AF388906

Honey Bee Health

Honey Bee Health

Editors

Giovanni Cilia
Antonio Nanetti

MDPI • Basel • Beijing • Wuhan • Barcelona • Belgrade • Manchester • Tokyo • Cluj • Tianjin

Editors
Giovanni Cilia
CREA Research Centre for Agriculture and Environment
Italy

Antonio Nanetti
CREA Research Centre for Agriculture and Environment
Italy

Editorial Office
MDPI
St. Alban-Anlage 66
4052 Basel, Switzerland

This is a reprint of articles from the Special Issue published online in the open access journal *Veterinary Sciences* (ISSN 2306-7381) (available at: https://www.mdpi.com/journal/vetsci/special_issues/bee-health).

For citation purposes, cite each article independently as indicated on the article page online and as indicated below:

LastName, A.A.; LastName, B.B.; LastName, C.C. Article Title. *Journal Name* **Year**, *Volume Number*, Page Range.

ISBN 978-3-0365-2680-5 (Hbk)
ISBN 978-3-0365-2681-2 (PDF)

Cover image courtesy of Giovanni Cilia.

Contents

About the Editors

Giovanni Cilia graduated *cum laude* with a Veterinary Biotechnological Sciences master's degree in 2016 at the University of Milan (Italy). He obtained a Ph.D. in Veterinary Sciences from the Department of Veterinary Sciences at the University of Pisa. Since 2021, he has been working at CREA Research Centre for Agriculture and Environment in Bologna, working on honey bee pathogens and molecular biology. He performs support teaching and seminar activities for the courses of "Infectious Bacterial Diseases", "Microbiology and Immunology", "Microbiology and Biotechnology Applied to Animal Production", and "Veterinary Hygiene" in the Department of Veterinary Sciences at the University of Pisa. During his activities, he applied for an internship at a different research institute, including the Biology of Spirochetes Unit, National Reference Center for Leptospirosis of Institute Pasteur (Paris, France). His research was focused on bacterial infectious diseases in domestic animals and wildlife. At present, his research topics include antibacterial activity of natural substances and infectious diseases in honey bee and insect molecular biology, with particular attention given to honey bee virus, nosemosis, and pests. He is the author/co-author of several scientific publications in international journals, the Editor of some infectious disease Special Issues, and the Topic Editor of "Veterinary Sciences".

Antonio Nanetti graduated *cum laude* with a degree in Science of Animal Production from the University of Bologna, Italy, in 1992. In 1979, he began working at the former National Institute of Beekeeping, the reference Italian Institution in apicultural research, which was recently incorporated into the Research Centre for Agriculture and Environment of the Italian Council for Agricultural Research and Economics (CREA). Within this institution, he is a researcher belonging to the apidological team, where he runs research projects focused on honey bees. His research covers the main honey bee pests and diseases, with a particular focus on exotic parasites and pathogens, such as *Varroa destructor*, *Nosema ceranae*, and *Aethina tumida*. His research activities have enabled considerable progress to be made in the control of bee diseases with natural compounds. He has also contributed to the ground-breaking research fields of honey bee welfare and the impact of climate change on pollinators. He has authored/co-authored more than 170 publications of different kinds, is the co-editor of a book on the welfare of managed honey bees, which is presently in preparation, and acted as a guest editor for a Special Issue on 'Honey Bee Health' in the journal Veterinary Science.

veterinary sciences

Editorial

Honey Bee Health

Giovanni Cilia * and Antonio Nanetti

CREA Research Centre for Agriculture and Environment (CREA-AA), Via di Saliceto 80, 40128 Bologna, Italy; antonio.nanetti@crea.gov.it
* Correspondence: giovanni.cilia@crea.gov.it; Tel.: +39-051-353103

Citation: Cilia, G.; Nanetti, A. Honey Bee Health. *Vet. Sci.* **2021**, *8*, 127. https://doi.org/10.3390/vetsci8070127

Received: 9 April 2021
Accepted: 26 June 2021
Published: 6 July 2021

Publisher's Note: MDPI stays neutral with regard to jurisdictional claims in published maps and institutional affiliations.

Honey bee health is a crucial issue that has recently received increased interest from researchers, stakeholders, and citizens.

To explore all possible features of honey bee health, this Special Issue, "Honey Bee health", aims to explore this topic through a series of research articles focused on different aspects of honey bee health at different levels, including molecular health, microbial health, and population genetic health. All the 21 published articles explore this theme and emphasize the importance of this issue.

Factors associates with honey bee colony losses were reviewed by Hristov et al. [1], and Amiri et al. analysed the interest in honey bee research [2]. Donkersley et al. report a One Health model to reverse honey bee decline [3].

Ribani et al. used environmental DNA (eDNA) to monitor honey bee pathogens and parasites, demonstrating that *V. destructor* is widespread and *L. passim* and *A. apis* with *N. ceranae* occurr frequently together [4].

V. destructor infections were modelled, suggesting that colony survival is sensitive to the hive grooming rate and reproductive rate of the mites, which is enhanced in drone-capped cells [5], whereas a new sampling and treatment of this pathogen was found to be a favourable and sustainable method of management [6].

Mendoza et al. found that *V. destructor* resistant honey bees have greater behavioural resistance than susceptible honey bees. At the end of the summer, resistant honey bees had fewer mites and a lower deformed wing virus type A (DWV-A) viral load than susceptible honey bees. Additionally, resistant honey bees were *A. mellifera scutellata* hybrids, whereas susceptible ones were closer to European subspecies [7].

The most promising molecular genetic markers for determining resistance to nosemosis in dark forest bees are microsatellite loci *AC117*, *Ap243*, and *SV185*, which were investigated by Ostroverkhova [8].

Emsen et al. evaluated the seasonality of *N. ceranae* and their relationship with honey bee survivorship, highlighting the highest infection rates, prevalence, and spore viability in the spring and summer, associated with reduced bee populations and food stores in colonies [9].

In the honey bee population in Asia, *Nosema* infection was found in 65% of apiaries by Ostroverkhova et al. Both *N. apis* and *N. ceranae* occur across subarctic and warm summer continental climates, but *N. ceranae* is more predominant in the latter, even if coinfections are predominant (36.3%) [10]. The presence of *N. apis* was also investigated by Naudi et al. in Estonia and Latvia. The results show that *N. apis* is dominant in Estonia (43%) and *N. ceranae* in Latvia (47%) [11].

Porrini et al. studied the effect of compounds commonly used to treat varroosis, evaluating the CHC profiles and EO production on *N. ceranae* infected and non-infected honey bees. The results indicate an absence of alteration in EO or CHC as a response to acaricides ingestion, suggesting that worker honey bees exposed to these highly ubiquitous drugs are hardly differentiated by nest-mates [12].

The efficacy of ApiHerb® and Api-Bioxal® as treatments against *N. ceranae* were investigated using two qPCR methods based on the *16S rRNA* and *Hsp70* genes. Both treatments reduced the abundance of *N. ceranae*, but ApiHerb also decreased the prevalence of infected bees. From the analysis, the qPCR method based on the *Hsp70* gene ensures a higher accuracy for the exact quantification of *N. ceranae* [13].

Dittes et al. discussed the veterinary approach to adult bee examination by analysing the differential diagnosis of the common virus diseases: Acute Bee Paralysis Virus (ABPV)-Kashmir Bee Virus (KBV)-Israeli Acute Paralysis Virus (IAPV)-Complex, Chronic Bee Paralysis Virus (CBPV), and DWV, as well as coinfections such as *Varroa* spp. and *Nosema* spp. [14].

The DWV-A transmission via hive products was investigated in a fully-crossed hoarding cage experiment, estimating the transmission risk by screening commercial products. The results show that DWV-A transmission via hive products is feasible, but the risk of introducing novel viruses and/or strains should be considered [15].

A case report highlights treatment and sanitary measures to save two *A. mellifera carnica* CBPV-infected colonies before the winter [16].

Bullock et al. proposed a silicone wristband as passive samplers in a beehive, developing a novel approach to passively sample honey bee hives. The silicone wristbands provide a simple, affordable, and passive method for sampling the chemical environment of honey bees [17].

Ludvigsen et al. evaluated the honey bee gut mycobiota cluster in different seasons and gut segments. The main finding was that bacteria cluster by gut segments, while fungi cluster by season [18]. Additionally, the administration of veterinary drugs, dietary supplements, and non-protein amino acids affected the ventriculum microbiological profile of *A. mellifera ligustica* [19].

Terenzi et al. reviewed the importance of the sound emitted by the hive. It is used by the bees to communicate within the hive, and its analysis can reveal useful information to understand the colony health status and to detect colony variations [20].

Power et al. found an innovative histological processing technique to analyse healthy drones of *A. mellifera ligustica*. The new approach can detect testes alterations, such as degenerated seminiferous tubules [21].

Author Contributions: G.C. and A.N. conceptualized, wrote, and reviewed the manuscript. All authors have read and agreed to the published version of the manuscript.

Funding: This research received no external funding.

Institutional Review Board Statement: Not applicable.

Informed Consent Statement: Not applicable.

Acknowledgments: We are grateful to all authors and reviewers who participated in this Special Issue.

Conflicts of Interest: The authors declare no conflict of interest.

References

1. Hristov, P.; Shumkova, R.; Palova, N.; Neov, B. Factors Associated with Honey Bee Colony Losses: A Mini-Review. *Vet. Sci.* **2020**, *7*, 166. [CrossRef]
2. Amiri, E.; Waiker, P.; Rueppell, O.; Manda, P. Using Manual and Computer-Based Text-Mining to Uncover Research Trends for Apis mellifera. *Vet. Sci.* **2020**, *7*, 61. [CrossRef]
3. Donkersley, P.; Elsner-Adams, E.; Maderson, S. A One-Health Model for Reversing Honeybee (Apis mellifera L.) Decline. *Vet. Sci.* **2020**, *7*, 119. [CrossRef]
4. Ribani, A.; Utzeri, V.J.; Taurisano, V.; Fontanesi, L. Honey as a Source of Environmental DNA for the Detection and Monitoring of Honey Bee Pathogens and Parasites. *Vet. Sci.* **2020**, *7*, 113. [CrossRef] [PubMed]
5. Torres, D.J.; Torres, N.A. Modeling the Influence of Mites on Honey Bee Populations. *Vet. Sci.* **2020**, *7*, 139. [CrossRef] [PubMed]
6. Floris, I.; Pusceddu, M.; Satta, A. How the Infestation Level of Varroa destructor Affects the Distribution Pattern of Multi-Infested Cells in Worker Brood of Apis mellifera. *Vet. Sci.* **2020**, *7*, 136. [CrossRef]
7. Mendoza, Y.; Tomasco, I.H.; Antúnez, K.; Castelli, L.; Branchiccela, B.; Santos, E.; Invernizzi, C. Unraveling Honey Bee–Varroa destructor Interaction: Multiple Factors Involved in Differential Resistance between Two Uruguayan Populations. *Vet. Sci.* **2020**, *7*, 116. [CrossRef]

8. Ostroverkhova, N.V. Association between the Microsatellite Ap243, AC117 and SV185 Polymorphisms and Nosema Disease in the Dark Forest Bee Apis mellifera mellifera. *Vet. Sci.* **2020**, *8*, 2. [CrossRef]
9. Emsen, B.; De la Mora, A.; Lacey, B.; Eccles, L.; Kelly, P.G.; Medina-Flores, C.A.; Petukhova, T.; Morfin, N.; Guzman-Novoa, E. Seasonality of Nosema ceranae Infections and Their Relationship with Honey Bee Populations, Food Stores, and Survivorship in a North American Region. *Vet. Sci.* **2020**, *7*, 131. [CrossRef] [PubMed]
10. Ostroverkhova, N.V.; Konusova, O.L.; Kucher, A.N.; Kireeva, T.N.; Rosseykina, S.A. Prevalence of the Microsporidian *Nosema* spp. in Honey Bee Populations (Apis mellifera) in Some Ecological Regions of North Asia. *Vet. Sci.* **2020**, *7*, 111. [CrossRef]
11. Naudi, S.; Šteiselis, J.; Jürison, M.; Raimets, R.; Tummeleht, L.; Praakle, K.; Raie, A.; Karise, R. Variation in the Distribution of Nosema Species in Honeybees (Apis mellifera Linnaeus) between the Neighboring Countries Estonia and Latvia. *Vet. Sci.* **2021**, *8*, 58. [CrossRef]
12. Porrini, M.P.; Garrido, P.M.; Umpiérrez, M.L.; Porrini, L.P.; Cuniolo, A.; Davyt, B.; González, A.; Eguaras, M.J.; Rossini, C. Effects of Synthetic Acaricides and Nosema ceranae (Microsporidia: Nosematidae) on Molecules Associated with Chemical Communication and Recognition in Honey Bees. *Vet. Sci.* **2020**, *7*, 199. [CrossRef]
13. Cilia, G.; Garrido, C.; Bonetto, M.; Tesoriero, D.; Nanetti, A. Effect of Api-Bioxal ®and ApiHerb ®Treatments against Nosema ceranae Infection in Apis mellifera Investigated by Two qPCR Methods. *Vet. Sci.* **2020**, *7*, 125. [CrossRef] [PubMed]
14. Dittes, J.; Aupperle-Lellbach, H.; Schäfer, M.O.; Mülling, C.K.W.; Emmerich, I.U. Veterinary Diagnostic Approach of Common Virus Diseases in Adult Honeybees. *Vet. Sci.* **2020**, *7*, 159. [CrossRef]
15. Schittny, D.; Yañez, O.; Neumann, P. Honey Bee Virus Transmission via Hive Products. *Vet. Sci.* **2020**, *7*, 96. [CrossRef]
16. Dittes, J.; Schäfer, M.O.; Aupperle-Lellbach, H.; Mülling, C.K.W.; Emmerich, I.U. Overt Infection with Chronic Bee Paralysis Virus (CBPV) in Two Honey Bee Colonies. *Vet. Sci.* **2020**, *7*, 142. [CrossRef]
17. Bullock, E.J.; Schafsnitz, A.M.; Wang, C.H.; Broadrup, R.L.; Macherone, A.; Mayack, C.; White, H.K. Silicone Wristbands as Passive Samplers in Honey Bee Hives. *Vet. Sci.* **2020**, *7*, 86. [CrossRef] [PubMed]
18. Ludvigsen, J.; Andersen, Å.; Hjeljord, L.; Rudi, K. The Honeybee Gut Mycobiota Cluster by Season versus the Microbiota which Cluster by Gut Segment. *Vet. Sci.* **2020**, *8*, 4. [CrossRef] [PubMed]
19. Cilia, G.; Fratini, F.; Tafi, E.; Turchi, B.; Mancini, S.; Sagona, S.; Nanetti, A.; Cerri, D.; Felicioli, A. Microbial Profile of the Ventriculum of Honey Bee (Apis mellifera ligustica Spinola, 1806) Fed with Veterinary Drugs, Dietary Supplements and Non-Protein Amino Acids. *Vet. Sci.* **2020**, *7*, 76. [CrossRef]
20. Terenzi, A.; Cecchi, S.; Spinsante, S. On the Importance of the Sound Emitted by Honey Bee Hives. *Vet. Sci.* **2020**, *7*, 168. [CrossRef] [PubMed]
21. Power, K.; Martano, M.; Altamura, G.; Maiolino, P. Histopathological Findings in Testes from Apparently Healthy Drones of Apis mellifera ligustica. *Vet. Sci.* **2020**, *7*, 124. [CrossRef] [PubMed]

Review

Factors Associated with Honey Bee Colony Losses: A Mini-Review

Peter Hristov [1,*], **Rositsa Shumkova** [2], **Nadezhda Palova** [3] **and Boyko Neov** [1]

[1] Department of Animal Diversity and Resources, Institute of Biodiversity and Ecosystem Research, Bulgarian Academy of Sciences, 1113 Sofia, Bulgaria; boikoneov@gmail.com
[2] Research Centre of Stockbreeding and Agriculture, Agricultural Academy, 4700 Smolyan, Bulgaria; rositsa6z@abv.bg
[3] Scientific Center of Agriculture, Agricultural Academy, 8300 Sredets, Bulgaria; nadejda_palova@abv.bg
[*] Correspondence: peter_hristoff@abv.bg; Tel.: +359-2-979-2327

Received: 15 September 2020; Accepted: 29 October 2020; Published: 30 October 2020

Abstract: The Western honey bee (*Apis mellifera* L., Hymenoptera: Apidae) is a species of crucial economic, agricultural and environmental importance. In the last ten years, some regions of the world have suffered from a significant reduction of honey bee colonies. In fact, honey bee losses are not an unusual phenomenon, but in many countries worldwide there has been a notable decrease in honey bee colonies. The cases in the USA, in many European countries, and in the Middle East have received considerable attention, mostly due to the absence of an easily identifiable cause. It has been difficult to determine the main factors leading to colony losses because of honey bees' diverse social behavior. Moreover, in their daily routine, they make contact with many agents of the environment and are exposed to a plethora of human activities and their consequences. Nevertheless, various factors have been considered to be contributing to honey bee losses, and recent investigations have established some of the most important ones, in particular, pests and diseases, bee management, including bee keeping practices and breeding, the change in climatic conditions, agricultural practices, and the use of pesticides. The global picture highlights the ectoparasitic mite *Varroa destructor* as a major factor in colony loss. Last but not least, microsporidian parasites, mainly Nosema ceranae, also contribute to the problem. Thus, it is obvious that there are many factors affecting honey bee colony losses globally. Increased monitoring and scientific research should throw new light on the factors involved in recent honey bee colony losses. The present review focuses on the main factors which have been found to have an impact on the increase in honey bee colony losses.

Keywords: honey bee losses; colony collapse disorder; *Varroa destructor*; viral diseases; nosematosis; negative pressures

---

1. Introduction

Managed honey bees are the most important pollinators for many crops and wild flowering species. Many countries worldwide, particularly in the Northern hemisphere, rely on the Western honey bee, *Apis mellifera*, for commercial pollination of certain crops, but over the recent years there has been an increase in losses in managed honey bee colonies in some regions of the world. Colony collapse disorder (CCD) has been reported for the first time in 2006 in the USA [1]. Although some bee losses have also been reported in China and Japan, published data from various investigations have shown that honey bee colony numbers have been stable for the past ten years in these regions [2,3]. The global picture has shown that there are no significant honey bee colony losses reported in Africa, Australia and South America. In the Middle East, the high temperatures and droughts in the summer are the main factor leading to colony losses because many plants, which are important sources for bee forage, suffer from heat stress [4]. Another factor aggravating the problem is the lack of comprehensive laws and legislations concerning the importation of bee colonies [5].

Indeed, bee colony losses are not a new phenomenon, and historical records show that extensive losses were not unusual in the past. Whilst recent problems may give the impression that there has been a massive decline, global research on honey bee colonies has shown that numbers actually increased between 1961 and 2007, mostly in Asia (426%), Africa (130%), South America (86%), and Oceania (39%) [6]. The most significant honey bee colony losses take place during overwintering, as shown by comparisons of colonies going into wintering and surviving the winter. The latter is a symptom of CCD, which has appeared in Europe, causing losses of up to 30% in some countries [7–9]. It has been difficult to determine a common pattern for the colony losses, but different investigations confirm that it is a phenomenon characteristic of the Western honey bee, while the Asiatic honey bee, present in southern, southeastern, and eastern Asia, appears to be more resistant to various pests and diseases [10].

2. Role of Pests and Diseases as Drivers Leading to Honey Bee Colony Losses

To understand the causes underlying the current decrease in honey bee colonies worldwide, it is important to shed light on the key pests and diseases that negatively affect bee health. Honey bees can be affected by various pests and diseases, including mites, different viruses, microsporidia, bacterial infections, and fungi (*Ascosphaera apis*) (Table 1). Due to the burden of infectious diseases and their agents, honey bee colonies may manifest significant weakness or even death. Only recently have scientists come to understand better the impact of the development and interactions of these pests and diseases.

Table 1. Some honey bee pests and diseases correlated with colony losses.

Type of Pathogen	Kind of Relationship	References
Varroa destructor	Ectoparasitic mite	[11,12]
Acarapis woodi	Tracheal mite	[13]
Varroa jacobsoni	Ectoparasitic mite	[14]
Tropilaelaps clareae	Ectoparasitic mite	[15]
Deformed wing virus A		
Deformed wing virus B (VDV1)		
Acute bee paralysis virus		
Kashmir bee virus	Viral pathogen	[16–20]
Israeli acute paralysis virus		
Chronic bee paralysis		
Sacbrood virus		
Black queen cell virus		
Nosema ceranae		
Nosema apis	Intestinal parasites	[21,22]
Nosema neumanni		
Ascosphaera apis	Fungal pathogen	[23]
Aspergillus spp.		
Aethina tumida	Beekeeping pest	[24–30]

2.1. Parasitic Mites

Honey bee hives can be a suitable habitat for various mites (Acari), including nonparasitic, omnivorous, pollen-feeding species, and parasites. Out of the different mite species associated with honey bees, *Varroa destructor*, *Acarapis woodi*, *Varroa jacobsoni* and *Tropilaelaps clareae* are economically significant pests of honey bees, and their infestation may lead to the destruction of the beekeeping industry in many cases [31,32]. *Varroa destructor* is the most serious pest of honey bee colonies around the world, as it is an obligate parasite which is able to attack different developmental stages and castes of *A. mellifera* [33]. It is interesting to note that Varroa mites have been established in New Zealand since 2000, but as of yet, Australia is still Varroa-free [34]. Additionally, in Africa, African honey bees seem to survive despite the presence of *Varoa destructor*, as do the Africanized honey bees in South America [35]. It is well known that the ectoparasitic mite *Varroa destructor* switched hosts from Eastern honey bees (*Apis cerana*) to Western honey bees (*Apis mellifera*) [36]. Thus, the Western honey bee has shown more susceptibility than *Apis cerana*. This increased resistance of the Africanized honey bees against *V. destructor* may be explained with their more aggressive behavior than the Western honey bee [37,38]. The association of *V. destructor* with the Western honey bee has led to a significant reduction of honey bee colonies.

V. destructor feeds on the fat bodies from adult bees, but not with hemolymph as previously believed [39]. For drones, it has been demonstrated that reduced weight due to *V. destructor* has resulted in decreased flight performance and sperm production [40]. There have also been reports of impaired orientation and homing ability of foraging bees infested with mite parasites, where the affected bees needed more time to return to the colony or did not return at all [41].

Therefore, a moderate Varroa infestation is considered to be one of the major factors leading to reduction of honey bee colonies, due to its influence on reproductive capacity and the general fitness of the colony. Moderate-to-severe infestations have been observed, especially during the autumn, when the mite population increases whereas the host population decreases [42]. The damage threshold is not correlated exactly with the number of mites per colony but is highly variable, depending on the bee and brood population, the season and its role as a very efficient vector of several honey bee-associated viruses [43].

Typical control of *V. destructor* involves the use of strips treated with fluvalinate, a pyrethroid, which are placed in the hive during times of no honey production. Intensive use of these strips has resulted in resistance to pyrethroid insecticides in some parts of Europe [44], the United States [45], Israel [46], and Mexico [47]. The spread of pyrethroid resistance in Europe seems to follow the initial spread of the mite according to bee movement, which suggests that resistance evolved once and spread thereafter [48]. Coumaphos, an organophosphate insecticide, was soon introduced for emergency use after control problems with fluvalinate. However, resistance to coumaphos is already present in Florida, USA [49] and northern Italy [50]. In Minnesota, USA [51] and in Mexico [47], there have been reports of resistance to both pyrethroids and amitraz, an amidine.

Thus, the increased resistance of *Varroa dectructor* against various insecticides creates a precondition for additional difficulty in combating mites and seeking alternative approaches.

2.2. Honey Bee-Associated Viruses

About 24 honey bee-associated viruses have been identified in the Western honey bee (*Apis mellifera*) [52]. Some of them generally persist in the bee's body, without causing a disease or manifestation of any clinical signs. In general, virus infestations were not considered to be a significant problem to honey bee health. On the other hand, some viruses are more virulent and infective, and thus may cause a significant loss in honey bee colonies as well as a decline in honey bees' health and production [53]. Some viruses show pathogenicity only under certain favorable environmental conditions.

Varroa mites *V. destructor* are considered to be the main transmitter of many honey bee viruses: deformed wing virus (DWV); acute bee paralysis virus (ABPV), Kashmir bee virus (KBV), and Israeli acute paralysis virus (IAPV) [39,54]. Furthermore, three viruses in the transmission of which Varroa seems to play no significant role, namely, chronic bee paralysis virus (CBPV), sacbrood virus (SBV), and black queen cell virus (BQCV) are also frequently surveyed [55,56]. This fact allows to us think that Varroa mites alone are not the (only) cause of honey bee losses. The negative influence of *V. destructor* results from its role as a viral reservoir and a transmitter of some honey bee-associated viruses [33]; the mite promotes replication of honey bee viruses like DWV [57]. Due to its feeding behavior, the Varroa mite injects directly viruses in the hemolymph, which has been associated with oral or sexual transmission of these viruses [58].

2.3. Microsporidia

Microsporidia are fungal, obligate intracellular parasites, infectious to honey bees. Microsporidia are possibly the smallest single-cell organisms which have a true nucleus. The genus *Nosema* is a parasitic fungus infecting insects such as honey bees, bumble bees and silkworms. Until now, only two species of microsporidia, namely, *Nosema ceranae* and *Nosema apis*, have been reported to parasitize on adult honey bees [59]. In 2017, a new species of Nosema, named *Nosema neumanni*, in honey bees from Uganda was reported [22]. It has been established that *N. apis* is specific for the Western honey bee, *Apis mellifera* L., whilst the Asiatic bee, *Apis cerana*, harbors *N. ceranae* [60]. For a long time, it was believed that *N. ceranae* and *N. apis* were species-specific. Since the beginning of this millennium (mainly post 2003), many investigations have revealed that *N. ceranae* has switched hosts and has become the dominant species in many countries [61–65]. Thus, it has been suggested that *N. ceranae* is possibly more virulent than *N. apis*.

It has been well documented that microsporidia invade the midgut epithelial cells of worker bees, queens and drones [66]. *Nosema* has adverse effects on the bee colony. The negative effect of nosemosis at the colony level is connected with the productivity and survival of honey bee colonies, including adult bee longevity, queen bees, brood rearing, bee biochemistry, pollen collection and other bee behaviors [67].

In contrast to *N. apis*, which rarely leads to the death of a diseased colony, since its emergence as a novel pathogen of the Western honey bee *A. mellifera*, *N. ceranae* has been generally associated with heavily diseased honey bee colonies [21]. Considering *N. ceranae* as a potential factor in CCD, we may summarize that almost any given disease organism has to persist over time (i.e., there has to be an increase in larval/adult incidence of infection) before causing colony mortality, and generally, *N. ceranae* acts simultaneously with other pathogens. For example, *N. ceranae* has been reported to be persistent for over an 18-month period in the colony before causing colony weakness [68]. Therefore, the possibility that *N. ceranae*, alone or in combination with other factors, causes CCD is still left open.

Bicyclohexylammonium fumagillin, an antibiotic isolated from the fungus *Aspergillus fumigatus*, has been the only widely used treatment for nosemosis, or "nosema disease", in Western honey bees, *Apis mellifera* [69]. The practice of periodic fumagillin treatment leads to decreasing yet nearly constant exposure of multiple generations of bees and pathogens to the drug. Although this practice appears to provide an environment conducive to selection of fumagillin-resistant Nosema strains, *N. apis* does not seem to have developed resistance to the drug. However, studies have shown that *N. ceranae* can reestablish pretreatment prevalence 6 months after treatments are terminated [70]. Research on protein profiles of bees treated with fumagillin has confirmed our hypothesis that fumagillin affects bee physiology at concentrations that no longer suppress *N. ceranae*. Thus, the use of fumagillin may increase the prevalence of *N. ceranae* and may be a factor in the replacement of *N. apis* by *N. ceranae* in US apiaries [71]. Besides having negative effects on host physiology, fumagillin presents some other drawbacks, e.g., increased management costs and risks to human health through honey consumption, as drug residues may persist in the hive [72]. Therefore, new alternatives to hard chemicals for Nosema spp. management are needed and sought.

2.4. Small Hive Beetle

The small hive beetle (SHB, *Aethina tumida*) is endemic to sub-Saharan Africa and alien for European honey bee pest [24]. During the last 25 years, SHBs were introduced into many countries worldwide, causing great damage to beekeeping as well as wild honey bee colonies [25]. The damages from SHBs occur when the honey bee population is insufficient to protect the honey combs from the scavenging beetle larvae [26]. Both adult and larval beetles are harmful to honey bee eggs and brood, but the larvae caused major damages as they funnel through comb with stored honey or pollen, damaging or destroying capping and comb [27]. Unlike wax moths, the SHB larvae do not necessarily damage the combs themselves, and do not produce extensive webbing. When large numbers of adult beetles defecate in the honey, they introduce yeasts, causing the honey to ferment and run out of the cells [28]. In this case, the queen bee may cease laying, and the entire colony may abscond. Weak colonies are particularly vulnerable to attack, but even strong colonies can be overwhelmed by large populations of beetles [24].

Beetles can create problems in cases when honey is removed from the hive, but not immediately extracted, i.e., beetles can invade the beekeeping room and quickly ruin a large portion of a honey harvest [29]. Honey contaminated by SHBs often is rejected by honey bees and is entirely unfit for human consumption [30].

2.5. Synergistic Effects of Various Diseases and Parasites

The interaction of Nosema spp. and viruses has been reported for bees co-infected with *N. ceranae* and chronic bee paralysis virus (CBPV) or DWV. One study showed that co-infection of bees with *N. ceranae* and CBPV led to increased replication of CBPV but did not result in mortality [18]. Costa et al. [73] found a significant negative correlation between N. ceranae spore load and DWV titer in midgut tissues of workers.

An interesting study has evaluated co-parasitism with Varroa (*Varroa destructor*) and Nosema (*Nosema ceranae* and *Nosema apis*) on honey bees (*Apis mellifera* L.) which have different defense levels [74]. The obtained results have shown that high-mite-mortality-rate (high-MMR) bees in the Nosema (−) group exhibited greater reductions in mean abundance of mites over time in comparison with low-mite-mortality-rate (low-MMR) bees, when inoculated with additional mites. However, high-MMR bees did not manage to reduce mite load as well as in the Nosema (−) group when fed with Nosema spores. As a whole, the study demonstrated differential mite mortality rates among different genotypes of bees, and also showed that Nosema infection inhibited the success of Varroa removal in high-MMR bees [74].

Other authors found considerable colony level variation regarding infection levels, and identified subtle differences between the microbiota of colonies with high infection levels and those with low infection levels [75]. Two exact sequence variants of Gilliamella, a core gut symbiont, previously associated with gut dysbiosis, were notably more abundant in bees from colonies with high Nosema loads in comparison with those with low Nosema loads.

An interesting study has been conducted to examine the association between *Varroa destructor* and DWV spreading from it in infected honey bees [76]. The study is a new approach to clarifying pathogen–parasite interactions in bees. The results obtained show that the experimental removal of increasing volumes of hemolymph from individual bees increases viral densities. The conducted research clearly demonstrates the defensive role of hemolymph in the bees and confirms that hemolymph has not only a bactericidal mode of action but also affects the regulation on viral replication in infected bees, which has been shown for the first time. The pathogen–parasite relationships observed in honey bees confirm their negative impact on colony losses [76].

Another interesting research study is related to a not so common symbiosis in bees [77]. The Varroa mite and DWV act in mutualistic symbiosis in infected honey bees, i.e., *Varroa destructor* vectored DWV, whereas the DWV-induced immunosuppression in honey bees is mediated by NF-κB signaling pathway, which facilitates mite feeding and reproduction. The decrease in the immune response of bees due to this symbiosis is another reason that leads to a significant weakening of bee colonies [77].

Other examples may be given about synergistic effects of various pathogens in honey bees, but we may conclude that they all lead to honey bee colony losses to a greater or lesser extent.

3. Anthropogenic Direct Drivers Associated with Honey Bee Colony Decline

In addition to different pest and diseases as direct natural drivers, there are many other drivers named anthropogenic, that lead to colony losses [78,79]. In many cases it is the interaction of these factors that leads to morbidity and mortality, and colony losses (Table 2). In this review, we will focus on the factors that are considered the most important.

Table 2. Environmental factors associated with honey bee colony losses.

Anthropogenic Direct Drivers that Cause Honey Bee Decline	Impact on Honey Bee	References
Pesticides	High rate of mortality, alteration of different biological processes	[80–83]
Climate change	Alteration of honey bee behavior, physiology and distribution, induced changes in flora for honey bees vitality	[84]
Introduction of alien species	Competition for food resources, decline of indigenous species, alteration of the new habitat	[85–88]
Genetically Modified Organisms (GMOs) crop	Alteration bees foraging behavior	[89]
Land use and management	Habitat and forage loss, honey bee and wild bee competition	[79,90]
Bee management	Hybridity of honey bees, migratory pollination	[91–94]
Environmental pollution	Imbalance in homeostasis, weakening of the immune system	[95]
Interactions between drivers	In many cases poorly studied	[96–105]

3.1. Pesticides

In recent decades, beekeepers have begun to use agrochemical pesticides not only for many crops, but in forests and other environments for the control of insect pests [106,107].

Pesticides are toxic chemicals with a specific mode of action, most often affecting a specific metabolic pathway in an organism [108]. Pesticides are applied via different pathways over the crop—sprays, seed coatings etc. [109]. The exposure of bees to pesticides is through ingestion of residues found in the pollen and nectar of contaminated plants (crop plants or the weeds around the fields) [110]. Insecticides pose the greatest danger to bees [80]. For this reason, beekeepers should be careful when using them near hives. In many countries, it is customary to inform beekeepers before treating crops in order to avoid their toxic effects on bees. These generally accepted rules are not observed in every country, which leads to mass mortality in bees and even the destruction of entire apiaries.

In recent years, the application of a new generation of pesticides—neonicotinoids—has been widely discussed among the scientific and beekeeping communities. They are used worldwide and are widely used for plant protection (crops, vegetables, fruits), veterinary products, and biocides to invertebrate pest control in fish farming [109]. Neonicotinoids represent a neurotoxic compound and their impact is more noticeable in insects than in mammals [111]. Their neurotoxic action is expressed as an agonist at nicotinic acetylcholine receptors (nAChRs) on the postsynaptic membrane, which plays a role in many cognitive processes [112].

One of the main marketed neonicotinoids is imidacloprid [113]. This neonicotinoid has various negative impacts on different biological processes in honey bees [81–83]. A recent study has shown that imidacloprid is involved in the reduction in protein biosynthesis and the decrease in the level of proteolytic activity (proteases) [114]. Taking into account the high protein content in royal jelly, this protein reduction is affected extremely adversely by the diet in larvae and queen, i.e., there is a subsequent decrease in honey bee populations [113]. In addition, it has been suggested that imidacloprid has a negative effect on antioxidant protection in young bees [115] as well as reduction in sperm motility (active mitochondria and sperm viability) [116].

These studies have shown the negative role of neonicotinoids on honey bee health and survival, which raises serious objections to their widespread use in agriculture.

3.2. Climate Change

Climate change may alter the synchrony between plant flowering and pollinator flight periods. Phenological mismatches probably contribute to pollinator losses that subsequently disrupt pollination of plants [117]. Climate is a crucial factor determining temperature and humidity. Humidity in hives must be maintained as low as possible, while the temperature of the brood must be maintained at 34 °C, and in winter the core temperature of the hive must not fall below 13 °C [118]. This is very important, and honey bee colonies must have sufficient access to carbohydrates to maintain these temperatures and survive. Extended periods of cold or wet weather or depletion of food sources can also have a negative impact on honey bee colony health. These can restrain flying activity and limit nectar and pollen supplies to the hive. In contrast to low temperature, if the brood temperature increases above 34.5 °C, bees display behavioral differences combined with learning and memory difficulties [119].

The effect of weather on bee colonies as a major factor in CCD has been reported in a survey of honey bee colony losses in the USA [120]. CCD has been linked to changes in bee habitats and malnutrition, both of which are indirectly caused by climate change. Moreover, climate change may allow invasive species to invade bee hives, causing disruption and further decline in bee populations [121,122].

3.3. Environmental Pollution

Honey bees interact with the environment especially while collecting pollen and nectar for feeding. Thus, they come into contact with some chemical substances and waste from the environment. In nature, waste and toxic substances (generally industrial gases, exhaust gases from vehicles, pesticides and insecticides) are absorbed and stored by plants. The largest part of air pollution is caused by anthropological factors (urbanization, industrialization, energy generation, mobile sources and other pollutants). One of the most important consequences of air pollution is heavy metal pollution [123,124]. Various heavy metal cations such as cadmium, cobalt, copper, zinc, lead, nickel and mercury are known to adversely affect both pollen (directly) and also the honey bees that feed on it (indirectly) [125,126]. Since bees collect pollen from different kinds of flowers, heavy metals found in large quantities within affected plants cause increased concentration of toxic heavy metals in bees' bodies and poison the latter. That is why honey bees and honey bee products are used as bioindicators of environmental pollution with heavy metals [123,127,128].

3.4. Bee Management

Specific peculiarities of beekeeping can be the direct cause or a supplement to the complex of stressors that can contribute to colony breakdown. These include artificial, unilateral feeding, use of antibiotics, acaricides and insecticides in hives, exposure to adverse temperatures and temperature fluctuations, infections and parasites, overexploitation of bee products, as well as unreliable sources of bees and queens [129,130]. One-sided selection of the honey bee results in genetic erosion in the species population and a lack of resistance to infectious diseases, mites, beekeeping acaricides applied in hives, etc. [130,131].

A pan-European epidemiological study aimed to clarify the role of beekeepers on honey bee losses [132]. The authors concluded that honey bees kept by professional beekeepers never showed signs of disease, unlike apiaries from hobbyist beekeepers that had symptoms of bacterial infection and heavy Varroa infestation. Moreover, they claimed that beekeepers' experience and apicultural practices are the major drivers of honey bee colony losses. To avoid these shortcomings in beekeeping, it is necessary to conduct trans-national monitoring schemes and keep improving beekeeper training [132].

Artificial feeding of bees is often necessary due to overcrowding in conditions of prolonged cold and rainy weather [133]. Feeding is found to be deficient in areas with intensive agricultural production, where the so-called stress from a monotonous or "monocultural" diet is observed [96]. This refers to the continuous foraging of bees on crops in mass flowering, grown in large fields, such as sunflower or rapeseed, and acacia, where the goal is honey production or just pollination of plants. Other factors related to insufficient nutrition include low-nutrient pollen and nectar, as well as certain plant species, including crops and flowers containing substances which are natural but toxic to bees. This is the case for the amygdalin glycoside found in almond flowers [134,135].

Pollen is one of the most important food sources for the proper development and functioning of honey bee colonies. Protein supplements are usually used in early spring to stimulate feeding and in mid-early autumn in order to successfully overwinter bee colonies. It is interesting to note that the addition of small amounts of pollen into protein supplements increases the resilience of honey bee colonies against DWV and Nosema infection [136]. Moreover, seasonal pollens are very important for honey bee feeding and development during their annual cycle [137]. For this reason, it is necessary to plan the presence of pollen annually, and in case of seasonal deficiency in nature, it is necessary to provide additional feeding to the bee colonies. Honey bee nutrition is also an important factor for increasing the resistance of honey bee colonies against various pathogens [138]. Proper and balanced nutrition throughout the year is a prerequisite for the effective function of the immune system at both the individual and social level. Consistently improving nutrition is a necessary condition for higher sustainability and resistance against pathogens (for example Varroa mites and honey bee-associated viruses) and diseases.

It is well known that, in most countries, large numbers of hives are transported by truck to multiple locations to pollinate seasonal fields and orchards. Transportation exposes colonies to many challenging stressors. The ecological conditions to which a hive is acclimated prior to transportation are often quite different from those of the destination. Bees are moved between locations at highway speeds and deployed in fields and orchards prior to the bloom. After transportation bees are exposed to changes in temperature, day length, and nutrient supplementation, which can result in increased foraging activity and brood production earlier than what would have occurred before relocation and in agricultural environments prior to floral bloom with low availability of resources [139]. Although transportation has been noted as a factor likely contributing to colony loss, the focus has been on changing forage quality and consistency rather than on stress endured during transportation [140]. Transportation stress has received less attention due to the difficulties of collecting data during shipping, yet it needs to be taken into consideration.

During transportation, colonies experience a number of stressors, including confinement, increased variation in temperature, air pressure, and vibration, frequent changes in elevation and latitude. Insufficient ventilation poses a serious risk of mortality due to overheating. Low-temperature stress also affects the bees, though less obviously. Transported colonies may experience extended periods of sub-lethal chill stress and loss of thermoregulation (LT), which affects long-term colony survival without proximate mortality by inducing developmental defects in new brood [141,142].

3.5. GMO Crops

Soybean and cotton varieties, followed by corn, with genetically incorporated genes for insecticide synthesis and herbicide tolerance, were first introduced in the USA in 1996. In 2007, 113 million hectares in different parts of the world (EU was among the exceptions) were sown with genetically modified crops [143]. With the expansion of the area planted with these crops, concerns have arisen about the safety of bees and other pollinators. Researchers have conducted a number of studies, involving dozens of plant species carrying Bt genes of *Bacillus thuringiensis* for resistance to insect pests [80,144,145]. It has been observed that there are no direct lethal effects of insect-resistant (IR) crops (e.g., producing *Bacillus thuringiensis* (Bt) toxins) on honey bees or other Hymenoptera, but some sub-lethal effects on honey bee behavior have been determined [139]. The impact of GMOs on honey bee health is contradictory. Some researchers have found that ingestion of high concentrations of Bt-toxins affected honey bee behavior [146], while others did not observe differences in the behavior and learning in honey bees [147]. There was no effect at lower toxin concentrations, such as those found in other transgenic varieties [148]. The presence of toxins in GMOS crops leads to reduced larval survival and body mass, and to increased developmental time in honey bees [149]. On the other hand, the introduction of varieties tolerant to certain herbicides and clean field technology have created conditions for growing maize, sunflower, etc., in the absence of weed vegetation in the crops. Since weeds are an alternative source of foraging, the widespread use of technology is considered to be one of the contributing factors to the starvation of bees—both wild and cultivated.

3.6. Interactions between Different Drivers

Except that multiple pressures individually impact the honey bee health, there are many evidences about the potential synergistic or additive effects among some drivers that leading to an overall negative impact on honey bee health and survival [12,150]. The most significant interactions among drivers are pointed out: 1. climate change and land-use—for instance, due to global warming, a pollinator species may migrate to a new geographical area, thus increasing the variety of pollinators of recipient region [97]. In many cases, this migration is unfavorable for the local fauna due to competition for food resources or transfer of various pests and diseases [98]; 2. pathogens and chemicals in the environment—the impact of pathogens and insecticides is another form of synergistic effect between drivers. Many authors have observed increased larval or worker honey bee mortality due to the additive or synergistic interactions between sub-lethal doses of neonicotinoid, infection by the *Nosema ceranae* and (BQCV) [99–101]; 3. bee nutrition and stress from disease and pesticides—honey bee colonies need proper and balanced nutrition to maintain their development and reproduction [102]. A large number of direct anthropogenic drivers produce alterations in diversity and may even lead to the extinction of many flowering plants which are the main food sources for honey bees [103,104]. These anthropogenic interventions may lead to malnutrition, i.e., reducing activity of immune system and potentially the function of some important detoxification enzymes; there is an elevated risk of the individual and combined impact of pesticides and pathogens on honey bees [96–105]. From what has been said so far, it is clear that the interaction between anthropogenic direct and natural direct drivers may represent a serious threat to honey bee health and survival.

4. Conclusions

Recent investigations have reported an increase in colony losses in some regions and have stimulated investment in more coordinated monitoring of bees and research on the impact of pests and diseases, bee diversity, bee-keeping practices and bee foraging environments on bee vitality. Factors such as land management and environmental conditions further affect the availability and quality of food sources as well as the conditions in the hive. Effective management of bee colonies under changing situations also depends on beekeeping practices and bee selection. All these diverse factors can affect bees' vitality and ability to overcome pests and diseases.

Author Contributions: All authors have equally contributed to the idea of this review. P.H. and B.N. prepared the original draft with considerable contributions from N.P. and R.S. All authors have substantially contributed to the writing of the final text. Moreover, all authors have read and agreed to the published version of the manuscript.

Funding: This work was funded by the National Scientific Fund of the Bulgarian Ministry of Education and Science, (grant numbers 06/10 17.12.2016). The funders had no role in study design, data collection and analysis, decision to publish, or preparation of the manuscript.

Conflicts of Interest: The authors declare that they have no conflict of interest.

References

1. Neumann, P.; Carreck, N.L. Honey bee colony losses. *J. Apic. Res.* **2010**, *49*, 1–6. [CrossRef]
2. Taniguchi, T.; Kita, Y.; Matsumoto, T.; Kimura, K. Honeybee Colony Losses during 2008~2010 Caused by Pesticide Application in Japan. *J. Apic.* **2012**, *27*, 15–27.
3. Liu, Z.; Chen, C.; Niu, Q.; Qi, W.; Yuan, C.; Su, S.; Liu, S.; Zhang, Y.; Zhang, X.; Ji, T.; et al. Survey results of honey bee (*Apis mellifera*) colony losses in China (2010–2013). *J. Apic. Res.* **2016**, *55*, 29–37. [CrossRef]
4. Awad, A.M.; Owayss, A.A.; Alqarni, A.S. Performance of two honey bee subspecies during harsh weather and Acacia gerrardii nectar-rich flow. *Sci. Agric.* **2017**, *74*, 474–480. [CrossRef]
5. Al-Ghamdi, A.; Adgaba, N.; Getachew, A.; Tadesse, Y. New approach for determination of an optimum honeybee colony's carrying capacity based on productivity and nectar secretion potential of bee forage species. *Saudi J. Biol. Sci.* **2016**, *23*, 92–100. [CrossRef] [PubMed]
6. Potts, S.G.; Roberts, S.P.; Dean, R.; Marris, G.; Brown, M.A.; Jones, R.; Neumann, P.; Settele, J. Declines of managed honey bees and beekeepers in Europe. *J. Apic. Res.* **2010**, *49*, 15–22. [CrossRef]
7. Aston, D. Honey bee winter loss survey for England, 2007–2008. *J. Apic. Res.* **2010**, *49*, 111–112. [CrossRef]
8. Topolska, G.; Gajda, A.; Pohorecka, K.; Bober, A.; Kasprzak, S.; Skubida, M.; Semkiw, P. Winter colony losses in Poland. *J. Apic. Res.* **2010**, *49*, 126–128. [CrossRef]
9. Gray, A.; Adjlane, N.; Arab, A.; Ballis, A.; Brusbardis, V.; Charrière, J.D.; Chlebo, R.; Coffey, M.F.; Cornelissen, B.; Amaro da Costa, C.; et al. Honey bee colony winter loss rates for 35 countries participating in the COLOSS survey for winter 2018–2019, and the effects of a new queen on the risk of colony winter loss. *J. Apic. Res.* **2020**, *59*, 744–751. [CrossRef]
10. Xu, P.; Shi, M.; Chen, X.X. Antimicrobial peptide evolution in the Asiatic honey bee *Apis cerana*. *PLoS ONE* **2009**, *4*, e4239. [CrossRef]
11. Clermont, A.; Pasquali, M.; Eickermann, M.; Kraus, F.; Hoffmann, L.; Beyer, M. Virus status, varroa levels and survival of 20 managed honey bee colonies monitored in Luxembourg between summer 2011 and spring 2013. *J. Apic. Sci.* **2015**, *59*, 59–73. [CrossRef]
12. Mõtus, K.; Raie, A.; Orro, T.; Chauzat, M.-P.; Viltrop, A. Epidemiology, risk factors and varroa mite control in the Estonian honey bee population. *J. Apic. Res.* **2016**, *55*, 396–412. [CrossRef]
13. Garrido-Bailón, E.; Bartolomé, C.; Prieto, L.; Botías, C.; Martínez-Salvador, A.; Meana, A.; Martín-Hernández, R.; Higes, M. The prevalence of *Acarapis woodi* in Spanish honey bee (*Apis mellifera*) colonies. *Exp. Parasitol.* **2012**, *132*, 530–536.
14. Roberts, J.; Anderson, D.; Tay, W. Multiple host shifts by the emerging honeybee parasite, Varroa jacobsoni. *Mol. Ecol.* **2015**, *24*, 2379–2391. [CrossRef]
15. Waghchoure-Camphor, E.S.; Martin, S.J. Population changes of *Tropilaelaps clareae* mites in *Apis mellifera* colonies in Pakistan. *J. Apic. Res.* **2009**, *48*, 46–49. [CrossRef]
16. Posada-Florez, F.; Childers, A.K.; Heerman, M.C.; Egekwu, N.I.; Cook, S.C.; Chen, Y.; Evans, J.D.; Ryabov, E.V. Deformed wing virus type A, a major honey bee pathogen, is vectored by the mite Varroa destructor in a non-propagative manner. *Sci. Rep.* **2019**, *9*, 12445. [CrossRef]
17. De Miranda, J.R.; Cordoni, G.; Budge, G. The acute bee paralysis virus–Kashmir bee virus–Israeli acute paralysis virus complex. *J. Invertebr. Pathol.* **2010**, *103*, S30–S47. [CrossRef]
18. Toplak, I.; Jamnikar Ciglenečki, U.; Aronstein, K.; Gregorc, A. Chronic bee paralysis virus and Nosema ceranae experimental co-infection of winter honey bee workers (*Apis mellifera* L.). *Viruses* **2013**, *5*, 2282–2297. [CrossRef]
19. Li, J.; Wang, T.; Evans, J.D.; Rose, R.; Zhao, Y.; Li, Z.; Li, J.; Huang, S.; Heerman, M.; Rodríguez-García, C.; et al. The phylogeny and pathogenesis of Sacbrood virus (SBV) infection in European honey bees, *Apis mellifera*. *Viruses* **2019**, *11*, 61. [CrossRef]
20. Spurny, R.; Přidal, A.; Pálková, L.; Kiem, H.K.T.; de Miranda, J.R.; Plevka, P. Virion structure of black queen cell virus, a common honeybee pathogen. *J. Virol.* **2017**, *91*, e02100-16. [CrossRef]
21. Vejsnaes, F.; Neilsen, S.L.; Kryger, P. Factors involved in the recent increase in colony losses in Denmark. *J. Apic. Res.* **2010**, *49*, 109–110. [CrossRef]
22. Chemurot, M.; De Smet, L.; Brunain, M.; De Rycke, R.; de Graaf, D.C. *Nosema neumanni* n. sp. (Microsporidia, Nosematidae), a new microsporidian parasite of honeybees, *Apis mellifera* in Uganda. *Eur. J. Protistol.* **2017**, *61*, 13–19. [CrossRef] [PubMed]

23. Sarwar, M. Fungal diseases of honey bees (Hymenoptera: Apidae) that induce considerable losses to colonies and protocol for treatment. *Int. J. Zool. Stud.* **2016**, *1*, 8–13.

24. Neumann, P.; Pettis, J.S.; Schäfer, M.O. *Quo vadis Aethina tumida*? Biology and control of small hive beetles. *Apidologie* **2016**, *47*, 427–466. [CrossRef]

25. Schäfer, M.O.; Cardaio, I.; Cilia, G.; Cornelissen, B.; Crailsheim, K.; Formato, G.; Lawrence, A.K.; Le Conte, Y.; Mutinelli, F.; Nanetti, A.; et al. How to slow the global spread of small hive beetles, *Aethina tumida*. *Biol. Invasions* **2019**, *21*, 1451–1459. [CrossRef]

26. Neumann, P.; Spiewok, S.; Pettis, J.; Radloff, S.E.; Spooner-Hart, R.; Hepburn, R. Differences in absconding between African and European honeybee subspecies facilitate invasion success of small hive beetles. *Apidologie* **2018**, *49*, 527–537. [CrossRef]

27. Huang, Q.; Lopez, D.; Evans, J.D. Shared and unique microbes between Small hive beetles (*Aethina tumida*) and their honey bee hosts. *Microbiol. Open* **2019**, *8*, e899. [CrossRef]

28. Mustafa, S.G.; Spiewok, S.; Duncan, M.; Spooner-Hart, R.; Rosenkranz, P. Susceptibility of small honey bee colonies to invasion by the small hive beetle, *Aethina tumida* (Coleoptera, Nitidulidae). *J. Appl. Entomol.* **2014**, *138*, 547–550. [CrossRef]

29. Ellis, J.D.; Hepburn, H.R. An ecological digest of the small hive beetle (*Aethina tumida*), a symbiont in honey bee colonies (*Apis mellifera*). *Insectes Sociaux* **2006**, *53*, 8–19. [CrossRef]

30. Bernier, M.; Fournier, V.; Eccles, L.; Giovenazzo, P. Control of *Aethina tumida* (Coleoptera: Nitidulidae) using in-hive traps. *Can. Entomol.* **2015**, *147*, 97–108. [CrossRef]

31. Sammataro, D.; Gerson, U.; Needham, G. Parasitic mites of honey bees: Life history, implications, and impact. *Annu. Rev. Entomol.* **2000**, *45*, 519–548. [CrossRef] [PubMed]

32. Dhooria, M.S. Parasitic Mites on Honeybees. In *Fundamentals of Applied Acarology*; Springer: Singapore, 2016.

33. Shen, M.; Cui, L.; Ostiguy, N.; Cox-Foster, D. Intricate transmission routes and interactions between picorna-like viruses (Kashmir bee virus and sacbrood virus) with the honeybee host and the parasitic varroa mite. *J. Gen. Virol.* **2005**, *86*, 2281–2289. [CrossRef] [PubMed]

34. Iwasaki, J.M.; Barratt, B.I.; Lord, J.M.; Mercer, A.R.; Dickinson, K.J. The New Zealand experience of varroa invasion highlights research opportunities for Australia. *Ambio* **2015**, *44*, 694–704. [CrossRef] [PubMed]

35. Strauss, U.; Dietemann, V.; Human, H.; Crewe, R.M.; Pirk, C.W. Resistance rather than tolerance explains survival of savannah honeybees (*Apis mellifera scutellata*) to infestation by the parasitic mite *Varroa destructor*. *Parasitology* **2016**, *143*, 374–387. [CrossRef]

36. Beaurepaire, A.L.; Truong, T.A.; Fajardo, A.C.; Dinh, T.Q.; Cervancia, C.; Moritz, R.F. Host specificity in the honeybee parasitic mite, Varroa spp. in *Apis mellifera* and *Apis cerana*. *PLoS ONE* **2015**, *10*, e0135103. [CrossRef]

37. Medina Flores, C.A.; Guzmán Novoa, E.; Hamiduzzaman, M.; Aréchiga Flores, C.F.; López Carlos, M.A. Africanized honey bees (*Apis mellifera*) have low infestation levels of the mite *Varroa destructor* in different ecological regions in Mexico. *Genet. Mol. Res.* **2014**, *13*, 7282–7293. [CrossRef]

38. Oddie, M.; Büchler, R.; Dahle, B.; Kovacic, M.; Le Conte, Y.; Locke, B.; de Miranda, J.R.; Mondet, F.; Neumann, P. Rapid parallel evolution overcomes global honey bee parasite. *Sci. Rep.* **2018**, *8*, 7704. [CrossRef]

39. Ramsey, S.D.; Ochoa, R.; Bauchan, G.; Gulbronson, C.; Mowery, J.D.; Cohen, A.; Lim, D.; Joklik, J.; Cicero, J.M.; Ellis, J.D.; et al. *Varroa destructor* feeds primarily on honey bee fat body tissue and not hemolymph. *Proc. Natl. Acad. Sci. USA* **2019**, *116*, 1792–1801. [CrossRef]

40. Rinkevich, F.D.; Danka, R.G.; Healy, K.B. Influence of Varroa Mite (*Varroa destructor*) Management Practices on Insecticide Sensitivity in the Honey Bee (*Apis mellifera*). *Insects* **2017**, *8*, 9. [CrossRef]

41. Peck, D.T.; Smith, M.L.; Seeley, T.D. *Varroa destructor* Mites Can Nimbly Climb from Flowers onto Foraging Honey Bees. *PLoS ONE* **2016**, *11*, e0167798. [CrossRef]

42. Oddie, M.; Dahle, B.; Neumann, P. Norwegian honey bees surviving *Varroa destructor* mite infestations by means of natural selection. *PeerJ* **2017**, *5*, e3956. [CrossRef] [PubMed]

43. Glenny, W.; Cavigli, I.; Daughenbaugh, K.F.; Radford, R.; Kegley, S.E.; Flenniken, M.L. Honey bee (*Apis mellifera*) colony health and pathogen composition in migratory beekeeping operations involved in California almond pollination. *PLoS ONE* **2017**, *12*, e0182814. [CrossRef] [PubMed]

44. Floris, I.; Cabras, P.; Garau, V.L.; Minelli, E.V.; Satta, A.; Troullier, J. Persistence and effectiveness of pyrethroids in plastic strips against *Varroa jacobsoni* (Acari: Varroidae) and mite resistance in a Mediterranean area. *J. Econ. Entomol.* **2001**, *94*, 806–810. [CrossRef] [PubMed]

45. Macedo, P.A.; Wu, J.; Ellis, M.D. Using inert dusts to detect and assess varroa infestations in honey bee colonies. *J. Apic. Res.* **2002**, *41*, 3–7. [CrossRef]

46. Mozes-Koch, R.; Slabezki, Y.; Efrat, H.; Kalev, H.; Kamer, Y.; Yakobson, B.A.; Dag, A. First detection in Israel of fluvalinate resistance in the varroa mite using bioassay and biochemical methods. *Exp. Appl. Acarol.* **2000**, *24*, 35–43. [CrossRef] [PubMed]

47. Rodríguez-Dehaibes, S.R.; Otero-Colina, G.; Sedas, V.P.; Jiménez, J.A.V. Resistance to amitraz and flumethrin in Varroa destructor populations from Veracruz, Mexico. *J. Apic. Res.* **2005**, *44*, 124–125. [CrossRef]

48. Büchler, R.; Berg, S.; Le Conte, Y. Breeding for resistance to *Varroa destructor* in Europe. *Apidologie* **2010**, *41*, 393–408. [CrossRef]

49. Elzen, P.J.; Westervelt, D. Detection of coumaphos resistance in Varroa destructor in Florida. *Am. Bee J.* **2002**, *142*, 291–292.

50. Spreafico, M.; Eördegh, F.R.; Bernardinelli, I.; Colombo, M. First detection of strains of *Varroa destructor* resistant to coumaphos. Results of laboratory tests and field trials. *Apidologie* **2001**, *32*, 49–55. [CrossRef]

51. Elzen, P.J.; Baxter, J.R.; Spivak, M.; Wilson, W.T. Control of *Varroa jacobsoni* Oud. resistant to fluvalinate and amitraz using coumaphos. *Apidologie* **2000**, *31*, 437–441. [CrossRef]

52. Gisder, S.; Genersch, E. Special issue: Honey bee viruses. *Viruses* **2015**, *7*, 5603–5608. [CrossRef]

53. Genersch, E.; Aubert, M. Emerging and re-emerging viruses of the honey bee (*Apis mellifera* L.). *Vet. Res.* **2010**, *41*, 54. [CrossRef] [PubMed]

54. Locke, B. Natural Varroa mite-surviving *Apis mellifera* honeybee populations. *Apidologie* **2016**, *47*, 467–482. [CrossRef]

55. Tentcheva, D.; Gauthier, L.; Zappulla, N.; Dainat, B.; Cousserans, F.; Colin, M.E.; Bergoin, M. Prevalence and seasonal variations of six bee viruses in *Apis mellifera* L. and *Varroa destructor* mite populations in France. *Appl. Environ. Microbiol.* **2004**, *70*, 7185–7291. [CrossRef]

56. Nielsen, S.L.; Nicolaisen, M.; Kryger, P. Incidence of acute bee paralysis virus, black queen cell virus, chronic bee paralysis virus, deformed wing virus, Kashmir bee virus and sacbrood virus in honey bees (*Apis mellifera*) in Denmark. *Apidologie* **2008**, *39*, 310–314. [CrossRef]

57. Levin, S.; Sela, N.; Chejanovsky, N. Two novel viruses associated with the *Apis mellifera* pathogenic mite *Varroa destructor*. *Sci. Rep.* **2016**, *6*, 37710. [CrossRef]

58. Francis, R.M.; Nielsen, S.L.; Kryger, P. Patterns of viral infection in honey bee queens. *J. Gen. Virol.* **2013**, *94*, 668–676. [CrossRef]

59. Paris, L.; El Alaoui, H.; Delbac, F.; Diogon, M. Effects of the gut parasite *Nosema ceranae* on honey bee physiology and behavior. *Curr. Opin. Insect. Sci.* **2018**, *26*, 149–154. [CrossRef]

60. Fries, I.; Feng, F.; Da Silva, A.; Slemenda, S.B.; Pieniazek, N.J. *Nosema ceranae n. sp.* (Microspora, Nosematidae), morphological and molecular characterization of a microsporidian parasite of the Asian honey bee *Apis cerana* (Hymenoptera, Apidae). *Eur. J. Protistol.* **1996**, *32*, 356–365. [CrossRef]

61. Klee, J.; Besana, A.M.; Genersch, E.; Gisder, S.; Nanetti, A.; Tam, D.Q.; Chinh, T.X.; Puerta, F.; Ruz, J.M.; Kryger, P.; et al. Widespread dispersal of the microsporidian *Nosema ceranae*, an emergent pathogen of the western honey bee, *Apis mellifera*. *J. Invertebr. Pathol.* **2007**, *96*, 1–10. [CrossRef]

62. Paxton, R.J.; Klee, J.; Korpela, S.; Fries, I. *Nosema ceranae* has infected *Apis mellifera* in Europe since at least 1998 and may be more virulent than *Nosema apis*. *Apidologie* **2007**, *38*, 558–565. [CrossRef]

63. Chen, Y.P.; Evans, J.D.; Smith, I.B.; Pettis, J.S. *Nosema ceranae* is a long-present and widespread microsporidean infection of the European honey bee (*Apis mellifera*) in the United States. *J. Invertebr. Pathol.* **2008**, *97*, 186–188. [CrossRef] [PubMed]

64. Invernizzi, C.; Abud, C.; Tomasco, I.H.; Harriet, J.; Ramallo, G.; Campa, J.; Katz, H.; Gardiol, G.; Mendoza, Y. Presence of *Nosema ceranae* in honeybees (*Apis mellifera*) in Uruguay. *J. Invertebr. Pathol.* **2009**, *101*, 150–153. [CrossRef]

65. Stevanovic, J.; Stanimirovic, Z.; Genersch, E.; Kovacevic, S.R.; Ljubenkovic, J.; Radakovic, M.; Aleksic, N. Dominance of *Nosema ceranae* in honey bees in the Balkan countries in the absence of symptoms of colony collapse disorder. *Apidologie* **2011**, *42*, 49–58. [CrossRef]

66. Papini, R.; Mancianti, F.; Canovai, R.; Cosci, F.; Rocchigiani, G.; Benelli, G.; Canale, A. Prevalence of the microsporidian *Nosema ceranae* in honeybee (*Apis mellifera*) apiaries in Central Italy. *Saudi J. Biol. Sci.* **2017**, *24*, 979–982. [CrossRef] [PubMed]

67. Botías, C.; Martín-Hernández, R.; Barrios, L.; Meana, A.; Higes, M. Nosema spp. infection and its negative effects on honey bees (*Apis mellifera iberiensis*) at the colony level. *Vet. Res.* **2013**, *44*, 25. [CrossRef]

68. Higes, M.; Martín-Hernandez, R.; Garrido-Bailon, E.; Gonzalez-Porto, A.V.; García-Palencia, P.; Meana, A.; Del Nozal, M.J.; Mayo, R.; Bernal, J.L. Honey bee colony collapse due to *Nosema ceranae* in professional apiaries. *Environ. Microbiol. Rep.* **2009**, *1*, 110–113. [CrossRef] [PubMed]

69. Higes, M.; Nozal, M.J.; Alvaro, A.; Barrios, L.; Meana, A.; Martín-Hernández, R.; Bernal, J.L.; Bernal, J. The stability and effectiveness of fumagillin in controlling *Nosema ceranae* (Microsporidia) infection in honey bees (*Apis mellifera*) under laboratory and field conditions. *Apidologie* **2011**, *42*, 364–377. [CrossRef]

70. Pajuelo, A.G.; Torres, C.; Bermejo, F.J.O. Colony losses: A double blind trial on the influence of supplementary protein nutrition and preventative treatment with fumagillin against *Nosema ceranae*. *J. Apic. Res.* **2008**, *47*, 84–86. [CrossRef]

71. Huang, W.F.; Solter, L.F.; Yau, P.M.; Imai, B.S. *Nosema ceranae* escapes fumagillin control in honey bees. *PLoS Pathog.* **2013**, *9*, e1003185. [CrossRef]

72. Van den Heever, J.P.; Thompson, T.S.; Curtis, J.M.; Pernal, S.F. Stability of dicyclohexylamine and fumagillin in honey. *Food Chem.* **2015**, *179*, 152–158. [CrossRef] [PubMed]

73. Costa, C.; Tanner, G.; Lodesani, M.; Maistrello, L.; Neumann, P. Negative correlation between *Nosema ceranae* spore loads and deformed wing virus infection levels in adult honey bee workers. *J. Invertebr. Pathol.* **2011**, *108*, 224–225. [CrossRef] [PubMed]

74. Bahreini, R.; Currie, R.W. The influence of Nosema (Microspora: Nosematidae) infection on honey bee (Hymenoptera: Apidae) defense against *Varroa destructor* (Mesostigmata: Varroidae). *J. Invertebr. Pathol.* **2015**, *132*, 57–65. [CrossRef] [PubMed]

75. Rubanov, A.; Russell, K.A.; Rothman, J.A.; Nieh, J.C.; McFrederick, Q.S. Intensity of *Nosema ceranae* infection is associated with specific honey bee gut bacteria and weakly associated with gut microbiome structure. *Sci. Rep.* **2019**, *9*, 3820. [CrossRef]

76. Annoscia, D.; Brown, S.P.; Di Prisco, G.; De Paoli, E.; Del Fabbro, S.; Frizzera, D.; Zanni, V.; Galbraith, D.A.; Caprio, E.; Grozinger, C.M.; et al. Haemolymph removal by Varroa mite destabilizes the dynamical interaction between immune effectors and virus in bees, as predicted by Volterra's model. *Proc. R. Soc. B.* **2019**, *286*, 28620190331. [CrossRef]

77. Di Prisco, G.; Annoscia, D.; Margiotta, M.; Ferrara, R.; Varricchio, P.; Zanni, V.; Caprio, E.; Nazzi, F.; Pennacchio, F. A mutualistic symbiosis between a parasitic mite and a pathogenic virus undermines honey bee immunity and health. *Proc. Natl. Acad. Sci. USA* **2016**, *113*, 3203–3208. [CrossRef]

78. Kovács-Hostyánszki, A.; Espíndola, A.; Vanbergen, A.J.; Settele, J.; Kremen, C.; Dicks, L.V. Ecological intensification to mitigate impacts of conventional intensive land use on pollinators and pollination. *Ecol. Lett.* **2017**, *20*, 673–689. [CrossRef]

79. IPBES. *The Assessment Report of the Intergovernmental Science-Policy Platform on Biodiversity and Ecosystem Services on Pollinators, Pollination and Food Production*; Potts, S.G., Imperatriz-Fonseca, V.L., Ngo, H.T., Eds.; Secretariat of the Intergovernmental Science-Policy Platform on Biodiversity and Ecosystem Services: Bonn, Germany, 2016; p. 552.

80. Al Naggar, Y.; Codling, G.; Vogt, A.; Naiem, E.; Mona, M.; Seif, A.; Giesy, J.P. Organophosphorus insecticides in honey, pollen and bees (*Apis mellifera* L.) and their potential hazard to bee colonies in Egypt. *Ecotoxicol. Environ. Saf.* **2015**, *114*, 1–8. [CrossRef]

81. Piiroinen, S.; Goulson, D. Chronic neonicotinoid pesticide exposure and parasite stress differentially affects learning in honey bees and bumblebees. *Proc. Royal Soc. B.* **2016**, *283*, 20160246. [CrossRef]

82. Schneider, C.W.; Tautz, J.; Gruenewald, B.; Fuchs, S. RFID tracking of sublethal effects of two neonicotinoids insecticides on the foraging behavior of *Apis mellifera*. *PLoS ONE* **2012**, *7*, e30023. [CrossRef]

83. Goulson, D. Review: An overview of the environmental risks posed by neonicotinoid insecticides. *J. Appl. Ecol.* **2013**, *50*, 977–987. [CrossRef]

84. Le Conte, Y.; Navajas, M. Climate change: Impact on honey bee populations and diseases. *Rev. Sci. Tech. OIE J.* **2008**, *27*, 499–510.

85. Laurino, D.; Lioy, S.; Carisio, L.; Manino, A.; Porporato, M. *Vespa velutina*: An Alien Driver of Honey Bee Colony Losses. *Diversity* **2020**, *12*, 5. [CrossRef]

86. Moritz, R.F.; Haddad, N.; Bataieneh, A.; Shalmon, B.; Hefetz, A. Invasion of the dwarf honeybee *Apis florea* into the near East. *Biol. Invasions* **2010**, *12*, 1093–1099. [CrossRef]

87. Brittain, C.; Williams, N.; Kremen, C.; Klein, A.M. Synergistic effects of non-Apis bees and honey bees for pollination services. *Proc. Royal Soc. B* **2013**, *280*, 20122767. [CrossRef]

88. Kenis, M.; Auger-Rozenberg, M.A.; Roques, A.; Timms, L.; Péré, C.; Cock, M.J.; Settele, J.; Augustin, S.; Lopez-Vaamonde, C. Ecological effects of invasive alien insects. *Biol. Invasions* **2009**, *11*, 21–45. [CrossRef]

89. Han, P.; Niu, C.; Lei, C.-L.; Cui, J.-J.; Desneux, N. Quantification of toxins in a Cry1Ac + CpTI cotton cultivar and its potential effects on the honey bee *Apis mellifera* L. *Ecotoxicology* **2010**, *19*, 1452–1459. [CrossRef] [PubMed]

90. Durant, J.L.; Otto, C.R. Feeling the sting? Addressing land-use changes can mitigate bee declines. *Land Use Policy* **2019**, *87*, 104005. [CrossRef]

91. Andrews, E. To save the bees or not to save the bees: Honey bee health in the Anthropocene. *Agric. Hum. Values* **2019**, *36*, 891–902. [CrossRef]

92. Dolezal, A.G.; Toth, A.L. Feedbacks between nutrition and disease in honey bee health. *Curr. Opin. Insect. Sci.* **2018**, *26*, 114–119. [CrossRef]

93. Brodschneider, R.; Crailsheim, K. Nutrition and health in honey bees. *Apidologie* **2010**, *41*, 278–294. [CrossRef]

94. Melicher, D.; Wilson, E.S.; Bowsher, J.H.; Peterson, S.S.; Yocum, G.D.; Rinehart, J.P. Long-Distance Transportation Causes Temperature Stress in the Honey Bee, *Apis mellifera* (Hymenoptera: Apidae). *Environ. Entomol.* **2019**, *48*, 691–701. [CrossRef] [PubMed]

95. Strachecka, A.; Gryzińska, M.; Krauze, M. The influence of environmental pollution on the protective proteolytic barrier of the honey bee *Apis mellifera mellifera*. *Pol. J. Environ. Stud.* **2010**, *19*, 855–859.

96. Goulson, D.; Nicholls, E.; Botías, C.; Rotheray, E.L. Bee declines driven by combined stress from parasites, pesticides, and lack of flowers. *Science* **2015**, *347*, 1255957. [CrossRef]

97. Giannini, T.C.; Boff, S.; Cordeiro, G.D.; Cartolano, E.A.; Veiga, A.K.; Imperatriz-Fonseca, V.L.; Saraiva, A.M. Crop pollinators in Brazil: A review of reported interactions. *Apidologie* **2015**, *46*, 209–223. [CrossRef]

98. Kerr, J.T.; Pindar, A.; Galpern, P.; Packer, L.; Potts, S.G.; Roberts, S.M.; Rasmont, P.; Schweiger, O.; Colla, S.R.; Richardson, L.L.; et al. Climate change impacts on bumblebees converge across continents. *Science* **2015**, *349*, 177–180. [CrossRef]

99. Aufauvre, J.; Biron, D.G.; Vidau, C.; Fontbonne, R.; Roudel, M.; Diogon, M.; Viguès, B.; Belzunces, L.P.; Delbac, F.; Blot, N. Parasite-insecticide interactions: A case study of Nosema ceranae and fipronil synergy on honeybee. *Sci. Rep.* **2012**, *2*, 326. [CrossRef]

100. Doublet, V.; Labarussias, M.; de Miranda, J.R.; Moritz, R.F.; Paxton, R.J. Bees under stress: Sublethal doses of a neonicotinoid pesticide and pathogens interact to elevate honey bee mortality across the life cycle. *Environ. Microbiol.* **2015**, *17*, 969–983. [CrossRef]

101. Retschnig, G.; Neumann, P.; Williams, G.R. Thiacloprid–*Nosema ceranae* interactions in honey bees: Host survivorship but not parasite reproduction is dependent on pesticide dose. *J. Invertebr. Pathol.* **2014**, *118*, 18–19. [CrossRef]

102. Paoli, P.P.; Donley, D.; Stabler, D.; Saseendranath, A.; Nicolson, S.W.; Simpson, S.J.; Wright, G.A. Nutritional balance of essential amino acids and carbohydrates of the adult worker honeybee depends on age. *Amino Acids* **2014**, *46*, 1449–1458. [CrossRef]

103. Barber, N.A.; Soper Gorden, N.L. How do belowground organisms influence plant–pollinator interactions? *J. Plant Ecol.* **2015**, *8*, 1–11. [CrossRef]

104. Hladun, K.R.; Parker, D.R.; Tran, K.D.; Trumble, J.T. Effects of selenium accumulation on phytotoxicity, herbivory, and pollination ecology in radish (*Raphanus sativus* L.). *Environ. Pollut.* **2013**, *172*, 70–75. [CrossRef] [PubMed]

105. Vanbergen, A.J.; Initiative, T.I.P. Threats to an ecosystem service: Pressures on pollinators. *Front. Ecol. Environ.* **2013**, *11*, 251–259. [CrossRef]

106. Moritz, R.F.; Erler, S. Lost colonies found in a data mine: Global honey trade but not pests or pesticides as a major cause of regional honeybee colony declines. *Agric. Ecosyst. Environ.* **2016**, *216*, 44–50. [CrossRef]

107. Hladik, M.L.; Main, A.R.; Goulson, D. Environmental Risks and Challenges Associated with Neonicotinoid Insecticides. *Environ. Sci. Technol.* **2018**, *52*, 3329–3335. [CrossRef]

108. Lushchak, V.I.; Matviishyn, T.M.; Husak, V.V.; Storey, J.M.; Storey, K.B. Pesticide toxicity: A mechanistic approach. *EXCLI J.* **2018**, *17*, 1101.

109. Simon-Delso, N.; Amaralrogers, V.; Belzunces, L.P.; Bonmatin, J.M.; Chagnon, M.; Downs, C.A.; Furlan, L.; Gibbons, D.W.; Giorio, C.; Girolami, V.; et al. Systemic insecticides (neonicotinoids and fipronil): Trends, uses, mode of action and metabolites. *Environ. Sci. Pollut. Res.* **2015**, *22*, 5–34. [CrossRef]

110. Sanchez-Bayo, F.; Goka, K. Pesticide residues and bees–a risk assessment. *PLoS ONE* **2014**, *9*, e94482. [CrossRef]

111. Tomizawa, M.; Casida, J.E. Selective toxicity of neonicotinoids attributable to specificity of insect and mammalian nicotinic receptors. *Annu. Rev. Entomol.* **2003**, *48*, 339–364. [CrossRef]

112. Van der Sluijs, J.P.; Simon-Delso, N.; Goulson, D.; Maxim, L.; Bonmatin, J.M.; Belzunces, L.P. Neonicotinoids, bee disorders and the sustainability of pollinator services. *Curr. Opin. Environ. Sustain.* **2013**, *5*, 293–305. [CrossRef]

113. Jeschke, P.; Nauen, R.; Schindler, M.; Elbert, A. Overview of the status and global strategy for neonicotinoids. *J. Agric. Food Chem.* **2011**, *59*, 2897–2908. [CrossRef] [PubMed]

114. Wilde, J.; Frączek, R.J.; Siuda, M.; Bąk, B.; Hatjina, F.; Miszczak, A. The influence of sublethal doses of imidacloprid on protein content and proteolytic activitity in honey bees (*Apis mellifera* L.). *J. Apic. Res.* **2016**, *55*, 212–220. [CrossRef]

115. Słowińska, M.; Nynca, J.; Wilde, J.; Bąk, B.; Siuda, M.; Ciereszko, A. Total antioxidant capacity of honey bee hemolymph in relation to age and exposure to pesticides, and comparison to antioxidant capacity of seminal plasma. *Apidologie* **2016**, *47*, 227–236. [CrossRef]

116. Ciereszko, A.; Wilde, J.; Dietrich, G.J.; Siuda, M.; Bąk, B.; Judycka, S.; Karol, H. Sperm parameters of honeybee drone exposed to imidacloprid. *Apidologie* **2017**, *48*, 211–222. [CrossRef]

117. Petanidou, T.; Kallimanis, A.S.; Sgardelis, S.P.; Mazaris, A.D.; Pantis, J.D.; Waser, N.M. Variable flowering phenology and pollinator use in a community suggest future phenological mismatch. *Acta Oecologica* **2014**, *59*, 104–111. [CrossRef]

118. Nürnberger, F.; Härtel, S.; Steffan-Dewenter, I. The influence of temperature and photoperiod on the timing of brood onset in hibernating honey bee colonies. *PeerJ* **2018**, *6*, e4801. [CrossRef]

119. Wang, Q.; Xu, X.; Zhu, X.; Chen, L.; Zhou, S.; Huang, Z.Y.; Zhou, B. Low-Temperature Stress during Capped Brood Stage Increases Pupal Mortality, Misorientation and Adult Mortality in Honey Bees. *PLoS ONE* **2016**, *11*, e0154547. [CrossRef] [PubMed]

120. VanEngelsdorp, D.; Speybroeck, N.; Evans, J.D.; Kim Nguyen, B.; Mullin, C.; Frazier, M.; Frazier, J.; Cox-Foster, D.; Chen, Y.; Tarpy, D.R.; et al. Weighing risk factors associated with bee colony collapse disorder by classification and regression tree analysis. *J. Econ. Entomol.* **2010**, *103*, 1517–1523. [CrossRef] [PubMed]

121. Memmott, J.; Craze, P.G.; Waser, N.M.; Price, M.V. Global warming and the disruption of plant-pollinator interactions. *Ecol. Lett.* **2007**, *10*, 710–717. [CrossRef]

122. Thomson, J.D. Flowering phenology, fruiting success and progressive deterioration of pollination in an early-flowering geophyte. *Philos. Trans. R. Soc. Lond. B Biol. Sci.* **2010**, *365*, 3187–3199. [CrossRef]

123. Bargańska, Ż.; Ślebioda, M.; Namieśnik, J. Honey bees and their products: Bioindicators of environmental contamination. *Crit. Rev. Environ. Sci. Technol.* **2016**, *46*, 235–248. [CrossRef]

124. Søvik, E.; Perry, C.J.; LaMora, A.; Barron, A.B.; Ben-Shahar, Y. Negative impact of manganese on honeybee foraging. *Biol. Lett.* **2015**, *11*, 20140989. [CrossRef]

125. Nikolić, T.V.; Kojić, D.; Orčić, S.; Batinić, D.; Vukašinović, E.; Blagojević, D.P.; Purać, J. The impact of sublethal concentrations of Cu, Pb and Cd on honey bee redox status, superoxide dismutase and catalase in laboratory conditions. *Chemosphere* **2016**, *164*, 98–105. [CrossRef] [PubMed]

126. Lazor, P.; Tomáš, J.; Tóth, T.; Tóth, J.; Čéryová, S. Monitoring of air pollution and atmospheric deposition of heavy metals by analysis of honey. *J. Microbiol. Biotechnol. Food Sci.* **2020**, *9*, 522–533.

127. Skorbiłowicz, E.; Skorbiłowicz, M.; Cieśluk, I. Bees as bioindicators of environmental pollution with metals in an urban area. *J. Ecol. Eng.* **2018**, *19*, 229–234. [CrossRef]

128. Goretti, E.; Pallottini, M.; Rossi, R.; La Porta, G.; Gardi, T.; Goga, B.C.; Elia, A.C.; Galletti, M.; Moroni, B.; Petroselli, C.; et al. Heavy metal bioaccumulation in honey bee matrix, an indicator to assess the contamination level in terrestrial environments. *Environ. Pollut.* **2020**, *256*, 113388. [CrossRef]

129. Schierow, L.-J.; Johnson, R.; Corn, M.L. Bee Health: The Role of Pesticides, Congressional Research Service (CRS) 2012, Reports for Congress. p. 26. Available online: https://www.fas.org/sgp/crs/misc/R42855.pdf (accessed on 28 April 2018).

130. Capri, E.; Marchis, A. *Bee Health in Europe: Facts and Figures 2013. Compendium of the Latest Information on Bee Health in Europe*; OPERA Research Centre, Università Cattolica del Sacro Cuore: Milan, Italy, 2013; p. 64.

131. Johnson, R.; Corn, M.L. Bee Health: The Role of Pesticides. Congressional Research Service (CRS) 2015. Reports for Congress. p. 47. Available online: http://fas.org/sgp/crs/misc/R43900.pdf (accessed on 26 June 2017).

132. Jacques, A.; Laurent, M.; EPILOBEE Consortium; Ribière-Chabert, M.; Saussac, M.; Bougeard, S.; Budge, G.E.; Hendrikx, P.; Chauzat, M.-P. A pan-European epidemiological study reveals honey bee colony survival depends on beekeeper education and disease control. *PLoS ONE* **2017**, *12*, e0172591. [CrossRef]

133. van Engelsdorp, D.; Evans, J.D.; Donovall, L.; Mullin, C.; Frazier, M.; Frazier, J.; Tarpy, D.R.; Hayes, J.; Pettis, J.S. Entombed pollen: A new condition in honey bee colonies associated with increased risk of colony mortality. *J. Invertebr. Pathol.* **2009**, *101*, 147–149. [CrossRef]

134. London-Shafir, I.; Shafir, S.; Eisikowitch, D. Amygdalin in almond nectar and pollen-facts and possible roles. *Plant Syst. Evol.* **2003**, *238*, 87–95. [CrossRef]

135. United Nations Environment Programme (UNEP). UNEP Emerging Issues: Global Honey Bee Colony Disorder and Other Threats to Insect Pollinators 2010. p. 16. Available online: http://www.unep.org/dewa/Portals/67/pdf/Global_Bee_Colony_Disorder_and_Threats_insect_pollinators.pdf (accessed on 29 May 2016).

136. Watkins de Jong, E.; DeGrandi-Hoffman, G.; Chen, Y.; Graham, H.; Ziolkowski, N. Effects of diets containing different concentrations of pollen and pollen substitutes on physiology, Nosema burden, and virus titers in the honey bee (*Apis mellifera* L.). *Apidologie* **2019**, *50*, 845–858. [CrossRef]

137. DeGrandi-Hoffman, G.; Gage, S.L.; Corby-Harris, V.; Carroll, M.; Chambers, M.; Graham, H.; deJong, E.W.; Hidalgo, G.; Calle, S.; Azzouz-Olden, F.; et al. Connecting the nutrient composition of seasonal pollens with changing nutritional needs of honey bee (*Apis mellifera* L.) colonies. *J. Insect Physiol.* **2018**, *109*, 114–124. [CrossRef] [PubMed]

138. Hoffman, G.D.; Chen, Y. Nutrition, immunity and viral infections in honey bees. *Curr. Opin. Insect. Sci.* **2015**, *10*, 170–176. [CrossRef] [PubMed]

139. Fewell, J.H.; Winston, M.L. Colony state and regulation of pollen foraging in the honey bee, *Apis mellifera* L. *Behav. Ecol. Sociobiol.* **1992**, *30*, 387–393. [CrossRef]

140. Oldroyd, B.P. What's killing American honey bees? *PLoS Biol.* **2007**, *5*, e168. [CrossRef] [PubMed]

141. Groh, C.; Tautz, J.; Rössler, W. Synaptic organization in the adult honey bee brain is influenced by brood-temperature control during pupal development. *Proc. Natl. Acad. Sci. USA* **2004**, *101*, 4268–4273. [CrossRef]

142. Jones, J.C.; Helliwell, P.; Beekman, M.; Maleszka, R.; Oldroyd, B.P. The effects of rearing temperature on developmental stability and learning and memory in the honey bee, *Apis mellifera*. *J. Comp. Physiol. A. Neuroethol. Sens. Neural. Behav. Physiol.* **2005**, *191*, 1121–1129. [CrossRef]

143. USDA-Biotech Crop Data. Adoption of Genetically Engineered Crops in the U.S. 2009. Available online: http://www.ers.usda.gov/Data/BiotechCrops/#2009-7-1 (accessed on 19 February 2018).

144. Johnson, R.M.; Ellis, M.D.; Mullin, C.A.; Frazier, M. Pesticides and honey bee toxicity—USA. *Apidologie* **2010**, *41*, 312–331. [CrossRef]

145. Johnson, R.M. Honey Bee Toxicology. *Annu. Rev. Entomol.* **2015**, *60*, 415–434. [CrossRef]

146. Fearing, P.L.; Brown, D.; Vlachos, D.; Meghji, M.; Privalle, L. Quantitative analysis of CryIA(b) expression in Bt maize plants, tissues, and silage and stability of expression over successive generation. *Mol. Breed.* **1997**, *3*, 169–176. [CrossRef]

147. Arpaia, S.; De Cristofaro, A.; Guerrieri, E.; Bossi, S.; Cellini, F.; Di Leo, G.M.; Germinara, G.S.; Iodice, L.; Maffei, M.E.; Petrozza, A.; et al. Foraging activity of bumblebees (*Bombus terrestris* L.) on Bt-expressing eggplants. *Arthropod-Plant Interact.* **2011**, *5*, 255–261. [CrossRef]

148. Ramirez-Romero, R.; Desneux, N.; Decourtye, A.; Chaffiol, A.; Pham-Delègue, M.H. Does Cry1Ab protein affect learning performances of the honey bee *Apis mellifera* L. (Hymenoptera, Apidae)? *Ecotoxicol. Environ. Saf.* **2008**, *70*, 327–333. [CrossRef]

149. Paula, D.P.; Andow, D.A.; Timbo, R.V.; Sujii, E.R.; Pires, C.S.; Fontes, E.M. Uptake and transfer of a Bt toxin by a Lepidoptera to its eggs and effects on its offspring. *PLoS ONE* **2014**, *9*, e95422. [CrossRef]

150. González-Varo, J.P.; Biesmeijer, J.C.; Bommarco, R.; Potts, S.G.; Schweiger, O.; Smith, H.G.; Steffan-Dewenter, I.; Szentgyörgyi, H.; Woyciechowski, M.; Vilà, M. Combined effects of global change pressures on animal-mediated pollination. *Trends Ecol. Evol.* **2013**, *28*, 524–530. [CrossRef] [PubMed]

Publisher's Note: MDPI stays neutral with regard to jurisdictional claims in published maps and institutional affiliations.

Perspective

A One-Health Model for Reversing Honeybee (*Apis mellifera* L.) Decline

Philip Donkersley [1,*], **Emily Elsner-Adams** [2] **and Siobhan Maderson** [3]

[1] Lancaster Environment Centre, Lancaster University, Lancaster LA14YQ, UK
[2] Elsner Research and Consulting, 8118 Zürich, Switzerland; emilyadams13@gmail.com
[3] Department of Geography & Earth Sciences, Aberystwyth University, Aberystwyth SY233FL, UK; sim30@aber.ac.uk
* Correspondence: donkersleyp@gmail.com

Received: 27 July 2020; Accepted: 26 August 2020; Published: 27 August 2020

Abstract: Global insect decline impacts ecosystem resilience; pollinators such as honeybees (*Apis mellifera* L.) have suffered extensive losses over the last decade, threatening food security. Research has focused discretely on in-hive threats (e.g., *Nosema* and *Varroa destructor*) and broader external causes of decline (e.g., agrochemicals, habitat loss). This has notably failed to translate into successful reversal of bee declines. Working at the interdisciplinary nexus of entomological, social and ecological research, we posit that veterinary research needs to adopt a "One-Health" approach to address the scope of crises facing pollinators. We demonstrate that reversing declines will require integration of hive-specific solutions, a reappraisal of engagement with the many stakeholders whose actions affect bee health, and recontextualising both of these within landscape scale efforts. Other publications within this special issue explore novel technologies, emergent diseases and management approaches; our aim is to place these within the "One-Health" context as a pathway to securing honeybee health. Governmental policy reform offers a particularly timely pathway to achieving this goal. Acknowledging that healthy honeybees need an interdisciplinary approach to their management will enhance the contributions of veterinary research in delivering systemic improvements in bee health.

Keywords: honeybee; *Apis mellifera*; One-Health; nexus; landscape; beekeeper; pathogens

---

1. Introduction

Honeybee (*Apis mellifera* L.) populations have been in decline in many parts of Europe and North America since the start of the millennium [1,2]. Much of the discussion around bee decline has focused on external factors like chemical pesticides, land use change, diseases and pests [3]. Other internal physiological traits such as host gut microbiota, immune response or genetic variation have also been implicated [4–6]. To date the vast majority of this research was carried out by entomologists, physiologists and ecologists, but not veterinary researchers. Veterinary engagement with honeybees has been limited, save for a handful of papers focusing on disease incidence and management [7,8]. As this special issue highlights, veterinary science is increasingly required to engage with honeybee health. Adopting an interdisciplinary approach both for practical animal health purposes like prescribing antibiotics for use in hives because they are food producing animals [9], for food security and because honeybees are an indicator species for the wider health of the environment [10,11]. Here we explore the potential for veterinary science to rise to the challenge of ongoing bee declines through the One-Health approach.

The One-Health concept has been discussed since the mid-2000s, as an initiative between the medical and veterinary fields, acknowledging the need for a greater focus on zoonoses in the wake of the SARS outbreak and avian influenza outbreaks. Adopted by the WHO, OIE and FAO, as well as veterinary and medical societies around the world [12], the concept emphasises the interconnectedness of human, animal and environmental health, noting in particular that human health depends on a healthy and functioning ecosystem [13]. Although there are discussions about the different terms describing this interconnectedness (for example 'One-Health', 'EcoHealth', 'Planetary Health', [14]), they share a common focus on encouraging disciplines like medicine, veterinary science, ecological and environmental science to start collaborating across disciplinary boundaries. Although the focus has been on zoonotic diseases, the One-Health approach emphasises a holistic understanding to tackle challenges [13], and is therefore highly relevant for discussions about honeybee health.

Honeybees are closely linked to human well-being, through pollination of wild and agricultural plants, and through honey production [15]. They provide an important example of the need for interdisciplinary engagement over health, as their health is determined by many factors within the landscape, within human society and within the hive. In the UK, where the research underpinning this article took place, honeybees are primarily managed by beekeepers: wild colonies predominantly died out from the 1980s following the invasion of the infectious mite *Varroa destructor* [16], although there are still feral honeybee colonies present in the landscape [17]. The UK landscape is heavily agricultural, with arable and grazing land use covering 72% of the landscape. Honeybees co-exist with diseases and pest species, ranging from viruses to bacteria and Acari (See Table A1).

In this regard, the co-existence of disease-causing organisms and a species of human importance is not unusual—all domestic animals, and indeed humans, have a high incidence of potentially pathogenic organisms on their body [18]. What is important from a health perspective is understanding the relationships between the species present, and the context in which they are co-habiting. Hinchliffe et al. [19] argue that farm animal health is not a simple binary between healthy and unhealthy, where the definition of 'healthy' implies an absence of disease, but rather a result of the interactions between host, micro-organisms, environment, host immune system, management practices, and so forth [19]. The distinction between healthy and unhealthy is perhaps better described as a 'borderland' rather than a 'borderline', acknowledging that outbreaks of disease are rarely the result of incursions of a parasite or pest from outside into a 'healthy' area, but rather caused by a shift in the relationships between host, microorganism, management practices, host immune system and so on. This understanding of disease emphasises that diseases are endemic and co-generated [20].

Hinchliffe [20] suggests that addressing animal health issues requires a transdisciplinary approach to understanding disease, drawing in particular on knowledge from veterinarians, stock-people and farmers. Such an understanding of disease is vital to understanding the concept of health in honeybees. This paper explores three different groups of factors, operating at different scales, which, independently or in combination, can tip the balance of bee health into a condition of 'healthy' or 'unhealthy'. We consider how bee health is managed in hive-scale factors, human-scale factors and landscape-scale factors (Figure A1). Following critiques of One-Health initiatives to date [13,21,22], we actively engage with the environmental factors affecting honeybee health, and with the social science analyses [23]. We conclude by reflecting on the future for honeybee health, arguing that a One-Health approach could provide a useful framework for the many, diverse researchers, policy-makers and practitioners currently seeking to support healthy honeybee populations.

2. Materials and Methods

This paper draws on three projects on honeybee health that all used mixed methods, and were inspired by a transdisciplinary research perspective (Figure A1). Mixed methods research engages with multiple data types and different worldviews; our research suggests that this is necessary for improving bee health. Authors 1 and 2 worked in Lancashire, UK, with local hobbyist beekeepers, and in Herefordshire, UK, with a commercial beekeeper. Author 3 worked across the UK with commercial and hobby beekeepers, many of whom were nationally respected writers and lecturers on bee health. Authors 1 and 3 also interviewed policy-makers, scientific and academic experts, and bee health experts. Methods used in the authors' research included interviews, participant observation [24], in-hive and laboratory experiments [25], archival analysis and document analysis [26]. Archives and documents can provide temporal and/or contextual richness to otherwise singular understandings of bee health and decline [27,28]. By drawing together research and understandings about bee health from these diverse perspectives, the authors are well placed to present bee health in its full complexity.

3. Hive-Scale Factors

Most direct human–honeybee contact occurs at the hive, which is a highly complex environment. Honeybee hives are rich in bacterial, viral and insect diversity, co-existing within and on honeybees (Table A1). Beekeeping practices oriented towards health management and 'biosecurity' are often focused on individual beekeepers' hives, and those within flying range of the bees—up to 10 km from the hive. Security against diseases and pests has predominantly relied on creating barriers and spatial separations between livestock, pathogens and vectors of pathogens, and dividing them into categories such as healthy and diseased and then endeavouring to control movement across barriers [12]. Barriers include physical management practices, for example the use of disinfectant to stop pathogens being transported on clothes or vehicles, and legal/policy barriers, including legislation on movement, and surveillance of at-risk sites or animals [29]. Current beekeeping biosecurity advice focuses on good husbandry to maintain strong colonies that can resist infection or invasion [30], and specific management techniques designed to avoid the arrival of, or spread of, diseases, i.e., techniques that create a barrier around colonies and apiaries, known as 'barrier management methods'.

However, honeybees pose three specific management challenges to beekeepers and others tasked with monitoring and managing their health. First, honeybees are semi-domesticated animals: though beekeepers manage honeybee colonies, they are reliant on the open environment for large parts of their lives. Foraging excursions are inherently hazardous: bees interact with individuals from other colonies, and potentially transmit diseases and pests by arriving back to the wrong hive [31]; robbing other hives for food [32]; visiting plants previously visited by infected bees [33], and simply through life history traits such as mass mating where many individuals from different hives congregate at once [34]. These communal behaviours emphasise the challenges facing beekeepers and others responsible for healthy honeybee populations, and make clear the need for more interdisciplinary, flexible approaches to understanding and managing health.

Second, unusually in a veterinary context, the unit of management and treatment in the case of honeybees is the colony, not the individual animal—with colonies being composed of between 10,000 and 40,000 individuals, depending on the time of year. Honeybee colonies are highly tolerant of diseases: their behaviour is characterised by temporal polyethism, or age-dependent division of labour [35], and the age at which they transition between tasks depends on the condition of the colony and the individuals. Bees suffering from disease outbreaks or other physical challenges transition more quickly than healthy bees through to the most hazardous tasks such as out-of-hive foraging, reflecting the shorter longevity of these tasks [36]. An example where this is an issue is the parasitic mite *Varroa destructor*, which has a well-documented negative effect on honeybees. Its direct parasitism causes loss of haemolymph and body fat, reduced longevity and other issues, and it is a vector for viruses (Table A1), which then further damage the individual bees [37]. There are treatments available for *Varroa*, although none are perfect and some can also weaken bees [38].

Managing Varroa and its effects on honeybees raises the third challenge: honeybee managers are seeking to manage a species with which we share few characteristics. Even for mammalian livestock, it is difficult to measure what animals experience, although veterinarians and animal scientists have developed a series of parameters to measure nutrition, health, physical comfort and behavior, which can be interpreted to give an idea of the mental state of an animal [39]. Communication with insects is even more challenging, and although some honeybee communications have been decoded [40], judging the physical and emotional states of honeybees remains a challenge. Parasite and pathogen challenges like the negative direct and indirect effects of *Varroa*, require beekeepers to be skilled observers of their colonies [24], using proxy measures within the hive such as amount of brood as signs of stress [41].

Managing bee health therefore involves a constant process of negotiation between actions that might benefit a colony and actions that might put it at risk. Rather than treating all pathogen and pest species present within a single hive, it may be more useful to work within a conceptual framework of "tipping points", where attention is focused only on levels of a disease or pest that pose sufficient threat [20]. This more holistic approach to health management is somewhat practiced within the beekeeping community under the term "integrated pest management" (IPM). IPM within a beekeeping context pushes for intense monitoring of pests and parasites, allowing action beyond prescribed tipping points of symptoms or density of parasites.

Beekeepers, and bee inspectors charged with managing bee health adopt IPM techniques that seek to treat all aspects of bee lives, rather than working in isolation on single pathogen or pest species. IPM retains a hive-level focus, and does not engage directly with factors operating at a landscape-scale, such as forage availability or pesticide use, nor with economic issues like trade in bees (a classic source of infection of novel parasites and pathogens). Beekeepers and inspectors seeking to create resilient, healthy colonies therefore struggle to create the ideal conditions for healthy bees—one with plentiful, nutritious food across a season, and with limited chemicals, and where threats of novel pests and parasites are kept to a minimum.

4. Human-Scale Factors

Many of the factors causing bee decline occur within the hive. Beekeepers are responsible for monitoring hive health, as well as liaising with others whose actions affect bees, including farmers and bee inspectors. Beekeepers' knowledge and practices are central to any efforts to reverse bees' decline. The role of beekeepers in monitoring and ensuring honeybee health serves as a linkage between hive-based factors, and wider landscape scale factors affecting bees. Beekeepers' observations must be a cornerstone of any efforts to maintain species wellbeing, if they are to succeed [42,43]. There is often a tension between scientific assessment of bee health, and beekeepers' practical engagement with their colonies. This is rooted in fundamentally different ways of gaining knowledge about bees and the wider world [44]. Much of our current knowledge of bee health is generated via an epidemiological model, while beekeepers' knowledge is the result of a highly situated, locally generated knowledge [45]. This combines practical husbandry and direct observation, often resulting from generations of beekeeping in the same local landscape, as well as the more formal scientific knowledge that is the cornerstone of veterinary analysis of animal health. Beekeepers note how their practice results in them 'seeing like a bee' (e.g., interpreting the landscape's challenges and opportunities in terms of their colonies) [26].

Within the UK, national bee inspectors are tasked with inspecting colonies for foulbrood diseases, as well as providing beekeepers with information and education on bee health. While the bee inspectorate is overseen by the governmental Animal and Plant Health Agency (APHA), and inspections are carried out according to strict epidemiological standards, which are primarily focused on observing colonies for foulbrood diseases (Table A1), the education and outreach aspect of the position is rooted in personal working relationships with individual beekeepers.

While beekeepers are responsible for the immediate health of their bees, some beekeeping practices may be ultimately contrary to the short, and long term, health of bees. Characteristics of bees that may be attractive to beekeepers, such as high productivity, a low tendency to swarm, and being easy to handle and manipulate, may be ultimately counter to bees' welfare [41,46]. Over the past decade, concern about the negative effects of various in-hive treatments for *Varroa* has led to an increase in beekeepers practicing treatment-free beekeeping [47,48]. This approach can be highly problematic for the wider beekeeping community, as *Varroa* infestation has been widely associated with colony death [49], and the spread of infection between neighbouring colonies through robbing and drifting [50], or even to non-*Apis* pollinators [51] (see Section 6).

Hobby and professional beekeepers often have opposing perspectives on what are acceptable risks to their bees, and the timescale and nature of necessary human interventions. While most beekeepers express a deep sense of stewardship and responsibility to their bees' welfare, they have differing interpretations of what this involves [49]. For example, some beekeepers carry out their own bee breeding programmes, in an effort to move away from a reliance on chemical treatments for *Varroa* [52]. Although the treatment of *Varroa* is a volatile debate amongst the beekeeping community [49], the question of how beekeepers can manage *Varroa* whilst minimising the damaging side effects of miticides [53] highlights the human scale of factors affecting bee health. While beekeepers stay informed of innovative developments in veterinary understandings of bee health, and new in-hive treatments and management practices to combat the relentless onslaught of various infections [26], they also note the inadequacy of relying on such an approach. Others argue that their own husbandry can only go so far; ultimately, systemic change in landscape management is necessary to ensure bee health [26]. To this end, some beekeepers focus on negotiating with other land managers, upon whose land bees may be placed, and often educating them about best practice for pollinators. Some interviewees noted what they perceived as a discrepancy between land managers' formal adherence to agri-environment schemes designed to enhance forage and habitat for bees, and actual significant landscape enhancement. Incorporating the locally situated knowledge of beekeepers into land management strategy will be necessary to assure bee health.

Formal scientific understanding of bee health is primarily relied upon by policy-makers in efforts to address pollinator decline, with beekeepers' experiential and observational knowledge often being dismissed as anecdotal [54]. Although it can be challenging to incorporate such diversity of views, it is central to successful conservation movements [55]. As bee decline affects biodiversity, strategies from other conservation efforts are relevant [56,57]. Scientific assessments of landscape effects have successfully been applied alongside beekeepers' knowledge, leading to a broader, more nuanced understanding of complex environmental synergies [46]. A One-Health perspective facilitates constructive engagement with seemingly disparate understandings of bee health, and enables the insights of the full, heterogeneous beekeeping community, as well as bee scientists, veterinarians, and farmers, to be combined. By considering bee decline within a One-Health framework, we can also address the shared challenges to human, bee, and wider environmental health created by the industrial food system [58]. It is clear that in-hive challenges, which are observed by both beekeepers, and by qualified external observers such as bee inspectors, are exacerbated by challenges in the wider environment [1,59]. The next section considers the landscape-level factors affecting bee health, and how the colony itself can and does work to avoid ill health.

5. Landscape-Scale Factors

A number of factors relevant to honeybee health operate at a landscape-scale. These factors are very large-scale, hard to see, slow-moving changes with an inherently shifting baseline [59]. We explore two key landscape-scale factors in the rest of this section: nutrition and agricultural intensification.

The first factor is nutrition, specifically the availability of forage (nectar and pollen sources) within the landscape. Foraging by pollinators is demonstrably affected by the land use composition around them [60,61]. Honeybees are generalists, employing a patch-based foraging strategy [62]. Landscape heterogeneity is directly related to the amount, richness and diversity of pollen that bees forage [63], and honeybees benefit from more diverse, heterogeneous landscapes [25,64].

All animals, even invertebrates, have an optimal diet that maximizes their fitness (or health), known as the "intake target" [65]. These diets are typically referred to in terms of a protein: carbohydrate ratio; research indicates that diets with an excess of protein (P:C ratio >5:1) can result in significantly shortened lifespans [66]. More moderate P:C ratios (~2:1) are generally associated with enhanced resistance to bacterial and viral pathogens. When given a choice of what to eat, insects are able to adapt what they eat depending on their state of health [67]. Honeybees are specially adapted to forage over a huge range (up to 10 km on a single flight), selecting what they bring back to the hive based on the colony's fitness. However, if this range is saturated with a single floral species, as can be the case in some agricultural land, then this choice may be taken away from the foragers, hindering their ability to reach their preferred intake target.

The link between nutritional diversity and richness and pollinator fitness is well established; high quality diets have benefits to immune responses, reproduction and adult survival [68]. Although there is a lack of direct experimental data supporting dietary adaptation in honeybees, our current understanding is that honeybees that are under duress from pests or pathogens will consume a higher protein diet, whereas healthy bees may eat higher carbohydrate diets to enable greater exploitation of their environment. The nutritional status of honeybee populations has been shown to vary consistently with landscape type, with, for example, urban and forest environments being positive for honeybee nutrition [25,64,69]. In the UK, rural landscapes are dominated by agricultural land uses. Here, honeybees are most important for human well-being given their role in providing pollination services, yet these are sites of significant pressure on pollinators [70]. Agricultural habitats provide significant but time-limited nutritional resources, with extensive areas of single-crop arable farming flowering at once but only for a short time, negatively impacting nutritional diversity [71].

The second landscape-scale factor is also a result of agricultural intensification: the effects on honeybee health of pesticide use. The breadth and scope of knowledge on the damaging effects of chemical pesticides on honeybees is remarkable. Not only do insecticidal pesticides cause significant damage to honeybees [72], we are now beginning to appreciate the effects of herbicides and fungicides [73,74]. Given the volumes of fungicides and herbicides applied in agricultural landscapes, it is likely we are only beginning to appreciate the importance of the results of this emergent research. While beekeepers are key stakeholders responsible for ensuring bee health [75], and have historically provided evidence of pesticide risks to bees [26], in practice agrochemical risk and application is assessed and controlled by scientists and farmers, who may not be assessing bee health in the same way as beekeepers, or have the same priorities in land management [44,54].

6. Towards an Integrated Approach to Honeybee Health

We have shown how honeybees live in a complex social-ecological system, influenced by highly varied forms of human management, from the hive- to landscape-scale. The One-Health concept can effectively bridge these different sectors and scales, focusing as it does on interdisciplinary approaches to understanding health. In this paper, we have considered research from animal and human geography, microbiology, ecology, entomology, nutrition and many other disciplines (Figure A1). When considered together, the result is a clearer understanding of the many factors that influence honeybee health. Zinsstag [76] argue for a systemic approach to understanding health within social–ecological systems at a time of global change. There are many other factors that influence bee health, which we have not explored here, especially seasonal and climatic changes. Disruption of seasonal patterns can have serious knock-on effects to pollinators, particularly through changing forage plant phenology [59]. As we have previously discussed, disruption to foraging on a landscape scale can have significant negative consequences for honeybee health; complicating this, disruption on a temporal scale through climate change would clearly have even further negative consequences. The factors we have discussed will continue to shift and interact with other influences, forcing honeybees to adapt and change to maintain good health. To maintain honeybee health, these adaptations must be understood and supported by those most closely involved with their daily management, i.e. beekeepers and bee inspectors.

Beekeepers' knowledge is applied at the interface between landscape and hive-scale factors, and has a great potential role in supporting a One-Health approach. One way to understand this role is as a form of citizen science (CS). CS projects have become an important method of increasing engagement between practitioners and researchers, and enriching policy [77,78]. The dominant model of CS with beekeepers was characterised by experiments that were designed by formal scientists, with beekeepers incorporated as data collectors, and requested to submit fairly basic information (e.g., annual colony losses, honey production) [43,79]. Several of these projects have been run for many years, generating key baseline information (BIP, BeeBase, and CoLOSS). As these projects have evolved, further questions have been incorporated into the surveys, exploring the knowledge and practices of beekeepers, and how these influence colony survival [49].

Alternative CS models emphasise collaborative and co-created approaches, which allow for a higher degree of scientific engagement with beekeepers' knowledge [80]. Volunteers and other 'amateurs' have a significant role to play in environmental monitoring; unfortunately, this is frequently underutilised due to their being outside formal scientific communities [81]. Beekeeping is often an intergenerational practice, carried out by highly skilled, environmentally observant individuals. This, coupled with the habit of keeping hive records, and notes of other relevant factors to bee health and productivity, can generate detailed information on landscape-level changes to forage. Sufficiently detailed records also represent a bank of data on in-hive factors, including medication histories, queen breeding histories, and more. These observations can contribute to a multiple evidence based approach to policy development [55].

The UK's exit from the European Union will bring changes to environmental and agricultural policy, with potentially far-reaching effects on honeybee health. Reflecting on how best to develop new policies that support honeybee health is important, especially given the investment to date in honeybee health in the UK through the Insect Pollinator Initiative and more [82]. While landscape-scale factors are beyond most veterinary medical research, honeybees, much like all wild animals, operate at this scale. Historically, the EU has used agri-environment schemes that support the development of rural areas, to reverse the decline in pollinator biodiversity and their associated agro-ecosystems, and to protect biodiversity and ecosystem function. The Common Agricultural Policy (CAP) supported localised greening measures, such as crop diversification, protection of permanent grassland from conversion to arable and the implementation of ecological focus areas. These aim to improve pollinator resilience and help combat losses [3]. Reviews of the overall success of CAP have suggested the results have been limited due to the nature of payments focusing on land ownership, rather than environmental intent. Consequently, the effects on landscapes have been less than effective [83]. New land management policies should incorporate factors supporting honeybee health at all scales.

The UK Government Agriculture Bill 2020 is centered on the new Environmental Land Management system (ELMs). This will determine subsidies based on "public pay for public goods" (UK Gov., 2020), adopting a natural capital valuation approach. When considering honeybees, due to the effective area to forage in and the context of the 'pollinator movement hypothesis' [84], cooperative habitat management at the landscape scale has more evident and immediate benefits to these mobile ecosystem service providers [85]. Future land management strategies should consider implementation at the inter-farm level, highlighting the importance of complementarity of resources. Practical impacts from national policy initiatives have consistently struggled to reach beyond localised effects [86]. Action to counter pollinator decline has frequently manifest in ways which are seen as piecemeal, and motivated by an effort to appeal to public interest, rather than fundamentally addressing landscape-scale challenges to bee health [87].

Cost-effectiveness in ecosystem conservation can be achieved through the implementation of multi-functionality or "stacking" services to maximise output from minimal input [88]. Applying a multi-functional ecosystem service framework may result in exponentially greater, synergistic and efficient use of limited resources [85,89], but equally, requires the understanding and approval of practitioners, and an incentive to elicit environmental change. Though lacking under CAP, ELMs may suitably address this necessity, though this proposal remains in the consultation phase and is unlikely to show demonstrable results for the near future.

There is a worrying dissonance between agricultural policies, and the ecological status of honeybees [90]. Wider research on conservation policy notes the importance, and benefits, of integrating the knowledge and concerns of disparate communities [91]. A top-down policy approach will not prove an effective pathway towards integrating a One-Health concept into bee health, and efforts at inclusion must be carefully assessed to assure they move beyond rhetoric [92]. The One-Health paradigm supports community-led, bottom-up approaches that link vets, beekeepers, landowners and policy makers. Such an approach has the capacity to engage with diverse factors affecting decision-making in animal husbandry, such as a sense of community responsibility, and a belief, or lack thereof, in individual capacity to affect change [93]. While this model has been gaining traction throughout the veterinary community, and raises important points about the transdisciplinary nature of contemporary zoonotic infections, there is a risk that this perspective still tends to overlook relational realities, as well as their socio-economic and cultural settings [20]. These are the locally specific, subjectively experienced factors that can give rise to differing interpretations of bee health and how best to ensure a colony's wellbeing.

The public and policy responses to honeybee decline in recent years reflect an awareness of the interwoven fates of humans and pollinators (albeit with some rhetorical misunderstandings, as noted by [94]). This awareness is emblematic of other interspecific global challenges that are being tackled via the One-Health approach [95]. A One-Health approach to bee health highlights the need to incorporate landscape-scale factors alongside other factors functioning at the hive-, and the human- scale. Unravelling their effects on honeybees requires research across many disciplines, including, but not limited to, nutrition, pollination ecology, microbiology, neuroscience, and agronomy. This emphasises that conceptualising "health", simply as presence/absence of disease, is insufficient for understanding pollinator fitness (Table A1). Successfully applying a One-Health model to bee health will require active engagement with diverse and, at times, antagonistic parties, many of whose actions are driven by wider economic forces. If applied within the One-Health model, veterinary knowledge can make a strong contribution to reversing honeybee decline.

Honeybee health cannot be separated from the health of the environment that surrounds them [25]. A single health issue for honeybees, e.g., the presence/absence/treatment of a mite, can make an entire agricultural and food system dependent on their pollinating and honey-producing activities vulnerable [59]. Furthermore, this risks adopting an "*Apis*-centric" perspective on pollinator conservation. Though honeybees are highly efficient pollinators, they are by no means the only active player in this system [96]. Honeybee pathogens have also demonstrated the potential for zoonotic transfer to non-*Apis* pollinators when visiting the same flower patches [51]. Consequently, as well as considering honeybee health in a holistic manner, we must also consider knock-on impacts of treatment regimens and fitness assessments on to non-target species. Adopting a One-Health approach, deliberately seeking to bridge across the diverse sectors that affect bee lives, is therefore crucial.

Author Contributions: All authors contributed to writing the manuscript. All authors have read and agreed to the published version of the manuscript.

Funding: Publication fees were supported by Lancaster University Library, Author 1 is in receipt of an EPSRC Impact Acceleration Account grant.

Conflicts of Interest: The authors declare no conflict of interest.

Appendix A

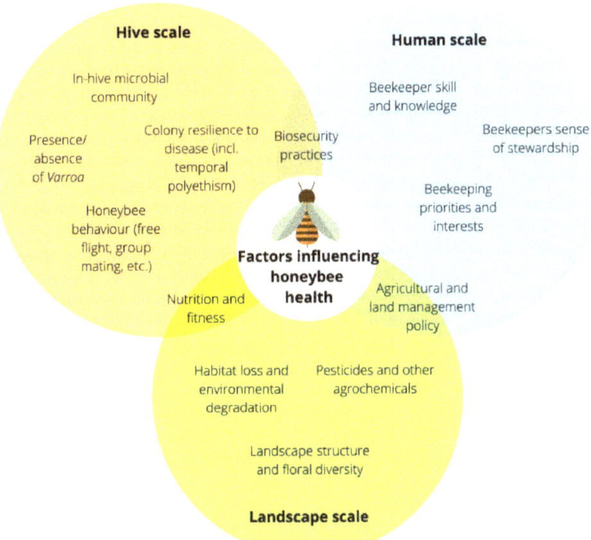

Figure A1. Factors influencing honey bee health at different scales. While factors at differing scales have predominant impacts on honeybee health, points of overlap illustrate the need for a One-Health approach to address challenges to honeybee health.

Table A1. Microbial, parasitic and commensal organisms associated with honeybees.

Organism	Type of Relationship	Reference
Lactobacillus spp. Firm-4		
Lactobacillus spp. Firm-5 (phylum Firmicutes)		
Bifidobacterium spp. (phylum Actinobacteria)		
Snodgrassella alvi	Symbiotic/commensal – collectively, these species form the 'core microbiome' of the honeybee's gut	[97,98]
Frischella perrara		
Gilliamella apicola		
Bartonella apis		
Alpha 2.1 (phylum Proteobacteria)		
Black queen cell virus		
Lake Sinai virus		
Deformed wing virus B (VDV1)		
Acute bee paralysis virus		
Chronic bee paralysis virus		
Sacbrood virus	Viral pathogen	[99,100]
Deformed wing virus A		
Aphid lethal paralysis virus		
Israeli acute paralysis virus		
Iridescent invertebrate virus IV		
Kashmir bee virus		

Table A1. *Cont.*

Organism	Type of Relationship	Reference
Varroa destructor	Parasitic mite, viral vector	[32]
Frischella perrara	Bacterial pathogen/commensal	[101,102]
Paenibacillus larvae (c.a. American Foulbrood)		
Melissococcus plutonius (c.a. European foul brood)		
Nosema ceranae	Intestinal parasites	[102]
Nosema apis		
Crithidia		
Acarapis woodi		

References

1. Potts, S.G.; Biesmeijer, J.C.; Kremen, C.; Neumann, P.; Schweiger, O.; Kunin, W.E. Global pollinator declines: Trends, impacts and drivers. *Trends Ecol. Evol.* **2010**, *25*, 345–353. [CrossRef] [PubMed]
2. Sánchez-Bayo, F.; Wyckhuys, K.A.G. Worldwide decline of the entomofauna: A review of its drivers. *Biol. Conserv.* **2019**. [CrossRef]
3. Scheper, J.; Holzschuh, A.; Kuussaari, M.; Potts, S.G.; Rundlöf, M.; Smith, H.G.; Kleijn, D. Environmental factors driving the effectiveness of European agri-environmental measures in mitigating pollinator loss—A meta-analysis. *Ecol. Lett.* **2013**, *16*, 912–920. [CrossRef] [PubMed]
4. Mattila, H.R.; Rios, D.; Walker-Sperling, V.E.; Roeselers, G.; Newton, I.L.G. Characterization of the active microbiotas associated with honeybees reveals healthier and broader communities when colonies are genetically diverse. *PLoS ONE* **2012**, *7*. [CrossRef]
5. Schmid, M.R.; Brockmann, A.; Pirk, C.W.W.; Stanley, D.W.; Tautz, J. Adult honeybees (*Apis mellifera* L.) abandon hemocytic, but not phenoloxidase-based immunity. *J. Insect Physiol.* **2008**. [CrossRef]
6. Saltykova, E.S.; Ben'kovskaya, G.V.; Gaifullina, L.R.; Novitskaya, O.P.; Poskryakov, A.V.; Nikolenko, A.G. Reaction of individual physiological barriers in bacterial infection in different races of the honeybee *Apis mellifera*. *J. Evol. Biochem. Physiol.* **2005**. [CrossRef]
7. vanEngelsdorp, D.; Meixner, M.D. A historical review of managed honeybee populations in Europe and the United States and the factors that may affect them. *J. Invertebr. Pathol.* **2010**. [CrossRef]
8. Amiri, E.; Strand, M.K.; Rueppell, O.; Tarpy, D.R. Queen quality and the impact of honeybee diseases on queen health: Potential for interactions between two major threats to colony health. *Insects* **2017**, *8*, 48. [CrossRef]
9. Mutinelli, F. Veterinary medicinal products to control *Varroa destructor* in honeybee colonies (*Apis mellifera*) and related EU legislation—An update. *J. Apic. Res.* **2016**. [CrossRef]
10. Narjes, M.E.; Lippert, C. The Optimal Supply of Crop Pollination and Honey from Wild and Managed Bees: An Analytical Framework for Diverse Socio-Economic and Ecological Settings. *Ecol. Econ.* **2019**. [CrossRef]
11. Croft, S.; Brown, M.; Wilkins, S.; Hart, A.; Smith, G.C. Evaluating European Food Safety Authority Protection Goals for Honeybees (*Apis mellifera*): What Do They Mean for Pollination? *Integr. Environ. Assess. Manag.* **2018**. [CrossRef] [PubMed]
12. Gibbs, E.P.J. The evolution of one health: A decade of progress and challenges for the future. *Vet. Rec.* **2014**. [CrossRef] [PubMed]
13. Destoumieux-Garzón, D.; Mavingui, P.; Boetsch, G.; Boissier, J.; Darriet, F.; Duboz, P.; Fritsch, C.; Giraudoux, P.; le Roux, F.; Morand, S.; et al. The one health concept: 10 years old and a long road ahead. *Front. Vet. Sci.* **2018**. [CrossRef]
14. Lerner, H.; Berg, C. A comparison of three holistic approaches to health: One health, ecohealth, and planetary health. *Front. Vet. Sci.* **2017**. [CrossRef] [PubMed]
15. Hanley, N.; Breeze, T.D.; Ellis, C.; Goulson, D. Measuring the economic value of pollination services: Principles, evidence and knowledge gaps. *Ecosyst. Serv.* **2015**. [CrossRef]
16. Potts, S.G.; Roberts, S.P.M.; Dean, R.; Marris, G.; Brown, M.A.; Jones, R.; Neumann, P.; Settele, J. Declines of managed honeybees and beekeepers in Europe. *J. Apic. Res.* **2010**. [CrossRef]

17. Thompson, C.E.; Biesmeijer, J.C.; Allnutt, T.R.; Pietravalle, S.; Budge, G.E. Parasite pressures on feral honeybees (*Apis mellifera* sp.). *PLoS ONE* **2014**. [CrossRef]
18. Barr, J.J. A bacteriophages journey through the human body. *Immunol. Rev.* **2017**. [CrossRef]
19. Hinchliffe, S.; Allen, J.; Lavau, S.; Bingham, N.; Carter, S. Biosecurity and the topologies of infected life: From borderlines to borderlands. *Trans. Inst. Br. Geogr.* **2013**. [CrossRef]
20. Hinchliffe, S. More than one world, more than one health: Re-configuring interspecies health. *Soc. Sci. Med.* **2015**. [CrossRef]
21. Stärk, K.D.C.; Arroyo Kuribreña, M.; Dauphin, G.; Vokaty, S.; Ward, M.P.; Wieland, B.; Lindberg, A. One Health surveillance—More than a buzz word? *Prev. Vet. Med.* **2015**. [CrossRef] [PubMed]
22. Lapinski, M.K.; Funk, J.A.; Moccia, L.T. Recommendations for the role of social science research in One Health. *Soc. Sci. Med.* **2015**. [CrossRef]
23. Craddock, S.; Hinchliffe, S. One world, one health? Social science engagements with the one health agenda. *Soc. Sci. Med.* **2015**. [CrossRef] [PubMed]
24. Adams, E.C. How to become a beekeeper: Learning and skill in managing honeybees. *Cult. Geogr.* **2018**. [CrossRef]
25. Donkersley, P.; Rhodes, G.; Pickup, R.W.; Jones, K.C.; Power, E.F.; Wright, G.A.; Wilson, K. Nutritional composition of honeybee food stores vary with floral composition. *Oecologia* **2017**, *185*, 749–761. [CrossRef]
26. Maderson, S.; Wynne-Jones, S. Beekeepers' knowledges and participation in pollinator conservation policy. *J. Rural Stud.* **2016**. [CrossRef]
27. Moritz, R.F.A.; Erler, S. Lost colonies found in a data mine: Global honey trade but not pests or pesticides as a major cause of regional honeybee colony declines. *Agric. Ecosyst. Environ.* **2016**. [CrossRef]
28. Watson, K.; Stallins, J.A. Honeybees and colony collapse disorder: A pluralistic reframing. *Geogr. Compass* **2016**. [CrossRef]
29. Donaldson, A. Biosecurity after event: Risk politics and animal disease. *Environ. Plan. A* **2008**. [CrossRef]
30. Jacques, A.; Laurent, M.; Ribière-Chabert, M.; Saussac, M.; Bougeard, S.; Budge, G.E.; Hendrikx, P.; Chauzat, M.P. A pan-European epidemiological study reveals honeybee colony survival depends on beekeeper education and disease control. *PLoS ONE* **2017**. [CrossRef]
31. Traver, B.E.; Fell, R.D. *Nosema ceranae* in drone honeybees (*Apis mellifera*). *J. Invertebr. Pathol.* **2011**. [CrossRef] [PubMed]
32. Frey, E.; Rosenkranz, P. Autumn invasion rates of *Varroa destructor* (Mesostigmata: Varroidae) into honeybee (Hymenoptera: Apidae) colonies and the resulting increase in mite populations. *J. Econ. Entomol.* **2014**. [CrossRef] [PubMed]
33. Anderson, K.E.; Sheehan, T.H.; Mott, B.M.; Maes, P.; Snyder, L.; Schwan, M.R.; Walton, A.; Jones, B.M.; Corby-Harris, V. Microbial ecology of the hive and pollination landscape: Bacterial associates from floral nectar, the alimentary tract and stored food of honeybees (*Apis mellifera*). *PLoS ONE* **2013**, *8*. [CrossRef] [PubMed]
34. Koeniger, N.; Koeniger, G.; Pechhacker, H. The nearer the better? Drones (*Apis mellifera*) prefer nearer drone congregation areas. *Insectes Soc.* **2005**. [CrossRef]
35. Johnson, B.R. Division of labor in honeybees: Form, function, and proximate mechanisms. *Behav. Ecol. Sociobiol.* **2010**. [CrossRef]
36. Perry, C.J.; Søvik, E.; Myerscough, M.R.; Barron, A.B. Rapid behavioral maturation accelerates failure of stressed honey bee colonies. *Proc. Natl. Acad. Sci. USA* **2015**. [CrossRef]
37. Gisder, S.; Aumeier, P.; Genersch, E. Deformed wing virus: Replication and viral load in mites (*Varroa destructor*). *J. Gen. Virol.* **2009**. [CrossRef]
38. Wu, J.Y.; Anelli, C.M.; Sheppard, W.S. Sub-lethal effects of pesticide residues in brood comb on worker honeybee (*Apis mellifera*) development and longevity. *PLoS ONE* **2011**. [CrossRef]
39. Stafford, K. Sheep veterinarians and the welfare of sheep: No simple matter. *Small Rumin. Res.* **2014**. [CrossRef]
40. Crist, E. Can an insect speak? The case of the honeybee dance language. *Soc. Stud. Sci.* **2004**. [CrossRef]
41. Smith, B.M.; Basu, P.C.; Chatterjee, A.; Chatterjee, S.; Dey, U.K.; Dicks, L.V.; Giri, B.; Laha, S.; Majhi, R.K.; Basu, P. Collating and validating indigenous and local knowledge to apply multiple knowledge systems to an environmental challenge: A case-study of pollinators in India. *Biol. Conserv.* **2017**. [CrossRef]

42. Hernández-Morcillo, M.; Hoberg, J.; Oteros-Rozas, E.; Plieninger, T.; Gómez-Baggethun, E.; Reyes-García, V. Traditional ecological knowledge in Europe: Status Quo and insights for the environmental policy agenda. *Environment* **2014**. [CrossRef]

43. Oteros-Rozas, E.; Ontillera-Sánchez, R.; Sanosa, P.; Gómez-Baggethun, E.; Reyes-García, V.; González, J.A. Traditional ecological knowledge among transhumant pastoralists in Mediterranean Spain. *Ecol. Soc.* **2013**. [CrossRef]

44. Suryanarayanan, S.; Kleinman, D.L. Be(e)coming experts: The controversy over insecticides in the honeybee colony collapse disorder. *Soc. Stud. Sci.* **2013**. [CrossRef]

45. Suryanarayanan, S. Balancing control and complexity in field studies of neonicotinoids and honeybee health. *Insects* **2013**, *4*, 153–167. [CrossRef]

46. Lehébel-Péron, A.; Sidawy, P.; Dounias, E.; Schatz, B. Attuning local and scientific knowledge in the context of global change: The case of heather honey production in southern France. *J. Rural Stud.* **2016**. [CrossRef]

47. Neumann, P.; Blacquière, T. The Darwin cure for apiculture? Natural selection and managed honeybee health. *Evol. Appl.* **2017**. [CrossRef]

48. Loftus, J.C.; Smith, M.L.; Seeley, T.D. How honeybee colonies survive in the wild: Testing the importance of small nests and frequent swarming. *PLoS ONE* **2016**. [CrossRef]

49. Thoms, C.A.; Nelson, K.C.; Kubas, A.; Steinhauer, N.; Wilson, M.E. Beekeeper stewardship, colony loss, and *Varroa destructor* management. *Ambio* **2019**. [CrossRef]

50. Peck, D.T.; Seeley, T.D. Mite bombs or robber lures? The roles of drifting and robbing in *Varroa destructor* transmission from collapsing honeybee colonies to their neighbors. *PLoS ONE* **2019**. [CrossRef]

51. Fürst, M.A.; McMahon, D.P.; Osborne, J.L.; Paxton, R.J.; Brown, M.J.F. Disease associations between honeybees and bumblebees as a threat to wild pollinators. *Nature* **2014**. [CrossRef]

52. Rinderer, T.E.; Harris, J.W.; Hunt, G.J.; de Guzman, L.I. Breeding for resistance to *Varroa destructor* in North America. *Apidologie* **2010**. [CrossRef]

53. De Mattos, I.M.; Soares, A.E.E.; Tarpy, D.R. Effects of synthetic acaricides on honeybee grooming behavior against the parasitic *Varroa destructor* mite. *Apidologie* **2017**. [CrossRef]

54. Sponsler, D.B.; Grozinger, C.M.; Hitaj, C.; Rundlöf, M.; Botías, C.; Code, A.; Lonsdorf, E.V.; Melathopoulos, A.P.; Smith, D.J.; Suryanarayanan, S.; et al. Pesticides and pollinators: A socioecological synthesis. *Sci. Total Environ.* **2019**. [CrossRef]

55. Tengö, M.; Brondizio, E.S.; Elmqvist, T.; Malmer, P.; Spierenburg, M. Connecting diverse knowledge systems for enhanced ecosystem governance: The multiple evidence base approach. *Ambio* **2014**. [CrossRef] [PubMed]

56. Brown, M.J.F.; Paxton, R.J. The conservation of bees: A global perspective. *Apidologie* **2009**. [CrossRef]

57. Hall, D.M.; Steiner, R. Insect pollinator conservation policy innovations: Lessons for lawmakers. *Environ. Sci. Policy* **2019**. [CrossRef]

58. Willett, W.; Rockström, J.; Loken, B.; Springmann, M.; Lang, T.; Vermeulen, S.; Garnett, T.; Tilman, D.; DeClerck, F.; Wood, A.; et al. Food in the Anthropocene: The EAT–Lancet Commission on healthy diets from sustainable food systems. *Lancet* **2019**. [CrossRef]

59. Vanbergen, A.J.; Garratt, M.P.; Vanbergen, A.J.; Baude, M.; Biesmeijer, J.C.; Britton, N.F.; Brown, M.J.F.; Brown, M.; Bryden, J.; Budge, G.E.; et al. Threats to an ecosystem service: Pressures on pollinators. *Front. Ecol. Environ.* **2013**. [CrossRef]

60. Kleijn, D.; van Langevelde, F. Interacting effects of landscape context and habitat quality on flower visiting insects in agricultural landscapes. *Basic Appl. Ecol.* **2006**. [CrossRef]

61. Klein, A.-M.; Vaissiere, B.E.; Cane, J.H.; Steffan-Dewenter, I.; Cunningham, S.A.; Kremen, C.; Tscharntke, T. Importance of pollinators in changing landscapes for world crops. *Proc. R. Soc. B Biol. Sci.* **2007**, *274*, 303–313. [CrossRef] [PubMed]

62. Nottebrock, H.; Schmid, B.; Mayer, K.; Devaux, C.; Esler, K.J.; Böhning-Gaese, K.; Schleuning, M.; Pagel, J.; Schurr, F.M. Sugar landscapes and pollinator-mediated interactions in plant communities. *Ecography Cop.* **2017**, *40*, 1129–1138. [CrossRef]

63. Odoux, J.F.; Feuillet, D.; Aupinel, P.; Loublier, Y.; Tasei, J.N.; Mateescu, C. Territorial biodiversity and consequences on physico-chemical characteristics of pollen collected by honeybee colonies. *Apidologie* **2012**. [CrossRef]

64. Donkersley, P.; Rhodes, G.; Pickup, R.W.; Jones, K.C.; Wilson, K. Honeybee nutrition is linked to landscape composition. *Ecol. Evol.* **2014**. [CrossRef] [PubMed]

65. Simpson, S.J.; Raubenheimer, D. *The Nature of Nutrition: A Unifying Framework from Animal Adaptation to Human Obesity*; Princeton University Press: Princeton, NJ, USA, 2012; ISBN 9781400842803.

66. Kwang, P.L.; Simpson, S.J.; Clissold, F.J.; Brooks, R.; Ballard, J.W.O.; Taylor, P.W.; Soran, N.; Raubenheimer, D. Lifespan and reproduction in Drosophila: New insights from nutritional geometry. *Proc. Natl. Acad. Sci. USA* **2008**. [CrossRef]

67. Lee, K.P.; Cory, J.S.; Wilson, K.; Raubenheimer, D.; Simpson, S.J. Flexible diet choice offsets protein costs of pathogen resistance in a caterpillar. *Proc. R. Soc. B Biol. Sci.* **2006**. [CrossRef]

68. Ruedenauer, F.A.; Raubenheimer, D.; Kessner-Beierlein, D.; Grund-Mueller, N.; Noack, L.; Spaethe, J.; Leonhardt, S.D. Best be(e) on low fat: Linking nutrient perception, regulation and fitness. *Ecol. Lett.* **2020**. [CrossRef]

69. Donkersley, P. Trees for bees. *Agric. Ecosyst. Environ.* **2019**. [CrossRef]

70. Deguines, N.; Jono, C.; Baude, M.; Henry, M.; Julliard, R.; Fontaine, C. Large-scale trade-off between agricultural intensification and crop pollination services. *Front. Ecol. Environ.* **2014**. [CrossRef]

71. Holzschuh, A.; Dainese, M.; González-Varo, J.P.; Mudri-Stojnić, S.; Riedinger, V.; Rundlöf, M.; Scheper, J.; Wickens, J.B.; Wickens, V.J.; Bommarco, R.; et al. Mass-flowering crops dilute pollinator abundance in agricultural landscapes across Europe. *Ecol. Lett.* **2016**. [CrossRef]

72. Woodcock, B.A.; Bullock, J.M.; Shore, R.F.; Heard, M.S.; Pereira, M.G.; Redhead, J.; Ridding, L.; Dean, H.; Sleep, D.; Henrys, P.; et al. Country-specific effects of neonicotinoid pesticides on honeybees and wild bees. *Science* **2017**. [CrossRef] [PubMed]

73. Motta, E.V.S.; Raymann, K.; Moran, N.A. Glyphosate perturbs the gut microbiota of honeybees. *Proc. Natl. Acad. Sci. USA* **2018**. [CrossRef]

74. Zaluski, R.; Justulin, L.A.; Orsi, R.D.O. Field-relevant doses of the systemic insecticide fipronil and fungicide pyraclostrobin impair mandibular and hypopharyngeal glands in nurse honeybees (*Apis mellifera*). *Sci. Rep.* **2017**. [CrossRef] [PubMed]

75. Reed, M.S.; Curzon, R. Stakeholder mapping for the governance of biosecurity: A literature review. *J. Integr. Environ. Sci.* **2015**. [CrossRef]

76. Zinsstag, J. Convergence of ecohealth and one health. *Ecohealth* **2012**. [CrossRef] [PubMed]

77. Donnelly, A.; Crowe, O.; Regan, E.; Begley, S.; Caffarra, A. The role of citizen science in monitoring biodiversity in Ireland. *Int. J. Biometeorol.* **2014**. [CrossRef] [PubMed]

78. Jalbert, K.; Kinchy, A.J.; Perry, S.L. Civil society research and Marcellus Shale natural gas development: Results of a survey of volunteer water monitoring organizations. *J. Environ. Stud. Sci.* **2014**. [CrossRef]

79. van der Steen, J.S.; Brodschneider, R. Public participation in bee science: C.S.I. pollen. *Bee World* **2014**. [CrossRef]

80. Bonney, R.; Cooper, C.B.; Dickinson, J.; Kelling, S.; Phillips, T.; Rosenberg, K.V.; Shirk, J. Citizen science: A developing tool for expanding science knowledge and scientific literacy. *Bioscience* **2009**. [CrossRef]

81. Ellis, R.; Waterton, C. Environmental citizenship in the making: The participation of volunteer naturalists in UK biological recording and biodiversity policy. *Sci. Public Policy* **2004**. [CrossRef]

82. Tang, M.; Hardman, C.J.; Ji, Y.; Meng, G.; Liu, S.; Tan, M.; Yang, S.; Moss, E.D.; Wang, J.; Yang, C.; et al. High-throughput monitoring of wild bee diversity and abundance via mitogenomics. *Methods Ecol. Evol.* **2015**. [CrossRef] [PubMed]

83. Lefebvre, M.; Espinosa, M.; Gomez y Paloma, S.; Paracchini, M.L.; Piorr, A.; Zasada, I. Agricultural landscapes as multi-scale public good and the role of the common agricultural policy. *J. Environ. Plan. Manag.* **2015**, *58*, 2088–2112. [CrossRef]

84. Hadley, A.S.; Betts, M.G. The effects of landscape fragmentation on pollination dynamics: Absence of evidence not evidence of absence. *Biol. Rev.* **2012**, *87*, 526–544. [CrossRef] [PubMed]

85. Eastburn, D.J.; O'Geen, A.T.; Tate, K.W.; Roche, L.M. Multiple ecosystem services in a working landscape. *PLoS ONE* **2017**, *12*, e0166595. [CrossRef]

86. Hall, M.A.; Nimmo, D.G.; Cunningham, S.A.; Walker, K.; Bennett, A.F. The response of wild bees to tree cover and rural land use is mediated by species' traits. *Biol. Conserv.* **2019**. [CrossRef]

87. Wilson, J.S.; Forister, M.L.; Carril, O.M. Interest exceeds understanding in public support of bee conservation. *Front. Ecol. Environ.* **2017**. [CrossRef]

88. Campbell, A.J.; Wilby, A.; Sutton, P.; Wäckers, F. Getting more power from your flowers: Multi-functional flower strips enhance pollinators and pest control agents in apple orchards. *Insects* **2017**, *8*, 101. [CrossRef]

89. Padilla, F.M.; Pugnaire, F.I. Species identity and water availability determine establishment success under the canopy of *Retama sphaerocarpa* shrubs in a dry environment. *Restor. Ecol.* **2009**, *17*, 900–907. [CrossRef]

90. Breeze, T.D.; Vaissière, B.E.; Bommarco, R.; Petanidou, T.; Seraphides, N.; Kozák, L.; Scheper, J.; Biesmeijer, J.C.; Kleijn, D.; Gyldenkærne, S.; et al. Agricultural policies exacerbate honeybee pollination service supply-demand mismatches across Europe. *PLoS ONE* **2014**. [CrossRef]

91. Bixler, P.R.; Dell'Angelo, J.; Mfune, O.; Roba, H. The political ecology of participatory conservation: Institutions and discourse. *J. Polit. Ecol.* **2015**. [CrossRef]

92. Rauschmayer, F.; van den Hove, S.; Koetz, T. Participation in EU biodiversity governance: How far beyond rhetoric? *Environ. Plan. C Gov. Policy* **2009**. [CrossRef]

93. Naylor, R.; Hamilton-Webb, A.; Little, R.; Maye, D. The 'good farmer': Farmer identities and the control of exotic livestock disease in England. *Sociol. Ruralis* **2018**. [CrossRef]

94. Smith, T.J.; Saunders, M.E. Honeybees: The queens of mass media, despite minority rule among insect pollinators. *Insect Conserv. Divers.* **2016**. [CrossRef]

95. Rousham, E.K.; Unicomb, L.; Islam, M.A. Human, animal and environmental contributors to antibiotic resistance in low-resource settings: Integrating behavioural, epidemiological and one health approaches. *Proc. R. Soc. B Biol. Sci.* **2018**. [CrossRef] [PubMed]

96. Ollerton, J. Pollinator diversity: Distribution, ecological function, and conservation. *Annu. Rev. Ecol. Evol. Syst.* **2017**. [CrossRef]

97. Engel, P.; Martinson, V.G.; Moran, N.A. Functional diversity within the simple gut microbiota of the honey bee. *Proc. Natl. Acad. Sci. USA* **2012**. [CrossRef] [PubMed]

98. Raymann, K.; Moran, N.A. The role of the gut microbiome in health and disease of adult honey bee workers. *Curr. Opin. Insect Sci.* **2018**. [CrossRef] [PubMed]

99. Johnson, R.M.; Evans, J.D.; Robinson, G.E.; Berenbaum, M.R. Changes in transcript abundance relating to colony collapse disorder in honey bees (*Apis mellifera*). *Proc. Natl. Acad. Sci. USA* **2009**, *106*, 14790–14795. [CrossRef]

100. DeGrandi-Hoffman, G.; Chen, Y. Nutrition, immunity and viral infections in honey bees. *Curr. Opin. Insect Sci.* **2015**, *10*, 170–176. [CrossRef]

101. Engel, P.; Kwong, W.K.; Moran, N.A. *Frischella perrara* gen. nov., sp. nov., a gammaproteobacterium isolated from the gut of the honeybee, *Apis mellifera*. *Int. J. Syst. Evol. Microbiol.* **2013**. [CrossRef]

102. D'Alvise, P.; Seeburger, V.; Gihring, K.; Kieboom, M.; Hasselmann, M. Seasonal dynamics and co-occurrence patterns of honey bee pathogens revealed by high-throughput RT-qPCR analysis. *Ecol. Evol.* **2019**. [CrossRef] [PubMed]

Article

Using Manual and Computer-Based Text-Mining to Uncover Research Trends for *Apis mellifera*

Esmaeil Amiri [1,*,†], Prashant Waiker [1,†], Olav Rueppell [1] and Prashanti Manda [2]

1   Department of Biology, University of North Carolina at Greensboro, Greensboro, NC 27402, USA;
    p_waiker@uncg.edu (P.W.); o_ruppel@uncg.edu (O.R.)
2   Department of Computer Science, University of North Carolina at Greensboro, Greensboro, NC 27402, USA;
    p_manda@uncg.edu
*   Correspondence: e.amiri79@gmail.com
†   These authors contributed equally to this work.

Received: 28 March 2020; Accepted: 2 May 2020; Published: 6 May 2020

Abstract: Honey bee research is believed to be influenced dramatically by colony collapse disorder (CCD) and the sequenced genome release in 2006, but this assertion has never been tested. By employing text-mining approaches, research trends were tested by analyzing over 14,000 publications during the period of 1957 to 2017. Quantitatively, the data revealed an exponential growth until 2010 when the number of articles published per year ceased following the trend. Analysis of author-assigned keywords revealed that changes in keywords occurred roughly every decade with the most fundamental change in 1991–1992, instead of 2006. This change might be due to several factors including the research intensification on the *Varroa* mite. The genome release and CCD had quantitively only minor effects, mainly on honey bee health-related topics post-2006. Further analysis revealed that computational topic modeling can provide potentially hidden information and connections between some topics that might be ignored in author-assigned keywords.

Keywords: text-mining; topic modeling; colony collapse disorder; genomics; *Varroa* mite; honey bee health; *Apis mellifera*

---

1. Introduction

Western honey bees, *Apis mellifera*, are of interest due to their beneficial products and impact on food security [1,2]. Their role as general pollinators and the easy mobility of large colonies with thousands of worker bees make them an indispensable component of modern agricultural systems [2,3]. Furthermore, they are an attractive scientific model to study caste development, haplo-diploidy, eusociality, symbolic language, and many other fundamental scientific topics [4,5].

Although the number of honey bee colonies has increased on a global scale, the demand for the pollination service of cultivated crops along with high overwintering colony mortality and changes in the political and socioeconomic system threaten the sustainability of local honey bee industries [6–10]. The decline in the population of pollinators has been of concern for stakeholders since the 1990s. In 2006, the description of a novel honey bee health problem called colony collapse disorder (CCD) allegedly caused a surge in research to understand and overcome this calamity [11,12]. Concomitantly, the release of the sequenced genome of *Apis mellifera* [13] facilitated new tools, which may have triggered a surge in fundamental research on honey bee health and biology [14]. However, the quantitative and qualitative consequences for the scientific output of honey bee research and sub-disciplines are unclear. A comprehensive aggregation of scientific knowledge to explore deep insights into apicultural research trends is lacking, which have been shown for other sub-fields of science [15,16]. It was estimated that the overall annual growth of scientific peer-reviewed articles was at a rate of 8–9 percent in recent years [15]. This volume of scientific output can exceed the capacity of researchers to keep track of every single publication and keep up with the pace of change in research within their field of expertise. Although review articles can effectively sum up the current state of research on a particular topic, they have been frequently criticized for presenting only a narrow and subjective view instead of a comprehensive update on the research field [17–19]. Moreover, connections and relationships between research topics are often overlooked by the reader because papers can contain valuable hidden knowledge beyond their key findings [20,21]. The development and application of text-mining tools and machine learning algorithms have been on the rise to address these limitations [15,16,19,20]. Text-mining tries to reduce the human role and instead utilize functional automated or semi-automated tools allowing researchers to evaluate the large volume of existing literature in an efficient manner [16,22,23].

In this study, we estimated overall research publication growth on *Apis mellifera* and tested the hypothesis that the release of the honey bee genome and the report of colony collapse disorder had a significant impact on shifting the focus of honey bee research. First, we extracted author-assigned keywords associated with the honey bee literature and analyzed them to understand usage and connections between different keywords over time. Next, we analyzed trends in honey bee research before and after 2006 to test our hypothesis. We also explored temporal trends for certain keywords of interest and investigated differences in keyword prevalence before and after 2006. We complemented these manual efforts with automated topic-modeling that used the abstract and title as additional data sources to compare manual and computer-based analyses of the literature.

2. Methods

2.1. Dataset

The Scopus database (www.scopus.com) was queried on 9 June 2018 using the search term "apis mellifera". All publications with the search term in the title, abstract, or keywords were retrieved as a subset of the honey bee literature. A minimum threshold of five publications per year led to the inclusion of all years from 1957 to 2017. Due to incomplete coverage and indexing at the time of retrieval, the data from 2018 was not included.

We further selected two different approaches to analyze this dataset. For author-assigned keyword analysis, we only used keywords, while for computer-based keyword analysis, we used the combination of keywords, abstracts, and titles.

2.2. Author-Assigned Keywords' Analysis

2.2.1. Pre-Processing the Dataset

Most of the retrieved publications from Scopus contained a set of author-assigned keywords. These keywords were pre-processed via stemming and manual synonymization to remove redundancies due to lexical differences across different publications. Initial manual exploration of keywords across publications revealed that authors often used synonyms of the same word. We manually assigned the keywords to a consistent synonym where applicable. For example, "honey-bee (apis mellifera)", "honey-bee apis mellifera", "honey bee a. mellifera", and "honey bee apis mellifera" were found as keywords in different publications. These keywords were all assigned "apis mellifera" as the synonym. We limited the manual synonymization process to the set of keywords that occurred more than once in the full dataset (publications from 1957–2017). Further, syntactic issues among the keywords such as different forms of the same keyword (e.g., singular and plural) were addressed by stemming keywords using the Porter Stemmer [24]. Stemming is a process that reduces a word to its etymological root, thereby removing redundancies due to different forms of the same root. For example, the words "foraging" and "forage" stem to" forag". Similarly, the words "colony" and "colonies" stem to "coloni". The final set of synonymized keywords were analyzed using multiple techniques, as described in subsequent sections.

2.2.2. Temporal Trends

The temporal trends of all keywords were examined to identify the change of interest in specific research areas over time. For each keyword, we analyzed the proportion of publications containing the particular keyword in each year of the dataset.

2.2.3. Cluster Analysis

Individual years in the data were clustered based on their similarity in keyword frequency to test whether 2006 represented a fundamental change in honey bee publication patterns. To analyze differences before and after 2006, the dataset of synonymized keywords was separated into two datasets corresponding to the time periods of (1) 1957–2005 and (2) 2006–2017. The top 50 most frequently occurring keywords were selected from each of the two time periods. After removal of the search term "apis mellifera", the remaining 49 keywords from each period were combined, resulting in 65 unique keywords that were used for the subsequent cluster analysis. Cluster analysis was performed by hierarchical clustering (R *hclust* function) with the Ward distance method (R "*ward.D*" method) [25]. To validate the clustering, we performed bootstrapping on our data for 1000 iterations using the R package "fpc" [26]. The keyword distribution over the resulting clusters was analyzed by creating a heatmap. The "RColorBrewer" package was further applied for better visualization and color enhancement. All the operations were performed in the RStudio v1.0.143 environment [27]. To identify dominant research topics across each cluster, we calculated the expected frequency for each keyword, based on the marginal sums (across all keywords for each year and for each keyword across all years) and then built a ratio (observed/expected). For each cluster, the geometric mean of these ratios was calculated (transformed by adding 0.001 to avoid division by zero) for each keyword across all years in the cluster. The three keywords with the highest observed/expected ratio were selected to illustrate dominant research topics of each cluster.

2.2.4. Network Analysis

Networks illustrating the representation and co-occurrence of keywords were built and analyzed to explore connectivity among the common keywords. Two networks were created using the Gephi visualization software v0.9.2 [28] based on the top 49 keywords from the two time periods; 1957–2005 and 2006–2017. Each node represents a unique keyword and connections (edges) between two keywords were drawn if they co-occurred together in more publications than expected by chance according to the following calculation:

Consider keywords i and j. Keyword pairs where the observed co-occurrence probability (O) is greater than the expected co-occurrence probability (E) are connected via an edge. O is defined as the number of publications containing both i and j as keywords. E is defined as $p(i) * p(j)$ where $p(i)$ is the probability of observing keywords i and j, respectively. Edge weights are proportional to $E - O$; the thickness of the connections represents the difference between the expected and observed probabilities of co-occurrence: a higher thickness depicts a higher observed co-occurrence.

The transitivity (known as the clustering coefficient) was computed on keyword networks generated from both time periods to explore differences in edge density. Transitivity is a measure of the degree to which nodes in a graph tend to cluster together [29]. It was computed using the density of triplets of nodes in a network graph [30]. Transitivity was calculated using the *transitivity()* function of the R package "igraph" [31]. To validate the results of transitivity, 95% CI were calculated by bootstrapping over 1000 iterations for each dataset using a custom R script.

2.3. Computer-Based Keywords' Analysis

2.3.1. Preparing and Pre-Processing the Corpora

According to the previously described hypothesis, two different corpora were built to conduct topic modeling from articles retrieved via Scopus: (1) pre-2006, which contained data (title, keywords, and abstract) for 5640 articles published between 1957 and 2005, and (2) post-2006, which contained data (title, keywords, and abstract) for 8473 articles published between 2006 and 2017. Each corpus was pre-processed using the steps of (1) corpus cleaning in which scientific notations, special characters, punctuation, and numbers were removed and (2) stemming, performed using the Porter Stemmer [24].

2.3.2. Applying Topic Modeling

Topic modeling is a text-mining technique to analyze large volumes of text to discover latent topics and patterns within texts. We applied latent Dirichlet allocation (LDA) [21] to both of the corpora. LDA assumes that the topic distribution has a sparse Dirichlet prior that supports the intuition that documents consist of a mixture of topics and that these topics can be described using sets of relevant words. One of the inputs to the LDA topic model is the number of topics to be generated from the corpus. Selecting the optimal number of topics from any corpus is a problem that has received extensive attention. As a result, several automated metrics such as perplexity [32] and coherence [33] have been developed to evaluate topic models. Several studies report that inferences of topic models' quality based on perplexity were negatively correlated with human perception [34,35]. Recent work [33] suggested coherence to be a measure that aligns better with the human perception of a model's quality.

We used a two-pronged approach to select the appropriate number of topics to be generated from the corpora. We developed models with 20, 50, 70, 90, 110, and 140 topics and evaluated the coherence scores using the c_v coherence metric [36]. Subsequently, the topics were manually inspected by domain scientists on our team to select the most representative model for the data. Based on these metrics, the model with 20 topics was selected for further analyses in this study. The topic models were analyzed/visualized as follows:

Overall view using LDAvis: First, the topic models were presented using LDAvis [37], a web-based interactive visualization tool. The visualization provided an overview of all topics while highlighting the important words associated with each topic. To confirm that this automated protocol yielded a meaningful grouping of the words, we evaluated the top ten words in each topic and manually labeled them to the most appropriate research focus they represented. LDAvis allows the user to glance over individual topics while keeping the entire topic landscape in view and is thus helpful to the user when interpreting and labeling topics.

Network analysis of topics: Next, we created two networks, using the visualization software, Gephi, populated by the top 5 words associated with each of the 20 topics in both time periods. These networks indicated the important words and scientific areas in each period as inferred by the topic modeling algorithm. Nodes in each network are important words for a topic, and edges connect words that co-occur in a topic. Further, to compare the density of interconnection in the network plots, transitivity analysis was performed as described in the previous section.

Comparison between the topic model and keyword networks: To make a comparison of human (keywords) versus automated methods (topic modeling), networks built using the two approaches for each time period were compared to estimate overlaps and differences. The networks were compared to identify overlaps in nodes and edges. An overlap in nodes was noted if the same node was present in both networks or if a node in one network was a substring in the other. Similarly, an overlapping edge was identified in one of three cases: (1) an edge between the same pair of nodes was present in both networks; (2) an edge between a node pair with one node matched exactly, and the other was a substring; and (3) an edge between a node pair where both nodes were related via substrings.

3. Results

3.1. Dataset

The Scopus search resulted in 14,113 articles with publication years ranging from 1957 to 2017. The publication growth could be approximated ($R^2 = 0.98$, n = 55, $p < 0.001$) by the exponential function ($y = 6.29\,e^{0.087x}$), with the number of publications doubling every 7.97 years. The exponential growth continued until 2010, after which the number of retrieved publications did not further increase consistently (Figure 1).

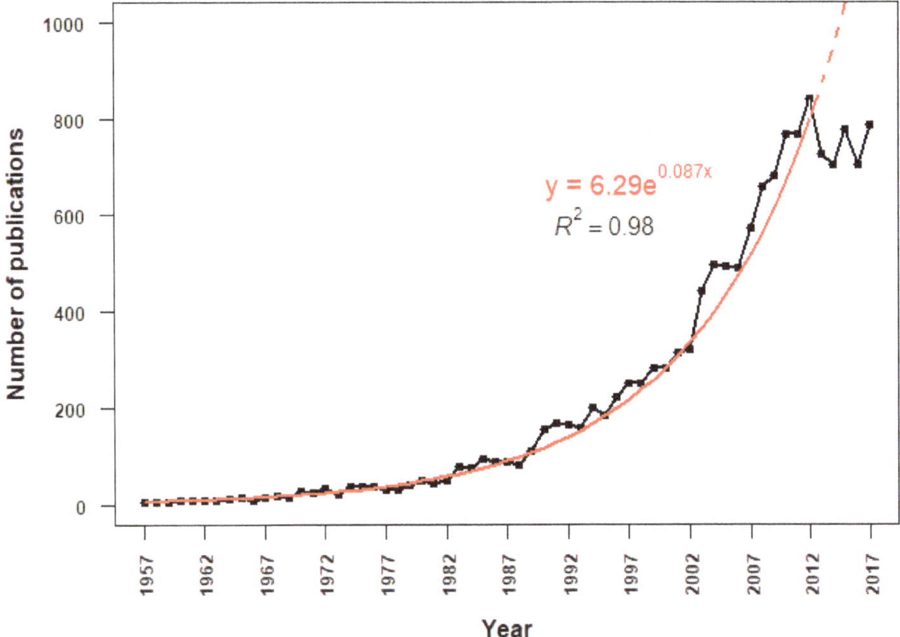

Figure 1. Publication trends over time for research articles retrieved from Scopus using the search Table 1957 to 2017. The publication growth fit an exponential regression curve until 2010, after which the growth leveled off with the exception of 2012.

3.2. Author-Assigned Keywords' Analysis

3.2.1. Temporal Trends

In general, the temporal trends of relative keywords abundance followed one of three trends: (1) general exponential increase, similar to the total number of publications; (2) abrupt increase without any prior occurrences; (3) increase with subsequent decrease. The average occurrence of most keywords increased with time, although the smaller number of publications per year in earlier years caused some large fluctuations. We selected a few keywords associated with two research foci "health" and "genomics" to highlight these temporal trends (Figure S1).

3.2.2. Cluster Analysis

Publication years were clustered based on similar usage of the 65 most common keywords prior to and after 2006 to evaluate the overall change in research focus over time (Figure 2). As shown in Figure 2, the most fundamental split among years occurred between 1991 and 1992 with the exception of 1993. The sub-clusters in both time periods corresponded roughly to decades. Moreover, the usage of words was plotted over the year cluster using a heatmap to explain the basis of clustering (Figure S2). Most top keywords were absent in the early years. The top keywords started to show up during the 1970s, and the years from 1992 to 2017 showed the highest frequency of these words (Figure S2).

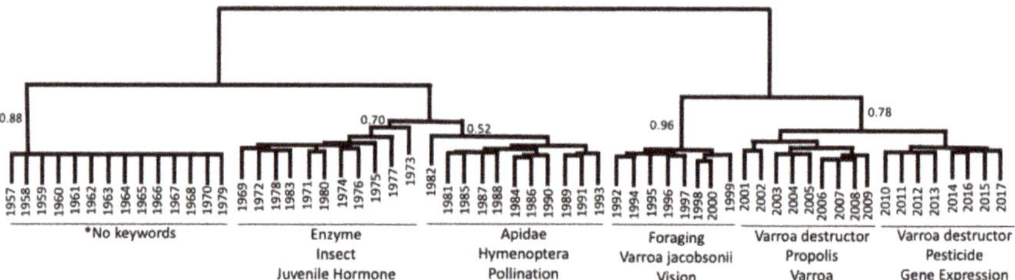

Figure 2. Cluster analysis of years based on the relative frequency of the overall most common aggregated keywords. Years clustered approximately into separate decades, but the most fundamental division in the dataset occurred between 1991 (and prior years) and 1992 (and following years). Bootstrap support is given for the major clusters. Furthermore, we show the three most enriched keywords for each cluster. * The most common keywords were not used in early articles, and therefore, the leftmost cluster did not have any over-enriched terms among the top keywords.

3.2.3. Network Analysis

Keyword networks were generated for the time periods of 1957–2005 (Figure 3) and 2006–2017 (Figure 4). Changes in the size of several words, signifying the degree (number of connections within the network) implied that the connections between topics in honey bee science were dynamic. Time changes were further confirmed by the presence of unique top keywords in each time period. For example, the words "behavior", "enzyme", and "brood" in Figure 3 had more connections to other words in the plot, while words such as "vision", "brain", and "neurotransmitter" had lesser connections. We noticed substantial changes in the same words in the post-2006 time period (Figure 4). For instance, words such as "immunity", "pollen", and "foraging" had more connections, while the words "venom", "hygienic", and "vision" had fewer connections. Most of the words were interconnected in both plots, which implied the significant co-occurrence of any set of two words more often than expected by chance. The measure of global clustering in the network (called transitivity) was higher in the period of 1957–2005 (transitivity, T = 0.53) suggesting higher connection density than in the later period (T = 0.44). The confidence intervals for these transitivity values were determined by bootstrapping over 1000 iterations (Table S1), and no overlap between confidence intervals of these transitivity values suggested that they were significantly different from each other. While the relative co-occurrence of some pairs did not change between time periods (e.g., "pollination - flowers" and "pollination - crops"), most highly over-represented connections changed (e.g., "behavior - division of labor" in Figure 3 and "pesticide - neonicotinoid" in Figure 4).

3.3. Computer-Based Topic Modeling

The 20 topics generated from computer-based topic modeling for the two time periods can be explored by the reader in detail using the interactive visualizations available within our data deposit (see the Supplementary Materials). A snapshot of the visualization for the pre-2006 period is shown in Figure S3.

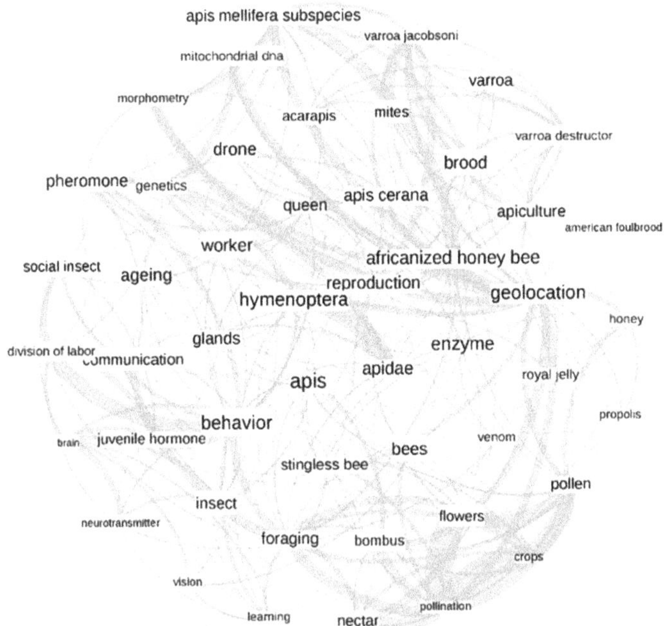

Figure 3. Network analysis plot of the top 49 keywords over the time period 1957–2005. The font size shows the degree of connections to that particular keyword, while each connection between words denotes a significant co-occurrence of the words in a research article. The edge thickness shows how much more keywords co-occur than expected by chance.

Manual analysis of the 20 topics indicated the presence of research themes such as "pollination", "genomics", "behavior", "reproduction", "apiculture", "varroa infestation", etc. The largest topic pre-2006 contained words such as "bee, forage, dance, test, model, pattern, fruit, communication, language, discrimination", indicative of research in behavior. In contrast, the largest topic post-2006 contained words such as "bee, geolocation, pollen, honey bee, apiculture, colony, hive, collect, year, nutrition", indicative of research in apiculture. The relevant words for each topic also were manually analyzed to label topics with research sub-fields where possible. For example, Topic 6 in the pre-2006 time period contained words such as "queen, pheromone, egg, reproduction, larvae, gland, ovary, cast, produce, cell" (Table S2). These words were labeled as "reproduction". Similarly, we identified topics corresponding to "behavior", "pollination", "varroa", and "genomics" pre-2006 and topics corresponding to "nosema infection", "population", and "virus infection" post-2006. Some sub-fields such as "genomics" were observed both pre- and post-2006. While certain topics were coherent and clearly indicative of a specific research area, other topics were deemed by the human observer (the authors) as a mixture of words from different areas and could not be labeled with a specific research theme. Tables S2 and S3 show five example topics corresponding to specific research foci from the two time periods.

Figure 4. Network analysis plot of the top 49 keywords over the time period 2006–2017. The font size shows the degree of connections to that particular keyword, while each connection between words denotes a significant co-occurrence of the words in a research article. The edge thickness shows how much more keywords co-occur than expected by chance.

The top five relevant words from each topic were used to create networks to shed light on the overall patterns of the co-occurrence and interconnection of computer-generated keywords before and after 2006 (Figures 5 and 6). The top five words from each topic were automatically connected to each other and formed a small sub-cluster. These small sub-clusters could be connected to each other through common words among themselves, forming larger clusters in the graph. For example, in Figure 5, words such as "bee", "forage", "dance", "test", and "model" from Topic 1 formed a small sub-cluster, which was connected to the small sub-cluster formed by the words "bee", "coloni", "hive", "day", and "group" from Topic 2 through the common word "bee" and even making a larger cluster by connecting to Topics 3, 4, and 16 by the same common word. Visual inspection revealed that important research connections in specific sub-fields were formed post-2006 as the field matured and new knowledge was gained. For example, the term "queen" was connected to only a few reproductive-related terms such as words "queen", "reproductive", "pheromon", "larva", and "egg" in the pre-2006 network (Figure 5), while neglecting the role of some important terms such as "drone" and "sperm", which appeared in the later time period (Figure 6). Further, we observed more small independent clusters not connected to the rest of the network post-2006 (Figure 6): for example, a cluster with words "brain", "neurotransmitt", "fli", "function", and "receptor" and another cluster with "mite", "varroa", "destructor", "neonicotinoid", and "pesticid" were not connected to the rest of the network, indicating more specialized research areas in the later time period.

Similar to the author-provided keywords, the computer-generated network graphs exhibited a difference in connection density among sub-clusters: it was higher in the 1957–2005 period (transitivity,

T = 0.80) than 2006–2017 (T = 0.71). The confidence intervals for these transitivity values were determined by bootstrapping over 1000 iterations (Table S1), and no overlap between confidence intervals of these transitivity values suggested they were significantly different from each other.

3.4. Author-Assigned vs. Computer-Based Networks

We compared networks from the author-assigned keyword (Figure 3 and Figure 4) to the topic modeling networks (Figure 5 and Figure 6) to explore similarities and differences between the human (keywords) and automated (topic models) methods. The overlap of network nodes between the topic network and keyword network was 44% and 51%, respectively, for pre- and post-2006, while only 23% and 21% of the edges in the topic model network overlapped with the keyword network pre- and post-2006, respectively. The higher topic overlap with keywords and only a fraction of edge overlap suggested that computer and human methods were more congruent in finding single topics than connections.

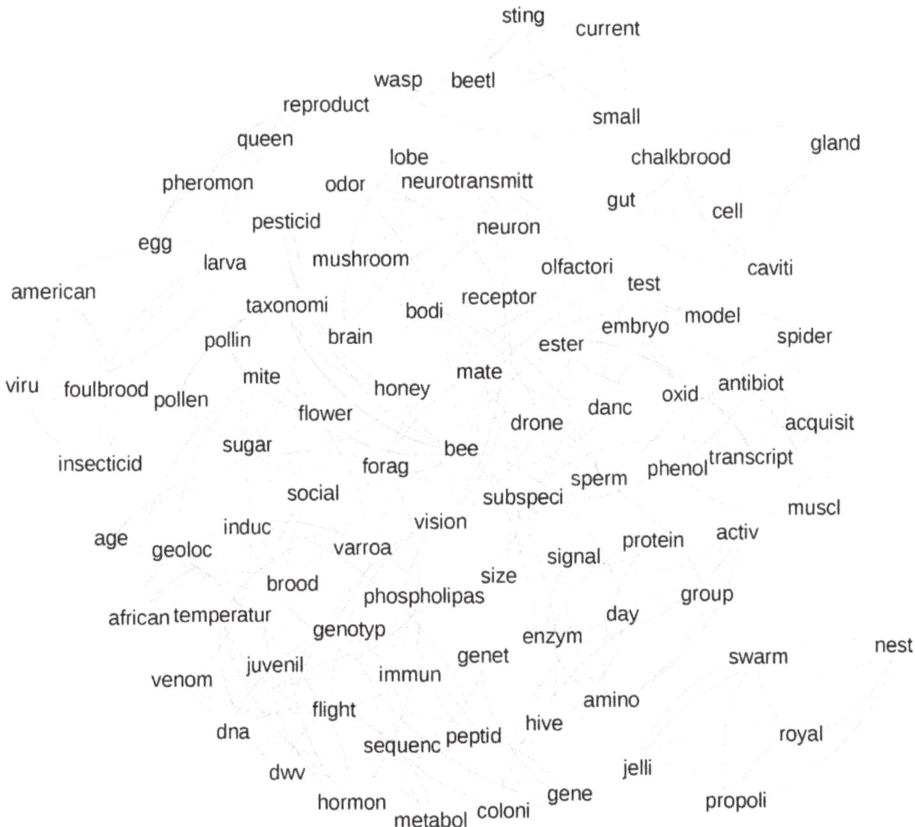

Figure 5. Topic modeling network graph of the top five words from each topic for the period of 1957–2005. The connection between words represent the co-occurrence of the words together in the title, keywords, and abstract of the analyzed articles.

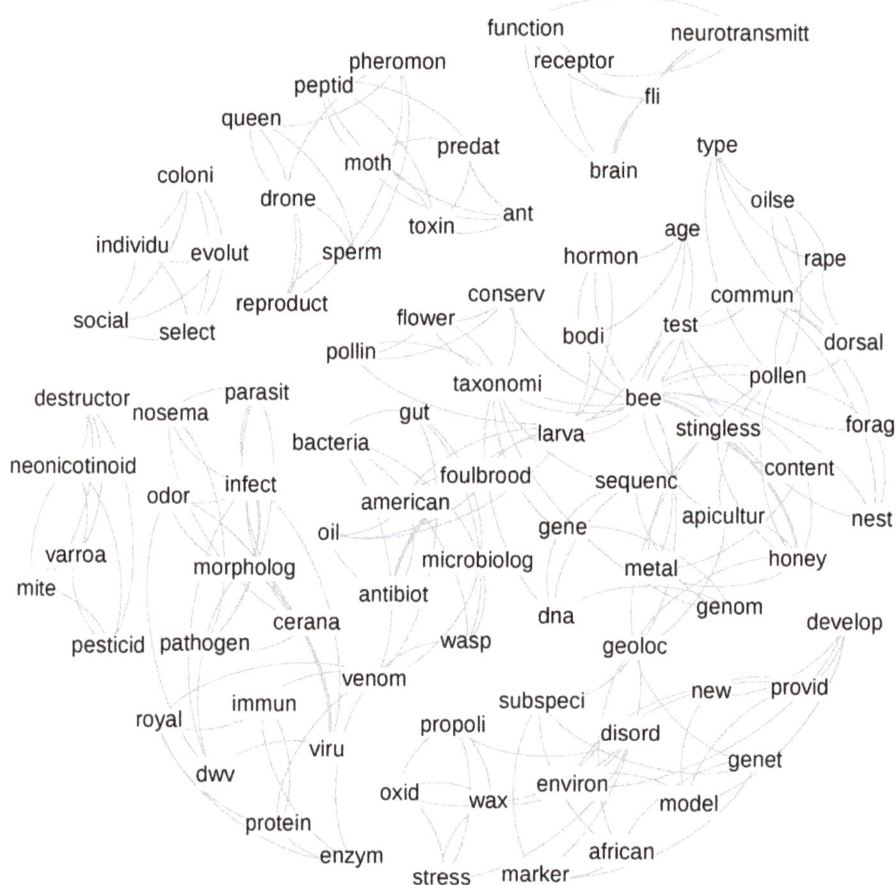

Figure 6. Topic modeling network graph of the top five words from each topic for the period of 2006–2017. The connection between words represents the co-occurrence of the words together in the title, keywords, and abstract of the analyzed articles.

4. Discussion

Continuous discoveries and publications have inundated scientific repositories with a tremendous volume of research articles [38]. The volume of articles threatens to exceed human capacity to read, understand, and comprehend the findings.

Here, we applied text-mining to aggregate the current knowledge and investigate potential research trends in the research field literature related to *Apis mellifera*, using over fourteen thousand scientific articles published between 1957 and 2017 from the Scopus database. Scopus is described as a source-neutral database that curates data from 24,600 journals/conference proceedings spanning across 5000 publishers. Scopus was selected as the source of data collection since we found it to be the most comprehensive resource available in addition to the ease of use for computational applications [39], compared to PubMed and Microsoft Academic Database. We are aware that using a single data source might introduce bias in the retrieved articles, which depend on where researchers choose to publish or a change in the portfolio of journals the database has selected to archive. We retrieved the articles by limiting our search to the term "apis mellifera" and avoided the generic word "honey bee" because the term "honey bee" has been used redundantly for several species, and our intent was to analyze specifically literature on Western honey bees, which are present all over the world [8]. It should be noted that the publications chosen here for our analysis were restricted to those published in English or available in multiple languages, one being English. Scientific literature, particularly before 1957 and shortly after, was predominantly published in German and French. English slowly became the language of scientific communication. We acknowledge that this choice omits a subsection of literature published in other languages portraying a geo-linguistic focus of Western literature culture.

We found that the number of research publications in honey bee science is growing similar to other biological sub-fields [19]. This exponential growth pattern of publications was consistent until the year 2010 with an exceptional of the year 2012. There could be multiple causes for such a growth rate. It could be the increase in the number of scientists during this period [40], technological advancement in the research field [41], or the administrative pressure to get or remain in research positions at academic institutions [42]. Furthermore, part of it can be explained by the onset of publications related to the honey bee genome sequencing project, which was conceptualized during 1998–2001 in several courses, workshops, and conferences [43]. In addition, honey bees have been experiencing higher colony losses in the U.S. and Western Europe, but relatively fewer in other parts of the world since the year 2006, which led to a surge in honey bee health research in the following years [10,44,45]. The saturation toward later years could be the result of the field maturation after publishing mostly descriptive research, leaving larger scale projects and investigations that require greater effort, time, and resources. Alternatively, the leveling off of the publication rate after 2010 might be an artifact of using Scopus as the sole data source. The result may simply reflect shifts in author choices of publication venues or changes in Scopus' publication sources.

We conducted cluster and network analyses on author-assigned keywords to understand trends, connectivity, and shifts in research sub-fields (Figure S2). In the scientific community and the public domain, it is commonly assumed that the discovery of CCD and release of the sequenced genome had a major impact on honey bee science, but this assumption was never tested. Our cluster analysis revealed that the usage of keywords experienced the biggest change in 1991–1992, and not in 2006. This observation might be due to several reasons including technological advancements [41], changes in political and socioeconomic systems [10], and the increasing impact of the parasitic *Varroa* mite [46,47]. There is no doubt that scientific advances depend not only on new ideas, conceptual leaps, and paradigm shifts, but also to a large extent on technology advances that make these steps possible. Technological advancements such as the emergence of PCR, genomic technologies [41], and the invention and use of the World Wide Web across academia are a notable few. These factors might have helped in elevating the research outputs and establishment of improved communications among researchers. Although the *Varroa* mite has been known and widespread in Europe since the 1970s [48], the final globalization of this honey bee health problem might be an explanation for our results [8,49]. It has been shown that research on the *Varroa* mite was by far the most studied bee threat in the early 1990s [50]. This was confirmed in our identification of major research topics across clusters that *Varroa* was a prominent research topic in all clusters from 1992 through 2017 (Figure 2). Our results also suggested that the main concepts and concerns of honey bee research had roughly changed every ten years (Figure 2). This might be due to the requirement and applicability of research; a shift in concepts and the development of advanced methodological tools, which often takes time to replace the earlier issues, and the duration may take up to a decade.

The co-occurrence of author-assigned keywords was explored using network analysis (Figures 3 and 4). The abundance of connections in both networks (pre- and post-2006), showing greater than randomly expected co-occurrence, indicated that many keywords were grouped and did not occur randomly. In part, this might be explained by the paucity of keyword usage in the early years of our dataset. However, tight clusters of correlated keywords, such as the clusters related to pollination ("crops", "flowers", "nectar", and "pollination"), colony products ("pollen", "honey", "royal jelly", "venom", and "propolis"), and taxonomical classifications ("insect", "hymenoptera", "bees", "apis mellifera subspecies", "apis cerana", and "africanized honey bee") were likely reflecting true connections. Overall, our findings indicated that core research related to apicultural science remained unchanged in both time periods. Words like "geolocation" (the term assigned to refer to geographical locations and environmental factors) acted as a connector between different research foci in both time periods, indicating that the interaction between the honey bee and the environment was related to many other concepts regardless of time period [51,52].

Although the year 2006 was not the most significant split in the cluster analysis, the network graphs (Figures 3 and 4) indicated connections between CCD and the genome release. For example, our post-2006 keyword network showed the co-occurrence of words related to genomic tools (RT-PCR and genome) and health (virus and immunity), which suggested that genomic tools were increasingly used to study honey bee health-related topics [14,53,54]. For example, genomic tools such as gene expression analysis facilitated understanding disease susceptibility, social immunity mechanisms, and response to environmental stressors [55,56]. Similarly, sequencing and RT-PCR have helped to sequence and detect pests and pathogens that impact honey bee health [14,57,58].

In addition to author-assigned keywords, the title and abstract provided more information, leading potentially to better insights and hidden information contained in the full articles. However, the increasing amount of information was more challenging to analyze manually. We used computer-based topic modeling to form the topics, which could be visualized conveniently by the LDAvis interactive tool. The results showed that many distinct, recognizable topics were extracted by the computer even though many keywords were shared among different topics.

The network plot generated from topic modeling showed subtle differences between the pre- and post-2006 time periods. The sharing of words in topics was more prominent in the earlier time period than in the later one, apparent as the connection density, which supported the findings of our keywords' network analysis. The topic modeling networks also suggested that the research had become more comprehensive and specialized after 2006. There may be multiple reasons for this observation. Increasing specialization has promoted the research topics related to colony losses in the USA and other countries [44,45], which may also have compartmentalized the science. Another reason could be the significant improvement and developments in genomic and molecular tools in science [14,54], opening new subfields. Our stringent selection of keywords in these network analyses (only 20 topics were chosen and further limited the words to the top five words in each topic) may also have obscured some connections between lower-tier keywords that were shared between these topics.

The comparison of networks obtained from author-assigned keywords to topic model keywords could provide information of the authors' adequacy in describing their overall research in a limited number of keywords. The considerable overlap between nodes was remarkable, but it was clear that topic modeling extracted additional information specifically regarding the connections between topics. Another extension of topic modeling for further improvement could be performed on the entire text rather than on select elements of the articles. Based on the current data, the node overlap increased from the pre-2006 to the post-2006 period, suggesting that authors may have become narrower in their research topics, as well as better at describing their research with relevant keywords over time.

5. Conclusions

In conclusion, our findings suggested that the author-assigned keywords were a decent representation of the research articles. However, computational methods could provide additional information, such as connections between some topics that might be missed by manual reading. This study showed that the year 1991–1992 had a major impact on shifting the honey bee research paradigm compared to the general perception of 2006 being the most influential year. However, CCD and honey bee genome release did fuel the bee research mostly comprised of immunity and health topics. The development of new tools and concepts and the need for practical applications induced clear transitions in honey bee research over time. These transitions were slower than anticipated and indicated that the majority of new research foci formed slowly, perhaps reflecting a healthy compromise between continuity and innovation.

Supplementary Materials: The following are available online at http://www.mdpi.com/2306-7381/7/2/61/s1: Figure S1: Temporal trends of selected keywords are shown over the period of 1957–2017, Figure S2: Heat map for the usage of top aggregated keywords over year cluster from the 1957–2005 and 2006–2017 time periods, Figure S3: Snapshot of the interactive LDAvis presentation of the topics (HTML file to explore and visualize topics and relevant keywords generated by topic modeling with LDAvis interactive tools for the time period 1957–2005 can be found at https://doi.org/10.5281/zenodo.3379018: "TopicModelingCorpusPre200620LDA", as well as for time period 2006–2017: "TopicModelingCorpusPost200620LDA".), Table S1: Confidence Intervals for transitivity values obtained through bootstrapping over 1000 iteration, Table S2: Example topics and associated salient words corresponding to specific research foci from the pre-2006 time period, Table S3: Example topics and associated salient words corresponding to specific research foci from the post-2006 time period.

Author Contributions: E.A., P.M., and O.R. designed the study; all authors performed the experiments; P.M. and P.W. analyzed the data; E.A. and P.W. wrote the manuscript; and O.R. and P.M. edited the manuscript. All authors read and approved the final manuscript.

Funding: This research was funded by the Giant Steps Research Development Grant from the University of North Carolina at Greensboro. This research was performed while Esmaeil Amiri held an NRC Research Associateship award.

Conflicts of Interest: The authors declare no conflict of interest. The funder had no role in the design of the study; in the collection, analyses, or interpretation of data; in the writing of the manuscript; nor in the decision to publish the results.

Data Availability: The data and code used to generate the results reported in this manuscript are publicly accessible via a Creative Commons Attribution 4.0 International license at https://doi.org/10.5281/zenodo.3379018.

Abbreviations

CCD	Colony Collapse Disorder
CI	Confidence Interval
LDA	Latent Dirichlet Allocation
PCR	Polymerase Chain Reaction
RT-PCR	Reverse Transcription Polymerase Chain Reaction

References

1. Morse, R.A.; Calderone, N.W. The value of honey bees as pollinators of US crops in 2000. *Bee Cult.* **2000**, *128*, 1–15.
2. Partap, U. The pollination role of honey bees. In *Honey Bees of Asia*; Hepburn, H., Radloff, S., Eds.; Springer: Berlin/Heidelberg, Germany, 2011; pp. 227–255. [CrossRef]
3. Aizen, M.A.; Garibaldi, L.A.; Cunningham, S.A.; Klein, A.M. Long-term global trends in crop yield and production reveal no current pollination shortage but increasing pollinator dependency. *Curr. Biol.* **2008**, *18*, 1572–1575. [CrossRef]
4. Menzel, R. The honey bee as a model for understanding the basis of cognition. *Nat. Rev. Neurosci.* **2012**, *13*, 758–768. [CrossRef]
5. Cridge, A.G.; Lovegrove, M.R.; Skelly, J.G.; Taylor, S.E.; Petersen, G.E.L.; Cameron, R.C.; Dearden, P.K. The honey bee as a model insect for developmental genetics. *Genesis* **2017**, *55*, e23019. [CrossRef] [PubMed]
6. van der Zee, R.; Pisa, L.; Andonov, S.; Brodschneider, R.; Charrière, J.-D.; Chlebo, R.; Coffey, M.F.; Crailsheim, K.; Dahle, B.; Gajda, A.; et al. Managed honey bee colony losses in Canada, China, Europe, Israel and Turkey, for the winters of 2008–9 and 2009–10. *J. Apic. Res.* **2012**, *51*, 100–114. [CrossRef]
7. Liu, Z.; Chen, C.; Niu, Q.; Qi, W.; Yuan, C.; Su, S.; Liu, S.; Zhang, Y.; Zhang, X.; Ji, T.; et al. Survey results of honey bee (*Apis mellifera*) colony losses in China (2010–2013). *J. Apic. Res.* **2016**, *55*, 29–37. [CrossRef]
8. vanEngelsdorp, D.; Meixner, M.D. A historical review of managed honey bee populations in Europe and the United States and the factors that may affect them. *J. Invertebr. Pathol.* **2010**, *103*, S80–S95. [CrossRef]
9. Potts, S.G.; Biesmeijer, J.C.; Kremen, C.; Neumann, P.; Schweiger, O.; Kunin, W.E. Global pollinator declines: Trends, impacts and drivers. *Trends Ecol. Evol.* **2010**, *25*, 345–353. [CrossRef]
10. Moritz, R.F.A.; Erler, S. Lost colonies found in a data mine: Global honey trade but not pests or pesticides as a major cause of regional honey bee colony declines. *Agric. Ecosyst. Environ.* **2016**, *216*, 44–50. [CrossRef]
11. Holden, C. Report warns of looming pollination crisis in North America. *Science* **2006**, *314*, 397. [CrossRef]

12. vanEngelsdorp, D.; Underwood, R.; Caron, D.; Hayes Jr, J. An estimate of managed colony losses in the winter of 2006-2007: A report commissioned by the apiary inspectors of America. *Am. Bee J.* **2007**, *147*, 599–603.

13. Honeybee Genome Sequencing Consortium. Insights into social insects from the genome of the honey bee *Apis mellifera*. *Nature* **2006**, *443*, 931–949. [CrossRef] [PubMed]

14. Grozinger, C.M.; Robinson, G.E. The power and promise of applying genomics to honey bee health. *Curr. Opin. Insect Sci.* **2015**, *10*, 124–132. [CrossRef] [PubMed]

15. Bornmann, L.; Mutz, R. Growth rates of modern science: A bibliometric analysis based on the number of publications and cited references. *J. Assoc. Inf. Sci. Technol.* **2015**, *66*, 2215–2222. [CrossRef]

16. Comeau, D.C.; Wei, C.H.; Islamaj Dogan, R.; Lu, Z. PMC text mining subset in BioC: About three million full-text articles and growing. *Bioinformatics* **2019**, *35*, 3533–3535. [CrossRef]

17. McAlister, F.A.; Clark, H.D.; van Walraven, C.; Straus, S.E.; Lawson, F.M.E.; Moher, D.; Mulrow, C.D. The medical review article revisited: Has the science improved? *Ann. Intern. Med.* **1999**, *131*, 947–951. [CrossRef]

18. Strube, M.J.; Gardner, W.; Hartmann, D.P. Limitations, liabilities, and obstacles in reviews of the literature: The current status of meta-analysis. *Clin. Psychol. Rev.* **1985**, *5*, 63–78. [CrossRef]

19. Pautasso, M. Publication growth in biological sub-fields: Patterns, predictability and sustainability. *Sustainability* **2012**, *4*, 3234–3247. [CrossRef]

20. Mooney, R.J.; Bunescu, R. Mining knowledge from text using information extraction. *SIGKDD Explor. Newsl.* **2005**, *7*, 3–10. [CrossRef]

21. Blei, D.M.; Ng, A.Y.; Jordan, M.I.; Lafferty, J. Latent dirichlet allocation. *J. Mach. Learn. Res.* **2003**, *3*, 993–1022.

22. Khalili, M.; Rahimi-Movaghar, A.; Shadloo, B.; Mojtabai, R.; Mann, K.; Amin-Esmaeili, M. Global scientific production on illicit drug addiction: A two-decade analysis. *Eur. Addict. Res.* **2018**, *24*, 60–70. [CrossRef] [PubMed]

23. Meyer, D.; Hornik, K.; Feinerer, I. Text mining infrastructure in R. *J. Stat. Softw.* **2008**, *25*, 1–54. [CrossRef]

24. Porter, M.F. Snowball: A language for Stemming Algorithms. Available online: http://snowball.tartarus.org/texts/introduction (accessed on 18 February 2019).

25. Murtagh, F.; Legendre, P. Ward"s hierarchical agglomerative clustering method: Which algorithms implement ward"s criterion? *J. Classif.* **2014**, *31*, 274–295. [CrossRef]

26. Hennig, C. Cluster-wise assessment of cluster stability. *Comput. Stat. Data Anal.* **2007**, *52*, 258–271. [CrossRef]

27. RStudio Team. *RStudio: Integrated Development for R*; RStudio, Inc.: Boston, MA, USA, 2015.

28. Bastian, M.; Heymann, S.; Jacomy, M. Gephi: An open source software for exploring and manipulating networks. In Proceedings of the International AAAI Conference on Web and Social Media, San Jose, CA, USA, 17–20 March 2009; pp. 361–362.

29. Wasserman, S.; Faust, K. *Social Network Analysis: Methods and Applications*; Cambridge University Press: Cambridge, UK, 1994; Volume 8.

30. Opsahl, T.; Panzarasa, P. Clustering in weighted networks. *Soc. Netw.* **2009**, *31*, 155–163. [CrossRef]

31. Csardi, G.; Nepusz, T. The igraph software package for complex network research. *InterJournal Complex Systems* **2006**, *1695*, 1–9. Available online: https://igraph.org (accessed on 5 March 2019).

32. Blei, D.M.; Lafferty, J.D. Correlated topic models. In *Proceedings of the 18th International Conference on Neural Information Processing Systems, Vancouver, BC, Canada, December 2006*; Weiss, Y., Schölkopf, B., Platt, J., Eds.; MIT Press: Cambridge, MA, USA, 2006; pp. 147–154.

33. Newman, D.; Lau, J.H.; Grieser, K.; Baldwin, T. Automatic evaluation of topic coherence. In *Human Language Technologies: The 2010 Annual Conference of the North American Chapter of the Association for Computational Linguistics*; Association for Computational Linguistics: Los Angeles, CA, USA, 2010; pp. 100–108.

34. Chang, J.; Boyd-Graber, J.; Gerrish, S.; Wang, C.; Blei, D.M. Reading tea leaves: How humans interpret topic models. In *Proceedings of the 22nd International Conference on Neural Information Processing Systems, Vancouver, BC, Canada, December 2009*; Bengio, Y., Schuumans, D., Lafferty, J.D., Williams, C.K.I., Culotta, A., Eds.; MIT Press: Cambridge, MA, USA, 2009; pp. 288–296.

35. Chang, J.; Blei, D. Relational topic models for document networks. In Proceedings of the Twelfth International Conference on Artificial Intelligence and Statistics, Clearwater Beach, FL, USA, 16–19 April 2009; pp. 81–88.

36. Roder, M.; Both, A.; Hinneburg, A. Exploring the space of topic coherence measures. In Proceedings of the Eighth ACM International Conference on Web Search and Data Mining, Shanghai, China, 31 January–2 February 2015; pp. 399–408.

37. Sievert, C.; Shirley, K.E. LDAvis: A method for visualizing and interpreting topics. In Proceedings of the Workshop on Interactive Language Learning, Visualization, and Interfaces, Baltimore, MD, USA, 27 June 2014; pp. 63–70.

38. Jinha, A.E. Article 50 million: An estimate of the number of scholarly articles in existence. *Learn. Publ.* **2010**, *23*, 258–263. [CrossRef]

39. Burnham, J.F. Scopus database: A review. *Biomed. Digit. Libr.* **2006**, *3*, e1. [CrossRef]

40. Lamb, C. Open access publishing models: Opportunity or threat to scholarly and academic publishers? *Learn. Publ.* **2004**, *17*, 143–150. [CrossRef]

41. Fields, S. The interplay of biology and technology. *Proc. Natl. Acad. Sci. USA* **2001**, *98*, 10051–10054. [CrossRef]

42. Lawrence, P.A. The politics of publication. *Nature* **2003**, *422*, 259–261. [CrossRef]

43. Robinson, G.E.; Evans, J.D.; Maleszka, R.; Robertson, H.M.; Weaver, D.B.; Worley, K.; Gibbs, R.A.; Weinstock, G.M. Sweetness and light: Illuminating the honey bee genome. *Insect Mol. Biol.* **2006**, *15*, 535–539. [CrossRef] [PubMed]

44. Pettis, J.S.; Delaplane, K.S. Coordinated responses to honey bee decline in the USA. *Apidologie* **2010**, *41*, 256–263. [CrossRef]

45. Moritz, R.F.A.; de Miranda, J.; Fries, I.; Le Conte, Y.; Neumann, P.; Paxton, R.J. Research strategies to improve honey bee health in Europe. *Apidologie* **2010**, *41*, 227–242. [CrossRef]

46. de Guzman, L.I.; Rinderer, T.E. Identification and comparison of *Varroa* species infesting honey bees. *Apidologie* **1999**, *30*, 85–95. [CrossRef]

47. Anderson, D.L.; Trueman, J.W. *Varroa jacobsoni* (Acari: Varroidae) is more than one species. *Exp. Appl. Acarol.* **2000**, *24*, 165–189. [CrossRef]

48. Ritter, W.; Leclercq, E.; Koch, W. Observations on bee and *Varroa* and mite populations in infested honey bee colonies. *Apidologie* **1984**, *15*, 389–399. [CrossRef]

49. Nazzi, F.; Le Conte, Y. Ecology of *Varroa destructor*, the major ectoparasite of the Western honey bee, *Apis mellifera*. *Annu. Rev. Entomol.* **2016**, *61*, 417–432. [CrossRef]

50. Decourtye, A.; Alaux, C.; Le Conte, Y.; Henry, M. Toward the protection of bees and pollination under global change: Present and future perspectives in a challenging applied science. *Curr. Opin. Insect Sci.* **2019**, *35*, 123–131. [CrossRef]

51. Paton, D.C. Honey bees in the Australian environment: Does *Apis mellifera* disrupt or benefit the native biota? *Bioscience* **1993**, *43*, 95–103. [CrossRef]

52. Hung, K.-L.J.; Kingston, J.M.; Albrecht, M.; Holway, D.A.; Kohn, J.R. The worldwide importance of honey bees as pollinators in natural habitats. *Proc. R. Soc. B* **2018**, *285*, e20172140. [CrossRef]

53. Grozinger, C.M.; Flenniken, M.L. Bee viruses: Ecology, pathogenicity, and impacts. *Annu. Rev. Entomol.* **2019**, *64*, 205–226. [CrossRef] [PubMed]

54. Trapp, J.; McAfee, A.; Foster, L.J. Genomics, transcriptomics and proteomics: Enabling insights into social evolution and disease challenges for managed and wild bees. *Mol. Ecol.* **2017**, *26*, 718–739. [CrossRef] [PubMed]

55. Boutin, S.; Alburaki, M.; Mercier, P.-L.; Giovenazzo, P.; Derome, N. Differential gene expression between hygienic and non-hygienic honey bee (*Apis mellifera* L.) hives. *BMC Genom.* **2015**, *16*, e500. [CrossRef] [PubMed]

56. Huang, Q.; Kryger, P.; Le Conte, Y.; Moritz, R.F.A. Survival and immune response of drones of a Nosemosis tolerant honey bee strain towards *N. ceranae* infections. *J. Invertebr. Pathol.* **2012**, *109*, 297–302. [CrossRef]

57. Cox-Foster, D.L.; Conlan, S.; Holmes, E.C.; Palacios, G.; Evans, J.D.; Moran, N.A.; Quan, P.-L.; Briese, T.; Hornig, M.; Geiser, D.M.; et al. A metagenomic survey of microbes in honey bee colony collapse disorder. *Science* **2007**, *318*, 283–287. [CrossRef]

58. Qin, X.; Evans, J.D.; Aronstein, K.A.; Murray, K.D.; Weinstock, G.M. Genome sequences of the honey bee pathogens *Paenibacillus larvae* and *Ascosphaera apis*. *Insect Mol. Biol.* **2006**, *15*, 715–718. [CrossRef]

veterinary sciences

Article

Honey as a Source of Environmental DNA for the Detection and Monitoring of Honey Bee Pathogens and Parasites

Anisa Ribani [1,2], Valerio Joe Utzeri [1,2], Valeria Taurisano [1] and Luca Fontanesi [1,*]

[1] Department of Agricultural and Food Sciences, University of Bologna, Viale Giuseppe Fanin 46, 40127 Bologna, Italy; anisa.ribani2@unibo.it (A.R.); valeriojoe.utzeri2@unibo.it (V.J.U.); valeria.taurisano2@unibo.it (V.T.)

[2] GRIFFA s.r.l., Viale Giuseppe Fanin 48, 40127 Bologna, Italy

[*] Correspondence: luca.fontanesi@unibo.it; Tel.: +39-051-2096535

Received: 29 July 2020; Accepted: 13 August 2020; Published: 15 August 2020

Abstract: Environmental DNA (eDNA) has been proposed as a powerful tool to detect and monitor cryptic, elusive, or invasive organisms. We recently demonstrated that honey constitutes an easily accessible source of eDNA. In this study, we extracted DNA from 102 honey samples (74 from Italy and 28 from 17 other countries of all continents) and tested the presence of DNA of nine honey bee pathogens and parasites (*Paenibacillus larvae*, *Melissococcus plutonius*, *Nosema apis*, *Nosema ceranae*, *Ascosphaera apis*, *Lotmaria passim*, *Acarapis woodi*, *Varroa destructor,* and *Tropilaelaps* spp.) using qualitative PCR assays. All honey samples contained DNA from *V. destructor*, confirming the widespread diffusion of this mite. None of the samples gave positive amplifications for *N. apis*, *A. woodi*, and *Tropilaelaps* spp. *M. plutonius* was detected in 87% of the samples, whereas the other pathogens were detected in 43% to 57% of all samples. The frequency of Italian samples positive for *P. larvae* was significantly lower (49%) than in all other countries (79%). The co-occurrence of positive samples for *L. passim* and *A. apis* with *N. ceranae* was significant. This study demonstrated that honey eDNA can be useful to establish monitoring tools to evaluate the sanitary status of honey bee populations.

Keywords: *Apis mellifera*; DNA analysis; epidemiology; health; *Lotmaria passim*; *Melissococcus plutonius*; *Nosema ceranae*; pathology; *Tropilaelaps*; *Varroa destructor*

---

1. Introduction

Environmental DNA (eDNA), defined as DNA extracted from environmental- or organismal-related specimens or matrixes, has been proposed as a powerful tool to detect and monitor cryptic, elusive, or invasive organisms, including parasites and many other pathogens that might be difficult to sample or to identify—also extending or facilitating the possibility to evaluate their distribution, even over time and geographic areas [1,2]. Environmental DNA approaches vary according to the objective, the organisms and, in turn, the methods of specimen collection, DNA isolation, or enrichment for targeted or metagenomics analyses [3–5]. The use of eDNA has been recently proposed in the apiculture sector for several purposes [6,7].

The problems determined by the global collapse of honey bee colonies, caused by the co-occurrence of debilitating factors, including environmental stressors and the increased frequency and intensity of chronic diseases, has raised the attention on disease monitoring approaches and pathogen identification methods [8,9]. Prognostic eDNA in the honey bee production systems is detectable in the debris collected from the bottom of the beehive. This material has been proposed as the source of simple and non-invasive collectable specimens useful to detect pathogens and assess the health status of honey bee colonies [10–13].

Honey constitutes another easily accessible source of eDNA useful for pathogen detection and honey bee epidemiological studies. For example, several authors proposed polymerase chain reaction (PCR) based-assays to identify the presence of *Paenibacillus larvae* and *Melissococcus plutonius* in naturally infected honey samples [14–17]. These are the two major infecting bacteria in honey bee colonies causing the American and the European foulbroods, respectively [18,19].

Giersch et al. [20] detected the presence of DNA of two microsporidia, *Nosema apis* and *N. ceranae*, in honey sampled in Australia and suggested that the later species was introduced in this continent through either contaminated honey or pollen. These two microsporidia have been significantly associated with the occurrence of colony losses [21–28].

Ascosphaera apis is a widespread fungal pathogen of honey bees causing the chalkbrood disease [29] that is responsible for colony losses, mainly when combined with other stressing conditions, derived by the co-presence of microsporidia or mite parasites and derived viruses [30].

Other parasitic microorganisms associated with increased colony mortalities in combination with other health problems and pathogens are two species of trypanosomatids, *Lotmaria passim* and *Crithidia mellificae*, which have been frequently confused and can be distinguished only using molecular methods [31–34].

Among the honey bee parasites, the tracheal tarsonemid mite *Acarapis woodi* (an internal parasite of the bee tracheal system, distributed in almost all continents; [35]) can causes significant economic damages in some conditions, even if it is not usually considered a real problem despite its quite high prevalence in apiaries of some countries [36].

The ectoparasite mite *Varroa destructor* is far more damaging than the tracheal mite [35]. Varroasis is acknowledged as the worst pest for *Apis mellifera* colonies all over the world, and one of the main responsible for death and reduced populations of overwintering colonies, and a potential factor of colony losses [37,38]. In addition to this mite species, other parasitic mites of the genus *Tropilaelaps* (parasites of *A. dorsata*) have recently raised concern after they have expanded their natural hosts to include *A. mellifera*, after this bee was introduced to Asia [39]. Invasion of *Tropilaelaps* mites in Europe and United States would likely be very damaging for the western apiculture, considering the higher reproduction rate and shorter life cycle that these parasites have than *V. destructor* [40]. To face this emerging threat, monitoring and detection systems have been developed to prevent the introduction of *Tropilaelaps* mites in areas outside Asia, the natural range of these species [41,42].

In this study, we used honey of different botanical and geographical origins as a source of eDNA to detect some of the most important pathogens and parasites of honey bees. We tested and developed qualitative PCR methods to identify two bacteria, *P. larvae* and *M. plutonius*, two microsporidia, *N. apis* and *N. ceranae*, one fungus, *A. apis*, one trypanosomatidae, *L. passim*, and three arthropod parasites, *A. woodi*, *V. destructor* and *Tropilaelaps* spp. The methods applied in this study expanded the possibility to evaluate the occurrence and prevalence of honey bee infecting and parasitizing agents by investigating an unconventional specimen (i.e., honey) than can be useful for epidemiological analyses and monitoring purposes.

2. Materials and Methods

2.1. Honey Samples

Honey samples were purchased from trade markets or were directly provided by beekeepers. These samples were not from a single hive as they derived by the routine procedures for their preparations and packaging into the final commercialized container or bin. A total of 102 honey samples were collected from European countries (n. 87), North, Central, and South American countries (n. 8), Asian countries (n. 5), African, and Oceanian countries (n. 1 for each continent). The European honey samples were from eight countries (n. 3 from Croatia, n. 4 from Finland, n. 2 from France, n. 1

from Greece, n. 1 from Hungary, n. 74 from Italy, n. 1 from Serbia, n. 1 from The Netherlands). The honey samples from the American regions were from USA (n. 3), Mexico (n. 1), Guatemala (n. 1), Brazil (n. 2), and Chile (n. 1). Asian samples were from China (n. 3), India (n. 1), and Japan (n. 1). The African sample was from Ethiopia and the Oceanian sample was from New Zealand. The year of production of the analysed honey samples ranged from 2004 to 2018 (Figure S1). For 13 samples, the precise year of production was not declared, but according to the collected information, it might be between 2015 and 2017. A total of 60 samples had monofloral and 32 had polyfloral origin, whereas 10 were honeydew honey samples. Table S1 reports the list of samples, including their origin and the year of production.

2.2. DNA Extraction from Honey Samples

DNA was extracted from each honey following the protocol previously described [43]. Briefly, for each sample, a total of 50 g of honey was divided into four 50 mL tubes (12.5 g for each tube) then 37.5 mL of ultrapure water was added to each tube. These tubes were vortexed and then incubated at 40 °C for 30 min. In the subsequent step, the tubes were centrifuged for 25 min at 5000× *g* at room temperature and the supernatant was discarded. The obtained pellet was resuspended in 5 mL of ultrapure water and the content of the four tubes was merged in one and then diluted again with ultrapure water. After another centrifugation step (25 min at 5000× *g* at room temperature), the supernatant was discarded and the pellet was resuspended in 0.5 mL of ultrapure water. Then, 1 mL of CTAB extraction buffer [2% (w/v) cetyltrimethylammonium bromide; 1.4 M NaCl; 100 mM Tris-HCl; 20 mM EDTA; pH 8] and 5 μL of RNase A solution (10 mg/mL) were added to each honey pre-treated sample and incubated for 10 min at 60 °C. After the incubation, 30 μL of proteinase K (20 mg/mL) were added and the mix was incubated at 65 °C for 90 min with gentle mixing. Cooled samples at room temperature were centrifuged for 10 min at 16,000× *g*. After the centrifugation, 700 μL of the supernatant was transferred in a tube containing 500 μL of chloroform/isoamyl alcohol (24:1) and mixed by vortexing. This step was followed by a centrifugation for 15 min at 16,000× *g* at room temperature. The supernatant was transferred in a new 1.5 mL tube and the DNA was precipitated with 500 μL of isopropanol and washed with 500 μL of ethanol 70%. Finally, pellets were rehydrated with 30 μL of sterile water and stored at −20 °C until PCR analyses.

Extracted DNA was quality checked using the NanoPhotometer IMPLEN P300 (Implen GmbH, Munchen, Germany) and visually evaluated by 1% agarose gel electrophoresis in TBE 1X buffer, after staining with 1X GelRed Nucleic Acid Gel Stain (Biotium Inc., Hayward, CA, USA).

2.3. PCR Analyses

PCR analyses were carried out using the PCR primer pairs reported in Table 1. To assess the possibility to successfully amplify DNA fragments from the extracted DNA, we first verified if amplification could occur for honey bee DNA using primers designed on the mitochondrial DNA (mtDNA) region [44]. Primer pairs used in this study that targeted sequences of *Paenibacillus larvae* (two primer pairs) *Melissococcus plutonius* (two primer pairs) *Nosema apis* (one primer pair), *N. ceranae* (one primer pair), *Ascosphaera apis* (one primer pair), *Lotmaria passim* (one primer pair), *Acarapis woodi* (two primer pairs), and *Varroa destructor* (two primer pairs) were described by other authors (Table 1). Primers to detect the presence of *Tropilaelaps* spp. were designed in this work on the mtDNA sequence of *Tropilaelaps clareae* to target a region of the mitochondrial DNA Cox1 gene that, according to the available sequence information in GenBank, was conserved only in the *Tropilaelaps* genus (Table 1). All selected primer pairs were expected to amplify DNA fragments of <300 bp as the highly degraded honey DNA limits the possibility to successful amplify larger fragments [45].

Table 1. Information on PCR primers and PCR conditions used in this study to amplify DNA extracted from the 102 honey samples reported in Table S1.

Species	Primer Name [1]	Primer Sequences (5′-3′): Forward and Reverse [2]	Size in Bp	Amplified Region	Ta °C [3]	Reference
Apis mellifera	AM_Forward AM_Reverse	GGCAGAATAAGTGCATTG TTAATATGAATTAAGTGGGG	C 85, M 139, A 153 [4]	mtDNA COI-COII	51	[44]
Paenibacillus larvae	AF_6f AF_7r	GCAAGTCGAGCGGACCTTGT TCGTCGCCTTGGTAAGC	237	16S rRNA	63	[14]
Paenibacillus larvae	Han233PaeLarv16S_F Han233PaeLarv16S_R	GTGTTTCCTTCGGGAGACG CTCTAGGTCGGCTACGCATC	233	16S rRNA	59	[46]
Melissococcus plutonius	MeliFORa MeliREVa	GTTAAAAGGCGCTTTCGGGT GAGGAAAACAGTTACTCTTTCCCCTA	281	16S rRNA	63	[36]
Melissococcus plutonius	Mp_Arai187_F Mp_Arai187_R	TGGTAGCTTAGGCGGAAAAC TGGAGCGATTAGAGTCGTTAGA	187	NapA	59	[47]
Nosema apis	Nose_apis_chen_F Nose_apis_chen_R	CCATTGCCGGATAAGAGAGT CCACCAAAAACTCCCAAGAG	269	SSUrRNA	54–60 [5]	[48]
Nosema ceranae	Nose_cera_chen_F Nose_cera_chen_R	CGGATAAAAGAGTCCGTTACC TGAGCAGGGTTCTAGGGAT	250	SSUrRNA	58	[48]
Lotmaria passim	LpCytb_F1 LpCytb_R1	CGAAGTGCACATATATGCTTTAC GCCAAACACCAATAACTGGTACT	247	mtDNA cyt b	59	[34]
Ascosphaera apis	AscosFORa AscosREVa	TGTGTCTGTGCGGCTAGGTG GCTAGCCAGGGGGGAACTAA	136	18S rRNA	61	[36]
Acarapis woodi	Acarap_cox1_F1 Acarap_cox1_R1	CAGTAGGGCTAGATATCGATACCCGAGCTT TGAGCTACAACATAATATCTGTCATGAAGA	247	mtDNA Cox 1	55–62 [5]	[49]
Acarapis woodi	Acarapis_cox1_F2 Acarapis_cox1_R2	CGGGCCCGAGCTTATTTTACTGCTG GCGCCTGTCAATCCACCTACAGAAA	162	mtDNA Cox 1	55–62 [5]	[50]
Varroa destructor	Varr_cox1_181_F Varr_cox1_181_R	GCCTTTATTTGTATGGTCTG TGGGTGTCCAAAAAATCA	181	mtDNA Cox 1	52	[45]
Varroa destructor	Varr_cox1_148_F Varr_cox1_148_R	TGGTATTATTTCTCATGTAATTTG ATCAATATCTATTCCTACTGTAAA	148	mtDNA Cox 1	55	[45]
Tropilaelaps clareae [6]	Tropi_cox1_F Tropi_cox1_R	TATTTGTATGATCTGTCCTA ATAATACCAAATCCTGGTA	214	mtDNA Cox 1	54–60 [5]	This study

[1] Internal name of the forward and reverse primers. [2] Sequence of the forward and reverse primers. [3] Annealing temperature used in the PCR amplifications (°C). [4] The amplified fragment can have different size according to the mitochondrial lineage (A, C or M) as described in Utzeri et al. [44]. [5] Range of the tested annealing temperature. [6] The PCR assay was designed on the mitochondrial DNA sequence of *Tropilaelaps clareae* (GenBank accession no. EF025461). However, it is not specific for this species, as according to the sequence alignment with the corresponding region of other *Tropilaelaps* species, these primers can also amplify other species of this genus (i.e., *T. mercedesae*, *T. thaii* and *T. koenigerum*) due to 100% identity of their mtDNA sequences in the primer regions. For this reason, the assay was indicated to amplify DNA of *Tropilaelaps* spp. The primer pair does not amplify *V. destructor* DNA.

PCR analyses used one primer pair for each reaction (single PCR analyses). PCR were performed on a 2700 Thermal Cycler (Life Technologies, Waltham, MA, USA) in a total volume of 20 µL including 50 ng of isolated DNA and using KAPA HiFi HotStart Mastermix (Kapa Biosystems, Wilmington, MA, USA) with 10 pmol of each primer and adopting the following PCR profile: initial denaturation step at 95 °C for 3 min, then 35 cycles of alternate temperatures (20 s at 98 °C, 15 s at the specific annealing temperature for the different primer pairs as indicated in Table 1, 30 s at 72 °C), followed by a final extension step at 72 °C for 1 min. Amplified DNA fragments were electrophoresed in 2.5% agarose gels in TBE 1X buffer and stained with 1X GelRed Nucleic Acid Gel Stain (Biotium Inc., Hayward, CA, USA). PCR was carried out at least twice for each primer pair/honey sample combination to confirm the results. Results of these PCR analyses are qualitative (presence or absence of amplification based on the expected DNA fragments).

2.4. Sequencing and Data Analysis

Three to seven PCR fragments obtained for each primer pairs from the respective positively amplified honey samples were sequenced. Amplified fragments (derived from 7 µL of PCR product) were purified with 1 µL of ExoSAP-IT® (USB Corporation, Cleveland, OH, USA) for 15 min at 37 °C and then sequenced using the BrightDye® Terminator Cycle Sequencing Kit (NIMAGEN, Nijmegen, The Netherlands). Sequencing reactions were purified using EDTA 0.125 M, Ethanol 100% and Ethanol 70%, following a standard protocol, and then were loaded on an ABI3100 Avant Genetic Analyzer sequencer (Life Technologies) for the detection of DNA sequences. Obtained electropherograms were visually inspected and analysed using MEGA 7 [51] and BioEdit Sequence Alignment Editor v7.0.5 [52] software. BLASTN (http://www.ncbi.nlm.nih.gov/BLAST/) was used to compare and validate the attribution of the obtained DNA sequences to the expected amplified region and organism.

A two-tailed Fisher exact test was used to compare the frequency of positive and negative honey samples between countries or groups of countries. Means were compared using t-Student tests.

3. Results

DNA extracted from all honey samples was amplified for the *Apis mellifera* mtDNA targeted region and the obtained fragments were the expected ones [44]. That means that for all samples, DNA was successfully isolated and PCR amplification was not inhibited by any contaminants eventually left in the extracted DNA aliquots (spectrophotometer analyses confirmed this result: A_{260}/A_{280} was always >1.6). Therefore, the DNA isolated from all honey samples was then amplified with the other 13 primer pairs in single PCR analysis to obtain information on the presence or absence of nine honey bee pathogens or parasites (*Paenibacillus larvae, Melissococcus plutonius, Nosema apis, N. ceranae, Ascosphaera apis, Lotmaria passim, Acarapis woodi, Varroa destructor,* and *Tropilaelaps* spp.). All sequenced fragments for the positively amplified honey samples corresponded to the targeted regions, confirming the specificity of the assays that were developed by other authors (Table 1).

Among the primers that produced positive results, the two primer pairs tested for *P. larvae* gave the same results (presence of amplification or absence of amplification) for 93% of the analysed samples, whereas the two primer pairs used to detect *M. plutonius* gave the same results for 95% of the investigated samples. As the final aim of this study was to obtain information on the presence of potential honey bee health threats, for these two pathogens, we then considered positive a honey sample if at least one or both primer pairs gave a positive amplification.

The two primer pairs tested for *V. destructor* and *A. woodi* were concordant in 100% of the analysed samples, but with opposite results. In the case of *V. destructors*, all 102 honey samples gave positive amplification, whereas in the case of *A. woodi* none of the 102 samples had any amplified fragment. Two other tested PCR assays, each designed for a different organism, did not evidence the presence of the targeted DNA in the extracted honey: *N. apis* and *Tropilaelaps* spp. were never amplified in all honey samples.

Figure 1 shows the percentage of honey samples that contained DNA of the other pathogens and parasites (i.e., *P. larvae*, *M. plutonius*, *N. ceranae*, *A. apis*, and *L. passim*). This figure reports the results considering all 102 honey samples together or dividing them into two other groups: only Italian samples (n. 74) and samples from all other countries (n. 28). *Melissococcus plutonius* was detected in 87% of the analysed samples whereas the other pathogens were detected in 43% to 57% of all samples. Comparing the results obtained for the Italian samples with the results obtained for all non-Italian samples, significant difference was observed for *P. larvae* ($p = 0.0074$; two tailed Fisher exact test). The frequency of positive samples was significantly lower in Italy (49%) than in all other countries considered all together (79%). This result was also confirmed dividing the non-Italian samples in other two groups: samples from all other European countries (n. 13) and samples from all other continents (n. 15) which both had a higher frequency of positive samples for *P. larvae* (77% with $p = 0.075$ and 80% with $p = 0.044$, respectively) when compared to the Italian samples. For the *M. plutonius*, *N. ceranae*, *A. apis*, and *L. passim* no significant differences between the Italian samples and the non-Italian samples were observed.

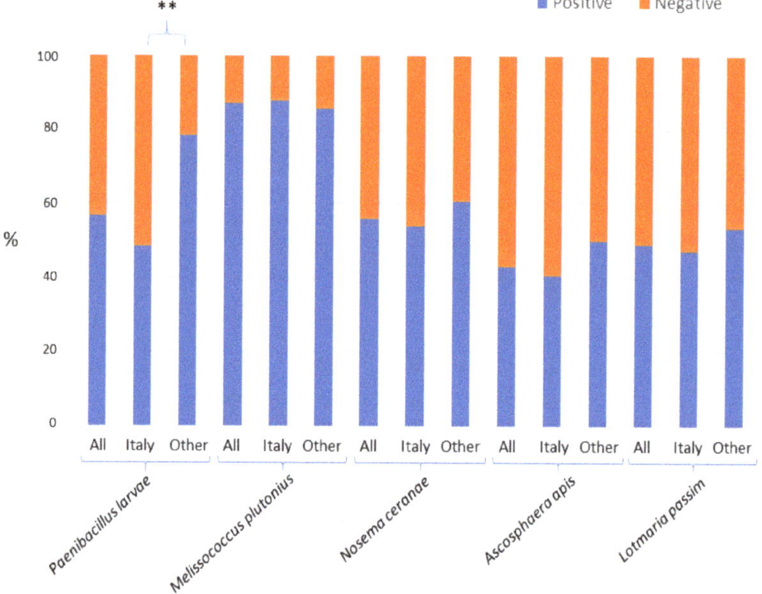

Figure 1. Frequency of the positive (presence of an amplified fragment, in blue) and negative (absence of an amplified fragment, in red) honey samples in the amplifications that targeted *Paenibacillus larvae*, *Melissococcus plutonius*, *Nosema ceranae*, *Ascosphaera apis*, and *Lotmaria passim*. "All" means that all 102 analysed samples were considered together; "Italy" means that only the 74 Italian samples were considered; "Other" means that only the remaining 28 non-Italian samples were considered. The two asterisks (**) mean that the difference of positive honey samples between the two contrasted groups was highly significant ($p < 0.01$).

Considering the stratification of the results based on north and south of Italy, including the Sicily and Sardinia islands in this later geographic area (n. 30 vs. 31 samples; we excluded the regions of Central Italy for the low number of remaining samples), *M. plutonius* resulted significantly less frequent in samples from the north of Italy ($p = 0.0125$) whereas *L. passim* resulted more frequent in samples from this part of Italy ($p = 0.074$).

A total of 14 honey samples (10 of which from Italy) were positive for all pathogens and parasites (excluding those that were not detected in any samples). Both *P. larvae* and *M. plutonius* were co-amplified in 50% of all samples. *Melissococcus plutonius* was always the pathogen that occurred with higher frequency together with all other pathogens or parasites, as expected from its general higher frequency of positive samples. High frequency of co-amplification was observed for *L. passim vs. N. ceranae* (80%; that means that 80% of the samples that were positive for *L. passim* were also positive for *N. ceranae*), for *A. apis vs. N. ceranae* (73%) and for *N. ceranae vs. L. passim* (70%). In these three cases, the co-occurrence of these pairs was statistically different from what would be expected considering the random distribution of each pathogen based on their general frequency of positive samples ($p < 0.001$, $p < 0.01$, and $p < 0.10$, respectively). Only three honey samples (all from Italy) did not show any amplification from all pathogens and parasites (except for *V. destructor*, that was detected in all samples). The distribution of the honey samples that gave amplified fragments for one to six pathogens/parasites is reported in Figure 2.

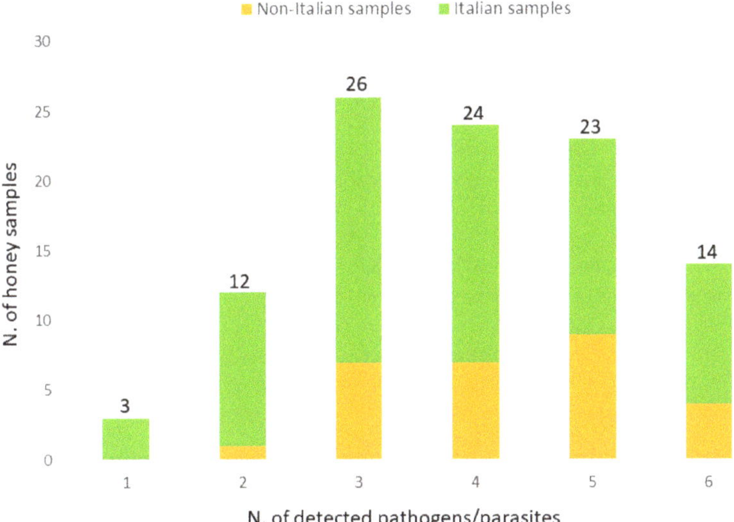

Figure 2. Number of honey samples that were positive for one to six pathogens/parasites differentiated between Italian samples (in green) and non-Italian samples (in orange). The total number of positive samples is reported at the top of each bar.

On average, the Italian samples were positive for 3.78 pathogens/parasites (standard deviation: 1.41) whereas for the non-Italian samples this mean was equal to 4.29 (±2.12). The two means differed at $p < 0.10$.

The two oldest honey samples that were analysed (produced in the year 2004) were positive for *V. destructor*, *P. larvae*, and *M. plutonius*. Considering the year of production, *N. ceranae*, *A. apis*, and *L. passim* were first detected in an Italian honey produced in the year 2012. Due to the unequal distribution of the year of production and the limited number of old samples among those that were analysed (only eight samples were produced in 2013 or earlier), it was not possible to test if there were any differences in the number of positive samples over the years.

4. Discussion

Honey is a complex matrix that is mainly made by sugars but that contains several other components that can be explored to disclose interesting information that could be used for several purposes. One of the components, usually neglected or not considered at all, is the DNA that derives from all organisms that directly or indirectly contributed to its productions or that were part of the production systems or environments where the honey has been produced [6,7,43–45]. This eDNA also contains information on pathogens and parasites that can be present in the beehive ecological niche and that can represent a threat for the honey bee health.

In this study, we used qualitative PCR-based assays that targeted several pathogens and parasites to identify their presence in honey samples that were produced in Italy and in other countries from all continents. This work should be considered a pilot study that wanted to set up a methodology and evaluate the potential strengths and limits that the use of honey extracted DNA could have for the detection and monitoring of honey bee pathogens and parasites.

DNA from the honey bees is always contained in the honey they produce [44]. Therefore, to evaluate the quality of the extracted DNA and the possibility to use it for other tests, we first amplified the extracted nucleic acid with primers that targeted a region of the honey bee mtDNA that can be also used to detect the mitochondrial lineage of the *A. mellifera* that produced the honey [44]. As all samples gave the expected amplified products for this *A. mellifera* DNA region, the presence or absence of amplification for all other tested primer pairs that targeted nine pathogens or parasites is indicative of the presence or absence of their DNA in the investigated honey samples. DNA could be extracted and then amplified independently by the year of production. DNA from honey produced more than 15 years ago was successfully amplified. Therefore, honey can be used for retrospective analyses and can be stored for this purpose.

All tested assays are qualitative, and in this first pilot study, we did not evaluate the limit of detection for these pathogens that, however, have been already tested in some of the studies from which we derived the primer pairs that we used to amplify honey DNA. It will be interesting to transform the qualitative assays in quantitative analyses that could provide a more precise information and better evaluate the limit of detection from this source of DNA.

It is also worth to mention that the honey samples that we analysed might derive from more than one colony or beehive and probably from more than one apiary. Commercial honey is prepared by mixing the honey derived by more than one beehive. Therefore, the information that we obtained should be referred to the presence of the pathogens and parasites at a higher level than that is usually considered in epidemiological or monitoring studies that usually use the single beehive/colony as the unit of their analyses. The use of honey (particularly, when it does not derive from a single colony, as in our case) might provide information based on the health situation in many colonies or at the apiary level or even at the regional level, depending on the way in which the honey is finally packaged and commercialized. This is an advantage for global monitoring purposes but it could not be considered useful for the punctual health evaluation of a single colony or beehive, that instead could be obtained by traditional approaches or, eventually, by analysing DNA of a single honeycomb (even if this later could not be considered practical in most cases).

The presence or absence of *Paenibacillus larvae*, *Melissococcus plutonius*, *Acarapis woodi*, and *Varroa destructor* in the 102 analysed honey samples was verified using two primer pairs for each targeted organism, obtaining always or for most samples the same results. Some differences in the results of the two assays were observed for the detection of *P. larvae* and *M. plutonius*, probably due to differences in the limits of detection of these tests or for the presence of different strains across samples, having mutations in the primer regions that could affect the amplification efficiency. It will be interesting to further explore this issue not only for these two pathogens but also for the other pathogens and parasites that were amplified in this study by using only one primer pair.

As expected from the global diffusion of *V. destructor*, the DNA of this mite was amplified from all honey samples. The positive amplification obtained for all samples can confirm the high quality of the extracted DNA, as already derived by the successful amplification of the honey bee DNA. We already demonstrated that the sequence information derived by the amplified Varroa DNA fragments from honey samples and obtained using next generation sequencing can provide additional details on the presence of several mite strains in different countries [45]. This approach can be also applied for all other pathogens and parasites investigated in this study if the amplified regions are informative, i.e., they are expected to include sequence differences among strains or lines that could add other information to the derived eDNA.

For two other arthropod parasites, *A. woodi* and *Tropilaelaps* spp., and one microsporidium, *Nosema apis*, we did not identify any positive honey samples. The negative results obtained in these cases could be due to the failure of amplification of the tested primer pairs, even if for two organisms (*A. woodi* and *N. apis*) we used primers that have been already extensively tested by other authors [48–50], two primer pairs were tested for *A. woodi* [49,50] with the same negative results, and for all used primers for these three organisms, several PCR conditions were tested (Table 1).

The negative results for *A. woodi* are unexpected, according to the supposed world-wide distribution of this parasite reported by a few studies [35,50,53,54]. However, other studies indicated that the prevalence and diffusion of this honey bee tracheal mite is limited and potentially related to local conditions and factors [55–57] and this picture might better match the results we obtained. Other studies are needed to confirm the negative results derived by our approach based of the amplification of honey DNA by two different PCR assays that have been specifically designed to detect this species [49,50].

On the other hand, the negative results for *Tropilaelaps* spp. were to some extent expected. That means that the absence of *Tropilaelaps* spp. amplified fragments indicates that these damaging mites [40] did not spread into the regions where we sampled the analysed honey. It could be also possible that the prevalence of mites of this genus was not relevant in these areas or it was under the detection limit of this analysis. It is also worth to mention that the assay we developed should be further tested using some positive honey samples. To be sure that honey DNA can be useful to capture the presence of these emerging parasites, it will be important to develop other PCR assays able to confirm these results. Anyway, even if the absence of *Tropilaelaps* spp. amplification is a good news, it further raises the attention to the usefulness of monitoring methods, like what we propose in this study, to prevent potential spreads in Europe or America where these mites would be probably very damaging to the apiculture sector [41].

Another pathogen that was never detected in the analysed samples was *N. apis*. The relevance of this microsporidium has been decreasing over the last decades that evidenced a corresponding increase of the infection prevalence and diffusion of *N. ceranae*, which has been suggested to replace *N. apis* [27,58–60]. Our results might confirm this general trend. *Nosema ceranae* was observed in more than 50% of the analysed samples further demonstrating the widespread and general prevalence of this obligate intracellular eukaryotic parasite.

Lotmaria passim, currently considered the predominant trypanosomatid species in *A. mellifera* host populations [33], was detected in about 50% of the honey samples, 80% of which were also positive for *N. ceranae* (and this co-occurrence was not derived by chance), confirming, to some extent, that its presence is usually associated with this microsporidium [34,61]. It is of course clear that, just by analysing honey DNA, is it not possible to establish a direct relationship between these two infection agents but this speculative deduction could open new opportunities to interpret the results when larger number of samples and specimens purposely collected for co-occurrence evaluations are investigated. It was also interesting to note that, stratifying the results of the Italian samples based on the two main geographic areas from which the largest number of samples were collected, *L. passim* occurred more frequently in the honey produced in the North of Italy than in that produced in the South of Italy. It will be important to increase the number of samples to confirm this preliminary evidence, that at present it does not have any clear explanations.

Ascosphaera apis, the agent of chalkbrood, was detected in more than 40% of all samples, confirming also for this entomopathogenic fungus the widespread distribution already known [29]. It will be important to understand if the DNA of *A. apis* derives from spores (which can have up to 15 years viability; [29]) or other materials, for the potential obvious implications derived by the possibility to further spread this disease through the exchange of honey. Similarly to what was discussed for *L. passim*, this fungus was detected in honey samples that in 73% of cases were also positive for

N. ceranae, further suggesting a possible relationship determining the co-occurrence of these pathogens (of course taking into account what already discussed above).

Melissococcus plutonius was the most frequent pathogen amplified in the 102 honey samples, with 87% of positive samples. This Gram-positive lanceolate coccus is the agent of the European foulbrood, one of the most serious brood diseases of *A. mellifera*, for which there is still a poor understanding of disease epidemiology. It is already known that this bacterium is also present in healthy colonies, variants of this agent have been described and that the disease can occur in combination with other health threats [62–64]. Anyway, the very high frequency of positive samples is quite puzzling and needs to be further investigated to better understand its meaning, by taking into account the methodology that was used for this detection. The lower frequency of positive samples produced in the North of Italy than those produced in the South of Italy is also another matter of further evaluation, if confirmed by analysing a larger number of samples.

The general quite high percentage of positive honey samples for *P. larvae* indicates that not in all cases the presence of this pathogen might cause American foulbrood, as already reported [65,66]. The results of the amplifications for the agent of this disease showed that this bacterium was less frequent in the Italian samples (49%) than in the non-Italian samples (79%) and this difference was highly significant. The interpretation of this result is not simple and might be derived by different management systems and practices or environmental conditions that would limit the diffusion of this Gram-positive pathogenic bacterium in Italy compared to other parts of the world. This is in line to the lower mean number of pathogens/parasites detected in the Italian samples than in the non-Italian samples. These first results might speculatively indicate general better healthy conditions of the colonies in Italy than in other countries—a hypothesis that should be supported by other evidences. Another explanation for this difference could be due to the fact that non-Italian samples might be derived by a larger number of beehives or apiaries (most of them were purchased and not directly provided by the beekeepers; Table S1) than the Italian samples (most of which were provided directly by the beekeepers; Table S1), increasing the possibility to include honey derived by positive beehives or beehives in which this pathogen was present. If this would be a possible explanation of this result, it is not clear, however, why this difference between Italian and non-Italian honey samples was observed only for *P. larvae* and not also for other pathogens.

5. Conclusions

This study reported for the first time an extensive use of honey eDNA to design epidemiological and monitoring approaches for pathogens and parasites with the final objective of obtaining a general picture of the sanitary status of the honey bee populations at macro-levels (i.e., apiary, beekeeper, regions, countries, continents). Using this unconventional approach, we also obtained, for the first time, a comprehensive analysis (even if preliminary) of the distribution and frequency of several pathogens and parasites in Italy. It will be useful (i) to refine and improve the applied assays, adding methods to detect other pathogens, (ii) to increase the number of the analysed honey samples to improve the interpretation of the results, and (iii) to correlate the results derived by the DNA analyses with the situations in the colony and/or apiary and epidemiological data and distribution of pathogens and parasites in a region that could be obtained from direct observations. The results of this study on the distribution, co-occurrence and prevalence of some of the targeted pathogens and parasites should be interpreted, considering that honey bee health threats cannot be regarded as local problems.

Supplementary Materials: The following are available online at http://www.mdpi.com/2306-7381/7/3/113/s1, Figure S1: Distribution of the analysed honey samples based on the years of production. Table S1: List and details of the investigated honey samples.

Author Contributions: Conceptualization, L.F.; methodology, L.F., A.R., V.J.U.; formal analysis, A.R., V.J.U., V.T.; resources, L.F., V.J.U.; data curation, L.F., A.R., V.J.U.; writing—Original draft preparation, L.F.; writing—Review and editing, L.F., A.R., V.J.U., V.T.; visualization, L.F.; supervision, L.F.; project administration, L.F.; funding acquisition, L.F. All authors have read and agreed to the published version of the manuscript.

Funding: This research was funded by University of Bologna RFO programme and by Regione Emilia-Romagna Misura F (DELIBERAZIONE DELL'ASSEMBLEA LEGISLATIVA DELLA REGIONE EMILIA-ROMAGNA 27 LUGLIO 2019, N. 216), BEE-RER project (2019-2020).

Acknowledgments: The authors thank Alberto Contessi and Gianfranco Naldi of Osservatorio Nazionale Miele, Lucia Piana of Lucia Piana srl and several beekeepers for the collaboration. This work was associated with the GRIFFA development activities.

Conflicts of Interest: The authors declare no conflict of interest. The funders had no role in the design of the study; in the collection, analyses, or interpretation of data; in the writing of the manuscript, or in the decision to publish the result.

References

1. Taberlet, P.; Coissac, E.; Hajibabaei, M.; Rieseberg, L.H. Environmental DNA. *Mol. Ecol.* **2012**, *21*, 1789–1793. [CrossRef] [PubMed]

2. Bohmann, K.; Evans, A.; Gilbert, M.T.; Carvalho, G.R.; Creer, S.; Knapp, M.; Yu, D.W.; De Bruyn, M. Environmental DNA for wildlife biology and biodiversity monitoring. *Trends Ecol. Evol.* **2014**, *29*, 358–367. [CrossRef]

3. Jerde, C.L.; Mahon, A.R.; Chadderton, W.L.; Lodge, D.M. "Sight-unseen" detection of rare aquatic species using environmental DNA. *Conserv. Lett.* **2011**, *4*, 150–157. [CrossRef]

4. Thomsen, P.F.; Kielgast, J.; Iversen, L.L.; Wiuf, C.; Rasmussen, M.; Gilbert, M.T.; Orlando, L.; Willerslev, E. Monitoring endangered freshwater biodiversity using environmental DNA. *Mol. Ecol.* **2012**, *21*, 2565–2573. [CrossRef] [PubMed]

5. Wilcox, T.M.; McKelvey, K.S.; Young, M.K.; Jane, S.F.; Lowe, W.H.; Whiteley, A.R.; Schwartz, M.K. Robust detection of rare species using environmental DNA: The importance of primer specificity. *PLoS ONE* **2013**, *8*, e59520. [CrossRef] [PubMed]

6. Bovo, S.; Ribani, A.; Utzeri, V.J.; Schiavo, G.; Bertolini, F.; Fontanesi, L. Shotgun metagenomics of honey DNA: Evaluation of a methodological approach to describe a multi-kingdom honey bee derived environmental DNA signature. *PLoS ONE* **2018**, *13*, e0205575. [CrossRef]

7. Bovo, S.; Utzeri, V.J.; Ribani, A.; Cabbri, R.; Fontanesi, L. Shotgun sequencing of honey DNA can describe honey bee derived environmental signatures and the honey bee hologenome complexity. *Sci. Rep.* **2020**, *10*, 9279. [CrossRef]

8. Dainat, B.; Evans, J.D.; Chen, Y.P.; Gauthier, L.; Neumann, P. Predictive markers of honey bee colony collapse. *PLoS ONE* **2012**, *7*, e32151. [CrossRef]

9. Lee, K.; Steinhauer, N.; Travis, D.A.; Meixner, M.D.; Deen, J. Honey bee surveillance: A tool for understanding and improving honey bee health. *Curr. Opin. Insect Sci.* **2015**, *10*, 37–44. [CrossRef]

10. Ward, L.; Brown, M.; Neumann, P.; Wilkins, S.; Pettis, J.; Boonham, N. A DNA method for screening hive debris for the presence of small hive beetle (*Aethina tumida*). *Apidologie* **2007**, *38*, 272–280. [CrossRef]

11. Ryba, S.; Titera, D.; Haklova, M.; Stopka, P. A PCR method of detecting American foulbrood (*Paenibacillus larvae*) in winter beehive wax debris. *Vet. Microbiol.* **2009**, *139*, 193–196. [CrossRef]

12. Ryba, S.; Kindlmann, P.; Titera, D.; Haklova, M.; Stopka, P. A new low-cost procedure for detecting nucleic acids in low-incidence samples: A case study of detecting spores of *Paenibacillus larvae* from bee debris. *J. Econ. Entomol.* **2012**, *105*, 1487–1491. [CrossRef] [PubMed]

13. Forsgren, E.; Laugen, A.T. Prognostic value of using bee and hive debris samples for the detection of American foulbrood disease in honey bee colonies. *Apidologie* **2014**, *45*, 10–20. [CrossRef]

14. Bakonyi, T.; Derakhshifar, I.; Grabensteiner, E.; Nowotny, N. Development and evaluation of PCR assays for the detection of *Paenibacillus larvae* in honey samples: Comparison with isolation and biochemical characterization. *Appl. Environ. Microbiol.* **2003**, *69*, 1504–1510. [CrossRef] [PubMed]

15. Lauro, F.M.; Favaretto, M.; Covolo, L.; Rassu, M.; Bertoloni, G. Rapid detection of *Paenibacillus larvae* from honey and hive samples with a novel nested PCR protocol. *Int. J. Food Microbiol.* **2003**, *81*, 195–201. [CrossRef]

16. McKee, B.; Djordjevic, S.; Goodman, R.; Hornitzky, M. The detection of *Melissococcus pluton* in honey bees (*Apis mellifera*) and their products using a hemi-nested PCR. *Apidologie* **2003**, *34*, 19–27. [CrossRef]

17. D'Alessandro, B.; Antúnez, K.; Piccini, C.; Zunino, P. DNA extraction and PCR detection of *Paenibacillus larvae* spores from naturally contaminated honey and bees using spore-decoating and freeze-thawing techniques. *World J. Microbiol. Biotech.* **2007**, *23*, 593–597. [CrossRef]

18. Forsgren, E. European foulbrood in honey bees. *J. Invertebr. Pathol.* **2010**, *103*, S5–S9. [CrossRef]
19. Genersch, E. American Foulbrood in honeybees and its causative agent, *Paenibacillus larvae*. *J. Invertebr. Pathol.* **2010**, *103*, S10–S19. [CrossRef]
20. Giersch, T.; Berg, T.; Galea, F.; Hornitzky, M. *Nosema ceranae* infects honey bees (*Apis mellifera*) and contaminates honey in Australia. *Apidologie* **2009**, *40*, 117–123. [CrossRef]
21. Cox-Foster, D.L.; Conlan, S.; Holmes, E.C.; Palacios, G.; Evans, J.D.; Moran, N.A.; Quan, P.L.; Briese, T.; Hornig, M.; Geiser, D.M.; et al. A metagenomic survey of microbes in honey bee colony collapse disorder. *Science* **2007**, *318*, 283–287. [CrossRef] [PubMed]
22. Vanengelsdorp, D.; Evans, J.D.; Saegerman, C.; Mullin, C.; Haubruge, E.; Nguyen, B.K.; Frazier, M.; Frazier, J.; Cox-Foster, D.; Chen, Y.; et al. Colony collapse disorder: A descriptive study. *PLoS ONE* **2009**, *4*, e6481. [CrossRef] [PubMed]
23. Cornman, R.S.; Tarpy, D.R.; Chen, Y.; Jeffreys, L.; Lopez, D.; Pettis, J.S.; Van Engelsdorp, D.; Evans, J.D. Pathogen webs in collapsing honey bee colonies. *PLoS ONE* **2012**, *7*, e43562. [CrossRef] [PubMed]
24. Higes, M.; Martín-Hernández, R.; Botías, C.; Bailón, E.G.; González-Porto, A.V.; Barrios, L.; Del Nozal, M.J.; Bernal, J.L.; Jiménez, J.J.; Palencia, P.G.; et al. How natural infection by *Nosema ceranae* causes honeybee colony collapse. *Environ. Microbiol.* **2008**, *10*, 2659–2669. [CrossRef] [PubMed]
25. Higes, M.; Martín-Hernández, R.; Garrido-Bailón, E.; González-Porto, A.V.; García-Palencia, P.; Meana, A.; Del Nozal, M.J.; Mayo, R.; Bernal, J.L. Honeybee colony collapse due to *Nosema ceranae* in professional apiaries. *Environ. Microbiol. Rep.* **2009**, *1*, 110–113. [CrossRef]
26. Klee, J.; Besana, A.M.; Genersch, E.; Gisder, S.; Nanetti, A.; Tam, D.Q.; Chinh, T.X.; Puerta, F.; Ruz, J.M.; Kryger, P.; et al. Widespread dispersal of the microsporidian *Nosema ceranae*, an emergent pathogen of the western honey bee, *Apis mellifera*. *J. Invertebr. Pathol.* **2007**, *96*, 1–10. [CrossRef]
27. Higes, M.; Meana, A.; Bartolomé, C.; Botías, C.; Martín-Hernández, R. *Nosema ceranae* (Microsporidia), a controversial 21st century honey bee pathogen. *Environ. Microbiol. Rep.* **2013**, *5*, 17–29. [CrossRef]
28. Branchiccela, B.; Arredondo, D.; Higes, M.; Invernizzi, C.; Martín-Hernández, R.; Tomasco, I.; Zunino, P.; Antúnez, K. Characterization of *Nosema ceranae* genetic variants from different geographic origins. *Microb. Ecol.* **2017**, *73*, 978–987. [CrossRef]
29. Aronstein, D.A.; Murray, K.D. Chalkbrood disease in honeybees. *J. Invertebr. Pathol.* **2010**, *103*, 20–29. [CrossRef]
30. Hedtke, K.; Jensen, P.M.; Jensen, A.B.; Genersch, E. Evidence for emerging parasites and pathogens influencing outbreaks of stress-related diseases like chalkbrood. *J. Invertebr. Pathol.* **2011**, *108*, 167–173. [CrossRef]
31. Ravoet, J.; Maharramov, J.; Meeus, I.; De Smet, L.; Wenseleers, T.; Smagghe, G.; De Graaf, D.C. Comprehensive bee pathogen screening in Belgium reveals Crithidia mellificae as a new contributory factor to winter mortality. *PLoS ONE* **2013**, *8*, e72443. [CrossRef] [PubMed]
32. Ravoet, J.; Schwarz, R.S.; Descamps, T.; Yañez, O.; Tozkar, C.O.; Martin-Hernandez, R.; Bartolomé, C.; De Smet, L.; Higes, M.; Wenseleers, T.; et al. Differential diagnosis of the honey bee trypanosomatids *Crithidia mellificae* and *Lotmaria passim*. *J. Invertebr. Pathol.* **2015**, *130*, 21–27. [CrossRef]
33. Schwarz, R.S.; Bauchan, G.; Murphy, C.; Ravoet, J.; De Graaf, D.C.; Evans, J.D. Characterization of two species of Trypanosomatidae from the honey bee *Apis mellifera*: *Crithidia mellificae* Langridge and McGhee, 1967 and *Lotmaria passim* n. gen., n. sp. *J. Eukaryot. Microbiol.* **2015**, *62*, 567–583. [CrossRef] [PubMed]
34. Stevanovic, J.; Schwarz, R.S.; Vejnovic, B.; Evans, J.D.; Irwin, R.E.; Glavinic, U.; Stanimirovic, Z. Species-specific diagnostics of *Apis mellifera* trypanosomatids: A nine-year survey (2007–2015) for trypanosomatids and microsporidians in Serbian honey bees. *J. Invertebr. Pathol.* **2016**, *139*, 6–11. [CrossRef] [PubMed]
35. Sammataro, D.; Gerson, U.; Needham, G. Parasitic mites of honey bees: Life history, implications, and impact. *Ann. Rev. Entomol.* **2000**, *45*, 519–548. [CrossRef] [PubMed]
36. Garrido-Bailón, E.; Higes, M.; Martínez-Salvador, A.; Antúnez, K.; Botías, C.; Meana, A.; Prieto, L.; Martín-Hernández, R. The prevalence of the honeybee brood pathogens *Ascosphaera apis*, *Paenibacillus larvae* and *Melissococcus plutonius* in Spanish apiaries determined with a new multiplex PCR assay. *Microb. Biotechnol.* **2013**, *6*, 731–739.
37. Guzmán-Novoa, E.; Eccles, L.; Calvete, Y.; Mcgowan, J.; Kelly, P.G.; Correa-Benítez, A. *Varroa destructor* is the main culprit for the death and reduced populations of overwintered honey bee (*Apis mellifera*) colonies in Ontario, Canada. *Apidologie* **2010**, *41*, 443–450. [CrossRef]

38. Le Conte, Y.; Ellis, M.; Ritter, W. Varroa mites and honey bee health: Can Varroa explain part of the colony losses. *Apidologie* **2010**, *41*, 353–363. [CrossRef]
39. Anderson, D.L.; Morgan, M.J. Genetic and morphological variation of bee-parasitic *Tropilaelaps* mites (Acari: Laelapidae): New and re-defined species. *Exp. Appl. Acarol.* **2007**, *43*, 1–24. [CrossRef]
40. De Guzman, L.I.; Williams, G.R.; Khongphinitbunjong, K.; Chantawannakul, P. Ecology, life history, and management of *Tropilaelaps* mites. *J. Econ. Entomol.* **2017**, *110*, 319–332. [CrossRef]
41. Traynor, K.S.; Rennich, K.; Forsgren, E.; Rose, R.; Pettis, J.; Kunkel, G.; Madella, S.; Evans, J.; Lopez, D.; Van Engelsdorp, D. Multiyear survey targeting disease incidence in US honey bees. *Apidologie* **2016**, *47*, 325–347. [CrossRef]
42. Pettis, J.S.; Rose, R.; Lichtenberg, E.M.; Chantawannakul, P.; Buawangpong, N.; Somana, W.; Sukumalanand, P.; Vanengelsdorp, D. A rapid survey technique for Tropilaelaps mite (Mesostigmata: Laelapidae) detection. *J. Econ. Entomol.* **2013**, *106*, 1535–1544. [CrossRef] [PubMed]
43. Utzeri, V.J.; Ribani, A.; Schiavo, G.; Bertolini, F.; Bovo, S.; Fontanesi, L. Application of next generation semiconductor based sequencing to detect the botanical composition of monofloral, polyfloral and honeydew honey. *Food Control* **2018**, *86*, 342–349. [CrossRef]
44. Utzeri, V.J.; Ribani, A.; Fontanesi, L. Authentication of honey based on a DNA method to differentiate *Apis mellifera* subspecies: Application to Sicilian honey bee (*A. m. siciliana*) and Iberian honey bee (*A. m. iberiensis*) honeys. *Food Control* **2018**, *91*, 294–301. [CrossRef]
45. Utzeri, V.J.; Schiavo, G.; Ribani, A.; Bertolini, F.; Bovo, S.; Fontanesi, L. A next generation sequencing approach for targeted Varroa destructor (Acari: Varroidae) mitochondrial DNA analysis based on honey derived environmental DNA. *J. Invertebr. Pathol.* **2019**, *161*, 47–53. [CrossRef]
46. Han, S.H.; Lee, D.B.; Lee, D.W.; Kim, E.H.; Yoon, B.S. Ultra-rapid real-time PCR for the detection of *Paenibacillus larvae*, the causative agent of American foulbrood (AFB). *J. Invertebr. Pathol.* **2008**, *99*, 8–13. [CrossRef]
47. Arai, R.; Miyoshi-Akiyama, T.; Okumura, Y.; Morinaga, K.; Wu, M.; Sugimura, Y.; Yoshiyama, M.; Okura, M.; Kirikae, T.; Takamatsu, D. Development of duplex PCR assay for detection and differentiation of typical and atypical *Melissococcus plutonius* strains. *J. Vet. Med. Sci.* **2014**, *76*, 491–498. [CrossRef]
48. Chen, Y.; Evans, J.D.; Zhou, L.; Boncristiani, H.; Kimura, K.; Xiao, T.; Litkowski, A.M.; Pettis, J.S. Asymmetrical coexistence of *Nosema ceranae* and *Nosema apis* in honey bees. *J. Invertebr. Pathol.* **2009**, *101*, 204–209. [CrossRef]
49. Kojima, Y.; Yoshiyama, M.; Kimura, K.; Kadowaki, T. PCR-based detection of a tracheal mite of the honey bee *Acarapis woodi. J. Invertebr. Pathol.* **2011**, *108*, 135–137. [CrossRef]
50. Garrido-Bailón, E.; Bartolomé, C.; Prieto, L.; Botías, C.; Martínez-Salvador, A.; Meana, A.; Martín-Hernández, R.; Higes, M. The prevalence of *Acarapis woodi* in Spanish honey bee (*Apis mellifera*) colonies. *Exp. Parasitol.* **2012**, *132*, 530–536. [CrossRef]
51. Kumar, S.; Stecher, G.; Li, M.; Knyaz, C.; Tamura, K. MEGA X: Molecular evolutionary genetics analysis across computing platforms. *Mol. Biol. Evol.* **2018**, *35*, 1547–1549. [CrossRef] [PubMed]
52. Hall, T.A. BioEdit: A user-friendly biological sequence alignment editor and analysis program for Windows 95/98/NT. *Nucl. Acids Symp. Ser.* **1999**, *41*, 95–98.
53. Ellis, J.D.; Munn, P.A. The worldwide health status of honey bees. *Bee World* **2005**, *86*, 88–101. [CrossRef]
54. Kojima, Y.; Toki, T.; Morimoto, T.; Yoshiyama, M.; Kimura, K.; Kadowaki, T. Infestation of Japanese native honey bees by tracheal mite and virus from non-native European honey bees in Japan. *Microb. Ecol.* **2011**, *62*, 895–906. [CrossRef] [PubMed]
55. Çakmak, I.; Aydın, L.; Gulegen, E.; Wells, H. Varroa (*Varroa destructor*) and tracheal mite (*Acarapis woodi*) incidence in the Republic of Turkey. *J. Apic. Res.* **2003**, *42*, 57–60. [CrossRef]
56. Adjlane, N.; Assia, H.C.; Haddad, N.; Ahcen, B. Prevalence of acariosis in honeybee colonies of *Apis mellifera intermissa* in Algeria. *J. Entomol. Res.* **2018**, *42*, 451–454. [CrossRef]
57. Ahn, A.J.; Ahn, K.S.; Noh, J.H.; Kim, Y.H.; Yoo, M.S.; Kang, S.W.; Yu, D.H.; Shin, S.S. Molecular prevalence of *Acarapis* mite infestations in honey bees in Korea. *Korean J. Parasitol.* **2015**, *53*, 315–320. [CrossRef]
58. Martín-Hernández, R.; Botías, C.; Bailón, E.G.; Martínez-Salvador, A.; Prieto, L.; Meana, A.; Higes, M. Microsporidia infecting *Apis mellifera*: Coexistence or competition. Is *Nosema ceranae* replacing *Nosema apis*. *Environ. Microbiol.* **2012**, *14*, 2127–2138. [CrossRef]

59. Natsopoulou, M.E.; McMahon, D.P.; Doublet, V.; Bryden, J.; Paxton, R.J. Interspecific competition in honeybee intracellular gut parasites is asymmetric and favours the spread of an emerging infectious disease. *Proc. R. Soc. B Biol. Sci.* **2015**, *282*, 20141896. [CrossRef]

60. Emsen, B.; Guzman-Novoa, E.; Hamiduzzaman, M.M.; Eccles, L.; Lacey, B.; Ruiz-Pérez, R.A.; Nasr, M. Higher prevalence and levels of *Nosema ceranae* than *Nosema apis* infections in Canadian honey bee colonies. *Parasitol. Res.* **2016**, *115*, 175–181. [CrossRef]

61. Tritschler, M.; Retschnig, G.; Yañez, O.; Williams, G.R.; Neumann, P. Host sharing by the honey bee parasites *Lotmaria passim* and *Nosema ceranae*. *Ecol. Evol.* **2017**, *7*, 1850–1857. [CrossRef] [PubMed]

62. Forsgren, E.; Lundhagen, A.C.; Imdorf, A.; Fries, I. Distribution of *Melissococcus plutonius* in honeybee colonies with and without symptoms of European foulbrood. *Microb. Ecol.* **2005**, *50*, 369–374. [CrossRef] [PubMed]

63. Belloy, L.; Imdorf, A.; Fries, I.; Forsgren, E.; Berthoud, H.; Kuhn, R.; Charrière, J.D. Spatial distribution of *Melissococcus plutonius* in adult honey bees collected from apiaries and colonies with and without symptoms of European foulbrood. *Apidologie* **2007**, *38*, 136–140. [CrossRef]

64. Haynes, E.; Helgason, T.; Young, J.P.W.; Thwaites, R.; Budge, G.E. A typing scheme for the honeybee pathogen *Melissococcus plutonius* allows detection of disease transmission events and a study of the distribution of variants. *Environ. Microbiol. Rep.* **2013**, *5*, 525–529. [CrossRef]

65. Antúnez, K.; D'Alessandro, B.; Piccini, C.; Corbella, E.; Zunino, P. *Paenibacillus larvae larvae* spores in honey samples from Uruguay: A nationwide survey. *J. Invertebr. Pathol.* **2004**, *86*, 56–58. [CrossRef]

66. Gillard, M.; Charriere, J.D.; Belloy, L. Distribution of *Paenibacillus larvae* spores inside honey bee colonies and its relevance for diagnosis. *J. Invertebr. Pathol.* **2008**, *99*, 92–95. [CrossRef]

veterinary sciences

Article

Modeling the Influence of Mites on Honey Bee Populations

David J. Torres [1],* and Nicholas A. Torres [2]

[1]   Department of Mathematics and Physical Science, Northern New Mexico College, Española, NM 87532, USA

[2]   Walker Department of Mechanical Engineering, The University of Texas at Austin, Austin, TX 78712, USA; natorres@utexas.edu

*   Correspondence: davytorres@nnmc.edu

Received: 20 July 2020; Accepted: 9 September 2020; Published: 21 September 2020

Abstract: The *Varroa destructor* mite has been associated with the recent decline in honey bee populations. While experimental data are crucial in understanding declines, insights can be gained from models of honey bee populations. We add the influence of the *V. destructor* mite to our existing honey bee model in order to better understand the impact of mites on honey bee colonies. Our model is based on differential equations which track the number of bees in each day in the life of the bee and accounts for differences in the survival rates of different bee castes. The model shows that colony survival is sensitive to the hive grooming rate and reproductive rate of mites, which is enhanced in drone capped cells.

Keywords: honey bee model; *Varroa destructor*; grooming; drones

---

1. Introduction

The decline of honey bees (*Apis mellifera*) has been associated with many factors, which include the *Varroa destructor* and *Varroa jacobsoni* mite and the viruses that they transmit, the microsporidia fungus *Nosema ceranae*, pesticides, including neonicotinoids, and bee management practices [1–3]. Here, we study the effect of the *V. destructor* mite with a mathematical model. The life cycle of the *V. destructor* mite is composed of a phoretic phase and a reproductive phase. In the phoretic phase, mites remain attached to adult bees where they feed on adult bee fat body tissue [4]. In the reproductive phase, the foundress female mite enters a worker brood or drone cell just before it is capped and lays eggs. Mites increase the mortality rate of bees, especially if they transmit a virus (e.g., the deformed wing virus and the acute paralysis virus) [5].

Models of honey bees have been developed by several authors [6–12]. Other authors have developed models to describe the growth of mites within honey bee colonies. Calis et al. [13] extends the model of Fries et al. [14] to study the growth of *V. jacobsoni* in colonies. Mites use available brood cells for reproduction but do not affect the bee population. Mite invasion rates are based on data from Boot et al. [15]. The model accounts for differences in mite mortality in the summer and winter as well as mite fertility in drone and worker cells. Wilkinson and Smith [16] use difference equations for adult bees and phoretic mites to study the growth rate of mites and perform sensitivity studies to study the effect of seasonal variation, the amount of drone brood, and post-capping times. Their model bases the growth of adult bees on the egg laying rate of the queen and similar to Calis et al. [13] accounts for differences in the invasion and reproductive rates of mites in worker and drone brood.

Many honey bee models include the effects of the *V. destructor* mite and the viruses that they transmit on the honey bee colony. DeGrandi-Hoffman and Curry [17] create a validated mathematical model that includes the effects of miticides. Martin [18] constructs a model of viral infection which uses mites as vectors. The model modifies the model developed by DeGrandi-Hoffman [8] and integrates meteorological conditions. Survival rates used are based on data from Fukuda and Sakagami [19]. Sumpter and Martin [20] create a model based on differential equations which tracks healthy hive bees, hive bees which acquire a virus, mites, and virus carrying mites. Hive bees are defined to be the adult worker bees that live inside the hive. Kang et al. [21] propose a model that includes parasitism, virus transmission terms, and allee effects. Ratti et al. [22] assemble a model of four differential equations, which tracks healthy and virus infested mites, and healthy and virus infested worker bees. Dénes and Ibrahim [23] use differential equations to model forager bees and three compartments of hive bees (susceptible, infested with virus free mites, and infested with virus-infected mites).

Torres et al. [24] develop a model based on differential equations that tracks each day in the life of the bee and uses different survival rates for each of the different bee castes. Th survival rates are affected by the relative number of bees in each caste. For example, the brood survival rate is decreased if there is insufficient hive bees. The model is unique in that it accounts for the brood and ethyl oleate pheromone by slowing or accelerating the maturation of hive bees into foraging bees. In this article, we add the *V. destructor* species to our existing model. This modification entails adding differential equations for the mite population, the drone population, and classes of infested pupae, drones, hive, and forager bees.

2. Mathematical Model

We begin with the fundamental Equation (1) from our transient model [24]

$$\frac{dB_i}{dt} = (S_{i-1}B_{i-1} - B_i)a_i, \quad 1 \leq i \leq 55 \tag{1}$$

where i refers to the age of the bee in days, S_i is the daily survival rate of a bee that is i days old, and B_i is the number of bees that are i days old. The term a_i accelerates or decelerates the maturation of hive bees into foraging bees due to the presence of pheromones. Equation (1) is solved using a forward Euler difference equation,

$$B_i^{n+1} = B_i^n + \triangle t S_{i-1}^n B_{i-1}^n a_i^n - \triangle t B_i^n a_i^n \tag{2}$$

where the subscript n refers to the time $n \triangle t$ and $\triangle t$ is the time step. While there are more accurate schemes, (2) benefits from simple conservation properties. The term $- \triangle t B_i^n$ accounts for the loss of $\triangle t B_i^n$ bees from the number of bees that are i days old due to natural aging in time $\triangle t$. Similarly, the term $\triangle t S_{i-1}^n B_{i-1}^n$ accounts for the movement of $\triangle t B_{i-1}^n$ bees that are $i-1$ days old into the number of bees that are i days old. However, the survival factor S_{i-1}^n reduces the number of bees $\triangle t B_{i-1}^n$ that become i days old.

Table 1 shows how an entire bee lifespan is divided into castes based on Schmickl et al. [12] and defines the total number of days a bee spends as an egg $E_t(i)$, brood or larvae $L_t(i)$, pupae $P_t(i)$, hive bee $H_t(i)$, and foraging bee $F_t(i)$. Each different caste has a different associated survival rate S_i.

Table 1. Day ranges used to calculate bee demographics.

Sum over B_i	Bee Caste	Number of Days in Caste
$E_t(i) \equiv \sum_{i=1}^{i=3} B_i$	Egg (E)	3
$L_t(i) \equiv \sum_{i=4}^{i=8} B_i$	Brood or larvae (L)	5
$P_t(i) \equiv \sum_{i=9}^{i=20} B_i$	Pupae (P)	12
$H_t(i) \equiv \sum_{i=21}^{i=41} B_i$	Hive (H)	21
$F_t(i) \equiv \sum_{i=42}^{i=55} B_i$	Forager (F)	14

We also add the castes for drones B_i^D, as shown in Table 2. The days spent in each drone caste are reported by DeGrandi-Hoffman and Curry [17] and used in the table.

Table 2. Day ranges used to calculate drone demographics.

Sum over B_i^D	Bee Caste	Number of Days in Caste
$E_t^D(i) \equiv \sum_{i=1}^{i=3} B_i^D$	Drone Egg (E^D)	3
$L_t^D(i) \equiv \sum_{i=4}^{i=10} B_i^D$	Drone Larvae (L^D)	7
$P_t^D(i) \equiv \sum_{i=11}^{i=24} B_i^D$	Drone Pupae (P^D)	14
$D_t(i) \equiv \sum_{i=25}^{i=45} B_i^D$	Adult Drone (D)	21

Table 3 shows the daily mortality (m) and survival rates $(S = 1 - m)$ based on data from Schmickl et al. [12] who base their rates on experimental data from Sakagami and Fukuda [25]. Hive bees also have a higher survival rate in the winter months, which we assume to be 0.9947 using the winter mortality rate that was provided by Sumpter and Martin [20]. If there are insufficient hive bees to tend the larvae, the survival rate of the larvae S_i^{reduce}, $4 \leq i \leq 8$ is reduced according to the equation

$$S_i^{reduce} = S_i r^\alpha, \quad r = \frac{R_L^H}{(R_L^H)_{ideal}}, \text{ if } R_L^H < (R_L^H)_{ideal}, \quad 4 \leq i \leq 8, \tag{3}$$

where $S_i = 0.99$ is the normal survival rate of larvae provided in Table 3, R_L^H is the ratio of hive bees to larvae

$$R_L^H = \frac{H_t(i)}{L_t(i)} = \frac{\sum_{i=21}^{i=41} B_i}{\sum_{i=4}^{i=8} B_i}, \tag{4}$$

and $(R_L^H)_{ideal}$ is the ideal ratio, which is assumed to be 2 [12]. The parameter $\alpha = \frac{1}{5}$ prevents the survival rate from declining too steeply if r is less than but close to 1. Martin [18] also modifies the survival rate of egg, larva, and pupae, but bases the change in survival on the total number of hive bees.

Table 3. Daily mortality (m) and survival $(S = 1 - m)$ rates.

m_{egg}	m_{larvae}	m_{pupae}	m_{hive}	$m_{forager}$
0.03	0.01	0.001	0.015	0.045
S_{egg}	S_{larvae}	S_{pupae}	S_{hive}	$S_{forager}$
0.97	0.99	0.999	0.985	0.955

The summer season is determined by the beginning S_{begin} and end days S_{end}. In the summer, hive bees mature into foraging bees, while in the winter they remain hive bees and survive at a higher daily survival rate of 0.9947. The daily egg laying rate is determined by the parameters $S_b < S_{egg} < S_{begin} < S_{peak} < S_{end} < S_e$. We found that the equation

$$L = L_{max} * \begin{cases} sin^2 \left[\frac{\pi}{2} \frac{(t - S_b)}{(S_{peak} - S_b)} \right] & \text{if } S_{egg} \leq t \leq S_{peak} \\ cos^2 \left[\frac{\pi}{2} \frac{(t - S_{peak})}{(S_e - S_{peak})} \right] & \text{if } S_{peak} < t \leq S_{end} \end{cases} \tag{5}$$

approximately matches the growth rate of the egg laying curve of Schmickl and Crailsheim [12] and can be used with adjustable beginning and ending summer dates. The parameters S_b, S_{egg}, and S_e allow the queen bee to begin and end the egg laying rate at a non-zero value in early spring and late summer. S_{peak} is the day when the egg laying rate peaks for the summer. L_{max} represents the maximum daily egg laying rate during the summer season.

2.1. Adding Mites to the Mathematical Model

In order to account for mites, we add equations to model the aging and survival of mites within the colony and assume mites survive an average of 27 days [21] in the summer when larvae brood is present

$$\frac{dM_i}{dt} = (S_{i-1}^M M_{i-1} - M_i) - \mathcal{D}_i, \quad 1 \le i \le 27. \tag{6}$$

Here, M_i represents the number of mites that are i days old, and S_{i-1}^M refers to the daily survival rate of mites. The survival rate of the mites is increased during the winter months. Martin provides the daily mortality rates of mites [26], which are 0.006 (corresponding to a survival rate of $S^M = 0.994$) in the summer and 0.002 (corresponding to a survival rate of 0.998) in the winter.

Phoretic mites are assumed to be equally distributed among the infested hive and drone bees. We assume that, if a bee is groomed, the mites on the bee die. Mites also die in a capped cell if the pupae dies within the cell. These reductions in the mite population are incorporated in the term $\mathcal{D}_i$ in (6). The product of the number of infested hive and drone bees times a maximum mite per bee value ζ determines the maximum number of phoretic mites that can be sustained by the colony.

2.2. Adding Drones and Infected Bee Populations to the Mathematical Model

An infested (denoted by an asterisk *) hive or forager bee is defined to be a bee with attached mites. In the case of pupae, an infested pupae harbors mites within its capped cell. We add additional equations for pupae infested with mites B_i^*, $9 \le i \le 20$, infested foraging bees B_i^*, $42 \le i \le 55$ (7), infested drone pupae B_i^{D*}, $11 \le i \le 24$ (8), and noninfested or "healthy" drone larvae and pupae B_i^D, $4 \le i \le 10$, $11 \le i \le 24$ (9),

$$\text{Infested pupae and foragers}: \quad \frac{dB_i^*}{dt} = S_{i-1}^* B_{i-1}^* - B_i^*, \; 9 \le i \le 20, \; 42 \le i \le 55, \tag{7}$$

$$\text{Infested drone pupae}: \quad \frac{dB_i^{D*}}{dt} = S_{i-1}^* B_{i-1}^{D*} - B_i^{D*}, \quad\quad\quad 11 \le i \le 24, \tag{8}$$

$$\text{Healthy drone larvae and pupae}: \quad \frac{dB_i^D}{dt} = S_{i-1} B_{i-1}^D - B_i^D, \; 4 \le i \le 10, \; 11 \le i \le 24. \tag{9}$$

For the healthy drone castes, we assume the same survival rates as their hive bee counterparts (see Table 3). However, we use different infested survival rates $S_i^* = 1 - m_i^*$ based on an increased mortality rate m_i^* for the infested castes.

The equations pertaining to noninfested or healthy B_i bees (10) and infested hive bees B_i^* (11), $21 \le i \le 41$, and healthy B_i^D (12), and infested drone bees B_i^{D*} (13), $25 \le i \le 45$, require additional terms to account for the movement of healthy bees B_i, B_i^D into the subpopulation of infested bees B_i^*, B_i^{D*} using the transmission rate β and the migration of infested bees back to the healthy subpopulation using the grooming rate γ

$$\text{Healthy hive bees}: \quad \frac{dB_i}{dt} = (S_{i-1} B_{i-1} - B_i)a_i - \beta \mathcal{R}(t) B_i + \gamma B_i^*, \quad 21 \le i \le 41, \tag{10}$$

$$\text{Infested hive bees}: \quad \frac{dB_i^*}{dt} = (S_{i-1}^* B_{i-1}^* - B_i^*)a_i + \beta \mathcal{R}(t) B_i - \gamma B_i^*, \quad 21 \le i \le 41, \tag{11}$$

$$\text{Healthy drones}: \quad \frac{dB_i^D}{dt} = S_{i-1} B_{i-1}^D - B_i^D - \beta \mathcal{R}(t) B_i^D + \gamma B_i^{D*}, \quad 25 \le i \le 45, \tag{12}$$

$$\text{Infested drones}: \quad \frac{dB_i^{D*}}{dt} = S_{i-1}^* B_{i-1}^{D*} - B_i^{D*} + \beta \mathcal{R}(t) B_i^D - \gamma B_i^{D*}, \quad 25 \le i \le 45, \tag{13}$$

where

$$\mathcal{R}(t) = \frac{H_t^*(t)}{H_t(i)}$$

is the ratio of the number of infested hive bees to the total number of hive bees. The term $\mathcal{R}$ is consistent with the SIR (Susceptible, Infected, Recovered) model of infectious disease [27]. These equations allow for a hive or drone bee to be infested with mites either through infested pupae maturation or as adults through the transmission term.

Because we assume that infested hive bees are less effective in tending to larvae, we modify the ideal hive to larvae ratio $(R_L^H)_{ideal}$ in (3) using the weighted mean

$$(R_L^H)_{ideal}^{mod} = \frac{[H_t(i) - H_t^*(t)](R_L^H)_{ideal} + H_t^*(i)(R_L^H)_{ideal}^*}{H_t(i)} \tag{14}$$

in the presence of infested hive bees, where $(R_L^H)_{ideal}^* = 3.0$.

2.3. Proportion of Drone Eggs

The proportion Z of eggs that are selected to be drone eggs is determined by the equation that was provided by Martin [18]

$$Z = 0.2263 \, log_{10}(0.1D_L) \, log_{10}[0.006(H_t(t) + H_t^*(t))] \tag{15}$$

where D_L is the day length which is determined by the day of the year and the latitude. We find that the equation produces fairly similar rates when compared to the rates of drone egg production used by Sumpter and Martin [20] in the spring 1%, summer 3.3%, and autumn 1%. The day length is computed using the equations that were provided by the CBM model in Forsythe et al. [28] and Brock [29].

2.4. Mite Reproduction

While the female foundress mite lays up to 5–6 eggs within the capped cell, surviving daughter mite offspring are produced at the rate of 1.3–1.45 for worker brood and 2.2–2.6 for drone brood [5]. The reproductive rate is higher in drone brood because drone brood remain capped for a longer period [5]. Martin [26] uses 1.01 and 2.91 for the reproductive rate for mites emerging from hive worker brood and drone brood, respectively, to account for a multitude of factors including mite infertility. DeGrandi-Hoffman et al. [30] use 1.5 and 2.6 for the number of surviving daughter mite offspring in worker and drone brood and we adopt these values in our model.

Sumpter and Martin [20] assume the percentage of time mites spend in capped pupae cells is 75%. While Kang et al. [21] state that the phoretic period may last 4.5–11 days when larvae are present, Martin notes that the mean number of days a mite spends as a phoretic mite is 4–6 [26], which slightly underestimates the 75% percentage if the mite lifespan is 27 days. Our model distributes 75% of total mites to available larvae cells on the day before they are capped. Let us refer to the mites that need to be distributed to available larvae as invading mites. Invading mites remain in capped cells for 12 days in worker pupae cells and 14 days in drone pupae cells. We assume that 85% of invading mites invade drone brood first up to four mites per cell due to the high propensity (5.5 to 12.1 times) of mites to invade drone cells compared to worker cells [14]. If there are insufficient drone brood to hold the invading drone mites, the remaining mites that did not invade a drone cell and 15% of the initial pool of invading mites are then distributed to available worker brood up to four mites per cell. The reduction in mite offspring due to multiple infestations of foundress mites is determined by using the values provided by Martin [26]. Specifically the number of offspring is reduced by 0.91, 0.86, and 0.60 in cells with two, three, and four foundress mites per cell in drone cells and 0.84, 0.65, and 0.66 in cells with two, three, and four foundress mites per cells in worker cells. The survival rates of the worker and drone pupae with multiple mite infestations are also reduced according to DeGrandi-Hoffman and Curry [17]. The survival rates are reduced 10%, 20%, and 40% for pupae with two, three, or four mites per cell respectively. A female mite usually has 2–3 reproductive cycles during her lifetime [5]. Martin [26] assumes the number of reproductive cycles for a mite to be 2.4. DeGrandi-Hoffman and Curry [17] only allow 60% of mites to invade a cell a second time. We only consider mites that are 16 days old or younger to be eligible for brood invasion, which allows for a maximum of two invasions.

2.5. Code and Computational Times

The code is written in MATLAB but also runs with Octave. Octave can be downloaded for free at gnu.org/software/octave. The free code can be downloaded at the github link: https://github.com/davytorres/beecode-with-mites/. Computational times vary on different computers and versions of MATLAB. A two-year simulation with a time step of 0.1 days runs in approximately 0.07 s on a Dell Ultrabook with a 2.6 GHz Intel Core i5 processor with 8 GB of RAM with MATLAB R2018b, while the same simulation runs in approximately 0.12 s on a MacBook Air with a 1.6 GHz Intel Core i5 processor with 8 GB of memory with MATLAB R2017a. Computational times can increase by orders of magnitude with earlier versions of MATLAB or Octave. The same simulation runs in approximately 4 s on a Dell Inspiron 3671 with a 2.9 GHz Intel Core i5 processor with 8 GB of installed memory with MATLAB R2014a and in 83.7 s with Octave 5.2.0 on the same machine. Despite the differences in computational times, the algorithm is fairly efficient and sensitivity studies can be conducted with two interacting variables.

3. Results

Figures 1 and 2 show a two-year simulation of a honey bee colony with mites with separate plots for eggs $E_t(i)$, larvae $B_t(i)$, total pupae $P_t(i)$, infested pupae $P_t^*(i)$, total hive bees $H_t(i)$, infested hive bees $H_t^*(i)$, total foraging bees $F_t(i)$, and infested foraging bees $F_t^*(i)$ (see Table 1). Note that the total caste subpopulation (e.g., $H_t(i)$) includes the infected caste (e.g., $H_t^*(i)$) subpopulation. The survival rates in Table 3 and the simulation parameters in Tables 4 and 5 are used. In the simulation, the mites do not affect the survival rates of any of the bee castes, and the hive is provided with an infinite supply of food. Differences in the number of bees in each caste and the height of each caste graph are mostly determined by the number of days bees spend in each caste, but are also determined to a lesser degree on the survival rate of each caste and pheromones in the hive caste. The number of drones $D_t(i)$ is also shown, but multiplied by 10 for plotting purposes, since the number of drones is relatively small. The location of the peaks are staggered beginning with the egg caste, and continuing with the larvae, pupae, hive, and foraging castes and reflect the maturation time spent in each caste. We note that the mite population peaks later than the peak of the foraging bee population. DeGrandi-Hoffman et al. [30] also show that mite populations are the largest in September through November. Figure 3 shows the percentage of eggs that become drones using (15) at latitude 35°. The graph has an average percent value of 2.5% over the two years during the summer. Figure 4 shows the positive values of the percent daily increase in the mite population,

$$100 * max \left\{ \left[\frac{M_t(i+1)}{M_t(i)} \right] - 1, 0.0 \right\}, \quad M_t(i) = \sum_{i=1}^{27} M_i, \tag{16}$$

corresponding to the mite population curve in Figure 2. After a period of oscillation, the rate relaxes. The average rate in the first year is 2.6% and the average rate in the second year is 1.3%. The reported percent rate of growth of the mite population in virus-free colonies ranges from 2.1% [31,32] to 2.2–2.5% [33]. Figure 5 shows the percentage of mites that spend their time in capped cells during the summer season. We see that approximately 70% of the mites reside within capped cells. Martin [26] predicts that 50% to 70% of mites reside within sealed brood with his model.

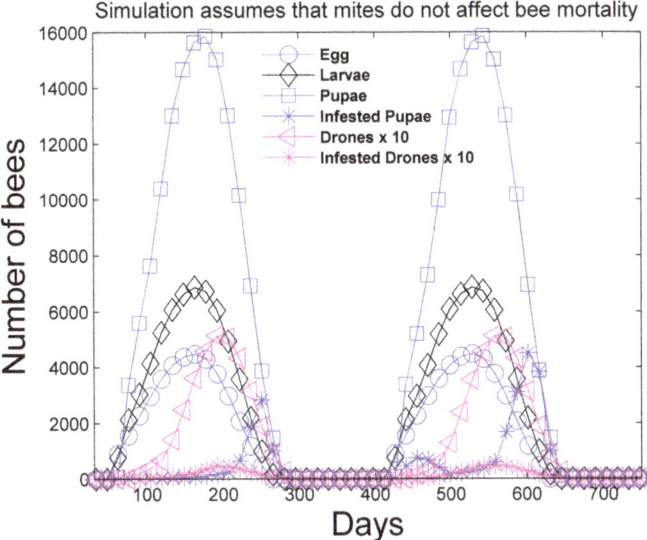

Figure 1. Two-year simulation of a honey bee colony with mites using the simulation parameters in Tables 4 and 5. The number of eggs, larvae, total pupae, infested pupae, total drones, and infested drones is shown. Mites do not affect the survival rate of any bee caste in this simulation. The number of drones is multiplied by 10 for plotting purposes.

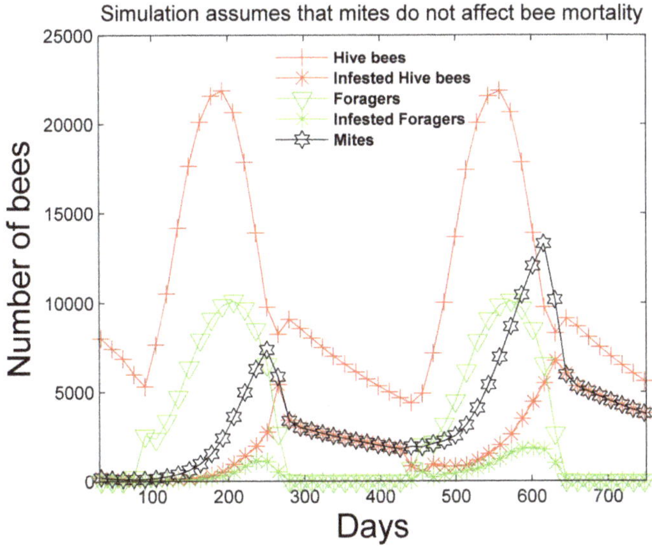

Figure 2. Two-year simulation of a honey bee colony with mites using the simulation parameters in Tables 4 and 5. The number of total hive bees, infested hive bees, total foraging bees, and infested foraging bees is shown. Mites do not affect the survival rate of any bee caste in this simulation. The number of mites closely matches the number of infested hive bees after the summer season ends when one mite remains attached to each infested hive bee.

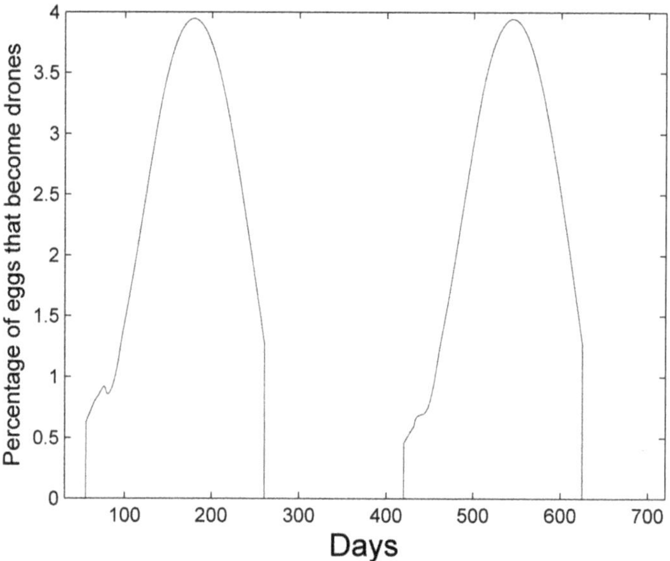

Figure 3. Percentage of eggs that become drone eggs using Equation (15) at latitude 35°. The average percent value over the two years is 2.5% during the summer.

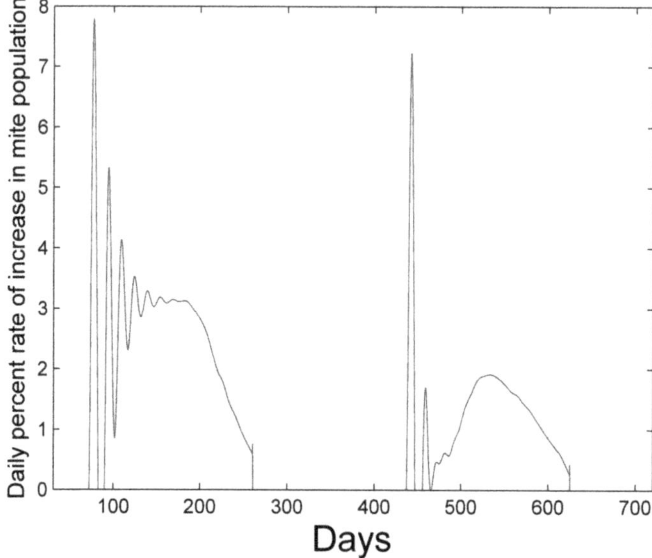

Figure 4. Daily percent rate of increase in mite population as defined by (16). After a period of oscillation, the rate relaxes. The average rate in the first year is 2.6% and the average rate in the second year is 1.3%.

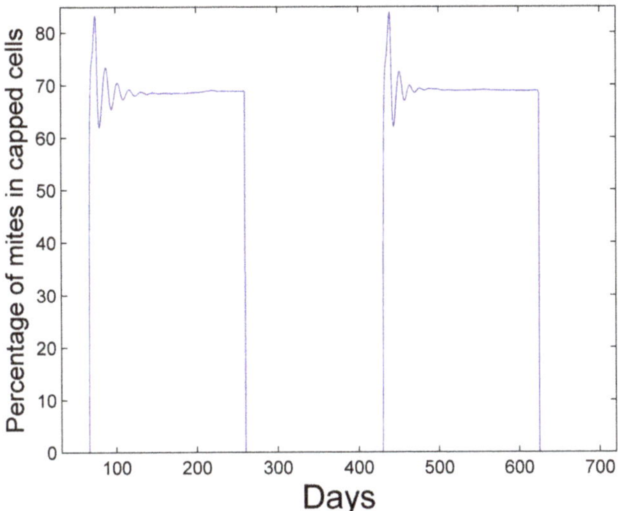

Figure 5. Percentage of mites in capped cells during summer season.

Table 4. Simulation parameters.

Initial number of hive bees	8000
Simulation start date	1 February
Maximum egg laying rate	1600
Ideal hive bee to brood ratio	2
Ideal hive bee to forager ratio	2.3
Hive bee daily survival rate in winter	0.9947
Latitude	35°
Time step $\triangle t$	0.1 day
Length of summer	6 March to 17 September
$S_b, S_{egg}, S_{peak}, S_e$	16, 55, 162, 272 days

Table 5. Simulation parameters with mites.

Initial number of mites	200
Lifetime of mite	27 days
Mite daily survival rate in summer	0.994
Mite daily survival rate in winter	0.998
Reproductive rate of foundress mite in drone brood	2.6
Reproductive rate of foundress mite in worker brood	1.5
Maximum number of mites in capped brood	4
Maximum number of mites per infested hive or drone bee ξ	6
Grooming rate γ	0.05
Transmission rate β	0.25

In Figures 6 and 7, a simulation is performed with mites under the same conditions as Figures 1 and 2, except the mites increase the mortality rates of the infested pupae, hive, drone, and forager castes shown in Table 3 by four times. We also apply (14), where $(R_L^H)^*_{ideal} = 3.0$ in order to require additional infested hive bees to effectively attend to larvae. This requirement is not enforced in Figures 1 and 2, where $(R_L^H)^*_{ideal} = (R_L^H)_{ideal} = 2$. The **differences** in the bee castes are plotted. For example, in regard to hive bees, the difference

$$H_t(i)[\text{with unchanged mortality}] - H_t(i)[\text{with increased mortality}]$$

is plotted in Figure 7. Figure 6 shows that the simulation with unchanged mortality rates (Figures 1 and 2) has larger numbers of larvae, pupae, and drones when compared to the simulation with increased bee mortality. However, the differences are less than 200 in each category. Figure 7 shows that more mites exist in the simulation with unchanged bee mortality, because the mites are negatively affected by increased bee mortality. In addition, the simulation with unchanged bee mortality has significantly larger numbers of hive and foraging bees. The number of hive bees that persist late into the summer season is significantly reduced with increased bee mortality and the reduced number of hive bees does not allow the colony to survive past the first year.

Figure 8 shows how the maximum number of hive bees varies in the second year when 32 different values of the grooming rate ranging from 0 to 0.15 and 32 different values of transmission rate ranging from 0 to 0.2 are used for a total of 1024 simulations. In this sensitivity study, mites increase the mortality rates of the infested pupae, hive, drone, and forager castes shown in Table 3 by four times. We also apply (14), where $(R_L^H)_{ideal}^* = 3.0$. With the exceptions noted above, the simulation parameters in Tables 3–5 are used. A sharp line separates the conditions under which the colony can be sustained (yellow) from the conditions under which the colony perishes (blue) beyond the first year. Figure 8 shows the importance of the grooming rate in sustaining the colony.

Figure 9 shows how the maximum number of hive bees varies in the second year when 32 different grooming rates ranging from 0 to 0.1 and 32 different reproductive rates ranging from 1.5 to 3.0 for the drone foundress mite are used. As with Figure 8, in this sensitivity study, mites increase the mortality rates of the infested pupae, hive, drone, and forager castes by four times. A transmission rate of $\beta = 0.2$ is used. We also apply (14) where $(R_L^H)_{ideal}^* = 3.0$. With the exceptions noted above, the simulation parameters in Tables 3–5 are used. Again, a sharp line separates the conditions under which the colony can be sustained. Figure 9 shows the importance of the higher reproductive level of mites in drone cells. Mite populations are significantly reduced in simulations of colonies without drones, despite the relatively small number of drones when compared to hive bees. We note that beekeepers do remove drone brood in practice to control mites [34].

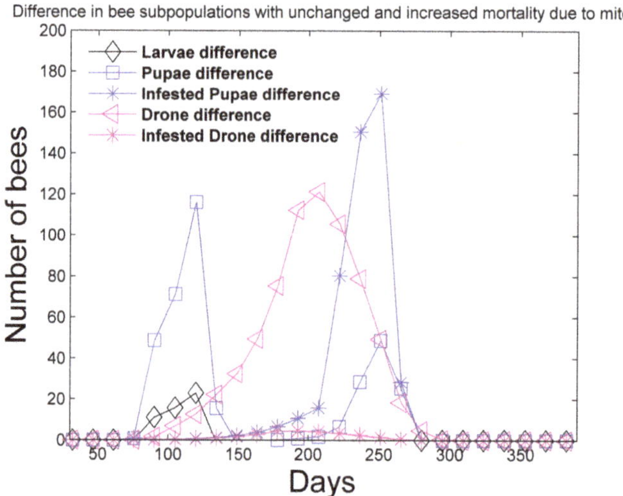

Figure 6. Two-year simulation of a honey bee colony with mites using the simulation parameters in Tables 4 and 5. Mites increase the mortality rates in Table 3 of infested pupae, hive, drone, and foragers by four times. Only the first year is shown since the colony does not survive past the first year. In this figure, the **difference** in the bee subpopulations (larvae, total pupae, infested pupae, total drones, infested drones) with unaffected mortality (Figures 1 and 2) and the bee subpopulations with increased mortality is shown.

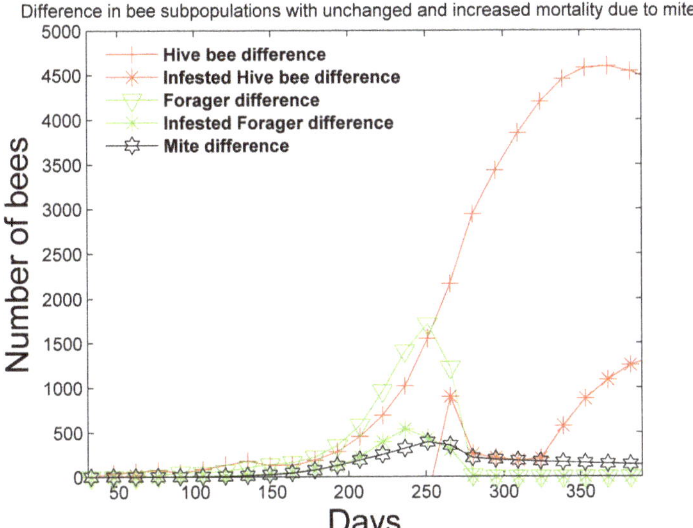

Figure 7. Two-year simulation of a honey bee colony with mites using the simulation parameters in Tables 4 and 5. Mites increase the mortality rates in Table 3 of infested pupae, hive, drone, and foragers by four times. Only the first year is shown since the colony does not survive past the first year. In this figure, the difference between the number of bee subpopulations (total hive bees, infested hive bees, total foraging bees, and infested foraging bees) with unchanged mortality (Figures 1 and 2) and the bee subpopulations with increased mortality is plotted.

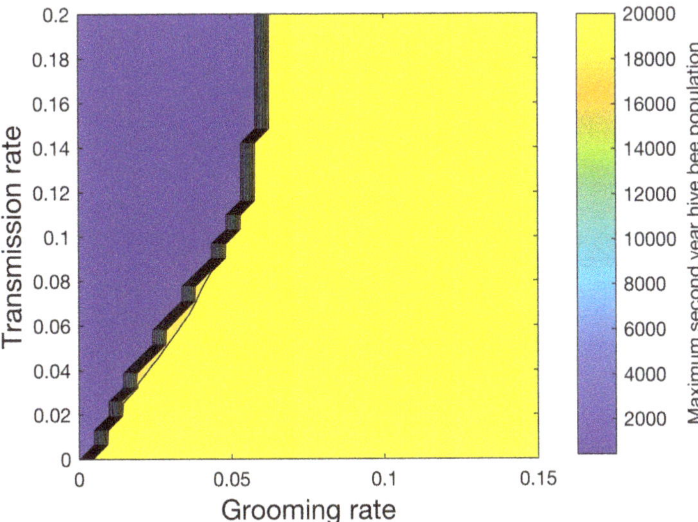

Figure 8. Effect of grooming rate γ and transmission rate β (10)–(13) on the second year maximum hive bee population. The transmission rate models the rate at which uninfested bees become infested. The grooming rate models the rate at which infested bees becomes uninfested. A sharp line divides the conditions under which the colony can be sustained (yellow) versus when the colony perishes (blue).

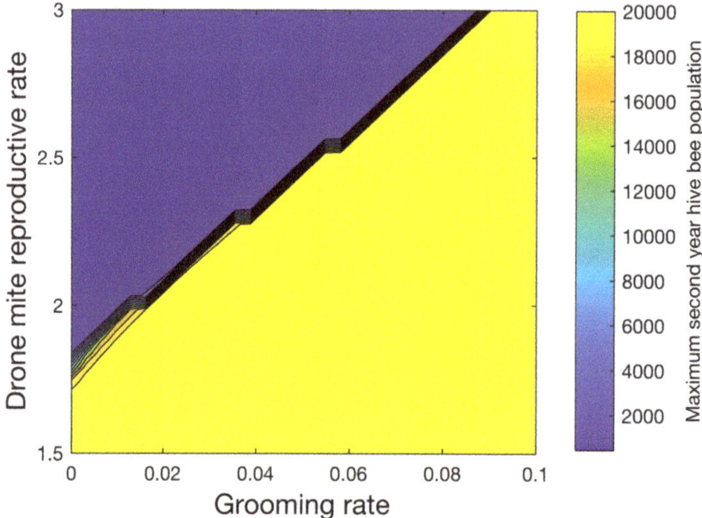

Figure 9. Effect of grooming rate γ (10)–(13) and drone reproductive rate on the second year maximum hive bee population. The grooming rate models the rate at which infested bees becomes uninfested. The transmission rate $\beta = 0.2$. A sharp line divides the conditions under which the colony can be sustained (yellow) versus when the colony perishes (blue). The figure shows that drones play an important role in sustaining the mite population due to the higher reproductive rate of mites in drone capped cells.

4. Discussion

We have developed a mathematical model of a honey bee colony with mites, which tracks healthy and infested pupae, hive bees, drones, and foragers. The model is based on differential equations that solve for the number of bees in each day in the life of the bee. Model simulations shed light on the effect of different parameters on colony sustainability. Specifically, the grooming rate, transmission rate, and reproductive rate of the foundress mites are all important parameters that determine whether a colony will survive or perish.

Our model shows that the mite population is significantly reduced in the absence of drones despite the relatively low number of drones compared to hive bees. The importance of drone brood to mite population growth has also been reported in other models [13,17,35]. Our model also shows that grooming is an effective mechanism for controlling the growth of mites. The absence of grooming in the model can lead to very high populations of mites and the importance of grooming to bee survival has been noted by other authors [13,36,37].

5. Conclusions

The observations from our mathematical model could lead to continued or new experimental investigations in order to test the impact drone reduction [23] and increased grooming rate have on colony sustainability and ultimately help guide bee management processes (e.g., removal of drone brood and selective breeding) to improve colony health.

Future research efforts would be aimed at adding the influence of viruses and the increased mortality that they bring to specific castes. A computational comparison of available models of mite growth would also be useful to quantitatively compare the differences and similarities of facets within each model.

Author Contributions: Conceptualization, D.J.T.; Writing, D.J.T.; Programming, D.J.T. and N.A.T.; Simulations, D.J.T. and N.A.T. Both authors have read and agreed to the published version of the manuscript.

Funding: This research is supported by an Institutional Development Award (IDeA) from the National Institute of General Medical Sciences of the National Institutes of Health under grant number P20GM103451.

Conflicts of Interest: The authors declare no conflict of interest.

References

1. Evans, J.D.; Spivak, M. Socialized medicine: Individual and communal disease barriers in honey bees. *J. Invertebr. Pathol.* **2010**, *103*, S62–S72. [CrossRef]
2. Johnson, R.M. Honey bee toxicology. *Annu. Rev. Entomol.* **2015**, *60*, 415–434. [CrossRef] [PubMed]
3. van Engelsdorp, D.; Evans, J.D.; Saegerman, C.; Mullin, C.; Haubruge, E.; Nguyen, B.K.; Frazier, M.; Frazier, J.; Cox-Foster, D.; Chen, Y.; et al. Colony Collapse Disorder: A Descriptive Study. *PLoS ONE* **2009**, *4*, E6481. [CrossRef] [PubMed]
4. Ramsey, S.D.; Ochoa, R.; Bauchan, G.; Gulbronson, C.; Mowery, J.D.; Cohen, A.; Lim, D.; Joklik, J.; Cicero, J.M.; Ellis, J.D.; et al. *Varroa destructor* feeds primarily on honey bee fat body tissue and not hemolymph. *Proc. Natl. Acad. Sci. USA* **2019**, *116*, 1792–1801. [CrossRef]
5. Rosenkranz, P.; Aumeier, P.; Ziegelmann, B. Biology and control of *Varroa destructor*. *J. Invertebr. Pathol.* **2010**, *103*, S96–S119. [CrossRef]
6. Becher, M.A.; Osborne, J.L.; Thorbek, P.; Kennedy, P.J.; Grimm, V. Towards a systems approach for understanding honeybee decline: A stocktaking and synthesis of existing models. *J. Appl. Ecol.* **2013**, *50*, 868–880. [CrossRef]
7. Becher, M.A.; Grimm, V.; Thorbek, P.; Horn, J.; Kennedy, P.J.; Osborne, J.L. BEEHAVE: A systems model of honeybee colony dynamics and foraging to explore multifactorial causes of colony failure. *J. Appl. Ecol.* **2014**, *51*, 470–482. [CrossRef]
8. DeGrandi-Hoffman, G.; Roth, S.A.; Loper, G.L.; Erickson, E.H. BEEPOP: A honeybee population dynamics simulation model. *Ecol. Model.* **1989**, *45*, 133–150. [CrossRef]
9. Khoury, D.S.; Myerscough, M.R.; Barron, A.B. A Quantitative Model of Honey Bee Colony Population Dynamics. *PLoS ONE* **2011**, *6*, E18491. [CrossRef]
10. Khoury, D.S.; Barron, A.B.; Myerscough, M.R. Modelling food and population dynamics in honey bee colonies. *PLoS ONE* **2013**, *8*, E59084. [CrossRef]
11. Russell, S.; Barron, A.B.; Harris, D. Dynamic modelling of honey bee (*Apis mellifera*) colony growth and failure. *Ecol. Model.* **2013**, *265*, 158–169. [CrossRef]
12. Schmickl, T.; Crailsheim, K. HoPoMo: A model of honeybee intracolonial population dynamics and resource management. *Ecol. Model.* **2007**, *204*, 219–245. [CrossRef]
13. Calis, J.N.M.; Fries, I.; Ryrie, S.C. Population modelling of *Varroa jacobsoni* Oud. *Apidologie* **1999**, *30*, 111–124. [CrossRef]
14. Fries, I.; Camazine, S.; Sneyd, J. Population dynamics of *Varroa jacobsoni*: A model and a review. *Bee World* **1994**, *75*, 5–28. [CrossRef]
15. Boot, W.J.; Beetsma, J.; Calis, J.N.M. Behaviour of *Varroa* mites invading honey bee brood cells. *Exp. Appl. Acarol.* **1994**, *18*, 371–379. [CrossRef]
16. Wilkinson, D.; Smith, G.C. A model of the mite parasite, *Varroa destructor*, on honeybees (*Apis mellifera*) to investigate parameters important to mite population growth. *Ecol. Model.* **2001**, *148*, 263–275. [CrossRef]
17. DeGrandi-Hoffman, G.; Curry, R.A. Mathematical Model of Varroa Mite (Varroa Destructor Anderson and Trueman) and Honeybee (*Apis Mellifera* L.) Population Dynamics. *Internate J. Acarol.* **2000**, *30*, 259–274. [CrossRef]
18. Martin, S.J. The role of *Varroa* and viral pathogens in the collapse of honeybee colonies: A modelling approach. *J. Appl. Ecol.* **2001**, *38*, 1082–1093. [CrossRef]
19. Fukuda, H.; Sakagami, S. Worker brood survival in honey bees. *Res. Popul. Ecol.* **1968**, *10*, 31–39. [CrossRef]
20. Sumpter, D.J.T.; Martin, S.J. The dynamics of virus epidemics in Varroa-infested honey bee colonies. *J. Anim. Ecol.* **2004**, *73*, 51–63. [CrossRef]
21. Kang, Y.; Blanco, K.; Davis, T.; Wang, Y.; DeGrandi-Hoffman, G. Disease dynamics of honeybees with Varroa destructor as parasite and virus vector. *Math. Biosci.* **2016**, *275*, 71–92. [CrossRef] [PubMed]

22. Ratti, V.; Kevan, P.G.; Eberl, H.J. A mathematical model for population dynamics in honeybee colonies infested with *Varroa destructor* and the Acute Bee Paralysis Virus. *Can. Appl. Math. Q.* **2013**, *21*, 63–93.

23. Dénes, A.; Ibrahim, M.A. Global dynamics of a mathematical model for a honeybee colony infested by virus-carrying *Varroa* mites. *J. Appl. Math. Comput.* **2019**. [CrossRef]

24. Torres, D.J.; Ricoy, U.M.; Roybal, S. Modeling honey bee populations. *PLoS ONE* **2015**, *10*, E0130966. [CrossRef] [PubMed]

25. Sakagami, S.F.; Fukuda, H. Life tables for worker honeybees. *Res. Popul. Ecol.* **1968**, *10*, 127–139. [CrossRef]

26. Martin, S. A population model for the ectoparasitic mite *Varroa jacobsoni* in honey bee *Apis mellifera* colonies. *Ecol. Model.* **1998**, *109*, 267–281. [CrossRef]

27. Kermack, W.O.; McKendrick, A.G. A contribution to the mathematical theory of epidemics. *Proc. R. Soc. Lond. A* **1927**, *115*, 700–721.

28. Forsythe, W.C.; Rykiel, E.J.; Stahl, R.S.; Wu, H.; Schoolfield, R.M. A model comparison for daylength as a function of latitude and day of year. *Ecol. Model.* **1995**, *80*, 87–95. [CrossRef]

29. Brock, T.D. Calculating solar radiation for ecological studies. *Ecol. Model.* **1981**, *14*, 1–19. [CrossRef]

30. DeGrandi-Hoffman, G.; Ahumada, F.; Zazueta, V.; Chambers, M.; Hidalgo, G.; DeJong, E.W. Population growth of *Varroa destructor* (Acari: Varroidae) in honey bee colonies is affected by the number of foragers with mites. *Exp. Appl. Acarol.* **2016**, *69*, 21–34. [CrossRef]

31. Calatayud, F.; Verdu, M.J. Number of adult female mites *Varroa jacobsoni* Oud. on hive debris from honey bee colonies artificially infested to monitor mite population increase. *Mesostigmata: Varroidae Exp. Appl. Acarol.* **1995**, *19*, 181–188. [CrossRef]

32. Kraus, B.; Page, R.E., Jr. Population growth of *Varroa jacobsoni* Oud. in Mediterranean climates of California. *Apidologie* **1995**, *26*, 149–157. [CrossRef]

33. Martin, S.J.; Kemp, D. Average number of reproductive cycles performed by *Varroa jacobsoni* in honey bee (*Apis mellifera*) colonies. *J. Apic. Res.* **1997**, *36*, 113–123. [CrossRef]

34. Charrière, J.D.; Indorf, A.; Bachofen, B.; Tschan, A. The removal of capped drone brood: An effective means of reducing the infestation of varroa in honey bee colonies. *Bee World* **2002**, *84*, 117–124. [CrossRef]

35. Wilkinson, D.; Smith, G.C. Modeling the efficiency of sampling and trapping *Varroa destructor* in the drone brood of honey bees (*Apis mellifera*). *Am. Bee J.* **2002**, *142*, 209–212.

36. Prichard, D.J. Grooming by honey bees as a component of varroa resistant behavior. *J. Apic. Res.* **2016**, *55*, 1–11. [CrossRef]

37. Locke, B. Natural *Varroa* mite-surviving *Apis mellifera* honeybee populations. *Apidologie* **2016**, *47*, 467–482. [CrossRef]

Article

How the Infestation Level of *Varroa destructor* Affects the Distribution Pattern of Multi-Infested Cells in Worker Brood of *Apis mellifera*

Ignazio Floris *, Michelina Pusceddu and Alberto Satta

Department of Agricultural Science, University of Sassari, Viale Italia 39, 07100 Sassari, Italy;
mpusceddu@uniss.it (M.P.); albsatta@uniss.it (A.S.)
* Correspondence: ifloris@uniss.it; Tel.: +39-079-229-369

Received: 31 July 2020; Accepted: 16 September 2020; Published: 17 September 2020

Abstract: The mite *Varroa destructor*, the main ectoparasite of honey bees, is a threat to apiculture worldwide. Understanding the ecological interactions between *Varroa* and honeybees is fundamental for reducing mite impact in apiaries. This work assesses bee colonies with various *Varroa* infestation levels in apiaries to determine: (1) the relationship between multi-infested brood cells and brood infestation level, (2) the damage caused by *Varroa* to parasitized honey bee pupae, and (3) mite reproduction rate at various infestation levels. Data were collected from 19 worker brood combs, each from a different colony, ranging from 160 to 1725 (mean = 706) sealed cells per comb. Mite distribution was aggregated, ranging from about 2% to 74% infested cells per comb. The percentage of cells invaded by one, two, three, four, or more than four foundress mites, as a function of infestation level, was estimated by five highly significant ($p < 0.0001$) second-degree polynomial regression equations. The correction factors found could increase the precision of prediction models. *Varroa* fertility and adult bee longevity decreased as multi-infestation levels increased, and the implications of this relationship are discussed. Finally, these findings could improve sampling methods and the timing of mite treatments in apiaries, thus favoring sustainable management strategies.

Keywords: mite; reproductive rate; worker brood; infestation level; longevity; distribution; model; *Apis mellifera*

---

1. Introduction

The parasitic mite *Varroa destructor* was originally confined to the Eastern honey bee *Apis cerana* [1]. After a shift to a new host, *A. mellifera*, and a worldwide dispersion, this mite has become the most serious threat to honeybees [1,2]. *Varroa* has had a fundamental role in the decline of honeybee colonies observed all over the Northern Hemisphere in the last few decades [3,4].

Many factors of the host and the parasite affect the population growth of *Varroa* in honeybee colonies [5,6], such as the worker brood availability. In fact, the number of brood cells and/or the fertility of the mites and population growth are significantly correlated [7–9]. Therefore, it is likely that the population dynamic of the bee colony significantly influences the development of *Varroa* infestation. It is known that the average number of adult female offspring produced by a single mother mite invading a worker brood cell ranges between 1.2–1.5, whereas this reproduction rate rises to 2.2–2.6 in drone brood cells due to their longer capping period [10–12]. Moreover, in multiple-infested brood cells, the reproductive rate per female mite is significantly reduced [12–15]. Consequently, the number of brood cells throughout the season, the temporal pattern of brood availability, and the percentage of drone brood, among other factors [1], can have an impact on the reproduction of the *Varroa* population. This is described in some population dynamic models of *Varroa* mites and honeybees [12,16,17]. Furthermore, in the honey bee brood, it is not uncommon to find sealed cells with bee larvae infested by two, three, or more female mites, while many other cells remain uninfested. This suggests that the distribution of the mite among brood cells is not random but, rather, aggregated [18–21]. However, it should also be pointed out that some other studies do not support the aggregation hypothesis [12,22]. Differences in the statistical approach and data collection methods adopted probably led to different interpretations in these studies. Despite that, what clearly emerges statistically is that the mean values of brood infestation are always lower than the respective variances, which is a basic assumption to demonstrate the tendency to aggregation from an ecological point of view [23,24].

An aspect that remains to be explained is the different attractiveness of the brood cells, which is at the basis of the aggregation behavior, such as possible chemical sources of attractiveness [2]. Aggregation could favor exogamy and may have an adaptive value for *Varroa*, but it is unknown whether this phenomenon is related to an aggregation pheromone, to the higher attractiveness of certain bee larvae, or to other biotic and abiotic factors [2].

One of the main sources of error in the application of sampling techniques is the aggregate distribution, especially when cutting honeycomb parts [25,26]. In fact, the variability of infestation between different areas of the same comb or between different combs [27,28], due to the irregular distribution of *Varroa*, may lead to substantial differences in sampling results. An accurate estimation of brood infestation, based on precise knowledge of the mite spatial distribution pattern and its interpretation by specific models (e.g., Iwao's regression method [20]), favors the development adoption of appropriate sampling plans (e.g., stratified random sampling, cross sampling or, for practical purposes, sequential sampling) [19,20,26]. Better understanding the basic ecology of the mite is useful for several other reasons: (1) the correct estimation of the infestation level of the brood can help to determine the most appropriate timing of treatments used for *Varroa* control, as done in the sustainable integrated pest management (IPM); (2) the damage caused by *Varroa* to larvae or pupae depends on the infestation level and the number of mites entering the cell, as demonstrated by the correlations between pathogen loads (positive) or colony strength (negative) and *Varroa* infestation rate [29,30]; thus, knowing the mite distribution pattern can be useful in determining the extent of brood damage as a function of infestation level; (3) an accurate estimate of the percentage of cells with a specific number of female mites allows us to correctly assess the *Varroa* reproduction rate because the increased competition among the offspring mites for food and space in a multi-infested cell can decrease mite fertility. In fact, in multiple-invaded drone and worker brood cells, the reproductive rate per female mite is significantly reduced [12,14,15,31].

The available predictive models of mite dynamics [11,12,16,17,32] do not consider the effects of multi-infested brood cells on the development of *Varroa* infestation, thus associating mite growth rates simply with brood availability (distinguishing only between male and female brood), or they refer to a Poisson distribution, which is not accurate enough to represent the real behavior of *Varroa* in bee brood [33]. Some studies have shown an aggregated distribution of *Varroa* in the brood [19,20], associated with increases in multi-infestation as population density grows, with consequent effects on mite development and reproduction rate, as well as on the extent of damage to bee colonies. It is important to highlight that in environmental conditions favorable for the constant presence of brood in the hives throughout seasons, such as in the Mediterranean area, it is crucial to correctly estimate the percentage of cells infested by one or more mites [34]. This allows for the definition of more realistic simulation models of the development dynamic of *Varroa*, in line with its statistical spatial distribution, thus promoting more sustainable and efficient mite control.

This work assesses bee colonies with various levels of natural infestations by *Varroa* in apiaries to determine: (1) the relationship between multi-infested brood cells and brood infestation level, (2) the damage caused by *Varroa* to parasitized honey bee pupae in terms of bee longevity, and (3) the effect of infestation level on mite reproduction rate. Based on data collected from numerous worker brood combs with percentages of infested cells ranging from about 2% to 74%, five second-degree polynomial regression equations were developed to estimate the percentage of cells invaded by one, two, three, four, and more than four foundress mites according to the infestation level. The work also discusses the implications of these relationships on the reduction of *Varroa* fertility and longevity of adult bees. Our findings could favor sustainable management strategies by improving predictive models, sampling methods, and timing of mite treatments in apiaries.

2. Materials and Methods

The work was carried out in the experimental apiary of the University of Sassari located in Ottava (40°46′23″ N; 8°29′34″ E), Province of Sassari, Italy, from late summer to early fall (September–October) in 2018. The apiary was composed of *A. mellifera* colonies kept in Dadant–Blatt standard hives naturally infested by *Varroa*, at various infestation levels.

In total, 19 combs containing worker sealed brood were taken from the central position of the nest of 19 hives, 3 days after cell sealing started, and maintained at −20 °C until inspection. For each honeycomb, all sealed brood cells were inspected, and the number of foundress mites present in each infested cell was recorded. The number of sealed brood cells in each comb ranged from 160 (comb no. 18) to 1725 (comb no. 2), with an average value of 706 (Table 1).

Table 1. The number of sealed and infested brood cells and foundress mites obtained by inspecting 19 combs sampled from 19 different colonies of honey bees naturally infested by *Varroa* in the apiary. The percentage of infestation and the mean number of foundress mites per infested cell were calculated for each comb.

Comb	Number of Sealed Brood Cells	Number of Infested Cells	Number of Foundress Mites	Percentage of Infestation [a]	Mean Number of Foundress Mites Per Infested Cell
1	483	8	8	1.7	1.0
2	1725	36	40	2.1	1.1
3	738	22	24	3.0	1.1
4	442	16	19	3.6	1.2
5	545	20	22	3.7	1.1
6	1378	51	53	3.7	1.0
7	374	26	28	7.0	1.1
8	445	40	49	9.0	1.2
9	738	89	116	12.1	1.3
10	1652	218	253	13.2	1.2
11	466	77	88	16.5	1.1
12	740	145	191	19.6	1.3
13	256	78	141	30.5	1.8
14	691	227	345	32.9	1.5
15	1040	368	597	35.4	1.6
16	1066	449	818	42.1	1.8
17	215	101	236	47.0	2.3
18	160	103	259	64.4	2.5
19	265	197	549	74.3	2.8

[a] Data are listed in ascending order according to the percentage of infestation.

The following descriptive variables were calculated for each comb: (a) the total number of sealed brood cells, (b) the total number of female (foundress) mites, (c) the average number of female mites per brood cell, and (d) the number and percentage of cells containing one, two, three, four, and more than four foundress mites per cell. These data were used to derive second-degree polynomial regression equations to correlate the average number of foundress mites per infested cell with the percentage of cells containing one, two, three, four, or more than four foundress mites per cells, separately.

The regression equations obtained were used to derive two other second-degree regression equations, useful to describe the mean reproduction rate (fecundity) of female mites and the percentage of longevity reduction of adult bees parasitized by mites in the preimaginal stage as a function of the mean number of mother mites per infested cell. In order to develop these two equations, we simulated

17 different distributions of foundress mites, considering a constant number of cells available for invasion (1000) and an increasing number of mites (from 10 to 2500). After that, the average fecundity of foundress mites for each mite distribution was calculated considering the following average values of mite offspring: 1.45, 1.32, 1.25, 0.87, and 0 mites in cells with one, two, three, four, and more than four foundress mites, respectively [12]. To calculate the average longevity of workers that have emerged from the infested cells, the following percentage reduction in lifespan was considered: 2%, 10%, 20%, 40%, and 80% in cells with one, two, three, four, and more than four foundress mites, respectively [16].

Data availability: The complete datasets analyzed during the current study are available from the corresponding author on reasonable request.

3. Results

The number of sealed brood cells, infested brood cells, and foundress mites in each of the 19 inspected combs, the percentage of infestation per comb, and the mean number of female mites per infested cell are given in Table 1. The infestation level of *Varroa* ranged from 1.7% to 74.3% in the inspected combs, and the mean number of foundress mites per infested cell varied from 1 to 2.8 (Table 1). Between these last two variables, a highly significant, linear and positive, relationship (Df = 1; F = 277.4; $p < 0.00001$) was observed (Figure 1), with approximately 94% of the variability of the mean number of foundress mites per infested cell explained by the infestation level ($R^2 = 0.942$). This relationship clearly shows that as the level of infestation increases, the phenomenon of multi-infestation increases.

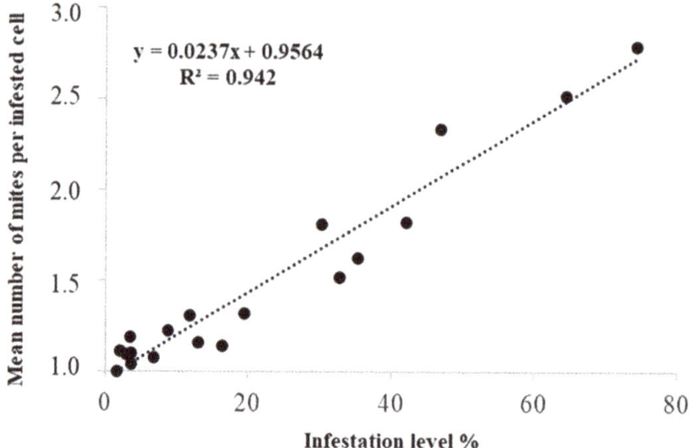

Figure 1. Relationship between the percentage of *Varroa* infestation level and the mean number of foundress mites per infested cells of worker brood.

From the raw data, we calculated, for each infestation level, the percentage of cells with one, two, three, four, and more than four foundress mites. Then, we correlated these parameters with the mean number of female mites per infested cells. The equations that better fitted the relationship between these variables were second-degree polynomials, as shown in Figure 2.

The second-degree polynomial curves (Figure 2A–E), all highly significant ($p < 0.0001$), allow us to evaluate the distribution of the foundress mites inside the brood cells for different infestation levels. In fact, we can observe that when the level of infestation increases, the progressive decrease in the percentage of cells containing only one mother mite ($R^2 = 0.9824$; Figure 2A) is compensated by the growth of cells containing two ($R^2 = 0.9045$; Figure 2B), three ($R^2 = 0.7795$; Figure 2C), four ($R^2 = 0.9031$; Figure 2D) or more than four ($R^2 = 0.9207$; Figure 2E) foundress mites.

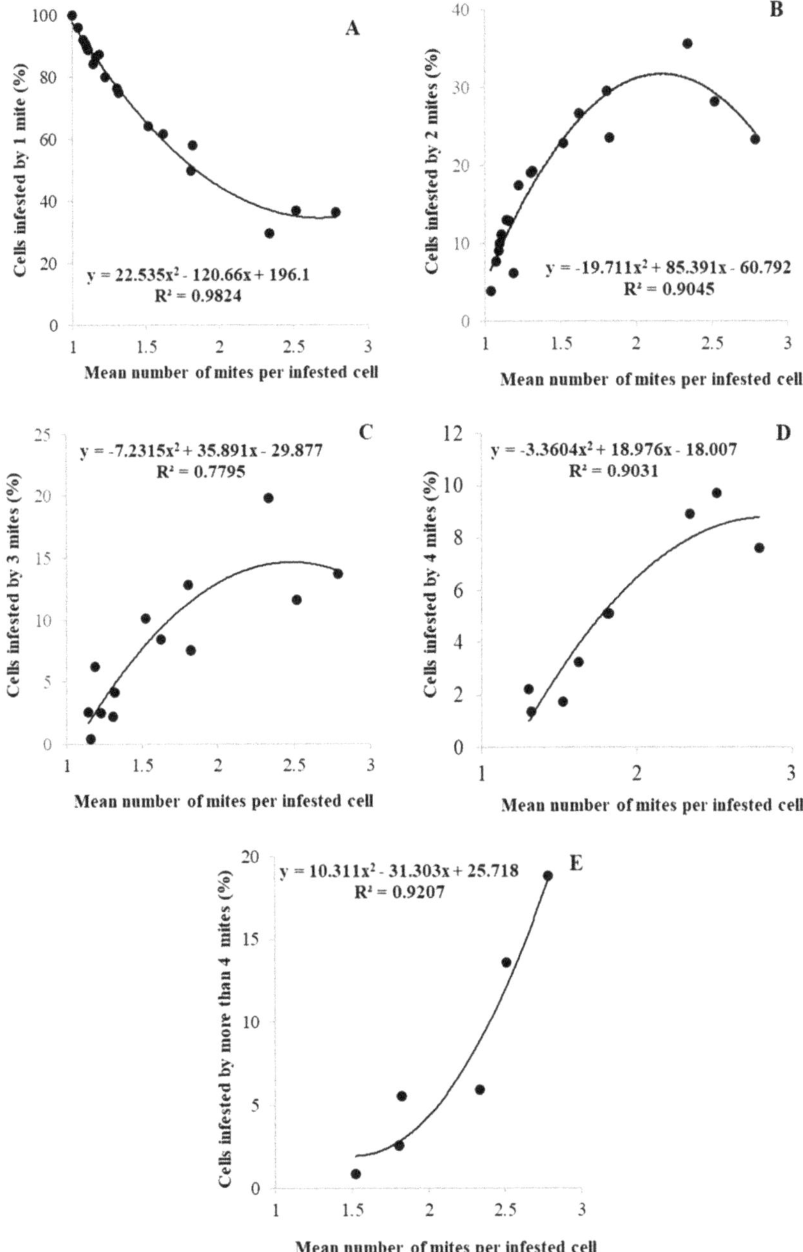

Figure 2. Relationship between the mean number of foundress mites per infested worker brood cells and the percentage of cells invaded by one (**A**), two (**B**), three (**C**), four (**D**), and more than four (**E**) foundress mite. In each graph, each dot represents a colony.

Considering that the *Varroa* reproduction rate decreases with increasing multi-infestation, if the distribution of multi-infested cells is known for each infestation level, we can also calculate the average reproduction rate of *Varroa*. For this purpose, by simulating a series of mite distributions at levels of growing infestation ($n = 17$), a regression curve was derived to express the negative relationship between the average fecundity per female mite and the average number of foundress mites for infested cell ($R^2 = 0.9989$; Figure 3). The same criterion used above was applied to derive a regression curve describing the decrease in the longevity of adult bees as a function of the average number of mites per infested cell ($R^2 = 0.9993$; Figure 4). In this case, the information obtained could be used to better understand the effects of the increasing infestation levels on the bee population dynamic.

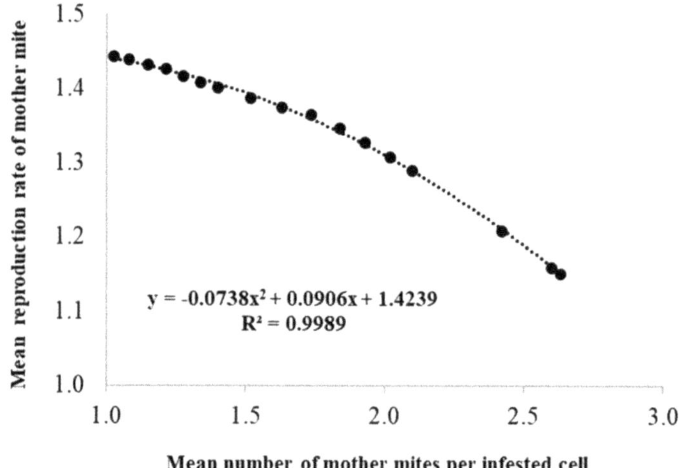

Figure 3. Relationship between the mean number of foundress mites per infested cells of worker brood and the reduction of their reproduction rate.

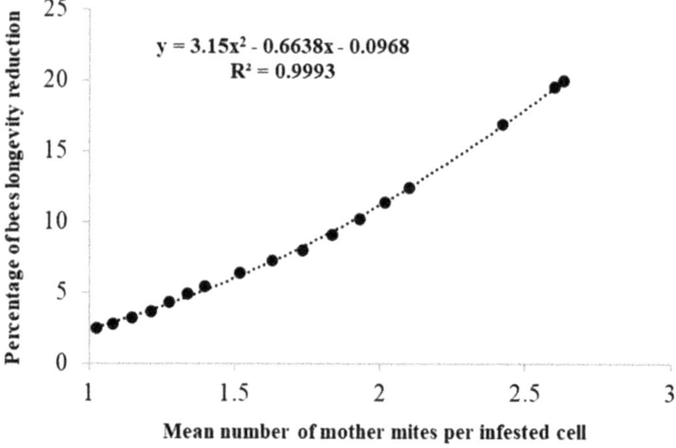

Figure 4. Relationship between the mean number of foundress mites per infested cells of worker brood and the longevity reduction of honeybee adults.

4. Discussion

Based on the data collected under our specific experimental conditions, our findings suggest applications of practical relevance in terms of improving *Varroa* control strategies. Differences due to seasonal or environmental variations are possible but likely limited to the range of infestation levels detected. In fact, these differences should not affect the relationship between infestation and multi-infestation levels nor the effects of multi-infestation because the latter is proven to be density-dependent [13]. This is clearly evidenced by the observed increase in multi-infestation as the infestation level increases. This phenomenon also has important effects on the genetic structure of the population of *Varroa*. In fact, as observed by Beaurepaire et al. [35], the increase in the co-infestation rate of brood cells also coincides with an increase in the recombined lines in the mite populations. Therefore, if treatments against *Varroa* are carried out before the recombination phase has taken place, the inbreeding will greatly promote the fixation of the alleles for acaricide resistance [35]. Furthermore, based on the model of De Grandi-Hoffmann and Curry [16], control treatments applied against *Varroa* in late summer provide the best chances for the survival of heavily infested colonies. Therefore, surveys conducted in late summer, as in our case, are particularly important in Mediterranean environments. However, it should also be considered that colony survival thresholds for mite populations and the effectiveness of miticides are dependent on the climate and the yearly brood dynamic.

The analysis of the data collected in our study revealed, according to the principles established by ecological methods [23,24], an aggregate distribution of *Varroa* in the brood. Such distribution can be interpreted by different models, such as Iwao's regression method, the negative binomial, and other interpretative models [19,20]. The aggregate distribution of the mite in the brood is associated with multi-infestation, with mites tending to concentrate in some areas of the brood comb, thus increasing the possibility that some of them may invade the same cell. However, the factors causing mite aggregation are still unknown [2].

Among the possible applications of our findings, the definition of more precise prediction models for the development of *Varroa* infestation is one of the most interesting aspects. In fact, based on our second-degree polynomial curves it could be possible to define the distribution of female mites in brood cells and their rate of reproduction for each level of infestation. Therefore, these equations provide significant correction factors, previously unknown in the literature, to define the evolution of *Varroa* infestation, better representing the behavior of the mite in apiary conditions. In addition, the information obtained in the reduction of adult bee longevity as a function of the average number of mites per infested cell, could be used to better understand the effects of the increasing infestation levels on bee population dynamic. Differently from previous studies, in our work, all the cells of the combs were examined according to their natural distribution, without preselection of combs or honeycomb areas with cells of the same age and so on, and the observations were not conducted under laboratory conditions. Therefore, our data reflect the natural behavior of the mite in honeybee colonies in their environment.

Our findings also have an important practical and scientific impact on the definition of more appropriate and precise sampling methods. In particular, the detected aggregated mite distribution suggests, from a statistical point of view, the need for a stratified sampling for a correct evaluation of brood infestation level. For practical purposes, sequential sampling [19,20], based on predefined infestation thresholds, is appropriate, whereas for experimental designs, crossed sampling [36] is recommended.

Finally, knowing the levels of *Varroa* infestation in apiaries and understanding the effects of their presence on beehives allows us to make correct decisions on treatment timing. This is particularly important in areas where bee brood are constantly present in the hives during the year, as in the case of Mediterranean environments, where chemical treatments are still the main means to control *Varroa* mite infestations, with the aim of limiting undesirable toxicological and pathological side effects of the applied products and of the mite infestation, respectively.

Author Contributions: Conceptualization and design of the work, I.F., M.P., and A.S.; investigation, data curation, and formal analysis, M.P. and A.S.; funding acquisition, writing—original draft preparation, and supervision, I.F.; writing—review and editing, M.P. and A.S. All authors have read and agreed to the published version of the manuscript.

Funding: This research was funded by Università degli Studi di Sassari "Fondo di Ateneo per la ricerca".

Acknowledgments: The authors thank Ana Helena Dias Francesconi from the University of Sassari for revising the manuscript.

Conflicts of Interest: The authors declare no conflict of interest.

References

1. Rosenkranz, P.; Aumeier, P.; Ziegelmann, B. Biology and control of *Varroa destructor*. *J. Invertebr. Pathol.* **2010**, *103*, S96–S119. [CrossRef]
2. Nazzi, F.; Le Conte, Y. Ecology of *Varroa destructor*, the major ectoparasite of the Western Honey Bee, *Apis mellifera*. *Annu. Rev. Entomol.* **2016**, *61*, 417–432. [CrossRef]
3. Le Conte, Y.; Ellis, M.; Ritter, W. *Varroa* mites and honey bee health: Can *Varroa* explain part of the colony losses? *Apidologie* **2010**, *41*, 353–363. [CrossRef]
4. Neumann, P.; Carreck, N.L. Honey bee colony losses. *J. Apic. Res.* **2010**, *49*, 1–6. [CrossRef]
5. Calis, J.N.M.; Fries, I.; Ryrie, S.C. Population modelling of *Varroa jacobsoni* Oud. *Apidologie* **1999**, *38*, 111–124. [CrossRef]
6. Fries, I.; Camazine, S.; Sneyd, J. Population Dynamics of *Varroa jacobsoni*: A Model and a Review. *Bee World* **1994**, *75*, 5–28. [CrossRef]
7. Arechavaleta-Velasco, M.E.; Guzman-Novoa, E. Relative effect of four characteristics that restrain the population growth of the mite *Varroa destructor* in honey bee (*Apis mellifera*) colonies. *Apidologie* **2001**, *32*, 157–174. [CrossRef]
8. Harris, J.W.; Harbo, J.R.; Villa, J.D.; Danka, R.G. Variable population growth of *Varroa destructor* (Mesostigmata: Varroidae) in colonies of honey bees (Hymenoptera: Apidae) during a 10-year period. *Environ. Entomol.* **2003**, *32*, 1305–1312. [CrossRef]
9. Lodesani, M.; Crailsheim, K.; Moritz, R.F.A. Effect of some characters on the population growth of mite *Varroa jacobsoni* in *Apis mellifera* L. colonies and results of a bi-directional selection. *J. Appl. Entomol.* **2002**, *126*, 130–137. [CrossRef]
10. Boot, W.J.; Schoenmaker, J.; Calis, J.N.M.; Beetsma, J. Invasion of *Varroa jacobsoni* into drone brood cells of the honey bee, *Apis mellifera*. *Apidologie* **1995**, *26*, 109–118. [CrossRef]
11. Martin, S.J. Ontogenesis of the mite *Varroa jacobsoni* Oud. in worker brood of the honeybee *Apis mellifera* L. under natural conditions. *Exp. Appl. Acarol.* **1994**, *18*, 87–100. [CrossRef]
12. Martin, S.J. Reproduction of *Varroa jacobsoni* in cells of *Apis mellifera* containing one or more mother mites and the distribution of these cells. *J. Apic. Res.* **1995**, *34*, 187–196.[CrossRef]
13. Fuchs, S.; Langenbach, K. Multiple infestation of *Apis mellifera* L. brood cells and reproduction in *Varroa jacobsoni* Oud. *Apidologie* **1989**, *20*, 257–266. [CrossRef]
14. Martin, S.J.; Medina, L.M. Africanized honey bees have unique tolerance to *Varroa* mites. *Trends Parasitol.* **2004**, *20*, 112–114. [CrossRef]
15. Mondragon, L.; Martin, S.; Vandame, R. Mortality of mite offspring: A major component of *Varroa destructor* resistance in a population of Africanized bees. *Apidologie* **2006**, *37*, 67–74. [CrossRef]
16. DeGrandi-Hoffman, G.; Curry, R. A mathematical model of *Varroa* mite (*Varroa destructor* Anderson and Trueman) and honeybee (*Apis mellifera* L.) population dynamics. *Int. J. Acarol.* **2004**, *30*, 259–274. [CrossRef]
17. Martin, S.J. A population model for the ectoparasitic mite *Varroa jacobsoni* in honey bee (*Apis mellifera*) colonies. *Ecol. Model.* **1998**, *109*, 267–281. [CrossRef]
18. Chiesa, F.; Milani, N. Some preliminary observations on the behaviour of *Varroa jacobsoni* Oud. on its natural host under laboratory conditions. In *European Research on Varroatosis Control*; Cavalloro, R., Ed.; Balkema: Rotterdam, The Netherlands, 1988; pp. 113–124.
19. Floris, I. Dispersion indices and sampling plans for the honeybee (*Apis mellifera ligustica* Spin.) mite *Varroa jacobsoni* Oud. *Apicoltura* **1991**, *7*, 161–170.
20. Floris, I. A sequential sampling technique for female adult mites of *Varroa jacobsoni* Oudemans in the sealed worker brood of *Apis mellifera ligustica* Spin. *Apidologie* **1997**, *28*, 63–70. [CrossRef]
21. Fuchs, S. Untersuchungen. zur quantitativen Abschägtzung des Befalls von Bienenvölkern mit *Varroa jacobsoni* Oudemans und zur Verteilung des Parasiten im Bienenvolk. *Apidologie* **1985**, *16*, 343–368. [CrossRef]
22. Salvy, M.; Capowiez, Y.; Le Conte, Y.; Clément, J.L. Does the spatial distribution of the parasitic mite *Varroa jacobsoni* Oud. (Mesostigmata: Varroidae) in worker brood of honey bee *Apis mellifera* L. (Hymenoptera: Apidae) rely on an aggregative process? *Naturwissenschaften* **1999**, *86*, 540–543. [CrossRef] [PubMed]
23. Southwood, T.R.E. *Ecological Methods*; Chapman and Hall: London, UK, 1970.

24. Taylor, L.R. Assessing and interpreting the spatial distributions of insect populations. *Ann. Rev. Entomol.* **2016**, *29*, 321–357. [CrossRef]

25. Accorti, M.; Barbattini, R.; Marchetti, S. La diagnosi ed il controllo di *Varroa jacobsoni* Oud. in campo: Proposta di unificazione delle metodologie nelle prove sperimentali. *Apicoltura* **1986**, *2*, 165–185.

26. Dietemann, V.; Nazzi, F.; Martin, S.J.; Anderson, D.L.; Locke, B.; Delaplane, K.S.; Wauquiez, Q.; Tannahill, C.; Frey, E.; Ziegelmann, B.; et al. Standard methods for varroa research. *J. Apic. Res.* **2015**, *52*, 1–54. [CrossRef]

27. Pappas, N.; Thrasyvoulou, A. Searching for an accurate method to evaluate the degree of *Varroa* infestation in honeybee colonies. In *European Research on Varroatosis*; Cavalloro, R., Ed.; Balkema: Rotterdam, The Netherlands, 1988; pp. 85–92.

28. Rosenkranz, P. Distribution of *Varroa* females within the honeybee brood nest and consequences for biological control. Experts meeting CCE. In Proceedings of the Research Activities on Varroatosis in the European Countries, Thessaloniki, Greece, 26–28 September 1984.

29. Natsopoulou, M.E.; McMahon, D.P.; Doublet, V.; Frey, E.; Rosenkranz, P.; Paxton, R.J. The virulent, emerging genotype B of *Deformed wing virus* is closely linked to overwinter honeybee worker loss. *Sci. Rep.* **2017**, *7*, 5242. [CrossRef]

30. Barroso-Arévalo, S.; Fernández-Carrión, E.; Goyache, J.; Molero, F.; Puerta, F.; Sánchez-Vizcaíno, J.M. High load of *Deformed wing virus* and *Varroa destructor* infestation are related to weakness of honey bee colonies in Southern Spain. *Front. Microbiol.* **2019**, *10*, 1331. [CrossRef]

31. Martin, S.J. The role of *Varroa* and viral pathogens in the collapse of honey bee colonies: A modeling approach. *J. Appl. Ecol.* **2001**, *38*, 1082–1093. [CrossRef]

32. Wilkinson, D.; Smith, G.C. A model of the mite parasite, *Varroa destructor*, on honeybees (*Apis mellifera*) to investigate parameters important to mite population growth. *Ecol. Model.* **2002**, *148*, 263–275. [CrossRef]

33. Fuchs, S. The distribution of *Varroa jacobsoni* on honeybee brood combs and within brood cells as a consequence of fluctuation infestation rates. In *European Research on Varroatosis Control*; Cavalloro, R., Ed.; Balkema: Rotterdam, The Netherlands, 1988; pp. 15–17.

34. Boot, W.J.; Sisselaar, D.J.A.; Calis, J.N.M.; Beetsma, J. Factors affecting invasion of *Varroa jacobsoni* (Acari: Varroaidae) into honey bee, *Apis mellifera* (Hymenoptera: Apidae), brood cells. *Bull. Entomol. Res.* **1994**, *84*, 3–10. [CrossRef]

35. Beaurepaire, A.L.; Krieger, K.J.; Moritz, R.F. Seasonal cycle of inbreeding and recombination of the parasitic mite *Varroa destructor* in honeybee colonies and its implications for the selection of acaricide resistance. *Infect. Genet. Evol.* **2017**, *50*, 49–54. [CrossRef]

36. Floris, I. Osservazioni sull'infestazione da *Varroa jacobsoni* Oud di covata femminile opercolata di *Apis mellifera ligustica* Spin. In Proceedings of the Convegno Stato attuale e sviluppo della ricerca in apicoltura, Sassari, Italy, 25–26 October 1991; pp. 87–98.

Article

Unraveling Honey Bee–*Varroa destructor* Interaction: Multiple Factors Involved in Differential Resistance between Two Uruguayan Populations

Yamandú Mendoza [1], Ivanna H. Tomasco [2], Karina Antúnez [3], Loreley Castelli [3], Belén Branchiccela [4], Estela Santos [1] and Ciro Invernizzi [1,*]

[1] Sección Etología, Facultad de Ciencias, Universidad de la República, Iguá 4225, Montevideo 11400, Uruguay; ymendozaspina@gmail.com (Y.M.); estelsantos@gmail.com (E.S.)

[2] Laboratorio de Evolución, Facultad de Ciencias, Universidad de la República, Iguá 4225, Montevideo 11400, Uruguay; ivanna@fcien.edu.uy

[3] Departamento de Microbiología, Instituto de Investigaciones Biológicas Clemente Estable, Av. Italia 3318, Montevideo 11600, Uruguay; kantunez03@gmail.com (K.A.); castelli.loreley@gmail.com (L.C.)

[4] Sección Apicultura, Programa de Producción Familiar, Instituto Nacional de Investigación Agropecuaria La Estanzuela, Ruta 50 km 11, Colonia 70002, Uruguay; bbranchiccela@inia.org.uy

* Correspondence: ciro@fcien.edu.uy

Received: 26 July 2020; Accepted: 17 August 2020; Published: 20 August 2020

Abstract: The ectoparasite *Varroa destructor* is the greatest biotic threat of honey bees *Apis mellifera* in vast regions of the world. Recently, the study of natural mite-resistant populations has gained much interest to understand the action of natural selection on the mechanisms that limit the mite population. In this study, the components of the *A. mellifera–V. destructor* relationship were thoroughly examined and compared in resistant and susceptible honey bee populations from two regions of Uruguay. Mite-resistant honey bees have greater behavioral resistance (hygienic and grooming behaviors) than susceptible honey bees. At the end of the summer, resistant honey bees had fewer mites and a lower deformed wing virus (DWV) viral load than susceptible honey bees. DWV variant A was the only detected variant in honey bees and mites. Molecular analysis by Short Tandem Repeat showed that resistant honey bees were Africanized (*A. m. scutellata* hybrids), whereas susceptible honey bees were closer to European subspecies. Furthermore, significant genetic differentiation was also found between the mite populations. The obtained results show that the natural resistance of honey bees to *V. destructor* in Uruguay depends on several factors and that the genetic variants of both organisms can play a relevant role.

Keywords: honey bees; mites; viruses; behavior; social immunity; Africanized bees; microsatellites; Uruguay

---

1. Introduction

The shift of the ectoparasitic mite *Varroa destructor* from the Asian honey bee *Apis cerana*, its first host, to *Apis mellifera* and its subsequent dispersal throughout the world, has created one of the largest biotic threats to honey bee populations and caused great damages to the beekeeping industry [1,2]. In *A. cerana*, the mite reproduces only in drone cells and it maintains a stable relationship with the host without causing significant damage [3–5]). However, in *A. mellifera* it can also reproduce in worker cells and might cause the death of the colonies if acaricides are not regularly applied [1].

In addition to the direct damage caused by *V. destructor* to *A. mellifera*, especially in the brood during the reproductive phase, it also acts as a vector for different RNA viruses and suppresses the immune response of honey bees [6,7]. One of the honey bee viruses that has received more attention due to its close association with *V. destructor* and the colony collapse is the deformed wing virus (DWV) [8,9].

Although *V. destructor* is currently the major threat for honey bees, the damage it causes to populations in different regions around the world varies significantly. For example, the mite has devastating effects in European countries, North America, and temperate regions of South America (Argentina, Chile, and Uruguay), where beekeepers must systematically use synthetic acaricides [2,10,11]. However, in tropical areas of South and Central America and broad regions of Africa, honey bees coexist with the mite without significant problems [11,12]. This difference in the impact of *V. destructor* depends on the time that both species have been interacting (<50 years in most countries), genetic aspects of both honey bees and mites, and the presence of other pathogens, particularly viruses [1,7,13]. Furthermore, the relationship between honey bees and *V. destructor* is strongly influenced by beekeeping practices. Thus, professional beekeepers group colonies into apiaries, which facilitates the horizontal transmission of mites and favors the selection of the most virulent variants [14–16]. Other common beekeeping practices, such as honey bee selection, the use of acaricides, movement of colonies, and honey bee trade, can have a strong impact on the interaction between honey bees and mites [15].

The resistance mechanisms of honey bees to *V. destructor* have been widely studied since the 1980s, focusing on hygienic and grooming behaviors [1,17–19]. Hygienic behavior (uncapping cells that contain dead, diseased, or parasitized brood and their subsequent removal) is a social behavior that helps control diseases of the offspring such as the American foulbrood (*Paenibacillus larvae*) and Ascosferiosis (*Ascosphaera apis*) and may interrupt the reproduction of *V. destructor* [20]. The selection of honey bees with improved hygienic behavior has had encouraging results in the control of mite populations [18,19]. In the USA, hygienic honey bees have been selected with a particular capacity to detect pupae parasitized by *V. destructor*, a trait known as Varroa Sensitive Hygiene (VSH) [21]. Grooming behavior, by which parasited honey bees can dislodge mites by themselves (autogrooming) or receiving help from other insects (allogrooming) [17], has been reported as an effective resistance mechanism against *V. destructor* [22–25]. Both hygienic and grooming behaviors are expressed very efficiently in *A. cerana* and would be key to controlling *V. destructor* populations [5,26]. In *A. mellifera*, these behaviors are expressed more highly in Africanized honey bees (hybrids of *A. m. scutellata*) than in European ones, which could partially explain the resistance to *V. destructor* in the former [13].

For several years, researchers from different countries have been working on the breeding and selection of *V. destructor*-resistant honey bees [18,19]. Although significant progress has been made, its impact on the beekeeping industry has been very limited and a situation where beekeepers can maintain their colonies without acaricide treatment is far from being reached. However, the possible successes of these initiatives are still under debate [27,28].

Given the limitations of artificial selection, the study of honey bee populations naturally resistant to *V. destructor* has gained increasing attention in recent years. The most studied populations have been those from Brazil, South Africa, Fernando de Noronha island (Brazil), Gotland island (Sweden), Avignon (France), and Arnost Forest (USA) [13]. It is interesting to observe how natural selection has shaped different responses in different honey bee populations to coexist with *V. destructor*. For example, in Brazil and South Africa, as well as in Primorsky's honey bees, behavioral resistance is important, whereas in the case of Gotland's bees, resistance is associated with a reduction in colony size and lower mite reproduction [13].

Unlike honey bees, differences in *V. destructor* that may affect the relationship with its host are less known. Among the several described haplotypes of *V. destructor*, Japanese and Korean are the only ones able to reproduce in *A. mellifera*. It is known that the Korean haplotype display a higher virulence and a wider geographic distribution than the Japanese one [29–31]. Strikingly, both haplotypes have almost no polymorphism and can, therefore, be considered as quasi-clonal populations [31]. However, genetic differences between mites at the population and colony levels, even within the colony, have been recently reported [32,33].

Uruguay is a South American country of 176,000 km^2 without significant geographical barriers, with a temperate climate where the spring–summer period presents marked differences compared to the autumn–winter period. In 1834, *A. m. mellifera* was introduced from France [34], but today most of the honey bee populations are hybrids with *A. m. scutellata* after decades of this subspecies entering from Brazil [35,36]. *Varroa destructor* entered the country in 1978 and until the late 1990s it did not cause significant problems, and colonies were able to survive without acaricide treatments. Afterward, colony losses associated with *V.* destructor increased and, in a few years, beekeepers had to use massively synthetic acaricides in almost all the country in order to ensure colony survival [37]. This change could be due to the introduction of more susceptible European honey bee subspecies, the entry of more virulent variants of *V. destructor*, a greater impact of viruses associated with the mite, among other factors. However, on the eastern side of the country there are still regions where honey bees coexist with *V. destructor* without acaricide treatments and with minimal colony losses [37]. It should be noted that in Uruguay only the K haplotype of *V. destructor* is present [38]. This is a striking situation considering the short distance where susceptible honey bees are found, and the increasing movement of colonies between regions. Thus, this is an interesting scenario to analyze the factors involved in differential resistance to *V. destructor* in Africanized bees.

The aim of this study was to analyze the role of different factors that could affect the *A. mellifera–V. destructor* interactions in two populations with marked mite resistance differences.

2. Materials and Methods

2.1. Overview

During the spring of 2013, one apiary comprising 21 colonies was installed in the experimental station of INIA La Estanzuela (34°20′48.60″ S; 57°41′29.02″ W), Colonia Department (western region of Uruguay), and another apiary with 23 colonies was installed in the experimental station INIA Treinta y Tres (33°15′06.60″ S; 54°25′40.63″ W), in the Treinta y Tres Department (eastern region of Uruguay). In both cases, colonies belonged to local honey bee populations with new queens and no symptoms of disease. Honey bee populations from Colonia received acaricide treatment at least once a year to survive the infestation by *V. destructor* ("mite-susceptible colonies"). On the contrary, colonies located in Treinta y Tres had not received acaricides for six years, showing average infestation levels lower than 5% and low annual colony losses (<15%, "mite-resistant colonies").

From the middle of the summer of 2014 (January) to the beginning of the autumn of 2014 (April), the evaluations and sampling described below were carried out in both apiaries. Later, in the summer of 2015, the estimation of mites that reproduced in drone and worker cells was evaluated.

2.2. Estimation of the Honey Bee Population and Brood Area

In order to estimate the honey bee population at the end of the summer (March), the number of frames covered by honey bees was recorded [39]. In the case of brood, the brood area occupied per frame was estimated [39]. During the winter, the apiaries were regularly visited to determine the survival of colonies.

2.3. Evaluation of Hygienic and Grooming Behaviors

The evaluation of hygienic behavior was carried out in 21 mite-susceptible colonies and 21 mite-resistant colonies. At least 100 pupae were killed by pricking them with an entomological pin through the cell cap and 24 h later the number of removed pupae was recorded. The result was expressed as a percentage of cleaned cells [40].

Grooming behavior was evaluated in 17 and 21 mite-susceptible and mite-resistant colonies, respectively. A petroleum jelly-smeared sheet was placed on the floor of the hives for 7 days so that the mites dislodged by the bees would remain attached. Mites were observed at 40× to determine if they had mutilated legs. Grooming behavior was expressed as a percentage of damaged mites [41].

2.4. Estimation of Mites in Honey Bees and in Brood Cells

The estimation of the mites' proportion in honey bees (phoretic mites) and in the brood cells was made in 21 mite-susceptible colonies and 21 mite-resistant colonies.

To estimate the percentage of mite-infected honey bees, a sample of approximately 300 workers collected in three combs was taken from each colony. The varroa mites were removed from the honey bees with ethanol 75%, and the percentage of infected honey bees was determined [42].

The brood infestation was determined by observing 400 capped cells with pupae older than 15 days of age (with purple eyes). When an infested cell was found, the number of adult mites (female founders) and offspring were recorded. Adult mites were differentiated from the immature forms (protonymph and deutonymph) by size, shape, and color. From these records, the fertility of the mites (cells infested with one varroa with offspring), the abundance (the average number of adult female mites per examined cells), the intensity (the average number of female mites per infested cell), and the prevalence (the percentage of infested cells) were estimated. To determine the fertility of *V. destructor*, only colonies with at least 10 infected pupae were considered. Thus, 21 mite-susceptible colonies and 11 mite-resistant colonies were analyzed. The relationship between the infestation level in adult bees and brood (abundance) was determined.

At the end of the summer of 2015, 8 mite-susceptible colonies and 5 mite-resistant colonies, the only ones that had at least 10 drone cells, were selected. None of these colonies had participated in the study carried out the previous year. The presence and quantity of founder females in the cells of drones and workers were recorded to determine the preference of mites to reproduce in the two types of cells.

2.5. Detection and Quantification of RNA Viruses in Honey Bees and Mites

Nurse honey bee samples were collected from the brood nest of 20 mite-susceptible and 20 mite-resistant colonies. At the same time, mite samples were collected from the infected pupae cells from 11 mite-susceptible and 10 mite-resistant colonies. All samples were immediately transported to the laboratory and stored at −80 °C until analysis.

Ten honey bees per colony were processed according to the method described by Antúnez et al. [43]. In the case of mites, 10 individuals per colony were subjected to mechanical homogenization in 50 uL of PBS using ceramic beads and a Fast Prep system (MP Biomedicals, Solon, OH, USA; 5 × 6.6 m/s for 30 s). Samples were centrifuged at 10,000 rpm for 5 min at 4 °C. In both cases (honey bees and mites), the supernatant was used for RNA extraction using a QIAamp Viral RNA Mini Kit (Qiagen, Germantown, MD, USA). Total RNA was subjected to reverse transcription using a QuantiTect Reverse Transcription Kit (Qiagen) according to the manufacturer's conditions. Real-time PCR reactions were performed using a QuantiTec SYBR PCR Kit (Qiagen) and specific primers for the amplification of the following honey bee viruses: acute bee paralysis virus (ABPV) [44], deformed wing virus (DWV) [45], black queen cell virus (BQCV) [45], and sacbrood bee virus (SBV) [44]. For the normalization of the results, the expression level of the gene encodes for the honey bee β-actin was used [46].

Real-time PCR reactions were performed as described by Anido et al. [47] using a Rotor Gene 6000 (Corbett Research-Qiagen). The reaction mixture consisted of 1× QuantiTec SYBR Green PCR Master Mix, 0.5 µM of each primer (one pair of primers for each reaction), RNA-free water, and 5 µL cDNA in a final volume of 50 µL.

The cycling program consisted of an initial activation at 50 °C for 2 min and 95 °C for 15 min, and 45 cycles of 94 °C for 15 s, 50 °C for 30 s, and 72 °C for 30 s. Fluorescence was measured at the elongation step and controls without DNA were included in each reaction. The specificity of the reactions was checked by melting curve analysis of the amplified products (from 65 to 95 °C).

The amplified cDNA of each virus as well as that of the β-actin were expressed as the threshold cycle value (Ct). Ct value represents the number of cycles required to generate fluorescence that exceeds a predefined threshold. The threshold and reaction efficiency were calculated automatically using the Rotor-Gene 6000 software 1.7 (Corbett Research, Qiagen).

To control the variation in mRNA levels between the different samples, the data were normalized by subtracting the Ct value of β-actin from the Ct value of each virus (ΔCt). Subsequently, the viral load of each sample was estimated using the relative quantification method [48]. The concentration of all samples was analyzed with respect to the sample with the lowest viral load ("calibrator"); thus, the ΔCt of the calibrator was subtracted from the ΔCt of each sample (ΔΔCt = ΔCt of the sample −ΔCt of the calibrator). Finally, the value of $2^{-\Delta\Delta Ct}$ was calculated to estimate the relative levels of cDNA.

2.6. DWV Variants

cDNA obtained from honey bees and mites from 2 mite-susceptible and 2 mite-resistant colonies was subjected to qPCR-high-resolution melting (HRM) in order to amplify a 144 bp of the replicase gene [49]. Amplified fragments were cloned using a TOPO® TA Cloning kit (Invitrogen), according to the manufacturer's instructions. Ten clones per sample (8 samples in total) were sequenced at Macrogen (Seoul, Korea). The nucleotide sequences were compared to the GenBank database of the National Center for Biotechnology Information (Bethesda, MD, USA) using the BLAST tool.

Furthermore, cDNA obtained from honey bees of the 21 mite-susceptible and 21 mite-resistant colonies was amplified by qPCR in order to evaluate the presence of DWV variants A/B/C [50].

2.7. Molecular Characterization of Honey Bees

Five mite-susceptible and five mite-resistant colonies were selected and eight pupae per colony were collected and stored in 95% ethanol until analysis. Half of each pupa were processed individually and total DNA was extracted using the modified protocol of Miller et al. [51]. Five STR loci (i.e., A88, A113, A28, A43, and A9) were selected for PCR amplification and subsequent genotyping [52–55]. PCR amplification was performed in a PxE thermocycler (Thermo Electron Corporation, Milford, MA, USA). A total volume of 10 µL containing 4 ng/µL of template DNA was used for the reaction, 0.2 mM of each primer, 0.2 mM of each deoxynucleotide (dNTPs), 2 mM of MgCl2, 1× of enzyme buffer, and 0.5 units of Taq polymerase (Thermo Fisher Scientific, Waltham, USA). Initial denaturation was performed at 95 °C for 3 min, followed by a series of variable cycles according to the microsatellite to be amplified with denaturation at 94 °C for 30 s, between 51 °C and 60 °C during 30 s and 73 °C for 30 s, ending with 72 °C for 5 min. The success of the reaction was verified with electrophoresis on a 5% polyacrylamide gel visualized through silver nitrate staining [56]. The amplification products were processed in an ABI3500 Genetic Analyzer (Applied Biosystems, Foster City, CA, USA), with a DS-33 matrix and Liz600 as molecular weight standard. Peak Scanner 2.0 software (Applied Biosystems, Foster City, CA, USA) was used for individual genotype determination. Furthermore, genotypes were obtained from three reference populations (Europe, Brazil, and Africa) provided by Alice Pinto [57].

2.8. Molecular Characterization of V. destructor

Eight mites per colony were collected from five mite-susceptible and five mite-resistant colonies and kept in 95% ethanol until analysis. For DNA extraction, each mite was processed individually and total DNA was extracted using a previously described protocol [51]. Five loci of variable STRs were amplified (i.e., VD112, VD001, VD114, VD016, and VJ295) [58,59]. PCR amplification was performed in a Thermo PxE thermocycler, in a total volume of 10 µL containing 8 ng/µL of template DNA, 0.2 mM of each primer, 0.2 mM of each deoxynucleotide (dNTPs), 2 mM of MgCl2, 1× of reaction buffer, and 1 rc unit of Taq polymerase (Thermo Scientific). An initial denaturation was carried out at 95 °C for 3 min and, subsequently, a series of variable cycles according to the microsatellite to be amplified with denaturation at 94 °C for 30 s, between 55 °C and 62 °C for 30 s and 73 °C for 30 s, ending with 72 °C for 5 min. The success of the reaction was verified with electrophoresis on a 5% polyacrylamide gel visualized through silver nitrate staining [56]. Finally, the amplification products were processed in an ABI3500 Genetic Analyzer, with a DS-33 and Liz600 matrix as the molecular weight standard individually for each sample. Peak Scanner 2.0 software (Applied Biosystems, Foster City, CA, USA) was used for individual genotype determination.

2.9. Statistical Analysis

To compare adult population, brood area, hygienic behavior, grooming behavior, level of *V. destructor* infection in adult honey bees, fertility, abundance, prevalence, and intensity of *V. destructor* in brood cells, the ratio of mites on adult honey bees/mites in brood cells (adding 1 to the numerator and denominator to avoid having 0 values), and the viral load between mite-susceptible and mite-resistant colonies, the Wilcoxon test was used, as variables did not fit the assumptions of parametric statistics.

To test differences between *V. destructor* infection of drone and worker cells between mite-susceptible and mite-resistant colonies, a generalized linear model (GLM) analysis with a logit function was used. The response variable was presence (1) or absence (0) of mites and the predictor variables were cell type (i.e., drone or worker), population, and an interaction term between them. Model selection was done with an Akaike information criterion (AIC) [60,61], and the best fit for the data was achieved for the model with the lowest AIC value (ΔAIC > 2).

The proportion of mite-susceptible and mite-resistant colonies infected by the four RNA viruses studied was compared using binomial tests to evaluate the presence of the virus in adult bees and mites.

All analyses were performed using the R statistical program (Vienna, Austria) [62]. The *p*-values lower than 0.05 were considered as statistically significant.

The genotypes obtained for each of the honey bee and mite samples were used to estimate population parameters. The allelic and genotypic observed and expected frequencies by yHardy–Weinberg equilibrium (HWE), the number of alleles (Na), observed (Ho) and expected (He) heterozygosity, and departures from HWE by exact test were estimated employing the GENEPOP v.4.1 package [63]. The genetic structuring was estimated with STRUCTURE 2.3.4 [64]. The program was instructed to test 1 to 10 K parameters using admixture ancestry model and correlated allele frequency for computing the Markov chain Monte Carlo (MCMC) simulation algorithm with a 10,000 burn-in length and a run length of 10,000. The simulation calculation was repeated 20 times for each K value. The K that best fit the data was chosen as the one that provided the highest likelihood values.

3. Results

3.1. Honey Bee Population and Brood Area

At the end of the summer, the population of honey bees in mite-susceptible and mite-resistant colonies was similar (10,010 ± 1540 and 9680 ± 2310 honey bees, respectively, W = 285.5; *p* = 0.297). In contrast, the brood area of mite-susceptible colonies was smaller than in the mite-resistant colonies (8096 ± 2816 and 10,560 ± 2464 cells, respectively, W = 121.5; *p* = 0.005).

None of the mite-susceptible colonies survived until the end of the autumn (June), whereas 82% of the mite-resistant colonies arrived in spring (September) in good condition.

3.2. Hygienic and Grooming Behaviors

Mite-resistant colonies displayed higher hygienic and grooming behaviors than the mite-susceptible colonies (Figure 1).

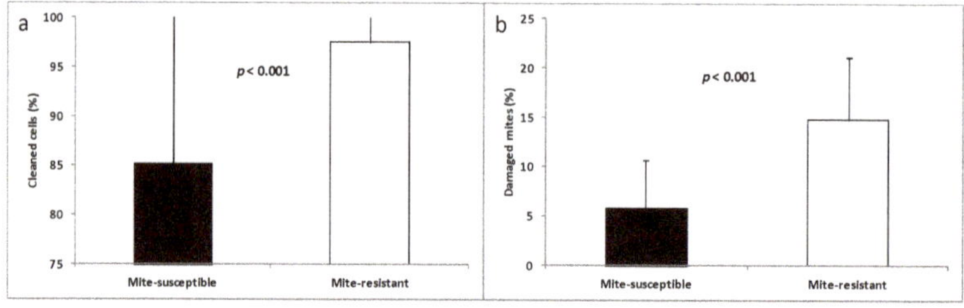

Figure 1. Hygienic (**a**) and grooming (**b**) behaviors in mite-susceptible and mite-resistant colonies.

3.3. Mites in Honey Bees and Brood Cells

At the end of the summer, mite-susceptible colonies showed more mites in honey bees and in the brood (reflected in values of abundance, prevalence, and intensity of infection), as well as a higher proportion of phoretic mites/mites in brood, than mite-resistant colonies. In contrast, the fertility of *V. destructor* was similar in both apiaries (Figure 2).

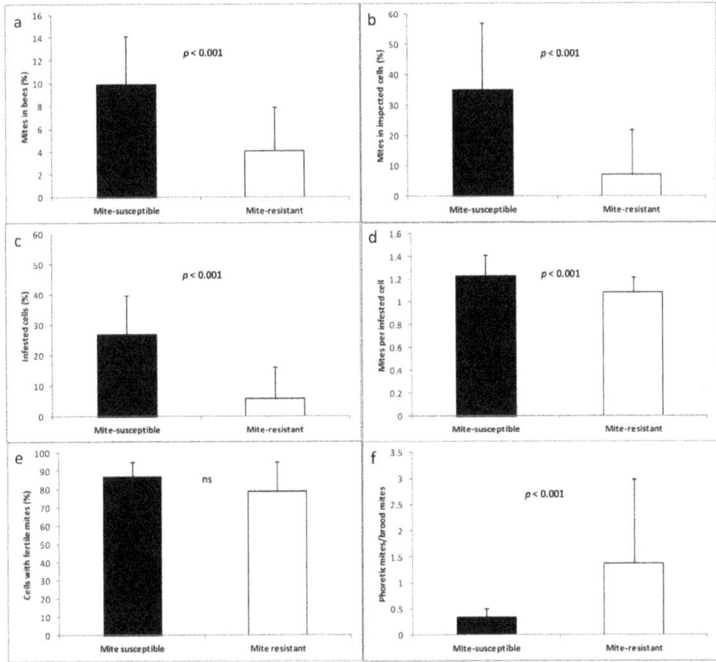

Figure 2. Presence of *V. destructor* in mite-susceptible and mite-resistant colonies. Infection in adult bees (**a**), abundance in brood cells (**b**), prevalence in brood cells (**c**), intensity of infection in brood cells (**d**), fertility (**e**), and relationship between phoretic and reproductive mites (**f**). ns: non significant.

Regarding the preference of the mites for the worker and drone cells, the best model explaining mite infection included all predictor variables (cell type, apiary, and the interaction between them) (Tables 1 and 2). Drone cells of mite-resistant colonies are three times more likely to be infected by the mite than those of mite-susceptible colonies. In contrast, worker cells have a low probability of being infected with no differences found between the two apiaries (Figure 3).

Table 1. Indicators of the level of infection by *V. destructor* in the totality of inspected drone cells and worker cells in 8 mite-susceptible colonies and 5 mite-resistant colonies.

Colonies		Mite-Susceptible	Mite-Resistant
	Inspected cells	887	282
	Mite-infested cells	176	101
Drones	Total mites	198	182
	Prevalence	19.8%	35.8%
	Abundance	22.3%	64.5%
	Inspected cells	1710	1210
	Mite-infested cells	67	60
Workers	Total mites	67	62
	Prevalence	3.9%	5.0%
	Abundance	3.9%	5.1%
Ratio of mite distribution between drone and worker cells		5.70	12.60

Table 2. Number of analyzed individuals (N), number of alleles found (Na), expected heterozygosity (He), observed heterozygosity (Ho), and probability of departure from Hardy–Weinberg equilibrium (*p*) for each microsatellite in mite-resistant honey bee populations of Treinta y Tres (Mite-R), mite-susceptible honey bee populations of La Estanzuela (Mite-S), and the reference honey bee populations from Europe (EU), Africa (AF), and Brazil (BR).

Locus		Mite-R	Mite-S	EU	AF	BR
A43	N	32	22	50	41	32
	Na	9	7	6	16	13
	Ho	0.063	0.682	0.640	0.854	0.875
	He	0.769	0.766	0.615	0.881	0.851
	p	<0.0001	<0.0001	0.2545	0.4865	0.9758
A88	N	32	23	50	41	32
	Na	7	3	6	14	12
	Ho	0.125	0.043	0.700	0.878	0.750
	He	0.771	0.463	0.629	0.878	0.853
	p	<0.0001	<0.0001	0.6378	0.8287	0.1182
A28	N	31	39	50	41	32
	Na	10	2	2	9	10
	Ho	0.065	0.000	0.380	0.805	0.813
	He	0.803	0.099	0.413	0.833	0.806
	p	<0.0001	<0.0001	0.2363	0.4356	0.4059
A8	N	38	37	50	41	32
	Na	8	4	6	9	9
	Ho	0.211	0.081	0.800	0.829	0.688
	He	0.780	0.681	0.801	0.838	0.802
	p	<0.0001	<0.0001	0.0825	0.6681	0.0440
A113	N	34	39	50	41	32
	Na	9	8	11	12	11
	Ho	0.382	0.769	0.640	0.854	0.875
	He	0.788	0.799	0.650	0.858	0.858
	p	<0.0001	<0.0001	0.8063	0.7393	0.1029

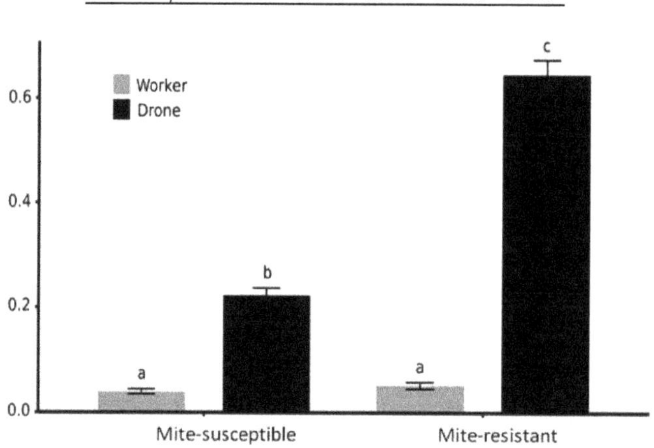

Figure 3. Estimated probability of *V. destructor* infestation in drone and worker cells in mite-susceptible and mite-resistant colonies according to generalized linear model (GLM) analysis with link logit. Different letters show significance level (*p* < 0.05) based on the odds ratio confidence intervals estimated in the model.

3.4. RNA Viruses in Honey Bees and Mites

The four analyzed viruses (ABPV, BQCV, DWV, and SBV) were detected in mite-susceptible and mite-resistant colonies, both in adult honey bees and in mites. BQCV and DWV were detected in honey bees from all colonies in the two apiaries (Figure 4). No statistical differences ($p > 0.10$) were found in the proportion of colonies infected by the four viruses studied between both groups, considering the presence of the viruses in bees and mites (Figure 4).

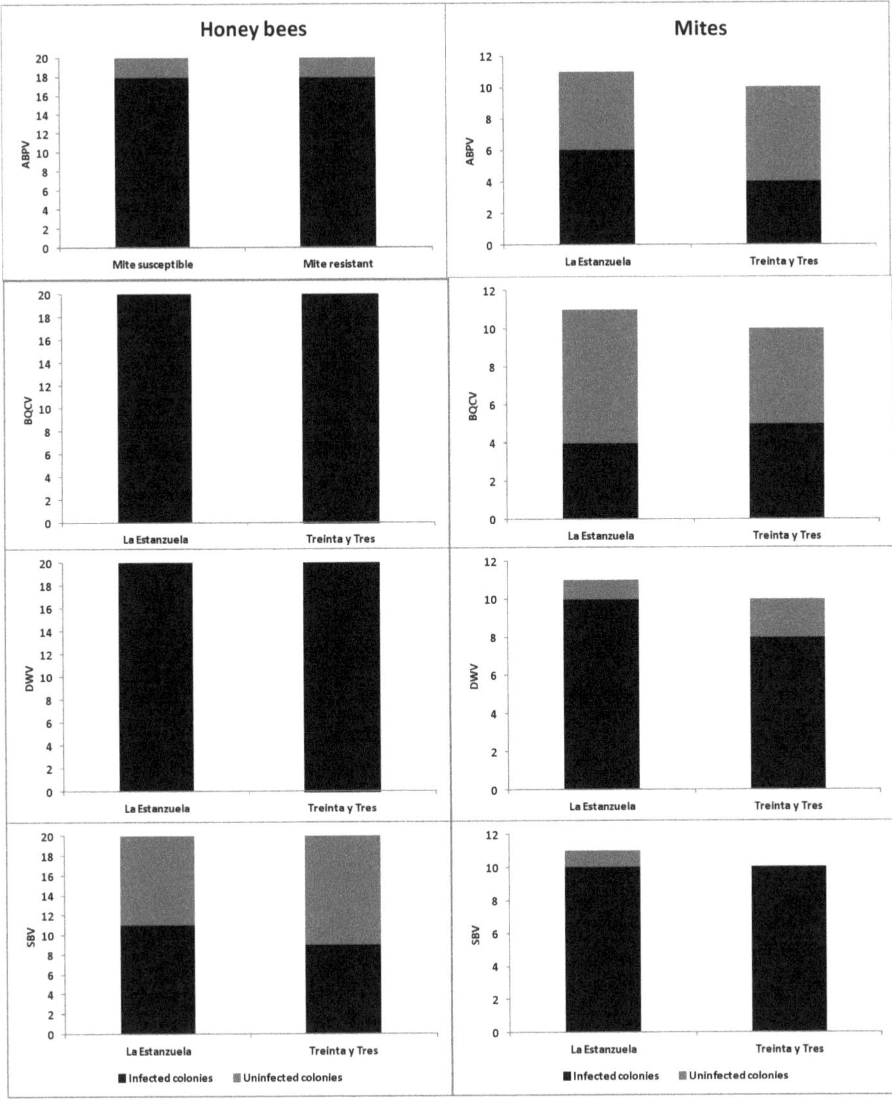

Figure 4. Percentage of colonies presenting honey bees or mites infected with acute bee paralysis virus (ABPV), black queen cell virus (BQCV), deformed wing virus (DWV), and sacbrood bee virus (SBV) in mite-susceptible and mite-resistant colonies.

The infection level of ABPV, BQCV, and SBV was similar in mite-susceptible and mite-resistant colonies ($p > 0.10$). However, in the case of DWV, mite-susceptible colonies showed a higher infection level than mite-resistant colonies (relative DWV level 2467 ± 5784 and 588 ± 2406, respectively, W = 294; $p = 0.011$).

Mites of colonies from both apiaries showed similar viral loads of the four analyzed viruses ($p > 0.10$).

According to the results of qPCR-HRM, clone sequencing, and A/B/C qPCR, only DWV variant A was detected in mites and honey bees from mite-susceptible and mite-resistant colonies.

3.5. Molecular Characterization of Honey Bees

The number of alleles found for each locus in the two honey bee populations is shown in Table 2. In all observed cases, heterozygosis was significantly lower than expected in all the studied loci (Table 2). This was not found in the reference populations. A lower number of alleles was recorded for Uruguayan samples than for references.

The distribution of allelic and genotypic frequencies was significantly different between mite-susceptible and mite-resistant colonies ($p < 0.0001$ in both cases). The differences between the two studied populations and between them and the reference populations can be graphically visualized in the figures generated by the STRUCTURE 2.3.4 software [64], which evaluates the combined variation of all microsatellites and assigns individuals to different theoretical populations. The model that better adjusts when analyzing only the samples from the two Uruguayan apiaries is that of two populations, coinciding with the samples from mite-resistant and mite-susceptible colonies, respectively (Figure S1, Supplementary Materials). When the reference population samples are included in the analysis (Figure 5), the model that fits better is the five-population model (K = 5), i.e., mite-resistant colonies, mite-susceptible colonies, Europe, and two hypothetical populations of mixed composition, with individuals from Brazil and Africa in similar proportions. When grouping into four populations (K = 4), these two hypothetical populations also appear, but include individuals from Africa, Brazil, and mite-resistant colonies. Some of the mite-resistant honey bees showed great affinity with mite-susceptible honey bees. In the model with three populations (K = 3), these hypothetical populations merge and form a single population made up of all individuals from Brazil and Africa, plus many individuals from mite-resistant honey bees. Again, some mite-resistant individuals showed greater affinity with the mite-susceptible individuals. The simplest grouping occurs when adjusting for two hypothetical populations (K = 2). In this case, one population is integrated by mite-resistant colonies, Brazil, and Africa, whereas the other corresponds to samples from mite-susceptible colonies and Europe. This model, while not being the best adjusted, shows the genetic affinities of honey bees from both Uruguayan apiaries with respect to those of reference samples.

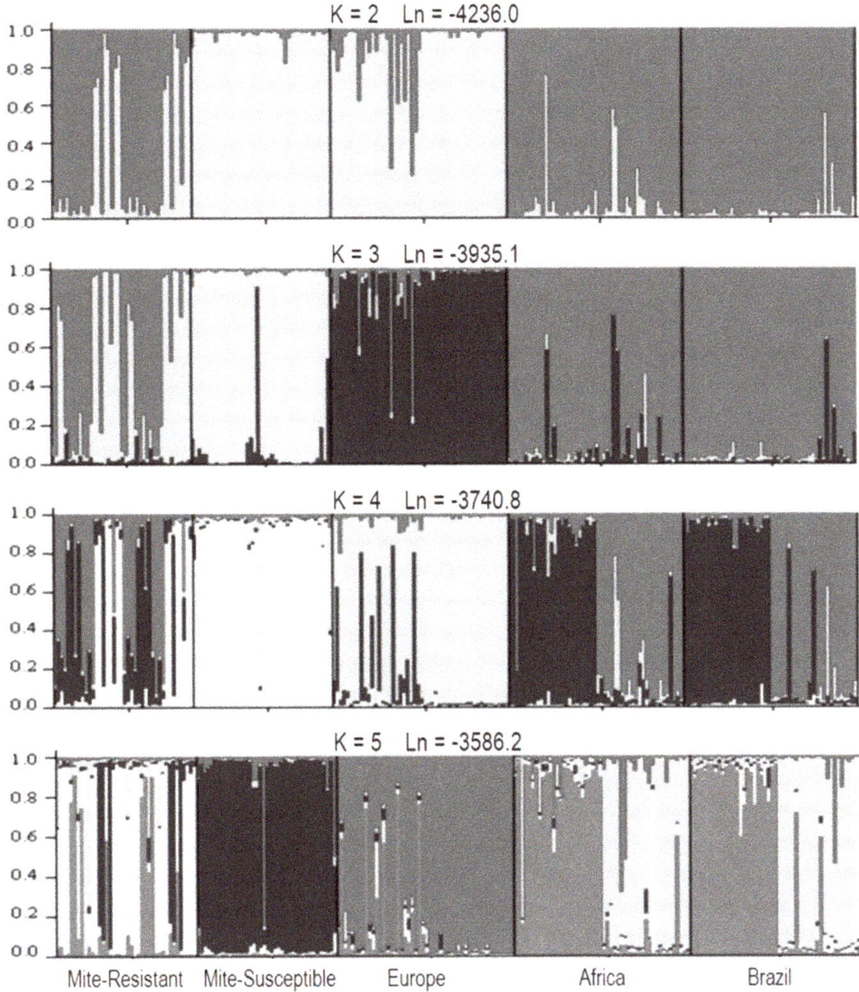

Figure 5. Honey bee population allocation by the STRUCTURE program based on the genotyping of 5 STR loci. Each individual is represented by a partitioned vertical bar, with partitions (represented by different colors) representing the posterior probability of each individual belonging to one or another population. K: number of populations assumed by the model; Ln: average likelihood value for all runs for the same K.

3.6. Molecular Characterization of V. destructor

The number of alleles found for each locus in the two mite populations ranged between 1 (VD016 monomorphic locus) and 4 (Table 3). In all studied loci, observed heterozygosis was lower than expected, and there was a very significant deviation from the HWE (Table 3).

Table 3. Number of analyzed individuals (N), number of alleles found (Na), expected heterozygosis (He), observed heterozygosis (Ho), and probability of departure from Hardy–Weinberg equilibrium (p) for each STR loci in populations of mites from mite-resistant colonies of Treinta y Tres (Mite-R) and mite-susceptible colonies of La Estanzuela (Mite-S).

Locus		Mite-R	Mite-S
VD112	N	29	33
	Na	2	3
	He	0.068	0.222
	Ho	0.000	0.242
	p	0.0175	1.000
VD001	N	29	26
	Na	3	2
	He	0.101	0.075
	Ho	0.034	0.000
	p	0.0175	0.0196
VD114	N	28	33
	Na	1	2
	He	0.000	0.060
	Ho	0.000	0.000
	p	NA	0.0154
VD016	N	29	30
	Na	1	1
	He	0.000	0.000
	Ho	0.000	0.000
	p	NA	NA
VJ295	N	28	28
	Na	4	3
	He	0.546	0.450
	Ho	0.500	0.500
	p	0.4031	0.0618

When comparing allelic and genotypic frequencies of mite populations of mite-resistant and mite-susceptible colonies, significant differences were found for the VJ295 locus ($p = 0.0006$ and $p = 0.0014$ for the allelic and genotypic frequencies, respectively). For the VD112 locus, marginal differences were found for allelic frequencies ($p = 0.0534$) and significant differences for genotypic frequencies ($p = 0.0014$). On the contrary, for VD001 and VD114 loci, no significant differences in allelic and genotypic frequencies were found between the two populations of mites studied ($p > 0.10$ in all cases).

4. Discussion

In Uruguay, after 40 years of interaction between honey bees and *V. destructor*, colonies in most of the country need acaricides to survive. A different scenario occurs in the eastern side of the country where the mite does not cause significant problems. The exhaustive analysis carried out of the factors that could explain the notable differences found in the *A. mellifera–V. destructor* relationship in two Uruguayan regions showed that this relationship is complex.

In the first place, this study confirmed the resistance to *V. destructor* of the honey bee population from the eastern side of the country, which was able to survive without acaricide treatment and showed only 18% colony mortality. In contrast, the western honey bee population showed extreme susceptibility to the mite, since none of the colonies managed to overcome autumn.

4.1. Resistance Behaviors to V. destructor

Behavioral resistance of honey bees appears as a critical factor in controlling the *V. destructor* population. *Varroa destructor*-resistant colonies showed higher hygienic behavior than mite-susceptible colonies. Numerous studies indicate that hygienic colonies display a better control of the *V. destructor* population [21,65–67]. Nevertheless, the role of hygienic behavior in limiting mite reproduction is still controversial [68,69].

The hygienic behavior exhibited by honey bees that are able to detect the onset of *V. destructor* reproduction and uncap the cell containing the infected pupae is called Varroa Sensitive Hygiene (VSH) [21]. A recent study, in which brood was artificially infected with *V. destructor*, showed that honey bees from Treinta y Tres presented higher VSH than honey bees from La Estanzuela (Alexis Beaurepaire, unpublished data). The interruption of *V. destructor* reproduction by hygienic honey bees, especially if the mites have already laid eggs, leads to a decrease in the mite population due to lost opportunities to reproduce [70,71].

Hygienic behavior may play a key role in honey bee populations with natural resistance to *V. destructor*, such as those from Brazil and South Africa. However, other factors explain mite resistance in honey bees from Fernando de Noronha island (Brazil) or those from Gotland island (Sweden) [13]. Recently, Oddie et al. [72] compared four populations of naturally surviving *V. destructor* bees with populations of susceptible local honey bees. They found that resistant honey bees uncap infested brood cells with higher frequency, and then recap the cells without the need to remove pupae. This behavior, a product of rapid evolution, avoids the cost of losing brood. These results were confirmed by Martin et al. [73] when comparing cell recapping in mite-resistant honey bees from Brazil and Africa (*A. m. scutellata*) with "mite-naive" honey bees from the United Kingdom and Australia.

Grooming behavior could also contribute to the better resistance to *V. destructor* presented by colonies from Treinta y Tres since this behavior was better expressed than in mite-susceptible colonies from La Estanzuela. In Uruguay, Invernizzi et al. [41] found that Africanized bees expressed more grooming than European bees (*A. m. ligustica*), both at colonial and individual levels. This behavior has already been reported as a valuable trait for the control of *V. destructor* [22–25]. However, we must be cautious when evaluating the importance of grooming behavior as resistance to *V. destructor* since a significant number of mites collected on the floor come from brood cells [74,75]. It is likely that part of the damaged mites in mite-resistant honey bees originated from the cleaning of the parasitized cells (hygienic behavior).

The differences in the expression of the two studied resistance behaviors could explain the lower infestation level by *V. destructor*, both in honey bees and brood, that the mite-resistant colonies presented when compared to mite-susceptible colonies at the beginning of the autumn. The selection of honey bees possessing these traits appears as a promising alternative to improve the resistance of honey bees to varroosis [18,19]. In this sense, progress has been made in the knowledge of genes associated with both behaviors, which would allow selection based on molecular markers [25,76,77].

4.2. Reproductive Aspects of V. destructor

The *V. destructor* fertility in both honey bee populations was similar, indicating that the difference in the mite population in both apiaries is not due to reproductive differences. This result contrasts with other studies that relate *V. destructor* fertility to the growth of mite populations in the colonies [23,69,78,79]. However, it cannot be ruled out that, with a larger sample, small differences would appear in the fertility of the mites of the two populations, having an impact on the level of infection in the colonies.

The phoretic mite/reproductive mite ratio was four times higher in mite-resistant colonies than in mite-susceptible colonies. This vital difference could be explained by the more remarkable hygienic behavior displayed by honey bees from the first apiary that allows interrupting the mites' reproduction in the cells, eliminating or forcing them to enter into the phoretic phase [72,80]. Whatever the fate of the mites found in the uncapped cells, the consequence is an increase in the phoretic mite population at the expense of the reproductive ones.

Another difference between both groups of colonies was the ratio of mites present in the drone and worker cells, since the probability of infecting drone cells was 3-fold greater in mite-resistant than in mite-susceptible colonies. *Varroa destructor* in its primary host, the Asian honey bee *A. cerana*, reproduces almost exclusively in drone cells [4,5]. Nevertheless, in *A. mellifera*, the mite maintains the preference to reproduce in drone cells but also in worker cells [1]. The preference ratio into the two cell types ranges from 12:1 to 8:1 [81–83]. The values found in mite-resistant colonies were similar to those mentioned (12.6:1), whereas in mite-susceptible colonies this ratio decreases markedly (5.7:1). This relationship has been poorly analyzed, although it may have significant consequences for the colonies' survival. If the worker population is harmed by increased mite reproduction into their cells, the colony viability and its chances of reproduction are compromised. In this sense, it has been found that the parasitization of worker pupae by *V. destructor* decreases honey bees' longevity [84,85] as well as their weight and flight capacity [86,87], increases the viral load of DWV [7,8], and suppresses the immune response exposing honey bees to infection by other organisms [6]. At the colony level, infected colonies produce fewer swarms [88,89]. A different virulence is possible among populations of *V. destructor* associated with its ability to reproduce in worker cells. This change in the reproductive biology of the mite may be a consequence of the colonies' density. Dynes et al. [16] tested the evolutionary hypothesis that mites from densely beekeeper-managed colonies would be more virulent than those from wild colonies (which rely more on vertical transmission to spread). When comparing the growth of the mite population of the two origins, the mites from wild colonies reproduced more slowly than those from commercial colonies. In line with this argument, it is essential to note that, in the Treinta y Tres Department (the location of the mite-resistant colonies), beekeeping activity is poorly developed and the density of colonies is 1 colony/km^2. At the same time, in the Colonia Department (the location of the mite-susceptible colonies), there are many beekeepers and the density of colonies is 10 colonies/km^2.

4.3. Presence of Viruses in Honey Bees and Mites

In both honey bees and mites from mite-resistant and mite-susceptible colonies, the viruses ABPV, BQCV, DWV, and SBV were found. The BQCV and DWV were present in honey bees from all colonies in both apiaries. There were no significant differences in the proportion of colonies with mites or honey bees infected by each virus in both honey bee populations. In addition, there were no significant differences between apiaries regarding the viral load of ABPV, BQCV, and SBV in mites and honey bees. However, mite-resistant honey bees showed a lower infection level by DWV than mite-susceptible colonies. It is possible that behavioral resistance to *V. destructor* shown by mite-resistant colonies limits their population and indirectly reduces DWV replication, preventing colony collapse. According to de Miranda and Genersch [8], the probability of pupae getting infected by DWV increases according to the *V. destructor* population. This situation can lead to the collapse of colonies. Similar results to those found in this study were obtained by Emsen et al. [90] comparing the load of several viruses in selected colonies for high and low growth of *V. destructor* population, finding that honey bees with higher resistance to the mite had less DWV load than the susceptible honey bees. In contrast, Locke et al. [91], who compared the evolution of RNA viruses in *V. destructor*-resistant honey bees (Gotland island population) to those in susceptible ones from summer to winter, found that DWV was the same in the two groups of colonies. Those authors suggest that *V. destructor*-resistant honey bees would have greater tolerance to DWV.

Regarding DWV variants, variant A was the only variant detected in both honey bee and mite populations. Martin et al. [92] showed that varroa facilitates the dominance of certain DWV strains, decreasing the viral diversity. In this sense, the low DWV genetic diversity detected is consistent with the long establishment of *V. destructor* in Uruguay. DWV variant A was also the dominant variant in the region, including Chile [93], Brazil [94], and Argentina [95], and seems to be more virulent than

DWV-B at the colony level. However, this point is under discussion, since other studies showed that variant DWV-B was more virulent [96,97].

4.4. Genetic Differences between Honey Bees

Microsatellite analysis in honey bees showed very marked allelic and genotypic differences between both populations. Some samples from Treinta y Tres had a higher posterior probability of belonging to the La Estanzuela honey bee population, indicating a certain degree of genetic exchange between both populations, undoubtedly due to the colony movement carried out by beekeepers. When incorporating the reference samples from Africa, Brazil, and Europe into the analysis [57] and asking STRUCTURE to form two groups (K = 2), clearly the honey bees of Treinta y Tres cluster with those from Brazil and Africa, and those from La Estanzuela with the European sample. Although 80% of honey bees in Uruguay belong to the African haplotype of *A. m. scutellata* [36], Africanization degree can still have a gradient from north and east to southeast, as described by Diniz et al. [35]. The genetic similarity of La Estanzuela honey bees to European honey bees can also be explained by the strong commercialization of European queens (especially *A. m. ligustica*) since beekeepers in the region highly value the gentleness of this subspecies. In contrast, in the eastern part of the country, the predominant honey bee populations are the Africanized ones that prevail in Brazil and the region [98,99]. When the program tested five groups, it was found, as expected, that the five analyzed bee populations were separated.

The fact that the two studied honey bee populations display genetic differences at the subspecies scale (the *V. destructor*-resistant populations of Treinta y Tres show a higher degree of Africanization, whereas the La Estanzuela colonies are more susceptible to the mites and more European-like) indicates that the differential resistance found to varroosis may reflect, at least in part, the resistance of each honey bee subspecies. Africanized honey bees are widely known to show good resistance to *V. destructor*, possibly due to greater hygienic and grooming behaviors than those of European honey bees [13].

4.5. Genetic Differences between Mites

The microsatellite analyses showed that the two populations of *V. destructor* displayed significant differences in allelic and genotypic frequencies. Until a few years ago, the population of *V. destructor* was thought to be almost genetically uniform, regardless of the analyzed regions [31]. The differences found coincide with recent studies that show significant genetic variability in *V. destructor* populations [32,33]. These results indicate that differences in the reproductive behavior of *V. destructor* between the two studied apiaries (ratio between phoretic and reproductive mites, and preference to reproduce in drone or worker cells) may be associated with genotypic variants. This is an aspect that will have to be studied in the future to understand the different *A. mellifera–V. destructor* relationships found worldwide.

4.6. Final Considerations

This study showed that the behavioral resistance of honey bees (hygienic and grooming behaviors) to *V. destructor* is critical to controlling mite populations. It is possible that the control of the *V. destructor* population reduces the DWV load in honey bees, the virus most associated with the mite, mitigating the damage it causes. The differences found in behavioral resistance may be associated with genetic differences in honey bees at the subspecies level. In any case, some aspects of the reproductive biology of *V. destructor* could be affecting the damage that this parasite causes to the colonies. These differences could be associated with genetic variants of the mite. The identification of a population of honey bees with clear resistance to *V. destructor* in Uruguay is an addition to those reported in other countries and contributes to the search for tools to improve mite control.

Supplementary Materials: The following are available online at http://www.mdpi.com/2306-7381/7/3/116/s1, Figure S1: Honey bee population allocation by the Structure program based on the genotyping of 5 STR loci.

Author Contributions: Conceptualization, Y.M., I.H.T., K.A., and C.I.; methodology, Y.M., I.H.T., K.A., and C.I.; software, Y.M. and I.H.T.; formal analysis, Y.M., I.H.T., K.A., and C.I.; investigation, Y.M., I.H.T., K.A., L.C., B.B., E.S., and C.I.; resources, Y.M., I.H.T., K.A., and C.I.; writing—original draft preparation, Y.M., I.H.T., and C.I.; writing—review and editing, all the authors; funding acquisition, Y.M. All authors have read and agreed to the published version of the manuscript.

Funding: This research was funded by the Instituto Nacional de Investigación Agropecuaria (INIA), grant number PF10, and the Agencia Nacional de Investigación e Innovación (ANII), grant number POS-NAC-2013-1-12259.

Acknowledgments: The authors wish to acknowledge Sebastian Díaz, Gustavo Ramallo, Daniela Arredondo, Sheena Salvarrey, Matías Maggi, and Pablo Juri for technical support and Sabrina Clavijo for her help in statistical analysis.

Conflicts of Interest: The authors declare no conflict of interest.

References

1. Rosenkranz, P.; Aumeier, P.; Ziegelmann, B. Biology and control of *Varroa destructor*. *J. Invertebr. Pathol.* **2010**, *103*, S96–S119. [CrossRef]
2. Le Conte, Y.; Ellis, M.; Ritter, W. Varroa mites and honey bee health: Can Varroa explain part of the colony losses? *Apidologie* **2010**, *41*, 353–363. [CrossRef]
3. Anderson, D.L.; Trueman, J.W.H. *Varroa jacobsoni* (Acari: Varroidae) is more than one species. *Exp. Appl. Acarol.* **2000**, *24*, 165–189. [CrossRef]
4. Boot, W.J.; Tan, N.Q.; Dien, P.C.; Huan, L.V.; Dung, N.V.; Long, L.T.; Beetsma, J. Reproductive success of *Varroa jacobsoni* in brood of its original host, *Apis cerana*, in comparison to that of its new host, *Apis mellifera*. *Bull. Entomol. Res.* **1996**, *87*, 119–126. [CrossRef]
5. Rath, W. Co-adaptation of *Apis cerana* Fabr. and *Varroa jacobsoni* Oud. *Apidologie* **1999**, *30*, 97–110. [CrossRef]
6. Yang, X.; Cox-Foster, D.L. Impact of an ectoparasite on the immunity and pathology of an invertebrate: Evidence for host immunosuppression and viral amplification. *Proc. Natl. Acad. Sci. USA* **2005**, *102*, 7470–7475. [CrossRef] [PubMed]
7. Beaurepaire, A.; Piot, N.; Doublet, V.; Antunez, K.; Campbell, E.; Chantawannakul, P.; Chejanovsky, N.; Gajda, A.; Heerman, M.; Panziera, D.; et al. Diversity and global distribution of viruses of the western honey bee. *Apis Mellifera Insects* **2020**, *11*, 239. [CrossRef]
8. De Miranda, J.R.; Genersch, E. Deformed wing virus. *J. Invertebr. Pathol.* **2010**, *103*, S48–S61. [CrossRef]
9. Martin, S.J.; Brettell, L.E. Deformed wing virus in honeybees and other insects. *Ann. Rev. Virol.* **2019**, *6*, 49–69. [CrossRef]
10. Guzmán-Novoa, E.; Eccles, L.; Calvete, Y.; McGowan, J.; Kelly, P.G.; Correa-Benítez, A. *Varroa destructor* is the main culprit for the death and reduced populations of overwintered honey bee (*Apis mellifera*) colonies in Ontario, Canada. *Apidologie* **2010**, *41*, 443–450. [CrossRef]
11. Maggi, M.; Antúnez, K.; Invernizzi, C.; Aldea, P.; Vargas, M.; Negri, P.; Brasesco, C.; De Jong, D.; Message, D.; Teixeira, E.W.; et al. Honeybee health in South America. *Apidologie* **2016**, *47*, 835–854. [CrossRef]
12. Pirk, C.W.W.; Strauss, U.; Yusuf, A.A.; Démares, F.; Human, H. Honey bee health in Africa—A review. *Apidologie* **2015**, *47*, 276–300. [CrossRef]
13. Locke, B. Natural *Varroa* mite-surviving *Apis mellifera* honeybee populations. *Apidologie* **2016**, *47*, 467–482. [CrossRef]
14. Fries, I.; Camazine, S. Implications of horizontal and vertical pathogen transmission for honey bee epidemiology. *Apidologie* **2001**, *32*, 199–214. [CrossRef]
15. Neumann, P.; Blacquière, T. The Darwin cure for apiculture? Natural selection and managed honeybee health. *Evol. Appl.* **2017**, *10*, 226–230. [CrossRef]
16. Dynes, T.L.; Berry, J.A.; Delaplane, K.S.; De Roode, J.C.; Brosi, B.J. Assessing virulence of *Varroa destructor* mites from different honey bee management regimes. *Apidologie* **2020**, *51*, 276–289. [CrossRef]
17. Boecking, O.; Spivak, M. Behavioral defenses of honey bees against *Varroa jacobsoni* Oud. *Apidologie* **1999**, *30*, 141–158. [CrossRef]
18. Büchler, R.; Berg, S.; Conte, Y.L. Breeding for resistance to *Varroa destructor* in Europe. *Apidologie* **2010**, *41*, 393–408. [CrossRef]
19. Rinderer, T.E.; Harris, J.W.; Hunt, G.J.; de Guzman, L.I. Breeding for resistance to *Varroa destructor* in North America. *Apidologie* **2010**, *41*, 409–424. [CrossRef]

20. Evans, J.D.; Spivak, M. Socialized medicine: Individual and communal disease barriers in honey bees. *J. Invertebr. Pathol.* **2010**, *103*, S62–S72. [CrossRef]

21. Harbo, J.R.; Harris, J.W. Suppressed mite reproduction explained by the behaviour of adult bees. *J. Apic. Res.* **2005**, *44*, 21–23. [CrossRef]

22. Arechavaleta-Velasco, M.E.; Guzman-Novoa, E. Relative effect of four characteristics that restrain the population growth of the mite *Varroa destructor* in honey bee (*Apis mellifera*) colonies. *Apidologie* **2001**, *32*, 157–174. [CrossRef]

23. Mondragón, L.; Spivak, M.; Vandame, R. A multifactorial study of the resistance of honeybees *Apis mellifera* to the mite *Varroa destructor* over one year in Mexico. *Apidologie* **2005**, *36*, 345–358. [CrossRef]

24. Guzman-Novoa, E.; Emsen, B.; Unger, P.; Espinosa Montaño, L.G.; Petukhova, T. Genotypic variability and relationships between mite infestation levels, mite damage, grooming intensity, and removal of *Varroa destructor* mites in selected strains of worker honey bees (*Apis mellifera* L.). *J. Invertebr. Pathol.* **2012**, *110*, 314–320. [CrossRef]

25. Morfin, N.; Given, K.; Evans, M.; Guzmán-Novoa, E.; Hunt, G.J. Grooming behavior and gene expression of the Indiana "mite-biter" honey bee stock. *Apidologie* **2020**, *51*, 267–275. [CrossRef]

26. Rosenkranz, P.; Tewarson, N.C.; Singh, A.; Engels, W. Differential hygienic behaviour towards *Varroa jacobsoni* in capped worker brood of *Apis cerana* depends on alien scent adhering to the mites. *J. Apicult. Res.* **1993**, *32*, 89–93. [CrossRef]

27. Dietemann, V.; Pflugfelder, J.; Anderson, D.; Charrière, J.-D.; Chejanovsky, N.; Dainat, B.; de Miranda, J.; Delaplane, K.; Dillier, F.-X.; Fuchs, S.; et al. *Varroa destructor*: Research avenues towards sustainable control. *J. Apicult. Res.* **2012**, *51*, 125–132. [CrossRef]

28. Danka, R.G.; Rinderer, T.E.; Spivak, M.; Kefuss, J. Comments on: "*Varroa destructor*: Research avenues towards sustainable control". *J. Apicult. Res.* **2013**, *52*, 69–71. [CrossRef]

29. De Guzman, L.I.; Rinderer, T.E. Identification and comparison of *Varroa* species infesting honey bees. *Apidologie* **1999**, *30*, 85–95. [CrossRef]

30. Garrido, C.; Rosenkranz, P.; Paxton, R.J.; Gonçalves, L.S. Temporal changes in *Varroa destructor* fertility and haplotype in Brazil. *Apidologie* **2003**, *34*, 535–541. [CrossRef]

31. Solignac, M.; Cornuet, J.-M.; Vautrin, D.; Le Conte, Y.; Anderson, D.; Evans, J.; Cros-Arteil, S.; Navajas, M. The invasive Korea and Japan types of *Varroa destructor*, ectoparasitic mites of the Western honey bee (*Apis mellifera*), are two partly isolated clones. *Proc. Roy. Soc. Lond. Ser. B Biol. Sci.* **2005**, *272*, 411–419. [CrossRef]

32. Dynes, T.L.; De Roode, J.C.; Lyons, J.I.; Berry, J.A.; Delaplane, K.S.; Brosi, B.J. Fine scale population genetic structure of *Varroa destructor*, an ectoparasitic mite of the honey bee (*Apis mellifera*). *Apidologie* **2017**, *48*, 9–101. [CrossRef]

33. Beaurepaire, A.L.; Moro, A.; Mondet, F.; Le Conte, Y.; Neumann, P.; Locke, B. Population genetics of ectoparasitic mites suggest arms race with honeybee hosts. *Sci. Rep.* **2019**, *9*, 11355. [CrossRef]

34. Cordara, J.J. *La Historia de La Apicultura en Uruguay*; Facultad de Ciencias Agrarias, Universidad de la Empresa: Montevideo, Uruguay, 2005.

35. Diniz, N.M.; Soares, A.E.E.; Sheppard, W.S.; Del Lama, M.A. Genetic structure of honeybee populations from southern Brazil and Uruguay. *Genet. Mol. Biol.* **2003**, *26*, 47–52. [CrossRef]

36. Branchiccela, B.; Aguirre, C.; Parra, G.; Estay, P.; Zunino, P.; Antúnez, K. Genetic changes in *Apis mellifera* after 40 years of Africanization. *Apidologie* **2014**, *45*, 752–756. [CrossRef]

37. Invernizzi, C.; Antúnez, K.; Campa, J.P.; Harriet, J.; Mendoza, Y.; Santos, E.; Zunino, P. Situación sanitaria de las abejas melíferas en Uruguay. *Veterinaria* **2011**, *47*, 15–27.

38. Mendoza, Y.; Gramajo, E.; Invernizzi, C.; Tomasco, I.H. Mitochondrial haplotype analyses of the mite *Varroa destructor* (Acari: Varroidae) collected from honeybees *Apis mellifera* (Hymenoptera, Apidae) in Uruguay. *Syst. Appl. Acarol.* **2020**, in press.

39. Delaplane, K.S.; van der Steen, J.; Guzmán-Novoa, E. Standard methods for estimating strength parameters of *Apis mellifera* colonies. *J. Apicult. Res.* **2013**, *52*, 1–12. [CrossRef]

40. Büchler, R.; Andonov, S.; Bienefeld, K.; Costa, C.; Hatjina, F.; Kezic, N.; Kryger, P.; Spivak, M.; Uzunov, A.; Wilde, J. Standard methods for rearing and selection of *Apismellifera* queens. *J. Apicult. Res.* **2013**, *52*, 1–30. [CrossRef]

41. Invernizzi, C.; Zefferino, I.; Santos, E.; Sánchez, L.; Mendoza, Y. Multilevel assessment of grooming behaviour against *Varroa destructor* in Italian and Africanized honey bees. *J. Apicult. Res.* **2016**, *54*, 1–7.

42. Dietemann, V.; Nazzi, F.; Martin, S.J.; Anderson, D.L.; Locke, B.; Delaplane, K.S.; Wauquiez, Q.; Tannahill, C.; Frey, E.; Ziegelmann, B.; et al. Standard methods for varroa research. *J. Apicult. Res.* **2013**, *52*, 1–54. [CrossRef]

43. Antúnez, K.; D'Alessandro, B.; Corbella, E.; Zunino, P. Detection of Chronic bee paralysis virus and Acute bee paralysis virus in Uruguayan honeybees. *J. Invertebr. Pathol.* **2005**, *90*, 69–72. [CrossRef] [PubMed]

44. Johnson, R.M.; Evans, J.D.; Robinson, G.E.; Berenbaum, M.R. Changes in transcript abundance relating to colony collapse disorder in honey bees (*Apis mellifera*). *Proc. Natl. Acad. Sci. USA* **2009**, *106*, 14790–14795. [CrossRef] [PubMed]

45. Kukielka, D.; Esperón, F.; Higes, M.; Sánchez-Vizcaíno, J.M. A sensitive one-step real-time RT-PCR method for detection of deformed wing virus and black queen cell virus in honeybee *Apis mellifera*. *J. Virol. Methods* **2008**, *147*, 275–281. [CrossRef] [PubMed]

46. Yang, X.; Cox-Foster, D. Effects of parasitization by *Varroa destructor* on survivorship and physiological traits of *Apis mellifera* in correlation with viral incidence and microbial challenge. *Parasitology* **2007**, *134*, 405–412. [CrossRef]

47. Anido, M.; Branchiccela, B.; Castelli, L.; Harriet, J.; Campá, J.; Zunino, P.; Antúnez, K. Prevalence and distribution of honey bee pathogens in Uruguay. *J. Apicult. Res.* **2016**, *54*, 532–540. [CrossRef]

48. Chen, Y.P.; Pettis, J.S.; Feldlaufer, M.F. Detection of multiple viruses in queens of the honey bee, *Apis mellifera* L. *J. Invertebr. Pathol.* **2005**, *90*, 118–121. [CrossRef]

49. Highfield, A.C.; El Nagar, A.; Mackinder, L.C.; Noel, L.M.; Hall, M.J.; Martin, S.J.; Schroeder, D.C. Deformed wing virus implicated in overwintering honeybee colony losses. *Appl. Environ. Microbiol.* **2009**, *75*, 7212–7220. [CrossRef]

50. Kevill, J.L.; Highfield, A.; Mordecai, G.J.; Martin, S.J.; Schroeder, D.C. ABC assay: Method development and application to quantify the role of three DWV master variants in overwinter colony losses of European honey bees. *Viruses* **2017**, *9*, 314. [CrossRef]

51. Miller, S.A.; Dykes, D.D.; Polesky, H.F.R.N. A simple salting out procedure for extracting DNA from human nucleated cell. *Nucleic Acids Res.* **1988**, *16*, 1215. [CrossRef]

52. Estoup, A.; Solignac, M.; Cornuet, J.-M. Precise assessment of the number of patrilines and of genetic relatedness in honeybee colonies. *Proc. R. Soc. Lond. B* **1994**, *258*, 1–7.

53. Estoup, A.; Garnery, L.; Solignac, M.; Cornuet, J.-M. Microsatellite variation in honey bee (*Apis mellifera* L.) populations: Hierarchical genetic structure and test of the infinite allele and stepwise mutation models. *Genetics* **1995**, *140*, 679–695.

54. Oldroyd, B.P.; Smolenski, A.J.; Cornuet, J.; Wongsiri, S.; Estoup, A.; Rinderer, T.E.; Crozier, R.H. Levels of polyandry and intracolonial genetic relationships in *Apis florea*. *Behav. Ecol. Sociobiol.* **1995**, *37*, 329–335. [CrossRef]

55. Rowe, D.; Rinderer, T.; Stelzer, J.; Oldroyd, B.P.; Crozier, R.H. Seven polymorphic microsatellite loci in honeybees (*Apis mellifera*). *Insectes Soc.* **1997**, *44*, 85–93. [CrossRef]

56. Sanguinetti, C.J.; Días Neto, E.; Simpson, A.J. Rapid silver staining and recovery of PCR products separated on polyacrylamide gels. *Biotechniques* **1994**, *17*, 914–921.

57. Pinto, M.A.; Rubink, W.L.; Patton, J.C.; Coulson, R.N.; Johnston, J.S. Africanization in the United States: Replacement of feral European honeybees (*Apis mellifera* L.) by an African hybrid swarm. *Genetics* **2005**, *170*, 1653–1665. [CrossRef]

58. Evans, J.D. Microsatellite loci in the honey bee parasitic mite *Varroa jacobsoni*. *Mol. Ecol.* **2000**, *9*, 1436–1438. [CrossRef]

59. Solignac, M.; Vautrin, D.; Pizzo, A.; Navajas, M.; Le Conte, Y.; Cornuet, J.-M. Characterization of microsatellite markers for the apicultural pest *Varroa destructor* (Acari: Varroidae) and its relatives. *Mol. Ecol. Notes* **2003**, *3*, 556–559. [CrossRef]

60. Akaike, H. A new look at the statistical model identification. *IEEE Trans. Autom. Contr.* **1974**, *19*, 716–723. [CrossRef]

61. Akaike, H. Autoregressive model fitting for control. *Ann. Inst. Statist. Math.* **1971**, *23*, 163–180. [CrossRef]

62. R Development Core Team. *R: A Language and Environment for Statistical Computing*; R Foundation for Statistical Computing: Vienna, Austria, 2012.

63. Raymond, M.; Rousset, F. GENEPOP Version 1.2: Population genetics software for exact tests and ecumenicism. *J. Hered.* **1995**, *86*, 248–249. [CrossRef]

64. Pritchard, J.K.; Stephens, M.; Donnelly, P. Inference of population structure using multilocus genotype data. *Genetics* **2000**, *155*, 945–959. [PubMed]

65. Boecking, O.; Drescher, W. The removal response of *Apis mellifera* L. colonies to brood in wax and plastic cells after artificial and natural infestation with *Varroa jacobsoni* Oud. and to freeze killed brood. *Exp. Appl. Acarol.* **1992**, *16*, 321–329. [CrossRef]

66. Spivak, M.; Reuter, G.S. *Varroa jacobsoni* infestation in untreated honey bee (Hymenoptera: Apidae) colonies selected for hygienic behavior. *J. Econ. Entomol.* **2001**, *94*, 326–331. [CrossRef]

67. Ibrahim, A.; Reuter, G.S.; Spivak, M. Field trial of honey bee colonies bred for mechanisms of resistance against *Varroa destructor*. *Apidologie* **2007**, *38*, 67–76. [CrossRef]

68. Leclercq, G.; Pannebakker, B.; Gengler, N.; Nguyen, B.K.; Francis, F. Drawbacks and benefits of hygienic behavior in honey bees (*Apis mellifera* L.): A review. *J. Apic. Res.* **2017**, *56*, 366–375. [CrossRef]

69. Nganso, B.; Fombong, A.; Yusuf, A.; Pirk, C.; Stuhl, C.; Torto, B. Low fertility, fecundity and numbers of mated female offspring explain the lower reproductive success of the parasitic mite *Varroa destructor* in African honeybees. *Parasitology* **2018**, *145*, 1633–1639. [CrossRef]

70. Fries, I.; Rosenkranz, P. Number of reproductive cycles of *Varroa jacobsoni* in honey-bee (*Apis mellifera*) colonies. *Exp. Appl. Acarol.* **1996**, *20*, 103–112. [CrossRef]

71. Martin, S.J.; Kemp, D. Average number of reproductive cycles performed by *Varroa jacobsoni* in honey bee (*Apis mellifera*) colonies. *J. Apicult. Res.* **1997**, *36*, 113–123. [CrossRef]

72. Oddie, M.A.Y.; Büchler, R.; Dahle, B.; Kovacic, M.; Le Conte, Y.; Locke, B.; de Miranda, J.; Mondet, F.; Neumann, P. Rapid parallel evolution overcomes global honey bee parasite. *Sci. Rep.* **2018**, *8*, 7704. [CrossRef]

73. Martin, S.J.; Hawkins, G.P.; Brettell, L.E.; Reece1, N.; Correia-Oliveira, M.E.; Allsopp, M.H. *Varroa destructor* reproduction and cell re-capping in mite-resistant *Apis mellifera* populations. *Apidologie* **2020**, *51*, 369–381. [CrossRef]

74. Rosenkranz, P.; Fries, I.; Boecking, O.; Stürmer, M. Damaged Varroa mites in the debris of honey bee (*Apis mellifera* L.) colonies with and without hatching brood. *Apidologie* **1997**, *28*, 427–437. [CrossRef]

75. Lobb, N.; Martin, S. Mortality of *Varroa jacobsoni* Oudemans during or soon after the emergence of worker and drone honeybees *Apis mellifera* L. *Apidologie* **1997**, *28*, 367–374. [CrossRef]

76. Tsuruda, J.M.; Harris, J.W.; Bourgeois, L.; Danka, R.G.; Hunt, G.J. High-resolution linkage analyses to identify genes that influence *Varroa* sensitive hygiene behavior in honey bees. *PLoS ONE* **2012**, *7*, e48276. [CrossRef] [PubMed]

77. Spötter, A.; Gupta, P.; Mayer, M.; Reinsch, N.; Bienefeld, K. Genome-wide association study of a *Varroa*-specific defense behavior in honeybees (*Apis mellifera*). *J. Hered.* **2016**, *107*, 220–227. [CrossRef] [PubMed]

78. Martin, S.; Holland, K.; Murray, M. Non-reproduction in the honeybee mite *Varroa jacobsoni*. *Exp. Appl. Acarol.* **1997**, *21*, 539–549. [CrossRef]

79. Martin, S.J.; Medina, L.M. Africanized honey bees have unique tolerance to *Varroa* mites. *Trends Parasitol.* **2004**, *20*, 112–114. [CrossRef]

80. Harris, J.W. Effect of brood type on *Varroa*-sensitive hygiene by worker honey bees (Hymenoptera: Apidae). *Ann. Entomol. Soc. Am.* **2008**, *101*, 1137–1144. [CrossRef]

81. Fuchs, S. Preference for drone brood cells by *Varroa jacobsoni* Oud. in colonies of *Apis mellifera carnica*. *Apidologie* **1990**, *21*, 193–196. [CrossRef]

82. Boot, W.J.; Schoenmaker, J.; Calis, J.N.M.; Beetsma, J. Invasion of *Varroa jacobsoni* into drone brood cells of the honey bee, *Apis mellifera*. *Apidologie* **1995**, *26*, 109–118. [CrossRef]

83. Calderone, N.W.; Kuenen, L.P.S. Effects of western honey bee (Hymenoptera: Apidae) colony, cell, type, and larval sex on host acquisition by female *Varroa destructor* (Acari: Varroidae). *J. Econ. Entomol.* **2001**, *94*, 1022–1030. [CrossRef] [PubMed]

84. Amdam, G.V.; Hartfelder, K.; Norberg, K.; Hagen, A.; Omholt, S.W. Altered physiology in worker honey bees (Hymenoptera: Apidae) infested with the mite *Varroa destructor* (Acari: Varroidae): A factor in colony loss during overwintering? *J. Econ. Entomol.* **2004**, *97*, 741–747. [CrossRef]

85. Aldea, P.; Bozinovic, F. The energetic and survival costs of *Varroa* parasitism in honeybees. *Apidologie* **2020**, in press. [CrossRef]

86. Duay, P.; de Jong, D.; Engels, W. Decreased flight performance and sperm production in drones of the honey bee (*Apis mellifera*) slightly infested by *Varroa destructor* mites during pupal development. *Genet. Mol. Res.* **2002**, *1*, 227–232. [PubMed]

87. Duay, P.; de Jong, D.; Engels, W. Weight loss in drone pupae (*Apis mellifera*) multiply infested by *Varroa destructor* mites. *Apidologie* **2003**, *34*, 61–65. [CrossRef]

88. Fries, I.; Hansen, H.; Imdorf, A.; Rosenkranz, P. Swarming in honey bees (*Apis mellifera*) and *Varroa destructor* population development in Sweden. *Apidologie* **2003**, *34*, 389–398. [CrossRef]

89. Villa, J.D.; Bustamante, D.M.; Dunkley, J.P.; Escobar, L.A. Changes in honey bee (Hymenoptera: Apidae) colony swarming and survival pre- and postarrival of *Varroa destructor* (Mesostigmata: Varroidae) in Louisiana. *Ann. Entomol. Soc. Am.* **2008**, *101*, 867–871. [CrossRef]

90. Emsen, B.; Hamiduzzaman, M.M.; Goodwin, P.H.; Guzmán-Novoa, E. Lower virus infections in *Varroa destructor*-infested and uninfested brood and adult honey bees (*Apis mellifera*) of a low mite population growth colony compared to a high mite population growth colony. *PLoS ONE* **2015**, *10*, e0118885. [CrossRef]

91. Locke, B.; Forsgren, E.; de Miranda, J.R. Increased tolerance and resistance to virus infections: A possible factor in the survival of *Varroa destructor*-resistant honey bees (*Apis mellifera*). *PLoS ONE* **2014**, *9*, e99998. [CrossRef]

92. Martin, S.J.; Highfield, A.C.; Brettell, L.; Villalobos, E.M.; Budge, G.E.; Powell, M.; Nikaido, S.; Schroeder, D.C. Global honey bee viral landscape altered by a parasitic mite. *Science* **2012**, *336*, 1304–1306. [CrossRef]

93. Riveros, G.; Arismendi, N.; Zapata, N.; Evans, D.; Pérez, I.; Aldea, P.; Vargas, M. Occurrence, prevalence and viral load of deformed wing virus variants in *Apis mellifera* colonies in Chile. *J. Apicult. Res.* **2020**, *59*, 63–68. [CrossRef]

94. de Souza, F.S.; Kevill, J.L.; Correia-Oliveira, M.E.; de Carvalho, C.A.L.; Martin, S.J. Occurrence of deformed wing virus variants in the stingless bee *Melipona subnitida* and honey bee *Apis mellifera* populations in Brazil. *J. Gen. Virol.* **2019**, *100*, 289–294. [CrossRef] [PubMed]

95. Brasesco, C.; Quintana, S.; Di Gerónimo, V.; Genchi García, M.L.; Sguazza, G.; Bravi, M.E.; Fargnoli, L.; Reynaldi, F.J.; Eguaras, M.; Maggi, M. Deformed wing virus type a and b in managed honeybee colonies of Argentina. *Bull. Entomol. Res.* **2020**, *29*, 1–11. [CrossRef] [PubMed]

96. McMahon, D.P.; Natsopoulou, M.E.; Doublet, V.; Fürst, M.; Weging, S.; Brown, M.J.; Gogol-Döring, A.; Paxton, R.J. Elevated virulence of an emerging viral genotype as a driver of honeybee loss. *Proc. Biol. Sci.* **2016**, *283*, 20160811. [CrossRef]

97. Kevill, J.L.; de Souza, F.S.; Sharples, C.; Oliver, R.; Schroeder, D.C.; Martin, S.J. DWV-A lethal to honey bees (*Apis mellifera*): A colony level survey of DWV variants (A, B, and C) in England, Wales, and 32 states across the US. *Viruses* **2019**, *11*, 426. [CrossRef]

98. Sheppard, W.S.; Rinderer, T.E.; Garnery, L.; Shimanuki, H. Analysis of Africanized honey bee mitochondrial DNA reveals further diversity of origin. *Gen. Mol. Biol.* **1999**, *22*, 73–75. [CrossRef]

99. Abrahamovich, A.H.; Atela, O.; De la Rúa, P.; Galián, J. Assessment of the mitochondrial origin of honey bees from Argentina. *J. Apicul. Res.* **2007**, *46*, 191–194. [CrossRef]

Article

Association between the Microsatellite Ap243, AC117 and SV185 Polymorphisms and *Nosema* Disease in the Dark Forest Bee *Apis mellifera mellifera*

Nadezhda V. Ostroverkhova [1,2]

1   Invertebrate Zoology Department, Biology Institute, National Research Tomsk State University, 36 Lenina Avenue, 634050 Tomsk, Russia; nvostrov@mail.ru; Tel.: +7-3822-529-461
2   Department of Biology and Genetics, Siberian State Medical University, 2 Moskovsky Trakt, 634055 Tomsk, Russia

Abstract: The microsporidian *Nosema* parasites, primarily *Nosema ceranae*, remain critical threats to the health of the honey bee *Apis mellifera*. One promising intervention approach is the breeding of *Nosema*-resistant honey bee colonies using molecular technologies, for example marker-assisted selection (MAS). For this, specific genetic markers used in bee selection should be developed. The objective of the paper is to search for associations between some microsatellite markers and *Nosema* disease in a dark forest bee *Apis mellifera mellifera*. For the dark forest bee, the most promising molecular genetic markers for determining resistance to nosemosis are microsatellite loci AC117, Ap243 and SV185, the alleles of which ("177", "263" and "269", respectively) were associated with a low level of *Nosema* infection. This article is the first associative study aimed at finding DNA loci of resistance to nosemosis in the dark forest bee. Nevertheless, microsatellite markers identified can be used to predict the risk of developing the *Nosema* disease.

Keywords: *Nosema* disease; dark forest bee; *Apis mellifera mellifera*; microsatellite loci; association

Citation: Ostroverkhova, N.V. Association between the Microsatellite Ap243, AC117 and SV185 Polymorphisms and *Nosema* Disease in the Dark Forest Bee *Apis mellifera mellifera*. Vet. Sci. **2021**, 8, 2. https://doi.org/10.3390/vetsci 8010002

Received: 9 December 2020
Accepted: 24 December 2020
Published: 29 December 2020

Publisher's Note: MDPI stays neutral with regard to jurisdictional claims in published maps and institutional affiliations.

1. Introduction

In the last few decades, negative processes such as massive losses of bee colonies and hybridization have been observed in honey bee populations worldwide. The honey bee colony losses, called colony collapse disorder (CCD), as a result of reduced adaptation of honey bees to environmental factors, pose a threat to beekeeping worldwide [1,2]. It has been suggested that CCD can be caused by many causes, including various diseases such as nosemosis, environmental pollution, exposure to pesticides, weather and agricultural and beekeeping practices [3–5]. On the one hand, in order to avoid a catastrophic population decline from pests and diseases, it is necessary to maintain a high level of genetic diversity in honey bee populations [6,7]. On the other hand, molecular technologies, for example marker-assisted selection (MAS) can be used to identify bee colonies carrying specific traits of interest (e.g., resistance to pathogens and parasites, gentleness and high honey productivity) or the lack of undesirable traits (e.g., aggression and swarming) [8–11]. Marker-assisted selection is a new technology in beekeeping, and no specific genetic markers that could be used in bee breeding have been proposed [12,13].

The search for informative DNA loci/genes associated with economically useful and other traits (associative mapping) is highly relevant. The preferred strategy is to genotype a high number of genetic markers in linkage or association studies in order to identify genomic regions and discover the causative genes [14]. The identification of genetic markers associated with the phenotype can also immediately be used to selectively breed colonies that are more resistant [15].

Currently, quantitative trait loci (QTLs) associated with queen fertility [16,17], resistance to chalkbrood [18,19] and varroosis [20,21], and various types of behavior [22–24]

have been identified. For example, hygiene behavior, which is a social behavior helps control various diseases of the offspring such as varroosis [25–27]. Varroosis is one of the most devastating diseases of the brood caused by the parasitic mite *Varroa destructor* [28–30]. Hygiene behavior has been shown to provide significant resistance against the *Varroa* mite [13,30]. QTL studies of *Varroa* resistance behavior in honey bees have identified over 20 suggestive QTLs in different genomic regions [14,20–22,31].

Like varroosis, nosemosis is also one of the most dangerous diseases [32–34], but the study of associations between molecular genetic markers and nosemosis is rare [35,36].

Nosemosis is a serious disease in adult honey bees caused by microsporidia, which are obligate intracellular eukaryotic parasites [37]. *Nosema* parasites multiply and develop within the host-cell cytoplasm causing extensive and even total destruction of the midgut epithelial layer [38,39].

In European honey bees, two species of microsporidia, *Nosema apis* and *Nosema ceranae*, have been described. *N. apis* Zander, 1909 [40] is an evolutionarily old parasite of the honey bee *A. mellifera*. The parasite causing type A nosemosis is moderately virulent, and bee colonies are able to resist disease under favorable environmental conditions [41–43]. *N. ceranae* Fries et al., 1996 [44], responsible for type C nosemosis, is a relatively new parasite for the *A. mellifera* [38,44–46]. This parasite was originally described in an Asian bee *Apis cerana* in the late 20th century [44]. Since 2006, it has been found in honey bee *A. mellifera* populations around the world [38,41–43,45,47–54]. Compared to *N. apis*, *N. ceranae* is considered a more virulent parasite, and in some countries, such as the Mediterranean countries, it is associated with the bee colony losses [32,55–58].

Using microsatellite markers, four QTLs associated with low spore load were revealed in a Danish selected *Nosema*-resistant honey bees line [35]. These Buckfast honey bee colonies have been selectively bred for the absence of *Nosema* over decades, resulting in a breeding line that is tolerant toward *Nosema* [59,60]. Unlike pure race breeding, the Buckfast breeding system mixes different stocks to establish a hybrid bee with desired characteristics. The Buckfast contains heritage from mainly *A. m. ligustica* and *A. m. mellifera* and from other subspecies. Since the Buckfast bee is a hybrid bee, the expression of its notable characteristics can vary greatly within the stock [61].

It is assumed that the honey bee subspecies (lines and colonies) differ in their resistance to disease, which may be determined by social immunity including hygienic and other types of behavior [13,15,30,62–70]. For example, in *A. mellifera*, hygienic and grooming behaviors are expressed more highly in Africanized honey bees than in European ones. Perhaps this explains the higher resistance of Africanized bees to *V. destructor* compared to European bees [71].

Although no significant effect of *N. ceranae* infections on hygienic behavior was detected [72], it is clear that natural resistance of honey bees to *Nosema* depends on many factors and the genetic variants of honey bees can play a relevant role. The purpose of this study was to identify associations between genetic variants of some microsatellite loci and *Nosema* infection/disease resistance in the dark forest bee *Apis mellifera mellifera*.

2. Materials and Methods

2.1. Bee Samples

In the present study, samples of a dark forest bee *Apis mellifera mellifera* from Siberian populations (longitude 81°29′–92°08′ E and latitude 50°44′–65°47′ N) were examined. The dark forest bee is a native bee that was introduced to Siberia about 230 years ago and has adapted well to the local climate and plant communities. In Siberia, the bee population is an artificial population; wintering of bees is controlled by people [73]. Honey bees were collected from 12 apiaries between the end of May 2016 and August of the same year. A total of 226 workers from twenty-eight bee colonies (from 8 to 10 bees from each bee colony) were examined.

For the diagnosis and detection of *Nosema* infection, the oldest honey bees (forager bees) were collected outside the entrance of hive, because they have the greatest infection and the highest proportion of infected bees [39]. Bee samples were stored in a freezer at $-20\ ^\circ\text{C}$ until further processing.

2.2. Study Design

The present research has conducted in several stages. At the first stage of the study, the presence of *Nosema* spp. in honey bees was investigated using both light microscopy and polymerase chain reaction (PCR).

At the second stage of the study, the genetic diversity of honey bees with different degrees of *Nosema* infestation was examined using polymorphic microsatellite loci. Earlier, the genetic diversity of local dark forest bees (Siberian population) was studied using a complex of microsatellite loci and identified polymorphic microsatellite loci [74]. In this study, 23 polymorphic microsatellite loci were used to search for genetic markers of *Nosema* disease resistance in honey bees.

Finally, the associations of polymorphic variants of microsatellite loci studied with nosemosis were analyzed using the odds ratio method (OR).

2.3. Experimental Procedures

To carry out individual analysis of the bees, for each sample, the bee's midgut was isolated and divided in two. One part of the midgut was used for light microscopy. For this, the midgut was ground in 0.5 mL of sterile, distilled water and the number of *Nosema* spores was counted using a Zeiss Axio Lab.1 light microscope.

DNA was extracted from another part of the midgut using a DNA purification kit, PureLink™ Mini (Invitrogen, Carlsbad, CA, USA) according to the manufacturer's protocol. PCR was performed using a thermal MyCycler T100 (BioRad, Foster City, CA, USA).

For the diagnosis of nosemosis, duplex-PCR was used [42]. The primer sequences used to amplify the 321 bp fragment corresponding to the 16S ribosomal gene of *N. apis* were 321APIS-FOR 5′-GGGGGCATGTCTTTGACGTACTATGTA-3′ and 321APIS-REV 5′-GGGGGGCGTTTAAAATGTGAAACAACTATG-3′. The primer sequences utilized to amplify the 218 bp fragment corresponding to the 16S ribosomal gene of *N. ceranae* were 218MITOC-FOR 5′-CGGCGACGATGTGATATGAAAATATTAA-3′ and 218MITOC-REV 5′-CCCGGTCATTCTCAAACAAAAAACCG-3′ [42]. PCR was performed in a reaction volume of 20 μL containing 5–10 ng of template DNA, $1 \times$ PCR buffer, 200 μM of each dNTP, 1.5 mM $MgCl_2$, 0.2 μM of each primer and 1U Taq polymerase (Fermentas, Thermo Fisher Scientific, Chelmsford, MA, USA). The routine consisted of an initial denaturation step at $94\ ^\circ\text{C}$ for 2 min, followed by 35 cycles of $94\ ^\circ\text{C}$ for 30 s, $58\ ^\circ\text{C}$ for 30 s and $72\ ^\circ\text{C}$ for 1 min, and a final extension step at $72\ ^\circ\text{C}$ for 5 min. PCR products were analyzed on 1.5% agarose gels. Gels were stained with ethidium bromide and visualized using UV illumination (Gel Doc XR+, BioRad, Foster City, CA, USA). For each PCR, positive control (reference *N. apis* and *N. ceranae* DNA extracts as template) was used. Negative control (ddH2O) was also included in each run of PCR amplification to detect possible contamination.

The variability of 23 polymorphic microsatellite loci (Ap066, K0457B, K1168, A007, A008, A028, A043, Ap049, Ap007, AC117, 6339, Ap068, Ap243, SV220, SV167, SV185, Ap226, H110, A024, AT139, A056, Ap249, and A113) was examined. When choosing loci, some factors such as a high level of polymorphism, maximum chromosomal coverage (the studied loci are located on 13 out of 16 honey bee chromosomes), and data from publications on DNA markers associated with bee resistance to diseases were considered. PCR was performed using specific fluorescence-labeled primers according to Solignac et al. [75]. The reactions were performed in 20 μL of a solution containing 5–10 ng DNA template, 0.4 μM of each primer, 60 μM of each dNTP, 1–2.5 mM $MgCl_2$ and $1 \times$ PCR buffer and 1U of Taq polymerase (Fermentas, Thermo Fisher Scientific, Chelmsford, MA, USA). After a denaturing step of 3 min at $94\ ^\circ\text{C}$, samples were processed through 35 cycles consisting of 30 s at $94\ ^\circ\text{C}$, 30 s at $55–60\ ^\circ\text{C}$ and 30 s at $72\ ^\circ\text{C}$. The final elongation step was 10 min at $72\ ^\circ\text{C}$ [75]. Amplification products were analyzed with ABI Prism 3730

Genetic Analyzer and GeneMapper Software (Applied Biosystems, Inc., Foster City, CA, USA) in the collective center Medical Genomics (Research Institute of Medical Genetics, Tomsk National Research Medical Center, Russian Academy of Sciences, Moscow, Russia). Two microliters of PCR products were mixed with GeneScan500-ROX size standards (Applied Biosystems, Inc.) and deionized formamide. Samples were run according to the manufacturer's recommendations.

2.4. Estimation of the Nosemosis Level

According to PCR, most of the bees examined were coinfected with two *Nosema* species *N. apis* and *N. ceranae*. In this regard, we considered the general infestation of honey bees with microsporidia, without division into *Nosema* species. A light microscope using 400× magnification was used for counting *Nosema* spores in macerated bee preparations.

Honey bees were divided into the following groups, uninfected (*Nosema*-negative) and *Nosema*-infected (*Nosema*-positive). Since our study did not require a high degree of precision, we used an arbitrary infection scale and divided the *Nosema*-positive bee group into two variants, *Nosema*-positive low and *Nosema*-positive high. *Nosema*-positive low bees were with a small amount of microsporidial spores on microscopic analysis (less than 100 spores in the field of view of the microscope). *Nosema*-positive high bees were with a significant number of microsporidial spores detected by microscopic analysis (more than 500 spores in the field of view of the microscope). An intermediate variant of the bee infection (100–500 *Nosema* spores in the field of view of the microscope) was not identified. In total, three groups of bees were formed: *Nosema*-negative, *Nosema*-positive low and *Nosema*-positive high.

2.5. Statistical Analysis

The genotypes obtained for each of the honey bee were used to estimate population parameters. The allelic and genotypic observed frequencies by Hardy–Weinberg equilibrium (HWE), the number of alleles, observed (Ho) and expected (He) heterozygosity were estimated employing the GENEPOP v.4.1 package [76]. To assess genetic diversity, for each infectious category, the observed and expected heterozygosity for each locus were compared using a Student's test. Comparison of allele and genotype frequencies between bee samples that differed in *Nosema* infestation was performed using the chi-square test. In the case of a small number of one of the comparison classes, the chi-square test with Yates correction was used.

To assess the association of polymorphic variants of the microsatellite loci studied with *Nosema* infection in bees, the odds ratio (OR) and 95% confidence interval (95% CI) corresponding to the p-value was calculated [77]. The association of the genetic marker with the tested trait was determined by the OR value when the differences in the allele frequency between the compared groups reached the level of statistical significance, $p < 0.05$. If the OR > 1, the assumption about the association of the analyzed genetic variant (allele/genotype) with the studied pathology is considered (increased chance of developing disease). When the OR < 1, a protective role of the corresponding genetic variant (allele/genotype) is assumed.

3. Results

3.1. Genetic Diversity of the A. m. mellifera Honey Bees in Siberia on the Microsatellite Loci

Honey bees with different degrees of *Nosema* infestation were genotyped using 23 microsatellite markers. When comparing the distribution of allele frequencies between three groups of the *A. m. mellifera* bees (*Nosema*-negative, *Nosema*-positive low and *Nosema*-positive high), several loci (AC117, A113, Ap243, A024, A007, Ap049 and SV185) were identified that are promising for further analysis. The parameters of genetic diversity of these loci, such as the frequencies of alleles and genotypes, expected heterozygosity and observed heterozygosity, are presented in Table 1.

Table 1. Genotype and allele frequencies and heterozygosity at 7 microsatellite loci in the dark forest bees with varying degrees of *Nosema* infection.

Locus	Genotype	Allele	Infection Categories of Honey Bees					
			Nosema-Negative		*Nosema*-Positive Low		*Nosema*-Positive High	
			Genotype	Allele	Genotype	Allele	Genotype	Allele
AC117	173–173	173	0.037	0.074 ± 0.036	0.024	0.079 ± 0.017		0.057 ± 0.021
	173–181	177	0.074	0.148 ± 0.048	0.110	0.075 ± 0.017	0.115	0.025 ± 0.014
	177–177	181	0.037	0.296 ± 0.062	0.008	0.260 ± 0.028		0.533 ± 0.045
	177–181	185		0.482 ± 0.068	0.039	0.587 ± 0.031		0.385 ± 0.044
	177–185		0.222		0.094		0.049	
	181–181		0.259		0.142		0.475	
	181–185				0.087			
	185–185		0.370		0.496		0.361	
	Ho/He		0.296 ± 0.088 **/0.653 ± 0.040		0.331 ± 0.042 **/0.577 ± 0.025		0.164 ± 0.047 **/0.564 ± 0.025	
	N		27		127		61	
A113	210–218	210			0.016	0.008 ± 0.006		
	212–212	212		0.107 ± 0.041	0.063	0.152 ± 0.023	0.016	0.063 ± 0.021
	212–214	214	0.036	0.018 ± 0.018	0.008	0.008 ± 0.006		
	212–218	218	0.143	0.518 ± 0.067	0.125	0.598 ± 0.031	0.094	0.719 ± 0.040
	212–220	220	0.036	0.339 ± 0.063	0.023	0.207 ± 0.025		0.219 ± 0.037
	212–222	222			0.008	0.008 ± 0.006		
	212–226	226			0.016	0.020 ± 0.009		
	214–226	228		0.018 ± 0.018	0.008			
	218–218		0.286		0.438		0.609	
	218–220		0.321		0.172		0.125	
	218–222				0.008			
	220–220		0.143		0.109		0.156	
	220–228		0.036					
	226–226				0.008			
	Ho/He		0.571 ± 0.094/0.605 ± 0.041		0.383 ± 0.043 **/0.577 ± 0.027		0.219 ± 0.052 */0.432 ± 0.043	
	N		28		128		64	
Ap243	253–260	253	0.043	0.022 ± 0.022			0.074	0.037 ± 0.026
	256–256	256	0.305	0.413 ± 0.073	0.284	0.419 ± 0.035	0.519	0.593 ± 0.067
	256–263	260	0.218	0.109 ± 0.046	0.233	0.111 ± 0.022	0.074	0.167 ± 0.051
	256–266	263		0.239 ± 0.063		0.283 ± 0.032	0.074	0.056 ± 0.031
	256–269	266		0.022 ± 0.022	0.030	0.010 ± 0.007		0.037 ± 0.026
	256–272	269		0.109 ± 0.046	0.010	0.096 ± 0.021		0.037 ± 0.026
	260–260	272	0.043	0.022 ± 0.022	0.081	0.035 ± 0.013	0.111	0.056 ± 0.031
	260–263	275	0.043	0.065 ± 0.036	0.010	0.046 ± 0.015		0.019 ± 0.018
	260–266		0.043		0.020			
	260–269				0.030		0.037	
	263–263		0.043		0.132			
	263–269		0.043		0.030			
	263–272		0.087		0.010		0.037	
	263–275				0.020			
	269–269		0.043		0.020			
	269–272				0.020		0.037	
	269–275		0.087		0.040			
	272–275				0.030		0.037	
	Ho/He		0.565 ± 0.103/0.743 ± 0.043		0.485 ± 0.050 **/0.719 ± 0.020		0.370 ± 0.093 */0.610 ± 0.067	
	N		23		99		27	

Table 1. *Cont.*

Locus	Genotype	Allele	Nosema-Negative		Nosema-Positive Low		Nosema-Positive High	
			Genotype	Allele	Genotype	Allele	Genotype	Allele
A024	92–92	92	0.500	0.712 ± 0.063	0.446	0.654 ± 0.030	0.313	0.578 ± 0.044
	92–100	96	0.231		0.177	0.008 ± 0.005	0.219	
	92–106	100	0.192	0.154 ± 0.050	0.238	0.181 ± 0.024	0.313	0.266 ± 0.039
	96–96	102			0.008	0.008 ± 0.005		
	100–100	106	0.038	0.135 ± 0.047	0.077	0.150 ± 0.022	0.156	0.156 ± 0.032
	100–102				0.015			
	100–106				0.015			
	106–106		0.038		0.023			
	Ho/He		0.423 ± 0.097/0.452 ± 0.070		0.446 ± 0.044/0.517 ± 0.029		0.531 ± 0.062/0.571 ± 0.032	
	N		26		130		64	
A007	104–108	104	0.185	0.093 ± 0.039	0.224	0.121 ± 0.021	0.364	0.182 ± 0.041
	104–113	108		0.815 ± 0.053	0.017	0.797 ± 0.026		0.818 ± 0.041
	108–108	113	0.704	0.093 ± 0.039	0.655	0.082 ± 0.018	0.636	
	108–113		0.037		0.060			
	113–113		0.074		0.043			
	Ho/He		0.222 ± 0.080/0.319 ± 0.075		0.302 ± 0.043/0.343 ± 0.036		0.364 ± 0.073/0.298 ± 0.052	
	N		27		116		44	
Ap049	120–120	120	0.036	0.161 ± 0.049	0.017	0.121 ± 0.021		0.057 ± 0.022
	120–127	127	0.250	0.714 ± 0.060	0.200	0.646 ± 0.031	0.113	0.745 ± 0.042
	120–130	130		0.054 ± 0.030	0.008	0.175 ± 0.025		0.085 ± 0.027
	127–127	139	0.536	0.071 ± 0.034	0.425	0.046 ± 0.014	0.585	0.085 ± 0.027
	127–130	152	0.036		0.192	0.013 ± 0.007	0.094	0.028 ± 0.016
	127–139		0.071		0.050		0.113	
	130–130		0.036		0.067		0.019	
	130–139						0.019	
	130–152				0.017		0.019	
	139–139		0.036		0.017		0.019	
	139–152				0.008			
	152–152						0.019	
	Ho/He		0.357 ± 0.091/0.456 ± 0.071		0.475 ± 0.046/0.535 ± 0.032		0.359 ± 0.066/0.426 ± 0.057	
	N		28		120		53	
SV185	253–253	253			0.009	0.022 ± 0.010		
	253–272	263		0.241 ± 0.058	0.027	0.313 ± 0.031		0.385 ± 0.050
	263–263	266	0.037	0.093 ± 0.039	0.134	0.094 ± 0.020	0.146	0.146 ± 0.036
	263–266	269		0.667 ± 0.064	0.045	0.549 ± 0.033	0.125	0.469 ± 0.051
	263–269	272	0.407		0.304	0.023 ± 0.010	0.354	
	263–272				0.009			
	266–266		0.074		0.045			
	266–269		0.037		0.054		0.167	
	269–269		0.444		0.366		0.208	
	269–272				0.009			
	Ho/He		0.444 ± 0.096/0.489 ± 0.061		0.446 ± 0.047 */0.591 ± 0.023		0.646 ± 0.069/0.611 ± 0.023	
	N		27		112		48	

N—the number of bee samples analyzed within each infection category; Ho—observed heterozygosity; He—expected heterozygosity according to Hardy–Weinberg equilibrium. In the table, the values of allele frequencies and parameters of heterozygosity with a standard error are given. Statistically significant differences in the observed heterozygosity from the expected heterozygosity are marked with (*) (* $p < 0.05$, ** $p < 0.001$).

All microsatellite loci studied were polymorphic: the minimum number of alleles was found for locus A007 (3 alleles), and the maximum number of alleles was for loci A113 and Ap243 (8 alleles); the average number of alleles per locus was 5 (Table 1).

Some studied loci differed in the variability in honey bees of different infectious categories (uninfected and *Nosema*-infected bees). For example, for the Ap243 locus, the frequency of the predominant allele "256" differs in bees of two *Nosema*-positive groups,

Nosema-positive low and *Nosema*-positive high bees (t = 2.30; $p < 0.05$). Interestingly, the frequency of the other allele ("263") of the Ap243 locus was also statistically significantly different between uninfected (*Nosema*-negative) and significantly *Nosema*-infected (*Nosema*-positive high) bees (t = 2.61; $p < 0.05$) and between two *Nosema*-positive groups (t = 5.10; $p < 0.001$). In addition, the frequency of alleles "177" and "181" of the AC117 locus differs between *Nosema*-negative and *Nosema*-positive high bees (t = 2.46; $p < 0.05$ and t = 3.09; $p < 0.05$, respectively), and between bees of two *Nosema*-positive groups (t = 2.27; $p < 0.05$ and t = 5.15; $p < 0.01$, respectively).

An assessment of the heterozygosity of most of the studied loci (except for loci A007 and SV185 in *Nosema*-positive high bees) revealed lower values of the observed heterozygosity (Ho) compared with the expected heterozygosity (He) (Table 1). A statistically significant level of differences between the values of the observed and expected heterozygosity is shown for the following loci: A113 (t = 3.82, $p < 0.001$ and t = 3.16, $p < 0.05$), Ap243 (t = 4.35, $p < 0.001$ and t = 2.09, $p < 0.05$) in *Nosema*-positive low and high bees, respectively; SV185 (t = 2.77, $p < 0.05$) in *Nosema*-positive low bees; AC117 (t > 3.69, $p < 0.001$) in all infection categories of bees.

Thus, the analysis of the variability of 23 microsatellite loci in the *A. m. mellifera* honey bees allowed us to identify the most promising loci for searching for associations of DNA markers with the *Nosema* disease.

3.2. Comparative Characteristics of the Genetic Diversity of A. m. mellifera Bees from Different Infectious Categories

The heterogeneity of bee groups differing in the degree of *Nosema* infection was evaluated. To do this, statistically significant differences between the bee groups in allele frequencies of microsatellite loci were determined.

For some microsatellite loci (AC117, A113, Ap243 and Ap049), differences in the distribution of allele frequencies (in total) between infectious categories of bees are shown. Mainly, differences were found between infected bees (*Nosema*-positive low and *Nosema*-positive high bees): locus AC117 ($\chi^2 = 44.61$, df = 7, $p < 0.01$); locus A113 ($\chi^2 = 12.76$, df = 5, $p < 0.05$); locus Ap243 ($\chi^2 = 19.77$, df = 7, $p < 0.01$) and locus Ap049 ($\chi^2 = 14.70$, df = 7, $p < 0.05$). In addition, for the AC117 locus, differences were revealed between uninfected and *Nosema*-positive high bees ($\chi^2 = 19.84$, df = 7, $p < 0.01$). No statistically significant differences were found between the groups of uninfected bees and *Nosema*-positive low bees.

For the locus AC117, the alleles "177", "181" and "185" make the largest contribution. The "177" allele is most often found in uninfected bees than *Nosema*-positive high bees ($\chi^2 = 9.59$, df = 1, $p < 0.01$). On the contrary, the "181" allele is more typical for *Nosema*-positive high bees than for uninfected ($\chi^2 = 8.66$, df = 1, $p < 0.01$) and *Nosema*-positive low ($\chi^2 = 26.56$, df = 1, $p < 0.01$) bees. For the "185" allele, differences were found between the two groups of infected bees ($\chi^2 = 14.17$, df = 1, $p < 0.01$).

For the locus A113, the allele "218" defines the differences between *Nosema*-negative and *Nosema*-positive high bees ($\chi^2 = 6.81$, df = 1, $p < 0.01$). In addition, significant differences were also found between *Nosema* infected (low and high) bees on the frequency of the alleles "218" ($\chi^2 = 6.12$, df = 1, $p < 0.05$) and "212" ($\chi^2 = 6.60$, df = 1, $p < 0.01$).

For the locus Ap243, the frequencies of two alleles "256" and "263" were significantly different in the groups of infected bees ($\chi^2 = 5.80$, df = 1, $p < 0.05$ and $\chi^2 = 12.45$, df = 1, $p < 0.01$, respectively). Allele "256" prevailed in *Nosema* infected high bees, while allele "263" was more common in *Nosema* infected low bees (Table 1). The allele "263" also determined differences between groups of uninfected and *Nosema*-positive high bees ($\chi^2 = 6.98$, df = 1, $p < 0.01$).

For the locus Ap049, alleles "127" and "130" determine the differences between groups of infected bees ($\chi^2 = 4.00$, df = 1, $p < 0.05$ and $\chi^2 = 5.12$, df = 1, $p < 0.05$, respectively), while allele "120" was between uninfected and *Nosema*-positive high bees ($\chi^2 = 4.69$, df = 1, $p < 0.05$).

Despite the fact that no significant differences were found in the distribution of all alleles (in total) of loci A007, A024 and SV185, there existed differences for some alleles of these loci. For example, allele "269" of the SV185 locus determined the differences between groups of *Nosema*-uninfected and *Nosema*-positive high bees ($\chi^2 = 5.65$, df = 1, $p < 0.05$).

3.3. Assessment of Associations of Genetic Markers with Nosema Infection/Resistance in the Dark Forest Bee A. m. mellifera

In order to search for alleles, possibly associated with *Nosema* disease in the dark forest bees (*A. m. mellifera*), the odds ratio, OR, was calculated (Table 2).

For loci AC117, A113, Ap243, A024, A007, Ap049 and SV185, statistically significant differences in allele and/or genotype frequencies between the compared groups (infection categories) were shown. However, according to OR calculations, only some genetic variants of these loci showed associations with *Nosema* infestation. Based on the calculated OR, it can be concluded that alleles "177" of the AC117 locus, "263" of the Ap243 locus and "269" of the SV185 locus have protective properties (that is, they reduce the risk of *Nosema* infection).

Table 2. Comparative analysis of the frequency of potentially significant genetic variants in the formation of nosemosis resistance in the dark forest bee *A. m. mellifera*.

Locus	Compared Alleles/Genotypes	Parameters	Compared Honey Bee Groups		
			Nosema-Negative–*Nosema*-Positive Low	*Nosema*-Negative–*Nosema*-Positive High	*Nosema*-Positive Low–*Nosema*-Positive High
AC117	Allele 177 vs. others	OR	0.46	**0.16**	0.35
		95% CI	0.18–1.24	**0.04–0.58**	0.11–1.13
		χ^2/p	2.15/0.17	**7.76/0.005**	2.92/0.09
	Homo- and heterozygous genotypes with an allele 177 vs. others	OR	0.47	**0.16**	0.35
		95% CI	0.16–1.43	**0.04–0.64**	0.11–1.16
		χ^2/p	1.48/0.22	**6.25/0.01**	2.69/0.10
A113	Allele 218 vs. others	OR	1.38	**2.38**	**1.72**
		95% CI	0.74–2.57	**1.18–4.80**	**1.04–1.39**
		χ^2/p	0.90/0.34	**6.12/0.01**	**4.91/0.03**
Ap243	Allele 263 vs. others	OR	1.25	**0.21**	**0.17**
		95% CI	0.27–2.83	**0.06–0.75**	**0.06–0.53**
		χ^2/p	0.17/0.68	**5.51/0.02**	**10.99/0.0009**
	Homo- and heterozygous genotypes with an allele 263 vs. others	OR	1.00	**0.18**	**0.19**
		95% CI	0.37–2.74	**0.05–0.73**	**0.06–0.51**
		χ^2/p	0.05/0.82	**5.19/0.02**	**8.22/0.004**
A024	Allele 92 vs. others	OR	0.77	0.56	0.73
		95% CI	0.38–1.53	0.26–1.17	0.46–1.15
		χ^2/p	0.41/0.52	2.25/0.13	1.80/0.18
	Allele 100 vs. others	OR	1.21	1.99	1.64
		95% CI	0.51–3.00	0.80–5.10	0.96–2.16
		χ^2/p	0.07/0.79	2.00/0.16	3.24/0.07

Table 2. *Cont.*

Locus	Compared Alleles/Genotypes	Parameters	Compared Honey Bee Groups		
			Nosema-Negative–*Nosema*-Positive Low	*Nosema*-Negative–*Nosema*-Positive High	*Nosema*-Positive Low–*Nosema*-Positive High
A007	Allele 104 vs. others	OR	1.35	2.18	1.62
		95% CI	0.46–4.19	0.69–7.32	0.78–3.32
		χ^2/p	0.12/0.73	1.47/0.23	1.53/0.22
Ap049	Allele 120 vs. others	OR	0.72	0.31	0.44
		95% CI	0.30–1.76	0.09–1.04	0.16–1.15
		χ^2/p	0.34/0.56	3.57/0.06	2.67/0.10
SV185	Allele 269 vs. others	OR	0.61	**0.44**	0.72
		95% CI	0.31–1.18	**0.21–0.93**	0.44–1.20
		χ^2/p	2.00/0.16	**4.68/0.03**	1.43/0.23

OR—odds ratio, 95% CI—limits of the 95% confidence interval, χ^2/p—χ^2 test and its level of significance, df = 1. Alleles for which the level of statistical significance on OR has been reached are indicated in bold.

It is worth pointing out that the frequency of homo- and heterozygous genotypes with an allele "177" of the AC117 locus decreased in the sequence: uninfected bees (25.9% of *Nosema*-negative individuals) were weakly infected (14.1% of *Nosema*-positive low bees) and significantly infected (4.9% of *Nosema*-positive high individuals) (Table 1). Between the extreme compared groups of bees (*Nosema*-negative and *Nosema*-positive high) differences in the frequency of individuals with genotypes having this allele reach the level of statistical significance ($\chi^2 = 16.61$, df = 2, $p < 0.01$). For the Ap243 locus, the frequency of genotypes with the "263" allele in *Nosema*-positive high bees (11.1%) differed significantly from both uninfected bees (43.3%; $\chi^2 = 13.45$, df = 6, $p < 0.05$) and *Nosema*-positive low individuals (43.5%; $\chi^2 = 18.86$, df = 7, $p < 0.01$) (Table 1). For the SV185 locus, there were no differences between the compared groups at the genotype level ($p > 0.05$). In addition, the results obtained suggest that the "218" allele of the A113 locus is associated with an increased risk of nosemosis in *A. m. mellifera* bees (no differences were recorded at the genotype level).

4. Discussion

The genetic diversity of the *A. m. mellifera* bees, differing in the degree of *Nosema* infection, was investigated, and a search for associations between genetic variants of microsatellite loci and *Nosema* disease was carried out. The present results show that some alleles of microsatellite loci are associated with nosemosis in the *A. m. mellifera* bees living in the Siberian region. For example, alleles "177" of the AC117 locus, "263" of the Ap243 locus and "269" of the SV185 locus reduce the risk of *Nosema* infection.

This study found associations of genetic markers with *Nosema* infection/disease resistance in honey bees and was an exploratory study. Therefore, it is necessary to understand whether the results obtained were random or reflected some general biological regularities. Possible solutions to this issue are expanding the bee samples and/or analyzing different bee subspecies and studying them in terms of the importance of microsatellite loci in determining resistance to nosemosis.

Previously, the associations of some microsatellite loci in the *Apis mellifera carpathica* bees were analyzed by us together with T. Kireeva (Tomsk State University, our unpublished data) [78]. Interestingly, for the Ap243 locus, statistically significant differences were also shown between uninfected and *Nosema*-infected *A. m. carpathica* bees ($\chi^2 = 22.93$, df = 7, $p < 0.01$). Alleles "253" and "256" determine the differences between groups of uninfected and *Nosema*-infected bees ($\chi^2 = 9.69$, df = 1, $p < 0.01$ and $\chi^2 = 7.03$, df = 1, $p < 0.01$, respectively). Based on the calculated OR, it can be concluded that allele "256" of the Ap243 locus has protective properties (OR = 0.29, 95% CI—0.10–0.84, χ^2/p—6.87/0.0088,

χ^2-Yeats/p—5.57/0.0182), while allele "253", on the contrary, increased the risk of *Nosema* infection (OR = 3.57, 95% CI—1.53–8.40, χ^2/p—11.10/0.00086, χ^2-Yeats/p—9.77/0.0018).

Thus, among the investigated microsatellite loci, the Ap243 locus located on chromosome 1 (group 1.1) was shared for two bee subspecies (*A. m. mellifera* and *A. m. carpathica*) and is probably associated with the incidence/resistance to nosemosis. However, in these subspecies, different alleles associated with *Nosema* disease were identified. For *A. m. mellifera* bees, allele "263" probably reduced the risk of infection with nosemosis (protective role), and statistically significant differences were also found at the genotype level in bees of different groups (*Nosema*-positive high bees differed from uninfected and *Nosema*-positive low bees). In *A. m. carpathica* bees, statistically significant differences were found for two alleles ("256" and "253") between uninfected bees and *Nosema*-infected individuals, and the allele "256" is likely to have a protective meaning, while the allele "253", on the contrary, is associated with a disease.

In addition, the A024 locus is also of interest for studying its association with nosemosis. Despite the fact that in the *A. m. mellifera* bees, no allele of the A024 locus showed association with *Nosema* disease (Table 2), in the *A. m. carpathica* bees, the allele "90" probably determines the resistance to nosemosis (OR = 0.09, 95% CI—0.04–0.023, χ^2/p—39.94/0.0000000, χ^2-Yeats/p—37.93/0.0000000).

It should be noted that for two bee subspecies, various microsatellite loci and alleles associated with the *Nosema* incidence/resistance have been identified, which may be determined by the different resistance of bee subspecies to nosemosis and/or by different habitats of bees (geographic, natural, climatic and nutritional conditions) [58,79–82]. Perhaps, in addition to the subspecies-specific features to *Nosema* resistance, the revealed differences can be determined by the structure of the chromosomal region where the QTL is located. For example, not this locus itself, but another, which is in linkage disequilibrium with it, may be involved in the determination of resistance to nosemosis, i.e., different alleles are in the same linkage group with a favorable variant in different bee subspecies.

The few hygienic behavior-related studies focused on finding quantitative trait loci controlling *Varroa* resistance hygienic behavior of honey bees have also identified different chromosomal regions related to hygienic behavior and reduced mite reproduction [14,20–22,31]. QTL studies of *Varroa* resistance behavior have identified over 20 suggestive chromosome regions associated with linkage groups 2, 3, 4, 5, 6, 7, 9, 10, 13, 15, 16 and 22. For example, using RAPD markers, Lapidge et al. (2002) found seven suggestive QTLs controlling hygienic behavior of honey bees [22]. Using microsatellite loci, Oxley et al. (2010) identified three significant and three suggestive QTLs that influence a honey bee worker's propensity to engage in hygienic behavior [31] whereas Behrens et al. (2011) identified three QTLs controlling reduced mite reproduction in the Swiss *Varroa* mite tolerant honey bee lineage [20]. In the QTL study presented by Spötter et al., six SNPs showed significant genome-wide associations with hygienic behavior against *Varroa* at the genotype level [14].

It is worth pointing out that the presented studies did not identify the same (common) QTLs associated with hygienic behavior of honey bees. For example, Oxley et al. (2010) and Behrens et al. (2011) identified two QTLs for the trait in different regions of chromosome 9 [20,31]. Additionally, Tsuruda et al. (2012) also reported one major QTL on chromosome 9, but in a different region, for hygienic behavior against *Varroa* using a small-scale SNP-Chip [21]. Thus, the identified genomic regions related to hygienic behavior are different, which can be explained by different bee materials (freeze-killed brood, brood and worker bees), different research methods used and DNA markers analyzed (RAPD markers, microsatellite loci or SNP), different genetic maps (map based on RAPD or microsatellite markers). Finally, the use of different bee subspecies can significantly affect the research outcome. For example, in QTL studies presented by Oxley et al. (2010) and Spötter et al. (2016), QTLs were identified in the same chromosomes (chromosomes 2 and 5), but at different ends of the chromosome [14,31]. It is assumed that different bee material used in these studies contributed to the differences in the genomic regions identified [14]. On the one hand, a freeze-killed brood instead of brood that was artificially infested with *Varroa* was used in Oxley et al.'s research [31]. On the other hand, two different bee subspecies (*Apis mellifera ligustica* and *Apis mellifera carnica*) have been investigated [14].

A comparison between the QTLs involved in *N. ceranae* infection tolerance according to Huang et al.'s study [35] and our own trait-associated regions found no agreement. So, four QTLs located on chromosomes 3, 10, 6 and 14 were significantly associated with low *Nosema* spore load, explaining 20.4% of total spore load variance in the selected *Nosema*-resistant Danish honey bee strain. The significant QTL on chromosome 14 explains 7.7% of the total variance and may be responsible for the resistance to nosemosis in the selected Danish honey bee. A candidate gene *Aubergine* (*Aub*) within this QTL region was significantly more overexpressed in drones with a low spore load than in those with a high spore load [35].

From the data presented, in the dark forest bee, microsatellite loci Ap243 (chromosome 1), SV185 (chromosome 5) and AC117 (chromosome 12) are associated with resistance to nosemosis. It is interesting to note that in the chromosome region 1.1, the microsatellite locus Ap243 and the honey bee microRNA (ame-miR-2b) are located. As shown, host microRNAs respond to infection by the parasite *N. ceranae* [36]. In honey bees, 17 miRNAs were differentially expressed during *N. ceranae* infection, which may target over 400 predicted genes for ion binding, signaling, the nucleus, transmembrane transport and DNA binding. MicroRNA ame-miR-2b is particularly interesting because 11 out of 27 enzymes were significantly correlated with its expression level [36]. In addition, in the same region on chromosome 1, a suggestive QTL associated with the performance of *Varroa* sensitive hygiene was identified [21]. This locus contains more than candidate 30 genes [21], including puromycin-sensitive aminopeptidase involved in proteolytic events essential for cell growth and viability [83], selenoprotein F located in the endoplasmic reticulum and regulated by cell stress conditions [84], transcription and splicing factors and other genes. Since Microsporidian *Nosema* species are intracellular parasite [37], of particular interest is the cell wall integrity and stress response component 1. In *Schizosaccharomyces pombe*, the homologous gene *wsc1* is responsible for such biological processes as a cell surface receptor signaling pathway, intracellular signal transduction and regulation of cell wall organization or biogenesis [85].

Despite the fact that numerous QTLs associated with the disease resistance of honey bee have been identified, it is assumed that the variations in this trait is controlled by a small number of loci. For example, a strong genetic component is involved in the control of hygienic behavior, although it may also be influenced by environmental factors to some extent [14]. Therefore, the identification of the gene variants that are responsible for disease resistance in honey bees, and then breeding resistant bee colonies is one promising approach in the fight against bee disease.

5. Conclusions

The present study examined the associations between several variants of microsatellite loci and *Nosema* disease. In the dark forest bee, genetic markers promising for the assessment of nosemosis resistance, such as the allele "177" of the locus AC117, the allele "263" of the locus Ap243, the allele "269" of the locus SV185 have been identified. At the same time, some issues, for example, differences in the spectrum of loci and/or alleles that determine resistance to nosemosis in different bee subspecies/breeds, remained unresolved. In this regard, additional research both for the same bee subspecies/breeds bred in other regions, and for other bee subspecies/breeds is needed. However, already at this stage, these markers can be used to predict the risk of developing nosemosis, but with the obligatory consideration of the specificity of diagnostic markers for different bee subspecies/breeds.

Funding: This research received no external funding.

Institutional Review Board Statement: Not applicable.

Informed Consent Statement: Not applicable.

Data Availability Statement: Data is contained within the article or Supplementary Material. The data presented in this study are available in [Ostroverkhova, N.V.; Kucher, A.N.; Konusova, O.L.; Kireeva, T.N.; Rosseykina, S.A.; Yartsev, V.V.; Pogorelov, Y.L. Genetic diversity of honey bee *Apis mellifera* in Siberia. In *Phylogenetics of Bees*; Ilyasov, R.A., Kwon, H.W., Eds.; CRC Press: Boca Raton, FL, USA, 2020; pp. 97–126 and Ostroverkhova, N.V.; Kucher, A.N.; Konusova, O.L.; Gushchina, E.S.; Yartsev, V.V.; Pogorelov, Y.L. Dark-Colored Forest Bee *Apis mellifera* in Siberia, Russia: Current State and Conservation of Populations. In *Selected Studies in Biodiversity*; IntechOpen: London, UK, 2018; pp. 157–180].

Acknowledgments: The work would have been impossible without the help and assistance of Tatyana N. Kireeva, a researcher of the Tomsk State University.

Conflicts of Interest: The author declares no conflict of interest.

References

1. Chauzat, M.P.; Jacques, A.; Laurent, M.; Bougeard, S.; Hendrikx, P.; Ribière-Chabert, M.; EPILOBEE Consortium. Risk indi-cators affecting honey bee colony survival in Europe: One year of surveillance. *Apidologie* **2016**, *47*, 348–378. [CrossRef]
2. Vanengelsdorp, D.; Meixner, M.D. A historical review of managed honey bee populations in Europe and the United States and the factors that may affect them. *J. Invertebr. Pathol.* **2010**, *103*, S80–S95. [CrossRef] [PubMed]
3. Gomes, T.; Feás, X.; Iglesias, A.; Estevinho, L.M. Study of Organic Honey from the Northeast of Portugal. *Molecules* **2011**, *16*, 5374–5386. [CrossRef] [PubMed]
4. Higes, M.; Meana, A.; Bartolomé, C.; Botías, C.; Martín-Hernández, R. *Nosema ceranae* (Microsporidia), a controversial 21st century honey bee pathogen: *N. ceranae* an emergent pathogen for beekeeping. *Environ. Microbiol. Rep.* **2013**, *5*, 17–29. [CrossRef] [PubMed]
5. Goulson, D.; Nicholls, E.; Botías, C.; Rotheray, E.L. Bee declines driven by combined stress from parasites, pesticides, and lack of flowers. *Science* **2015**, *347*, 1255957. [CrossRef] [PubMed]
6. Bilodeau, L.; Sylvester, A.; Danka, R.; Rinderer, T. Comparison of microsatellite DNA diversity among commercial queen breeder stocks of Italian honey bees in the United States and Italy. *J. Apic. Res.* **2008**, *47*, 93–98. [CrossRef]
7. Bilodeau, L.; Rinderer, T.E.; Sylvester, H.A.; Holloway, B.; Oldroyd, B.P. Patterns of *Apis mellifera* infestation by *Nosema ceranae* support the parasite hypothesis for the evolution of extreme polyandry in eusocial insects. *Apidologie* **2012**, *43*, 539–548. [CrossRef]
8. Bilodeau, L.; Villa, J.D.; Holloway, B.; Danka, R.G.; Rinderer, T.E. Molecular genetic analysis of tracheal mite resistance in honey bees. *J. Apic. Res.* **2015**, *54*, 1–7. [CrossRef]
9. Büchler, R.; Berg, S.; Le Conte, Y. Breeding for resistance to *Varroa destructor* in Europe. *Apidologie* **2010**, *41*, 393–408. [CrossRef]
10. Spötter, A.; Gupta, P.; Nürnberg, G.; Reinsch, N.; Bienefeld, K. Development of a 44K SNP assay focussing on the analysis of a *Varroa*-specific defence behaviour in honey bees (*Apis mellifera carnica*). *Mol. Ecol. Resour.* **2011**, *12*, 323–332. [CrossRef]
11. Bixby, M.; Baylis, K.; Hoover, S.E.; Currie, R.W.; Melathopoulos, A.P.; Pernal, S.F.; Foster, L.J.; Guarna, M.M. A Bio-Economic Case Study of Canadian Honey Bee (Hymenoptera: Apidae) Colonies: Marker-Assisted Selection (MAS) in Queen Breeding Affects Beekeeper Profits. *J. Econ. Èntomol.* **2017**, *110*, 816–825. [CrossRef] [PubMed]
12. Yunusbaev, U.B.; Kaskinova, M.D.; Ilyasov, R.A.; Gaifullina, L.R.; Saltykova, E.S.; Nikolenko, A.G. The Role of Whole-Genome Studies in the Investigation of Honey Bee Biology. *Russ. J. Genet.* **2019**, *55*, 815–824. [CrossRef]
13. Maucourt, S.; Fortin, F.; Robert, C.; Giovenazzo, P. Genetic parameters of honey bee colonies traits in a Canadian Selection Program. *Insects* **2020**, *11*, 587. [CrossRef] [PubMed]
14. Spötter, A.; Gupta, P.; Mayer, M.; Reinsch, N.; Bienefeld, K. Genome-Wide Association Study of a *Varroa*-Specific Defense Behavior in Honeybees (*Apis mellifera*). *J. Hered.* **2016**, *107*, 220–227. [CrossRef] [PubMed]
15. Broeckx, B.J.G.; De Smet, L.; Blacquière, T.; Maebe, K.; Khalenkow, M.; Van Poucke, M.; Dahle, B.; Neumann, P.; Nguyen, K.B.; Smagghe, G.; et al. Honey bee predisposition of resistance to ubiquitous mite infestations. *Sci. Rep.* **2019**, *9*, 7794. [CrossRef]
16. Graham, A.M.; Munday, M.D.; Kaftanoglu, O.; Page, R.E., Jr.; Amdam, G.V.; Rueppell, O. Support for the reproductive ground plan hypothesis of social evolution and major QTL for ovary traits of Africanized worker honey bees (*Apis mellifera* L.). *BMC Evol. Biol.* **2011**, *11*, 95. [CrossRef]
17. Rueppell, O.; Metheny, J.D.; Linksvayer, T.; Fondrk, M.K.; Page, R.E., Jr.; Amdam, G.V. Genetic architecture of ovary size and asymmetry in European honeybee workers. *Heredity* **2011**, *106*, 894–903. [CrossRef]
18. Holloway, B.; Sylvester, H.A.; Bourgeois, L.; Rinderer, T.E. Association of single nucleotide polymorphisms to resistance to chalkbrood in *Apis mellifera*. *J. Apic. Res.* **2012**, *51*, 154–163. [CrossRef]
19. Holloway, B.; Tarver, M.R.; Rinderer, T.E. Fine mapping identifies significantly associating markers for resistance to the honey bee brood fungal disease, Chalkbrood. *J. Apic. Res.* **2013**, *52*, 134–140. [CrossRef]
20. Behrens, D.; Huang, Q.; Geßner, C.; Rosenkranz, P.; Frey, E.; Locke, B.; Moritz, R.F.A.; Kraus, F.B. Three QTL in the honey bee *Apis mellifera* L. suppress reproduction of the parasitic mite *Varroa destructor*. *Ecol. Evol.* **2011**, *1*, 451–458. [CrossRef]
21. Tsuruda, J.M.; Harris, J.W.; Bourgeois, L.; Danka, R.G.; Hunt, G.J. High-Resolution Linkage Analyses to Identify Genes That Influence *Varroa* Sensitive Hygiene Behavior in Honey Bees. *PLoS ONE* **2012**, *7*, e48276. [CrossRef]
22. Lapidge, K.L.; Oldroyd, B.P.; Spivak, M. Seven suggestive quantitative trait loci influence hygienic behavior of honey bees. *Naturwissenschaften* **2002**, *89*, 565–568. [CrossRef] [PubMed]

23. Rüppell, O.; Pankiw, T.; Page, R.E., Jr. Pleiotropy, epistasis and new QTL: The genetic architecture of honey bee foraging behavior. *J. Hered.* **2004**, *95*, 481–491. [CrossRef] [PubMed]

24. Shorter, J.R.; Arechavaleta-Velasco, M.; Robles-Rios, C.; Hunt, G.J. A Genetic Analysis of the Stinging and Guarding Behaviors of the Honey Bee. *Behav. Genet.* **2012**, *42*, 663–674. [CrossRef] [PubMed]

25. Wilson-Rich, N.; Spivak, M.; Fefferman, N.H.; Starks, P.T. Genetic, individual, and group facilitation of disease resistance in insect societies. *Annu. Rev. Entomol.* **2007**, *54*, 405–423. [CrossRef] [PubMed]

26. Locke, B.; Forsgren, E.; De Miranda, J.R. Increased tolerance and resistance to virus infections: A possible factor in the survival of *Varroa destructor* resistant honey bees (*Apis mellifera*). *PLoS ONE* **2014**, *9*, e99998. [CrossRef] [PubMed]

27. Al Toufailia, H.; Evison, S.E.F.; Hughes, W.O.H.; Ratnieks, F.L.W. Both hygienic and non-hygienic honeybee, *Apis mellifera*, colonies remove dead and diseased larvae from open brood cells. *Philos. Trans. R. Soc. B Biol. Sci.* **2018**, *373*, 20170201. [CrossRef]

28. Rosenkranz, P.; Aumeier, P.; Ziegelmann, B. Biology and control of *Varroa destructor*. *J. Invertebr. Pathol.* **2010**, *103*, S96–S119. [CrossRef]

29. Nazzi, F.; Brown, S.P.; Annoscia, D.; Del Piccolo, F.; Di Prisco, G.; Varricchio, P.; Della Vedova, G.; Cattonaro, F.; Caprio, E.; Pennacchio, F. Synergistic parasite-pathogen interactions mediated by host immunity can drive the collapse of honeybee colonies. *PLoS Pathog.* **2012**, *8*, e1002735. [CrossRef]

30. Wagoner, K.M.; Spivak, M.; Rueppell, O. Brood Affects Hygienic Behavior in the Honey Bee (Hymenoptera: Apidae). *J. Econ. Entomol.* **2018**, *111*, 2520–2530. [CrossRef]

31. Oxley, P.R.; Spivak, M.; Oldroyd, B.P. Six quantitative trait loci influence task thresholds for hygienic behaviour in honeybees (*Apis mellifera*). *Mol. Ecol.* **2010**, *19*, 1452–1461. [CrossRef] [PubMed]

32. Higes, M.; Martín-Hernández, R.; Garrido-Bailon, E.; Gonzalez-Porto, A.V.; Garcia-Palencia, P.; Meana, A.; Del Nozal, M.J.; Mayo, R.; Bernal, J.L. Honeybee colony collapse due to *Nosema ceranae* in professional apiaries. *Environ. Microbiol. Rep.* **2009**, *1*, 110–113. [CrossRef] [PubMed]

33. Vanengelsdorp, D.; Evans, J.D.; Saegerman, C.; Mullin, C.; Haubruge, E.; Nguyen, B.K.; Frazier, M.; Frazier, J.; Cox-Foster, D.; Chen, Y.; et al. Colony Collapse Disorder: A Descriptive Study. *PLoS ONE* **2009**, *4*, e6481. [CrossRef] [PubMed]

34. Goblirsch, M. *Nosema ceranae* disease of the honey bee (*Apis mellifera*). *Apidologie* **2017**, *49*, 131–150. [CrossRef]

35. Huang, Q.; Kryger, P.; Le Conte, Y.; Lattorff, H.M.G.; Kraus, F.B.; Moritz, R.F.A.; Lattorff, M. Four quantitative trait loci associated with low *Nosema ceranae* (Microsporidia) spore load in the honeybee *Apis mellifera*. *Apidologie* **2013**, *45*, 248–256. [CrossRef]

36. Huang, Q.; Chen, Y.; Wang, R.W.; Schwarz, R.S.; Evans, J.D. Honey bee microRNAs respond to infection by the microsporidian parasite *Nosema ceranae*. *Sci. Rep.* **2015**, *5*, 17494. [CrossRef]

37. Fries, I. *Nosema apis*—A Parasite in the Honey Bee Colony. *Bee World* **1993**, *74*, 5–19. [CrossRef]

38. Higes, M.; Martín, R.; Meana, A. *Nosema ceranae*, a new microsporidian parasite in honeybees in Europe. *J. Invertebr. Pathol.* **2006**, *92*, 93–95. [CrossRef]

39. Fries, I.; Chauzat, M.-P.; Chen, Y.P.; Doublet, V.; Genersch, E.; Gisder, S.; Higes, M.; McMahon, D.P.; Martín-Hernández, R.; Natsopoulou, M.; et al. Standard methods for *Nosema* research. *J. Apic. Res.* **2013**, *52*, 1–28. [CrossRef]

40. Zander, E. Tierische Parasiten als Krankheitserreger bei der Biene. *Münchener Bienenztg.* **1909**, *31*, 196–204.

41. Klee, J.; Besana, A.M.; Genersch, E.; Gisder, S.; Nanetti, A.; Tam, D.Q.; Chinh, T.X.; Puerta, F.; Ruz, J.M.; Kryger, P.; et al. Widespread dispersal of the microsporidian *Nosema ceranae*, an emergent pathogen of the western honey bee, *Apis mellifera*. *J. Invertebr. Pathol.* **2007**, *96*, 1–10. [CrossRef] [PubMed]

42. Martín-Hernández, R.; Meana, A.; Prieto, L.; Salvador, A.M.; Garrido-Bailoón, E.; Higes, M. Outcome of Colonization of *Apis mellifera* by *Nosema ceranae*. *Appl. Environ. Microbiol.* **2007**, *73*, 6331–6338. [CrossRef] [PubMed]

43. Chen, Y.; Evans, J.D.; Smith, I.B.; Pettis, J.S. *Nosema ceranae* is a long-present and wide-spread microsporidian infection of the European honey bee (*Apis mellifera*) in the United States. *J. Invertebr. Pathol.* **2008**, *97*, 186–188. [CrossRef] [PubMed]

44. Fries, I.; Feng, F.; Da Silva, A.; Slemenda, S.B.; Pieniazek, N.J. *Nosema ceranae* n. sp. (Microspora, Nosematidae), morphological and molecular characterization of a microsporidian parasite of the Asian honey bee *Apis cerana* (Hymenoptera, Apidae). *Eur. J. Protistol.* **1996**, *32*, 356–365. [CrossRef]

45. Fries, I.; Martin, R.; Meana, A.; García-Palencia, P.; Higes, M. Natural infections of *Nosema ceranae* in European honey bees. *J. Apic. Res.* **2006**, *47*, 230–233. [CrossRef]

46. Fries, I. *Nosema ceranae* in European honey bees (*Apis mellifera*). *J. Invertebr. Pathol.* **2010**, *103*, S73–S79. [CrossRef]

47. Williams, G.R.; Shafer, A.B.; Rogers, R.E.; Shutler, D.; Stewart, D.T. First detection of *Nosema ceranae*, a microsporidian parasite of European honey bees (*Apis mellifera*), in Canada and central USA. *J. Invertebr. Pathol.* **2008**, *97*, 189–192. [CrossRef]

48. Giersch, T.; Berg, T.; Galea, F.; Hornitzky, M. *Nosema ceranae* infects honey bees (*Apis mellifera*) and contaminates honey in Australia. *Apidologie* **2009**, *40*, 117–123. [CrossRef]

49. Chen, Y.P.; Huang, Z.Y. *Nosema ceranae*, a newly identified pathogen of *Apis mellifera* in the USA and Asia. *Apidologie* **2010**, *41*, 364–374. [CrossRef]

50. Gisder, S.; Schüler, V.; Horchler, L.L.; Groth, D.; Genersch, E. Long-term temporal trends of *Nosema* spp. infection prevalence in northeast Germany: Continuous spread of *Nosema ceranae*, an emerging pathogen of honey bees (*Apis mellifera*), but no general replacement of *Nosema apis*. *Front. Cell. Infect. Microbiol.* **2017**, *7*, 301. [CrossRef]

51. Ostroverkhova, N.; Kucher, A.; Golubeva, E.; Rosseykina, S.; Konusova, O. Study of *Nosema* spp. in the Tomsk region, Siberia: Co-infection is widespread in honeybee colonies. *Far East. Entomol.* **2019**, *378*, 12–22. [CrossRef]

52. Ostroverkhova, N.V. Prevalence of *Nosema ceranae* (Microsporidia) in the *Apis mellifera mellifera* bee colonies from long time isolated apiaries of Siberia. *Far East. Èntomol.* **2020**, *407*, 8–20. [CrossRef]

53. Ostroverkhova, N.V.; Konusova, O.L.; Kucher, A.N.; Sharakhov, I.V. A comprehensive characterization of the honeybees in Siberia (Russia). In *Beekeeping and Bee Conservation—Advances in Research*; Chambo, D.E., Ed.; InTech: Rijeka, Croatia, 2016; pp. 1–37.

54. Ostroverkhova, N.V.; Konusova, O.L.; Kucher, A.N.; Kireeva, T.N.; Rosseykina, S.A. Prevalence of the Microsporidian *Nosema* spp. in Honey Bee Populations (*Apis mellifera*) in Some Ecological Regions of North Asia. *Vet. Sci.* **2020**, *7*, 111. [CrossRef]

55. Higes, M.; Martín-Hernández, R.; Martínez-Salvador, A.; Garrido-Bailón, E.; González-Porto, A.V.; Meana, A.; Bernal, J.L.; Del Nozal, M.J. A preliminary study of the epidemiological factors related to honey bee colony loss in Spain. *Environ. Microbiol. Rep.* **2010**, *2*, 243–250. [CrossRef]

56. Bacandritsos, N.; Granato, A.; Budge, G.; Papanastasiou, I.; Roinioti, E.; Caldon, M.; Falcaro, C.; Gallina, A.; Mutinelli, F. Sudden deaths and colony population decline in Greek honey bee colonies. *J. Invertebr. Pathol.* **2010**, *105*, 335–340. [CrossRef]

57. Soroker, V.; Hetzroni, A.; Yakobson, B.; David, D.; David, A.; Voet, H.; Slabezki, Y.; Efrat, H.; Levski, S.; Kamer, Y.; et al. Evaluation of colony losses in Israel in relation to the incidence of pathogens and pests. *Apidologie* **2011**, *42*, 192–199. [CrossRef]

58. Martín-Hernández, R.; Bartolomé, C.; Chejanovsky, N.; Le Conte, Y.; Dalmon, A.; Dussaubat, C.; García-Palencia, P.; Meana, A.; Pinto, M.A.; Soroker, V.; et al. *Nosema ceranae* in *Apis mellifera*: A 12 years post-detection perspective. *Environ. Microbiol.* **2018**, *20*, 1302–1329. [CrossRef]

59. Traynor, K. Bee breeding around the world. *Am. Bee J.* **2008**, *148*, 135–139.

60. Huang, Q.; Kryger, P.; Le Conte, Y.; Moritz, R.F.A. Survival and immune response of drones of a Nosemosis tolerant honey bee strain towards *N. ceranae* infections. *J. Invertebr. Pathol.* **2012**, *109*, 297–302. [CrossRef]

61. Ellis, J. Stocks of bees in the United State. *Am. Bee J.* **2015**, *1*, 141–148.

62. Masterman, R.; Ross, R.; Mesce, K.; Spivak, M. Olfactory and behavioral response thresholds to odors of diseased brood differ between hygienic and non-hygienic honey bees (*Apis mellifera* L.). *J. Comp. Physiol. A* **2001**, *187*, 441–452.

63. Gerdts, J.; Dewar, R.L.; Simone-Finstrom, M.D.; Edwards, T.; Angove, M. Hygienic behaviour selection via freeze-killed honey bee brood not associated with chalkbrood resistance in eastern Australia. *PLoS ONE* **2018**, *13*, e0203969. [CrossRef]

64. Leclercq, G.; Francis, F.; Gengler, N.; Blacquière, T. Bioassays to Quantify Hygienic Behavior in Honey Bee (*Apis mellifera* L.) Colonies: A Review. *J. Apic. Res.* **2018**, *57*, 663–673. [CrossRef]

65. Guichard, M.; Neuditschko, M.; Soland, G.; Fried, P.; Grandjean, M.; Gerster, S.; Dainat, B.; Bijma, P.; Brascamp, E.W. Estimates of genetic parameters for production, behaviour, and health traits in two Swiss honey bee populations. *Apidologie* **2020**, 1–16. [CrossRef]

66. Rothenbuhler, W.C. Behavior genetics of nest cleaning in honey bees. IV. Responses of F1 and backcross generations to disease-killed brood. *Am. Zool.* **1964**, *4*, 111–123. [CrossRef]

67. Gilliam, M.; Taber, S.; Richardson, G.V. Hygienic behavior of honey bees in relation to chalk brood disease. *Apidologie* **1983**, *14*, 29–39. [CrossRef]

68. Spivak, M.; Gilliam, M. Facultative expression of hygienic behavior of honey bees in relation to disease resistance. *J. Apic. Res.* **1993**, *32*, 147–157. [CrossRef]

69. Arechavaleta-Velasco, M.E.; Guzman-Novoa, E. Relative effect of four characteristics that restrain the population growth of the mite *Varroa destructor* in honey bee (*Apis mellifera*) colonies. *Apidologie* **2001**, *32*, 157–174. [CrossRef]

70. Ibrahim, A.; Spivak, M. The relationship between hygienic behavior and suppression of mite reproduction as honey bee (*Apis mellifera*) mechanisms of resistance to *Varroa destructor*. *Apidologie* **2005**, *37*, 31–40. [CrossRef]

71. Locke, B. Natural *Varroa* mite-surviving *Apis mellifera* honeybee populations. *Apidologie* **2016**, *47*, 467–482. [CrossRef]

72. Valizadeh, P.; Guzman-Novoa, E.; Goodwin, P.H. Effect of Immune Inducers on *Nosema ceranae* Multiplication and Their Impact on Honey Bee (*Apis mellifera* L.) Survivorship and Behaviors. *Insects* **2020**, *11*, 572. [CrossRef] [PubMed]

73. Ostroverkhova, N.V.; Kucher, A.N.; Konusova, O.L.; Kireeva, T.N.; Rosseykina, S.A.; Yartsev, V.V.; Pogorelov, Y.L. Genetic diversity of honey bee *Apis mellifera* in Siberia. In *Phylogenetics of Bees*; Ilyasov, R.A., Kwon, H.W., Eds.; CRC Press: Boca Raton, FL, USA, 2020; pp. 97–126.

74. Ostroverkhova, N.V.; Kucher, A.N.; Konusova, O.L.; Gushchina, E.S.; Yartsev, V.V.; Pogorelov, Y.L. Dark-colored forest bee *Apis mellifera* in Siberia, Russia: Current state and conservation of populations. In *Selected Studies in Biodiversity*; IntechOpen: London, UK, 2018; pp. 157–180.

75. Solignac, M.; Vautrin, D.; Loiseau, A.; Mougel, F.; Baudry, E. Five hundred and fifty microsatellite markers for the study of the honey bee (*Apis mellifera* L.) genome. *Mol. Ecol. Notes* **2003**, *3*, 307–311. [CrossRef]

76. Raymond, M.; Rousset, F. GENEPOP (Version 1.2): Population Genetics Software for Exact Tests and Ecumenicism. *J. Hered.* **1995**, *86*, 248–249. [CrossRef]

77. Morris, J.A.; Gardner, M.J. Statistics in Medicine: Calculating confidence intervals for relative risks (odds ratios) and standardised ratios and rates. *BMJ* **1988**, *296*, 1313–1316. [CrossRef]

78. Ostroverkhova, N.V.; Kireeva, T.N. Genetic diversity of *Apis mellifera carpathica* honey bee subspecies. *Far East. Èntomol.*. In Preparation.

79. Muli, E.; Patch, H.; Frazier, M.; Frazier, J.; Torto, B.; Baumgarten, T.; Kilonzo, J.; Kimani, J.N.; Mumoki, F.; Masiga, D.; et al. Evaluation of the Distribution and Impacts of Parasites, Pathogens, and Pesticides on Honey Bee (*Apis mellifera*) Populations in East Africa. *PLoS ONE* **2014**, *9*, e94459. [CrossRef]

10. Fleming, J.C.; Schmehl, D.R.; Ellis, J.D. Characterizing the Impact of Commercial Pollen Substitute Diets on the Level of *Nosema* spp. in Honey Bees (*Apis mellifera* L.). *PLoS ONE* **2015**, *10*, e0132014. [CrossRef]
11. Azzouz-Olden, F.; Hunt, A.G.; DeGrandi-Hoffman, G. Transcriptional response of honey bee (*Apis mellifera*) to differential nutritional status and *Nosema* infection. *BMC Genom.* **2018**, *19*, 628. [CrossRef]
12. Mendoza, Y.; Tomasco, I.; Antúnez, K.; Castelli, L.; Branchiccela, B.; Santos, E.; Invernizzi, C. Unraveling Honey Bee–*Varroa destructor* Interaction: Multiple Factors Involved in Differential Resistance between Two Uruguayan Populations. *Vet. Sci.* **2020**, *7*, 116. [CrossRef]
13. Bhutani, N.; Venkatraman, P.; Goldberg, A.L. Puromycin-sensitive aminopeptidase is the major peptidase responsible for digesting polyglutamine sequences released by proteasomes during protein degradation. *EMBO J.* **2007**, *26*, 1385–1396. [CrossRef]
14. Ren, B.; Liu, M.; Ni, J.; Tian, J. Role of Selenoprotein F in Protein Folding and Secretion: Potential Involvement in Human Disease. *Nutrients* **2018**, *10*, 1619. [CrossRef] [PubMed]
15. Davì, V.; Tanimoto, H.; Ershov, D.; Haupt, A.; De Belly, H.; Le Borgne, R.; Couturier, E.; Boudaoud, A.; Minc, N. Mechanosensation dynamically coordinates polar growth and cell wall assembly to promote cell survival. *Dev. Cell* **2018**, *45*, 170–182. [CrossRef] [PubMed]

Article

Seasonality of *Nosema ceranae* Infections and Their Relationship with Honey Bee Populations, Food Stores, and Survivorship in a North American Region

Berna Emsen [1,2], Alvaro De la Mora [1], Brian Lacey [3], Les Eccles [3], Paul G. Kelly [1], Carlos A. Medina-Flores [4], Tatiana Petukhova [5], Nuria Morfin [1] and Ernesto Guzman-Novoa [1,*]

[1] School of Environmental Sciences, University of Guelph, 50 Stone Road East, Guelph, ON N1G 2W1, Canada; bernaemsen@gmail.com (B.E.); alvarodlmr@gmail.com (A.D.l.M.); pgkelly@uoguelph.ca (P.G.K.); nmorfinr@uoguelph.ca (N.M.)

[2] Department of Animal Science, Agricultural Faculty, Ataturk University, Erzurum 25240, Turkey

[3] Ontario Beekeepers' Association Technology Transfer Program, 185, 5420 Hwy 6 N, Orchard Park Office, Guelph, ON N1H 6J2, Canada; brianwlacey@gmail.com (B.L.); les.eccles@ontariobee.com (L.E.)

[4] Unit of Veterinary Medicine and Animal Science, University of Zacatecas, Carr. Panamericana Km 35.5, El Cordovel 98500, Zac., Mexico; carlosmedina@uaz.edu.mx

[5] Department of Population Medicine, University of Guelph, 50 Stone Road East, Guelph, ON N1G 2W1, Canada; tpetukho@uoguelph.ca

* Correspondence: eguzman@uoguelph.ca

Received: 13 August 2020; Accepted: 6 September 2020; Published: 8 September 2020

Abstract: *Nosema ceranae* is an emerging pathogen of the western honey bee (*Apis mellifera* L.), and thus its seasonality and impact on bee colonies is not sufficiently documented for North America. This study was conducted to determine the infection intensity, prevalence, and viability of *N. ceranae* in >200 honey bee colonies during spring, summer, and fall, in a North American region. We also determined the relationship of *N. ceranae* infections with colony populations, food stores, bee survivorship, and overwinter colony mortality. The highest rates of *N. ceranae* infection, prevalence, and spore viability were found in the spring and summer, while the lowest were recorded in the fall. *N. ceranae* spore viability was significantly correlated with its prevalence and infection intensity in bees. Threshold to high levels of *N. ceranae* infections (>1,000,000 spores/bee) were significantly associated with reduced bee populations and food stores in colonies. Furthermore, worker bee survivorship was significantly reduced by *N. ceranae* infections, although no association between *N. ceranae* and winter colony mortality was found. It is concluded that *N. ceranae* infections are highest in spring and summer and may be detrimental to honey bee populations and colony productivity. Our results support the notion that treatment is justified when infections of *N. ceranae* exceed 1,000,000 spores/bee.

Keywords: *Nosema ceranae*; viability; prevalence; infection intensity; seasonality; bee longevity; bee population; honey stores; CCD

---

1. Introduction

The microsporidian parasite *Nosema ceranae* is an emerging pathogen of the western honey bee (*Apis mellifera* L.), and thus many aspects of its biology, pathology, and impact on colony conditions are unknown or not sufficiently documented. It is believed that the original host of *N. ceranae* is the Asiatic bee, *Apis cerana* [1], and it recently jumped to *A. mellifera* [2,3], which has been frequently infected by another microsporidian, *N. apis*, for more than a century [4]. *N. ceranae* and *N. apis* cause nosema disease in honey bees by parasitizing their midgut epithelial cells and impairing the digestion and absorption of nutrients in the infected individuals [5,6].

Spores of *Nosema* spp. can be observed and quantified by microscopy, but to identify the *Nosema* species observed, it is necessary to use molecular methods of detection. The most common molecular methods of *Nosema* spp. detection and quantification are those that use primers for the *16s rRNA* gene [1], which has multiple copies in the *Nosema* genome. More reliable methods of detection and quantification based on the single-copy gene *Hsp70* have been recently developed [7].

Nowaday, *N. ceranae* is distributed worldwide and is the microsporidian that is most frequently found in honey bee colonies, even on islands distant from mainland countries [6,8,9]. However, information on seasonal variation of *N. ceranae* infections, prevalence, and spore viability, as well as about its impact on colony conditions and bee survivorship, is particularly scarce for North America. Surveys in Europe have shown that the prevalence and intensity of *N. ceranae* infections are not constant throughout the year, and fluctuate between seasons and geographical regions [10]. However, the prevalence of *N. ceranae* in honey bee colonies does not show seasonality in Spain [11,12]. In the USA, a 12-month study showed that the intensity of *N. ceranae* infections in honey bee colonies vary seasonally in a similar way *N. apis* infections vary, with the highest spore counts of the parasite observed in the spring, and the lowest in the fall and winter [13]. Likewise, in Western Canada, Ibrahim et al. [14] reported that *N. ceranae* and *N. apis* showed a similar pattern of infection as that described for honey bee colonies in the USA [13]. However, in North America there is no information on whether the viability of *N. ceranae* spores shows a seasonal pattern and if this pattern is related to the intensity and prevalence of infections within colonies. It is also unknown how these patterns affect disease morbidity and colony conditions. This information is necessary for designing control strategies against the microsporidian.

Most of what is known about the epidemiology of nosema disease has been derived from studies with *N. apis*, and it is assumed that it must be similar for *N. ceranae* infections. However, this information needs to be corroborated with studies conducted with *N. ceranae* in different geographical locations. Therefore, in addition to determining seasonal changes in the intensity of *N. ceranae* infections, it is important to determine if these differences are related to the prevalence of infected bees within a colony, as well as to the viability of the spores at different times of the year. The viability of *Nosema* spp. spores influences their ability to infect bees and to multiply [15]. Therefore, determining the viability of the parasite's spores provide an indication of its ability to cause disease. This information will help to better understand the seasonality of infections in different locations and to determine the optimal treatment times for the control of nosema disease caused by *N. ceranae* in honey bee colonies.

The pathogenicity and virulence of *N. ceranae* to honey bees has been analyzed in numerous studies. In the majority of them, *N. ceranae* has been found to significantly shorten the lifespan of bees [16–20], although some have not found such effect (see, e.g., Huang et al. [21]). However, most of these studies have been conducted under laboratory conditions using caged bees, which are useful to study certain aspects of nosema disease, but cage experiments are probably not very suitable to study the effects of the parasite on bee longevity because they are conducted under artificial conditions. Studies under the natural environment of colonies within hives are thus required to confirm the impact of *N. ceranae* on bee longevity.

At the hive level, several studies conducted in Spain have related *N. ceranae* infections to colony collapse [22–24]. Similarly, another study suggested that *N. ceranae* was one of the factors that predicted seasonal colony mortality in Switzerland [25]. However, in other European countries, no clear causal link between the microsporidian and colony failure has been evidenced [10,26–28]. Likewise, the studies conducted in Canada, continental USA, and Hawaii did not find a relationship between *N. ceranae* infection and winter colony losses [29–32]. Nevertheless, the levels of *N. ceranae* infections were higher than those of *N. apis* infections in two Canadian provinces [33]. Similarly, *N. ceranae* has not been associated to colony collapse in Mexico and South America [34–36]. Apparently, data on the virulence and impact of *N. ceranae* at the colony level are contradictory and focused on the collapse of colonies rather than on the effect of the parasite on colony strength and food stores. The conflicting findings between studies conducted in different regions could be due to differences in virulence of *N. ceranae* isolates, differences in susceptibility of bee strains, or differences in other unknown factors [37].

Because seasonal and regional differences may be important for the pathogenicity and virulence of *N. ceranae*, additional studies on the epidemiology and impact of the microsporidian need to be conducted at the colony level in different regions of the world. Therefore, this study was conducted to determine the seasonal patterns of *N. ceranae* infection, prevalence, and viability in honey bees from commercial colonies in Ontario Canada. We also studied the relationship of *N. ceranae* infections with colony populations, food stores, and overwinter survival, as well as how the parasite affects the lifespan of worker bees in colonies.

2. Materials and Methods

2.1. Screening and Selection of Experimental Colonies

Approximately 100 worker bees per sample were initially collected with a bee vacuum [38] from the hive entrance of more than 300 colonies of commercial beekeepers from Southern Ontario, Canada in early spring. These colonies had not been treated against nosema disease during the last 12 months. The samples were stored at −20 °C and later microscopically analyzed for the presence of *Nosema* spp. spores [39]. Randomly chosen subsamples of 10 bees from each of the positive samples were subsequently subjected to PCR tests to detect *N. ceranae* infections [40]. Colonies infected with *N. ceranae* only (*n* = 233) were selected and sampled during different seasons (mid-spring, mid-summer, and mid-fall) to determine infection intensity of *N. ceranae* in all of them. Ten of the selected colonies were randomly chosen to provide subsamples of 30 bees/colony for assessments of *N. ceranae* spore prevalence and viability during each season. The selected 233 colonies in this study were managed similarly and treated against *Varroa destructor* mites (Apivar®, Veto-Pharma) and preventatively against foulbrood diseases (Oxytet-25®, Medivet) in the fall, according to the instructions of the manufacturers. No visual clinical signs of brood diseases were detected in the colonies during the course of the study. Treatments against nosema disease were not applied.

2.2. Infection Intensity

The intensity of *N. ceranae* infections was determined in samples of 60 forager bees as recommended by the OIE [41]. The bees were collected from the selected colonies (*n* = 233, 227, and 196, for spring, summer, and fall, respectively) as above described. The microscopic assessment of *N. ceranae* infection intensity was done using a phase contrast microscope (Olympus BX41; Olympus, Markham, ON, Canada) and a hemocytometer to obtain *Nosema* spores counts, which were used for calculations and recorded as mean number of spores per bee [39]. Spore counts are a reliable method of determining the intensity of *N. ceranae* infections because they correlate with PCR quantifications [40,42], another way of determining *Nosema* infection levels. Previously, we had demonstrated a high and significant linear relationship (R^2 = 0.95) between *N. ceranae* copies quantified by PCR, and number of spores from samples of bees with different infection levels of *N. ceranae* [33].

2.3. Spore Viability

To determine the viability of *N. ceranae* spores, three replicates of fresh spores were obtained from 30 bees of each of the 10 positive colonies that were chosen for these assessments. The spores were extracted and purified as per McGowan et al. [43]. After extraction, 10 µL of the spore suspension was used to count spores to determine their concentration in the suspension [39]. The *N. ceranae* spore concentrations were adjusted to obtain a final concentration of 1×10^5 spores/µL and kept in 1.5 mL microcentrifuge tubes at 5 °C until needed. Spores were stained with fluorescent 4′,6-diamino-2-phenylindole (DAPI; Boheringer Manheim Biochemicals, Indianapolis, USA [44]) and propidium iodide (PI; Boheringer Manheim Biochemicals, Indianapolis, USA [45]) dyes as described by McGowan et al. [43]. Briefly, 200 µL of spore suspension was placed in a microcentrifuge tube and incubated simultaneously with 20 µL of each of the dye solutions (1 mg/mL in ddH$_2$O) at room temperature for 20 min. After incubation, the spores were rinsed twice by centrifugation at 800 rpm for 6 min in 100 µL of ddH$_2$O. The final pellet was suspended in 100 µL of ddH$_2$O. A smear was prepared on a microscope slide with 5 µL of the spore suspension and 5 µL of ProLong® Gold reagent (Antifade, Invirtrogen, Burlington, ON, Canada) to protect the fluorescent dyes from fading. Subsequently, the smear was allowed to partially dry for 5 to 10 min, and a cover slip was placed on the smear. The slides were kept in the dark overnight and were sealed the next day using clear nail varnish.

Stained spores were observed at 400× magnification under a fluorescent microscope (Leica DM5500B, Leica Microsystems, Wetzlar, Germany). Between 25 and 30 different fields, containing approximately 30–40 spores per field, were observed using the bright field and the fluorescent light filters for DAPI and PI, and were photographed with a digital camera mounted on the microscope (Hamamatsu Photonics K.K., Hamamatsu, Japan). The photographs were analyzed with an imaging program (Volocity® version 5.1, Improvision, Coventry, UK) to compare the total number of spores between the bright field and the fluorescent filters. The bright light field provided the total number of spores, which are seen as refracting oval corpuscles. The DAPI fluorescent filter provided the number of total spores, which are blue-stained, indicating that they are indeed spores because their walls have been stained. The PI fluorescent filter provided the number of nonviable spores, which are observed red, indicating that the cell wall of the spore is compromised, allowing the dye to penetrate and stain its DNA. The percentage of viable spores was determined by subtracting the number of red spores counted under the PI fluorescent filter (nonviable spores), from the total number of spores counted under the bright field (which is roughly equal to the number of spores counted with the DAPI filter), and dividing this figure by the total number of spores. The resulting figure was then, multiplied by 100.

2.4. Nosema ceranae Prevalence in Bees (Percentage of Infected Bees)

The presence or absence of *N. ceranae* spores was determined individually in 30 bees per sample using microscopy as previously described. The percentage of infected bees was calculated by dividing the number of positive bees (having observable *Nosema* spores) by the total number of bees in the sample (30) and the result multiplied by 100.

2.5. Colony Evaluations and Relationship with Nosema ceranae Infections

Each of the selected colonies was assessed in the following manner. The colonies were opened and a frame by frame visual estimate to the 0.25 fraction level of the number of frames covered with bees or having brood, stored pollen, or stored honey was recorded [46]. Additionally, to detect possible relationships between different levels of *N. ceranae* infections and colony conditions, the colonies were arbitrarily classified as having low (<1,000,000 spores/bee), threshold (>1,000,000 <2,000,000 spores/bee), and high (>2,000,000 spores/bee) *N. ceranae* infection levels using the data from summer evaluations (the most productive season for the bees). This classification was partially based on estimations of the severity of *N. apis* infections by studies establishing the treatment threshold in 1,000,000 spores/bee [47–49], as this threshold has not been established for *N. ceranae* yet. Then, comparisons between the three categories of colonies were done for bee populations, brood amount, pollen stores and honey stores. Additionally, the population growth of colonies between spring and summer was determined by subtracting the number of frames covered by bees or containing brood of the summer assessments from that of the spring assessments. Data on colony growth rates were correlated with *N. ceranae* infections and colony conditions. In the following spring, colony evaluations were conducted as above for overwintered colonies, and the percentage of surviving colonies was determined.

2.6. Effect of Nosema ceranae Infections on Bee Survivorship in Colonies

The size of bee populations and colony productivity depend highly on how long bees live [50,51]. Therefore, it seemed important to determine how *N. ceranae* infections affect the survivorship of bees in the natural environment of a colony, rather than in cages within a laboratory, as most previous studies on bee longevity have been conducted. However, determining bee survivorship of marked bees in a colony is challenging. Therefore, three observation hives were used to facilitate daily censuses of the experimental bees. Each observation hive ($47.0 \times 4.1 \times 96.5$ cm) was installed in a laboratory and was connected to the exterior of the building to allow the bees to forage. Four newly drawn combs with ~3000 worker bees of variable ages and a queen unrelated to the experimental bees were installed into each hive. Newly emerged bees that were inoculated or that were not inoculated with *N. ceranae* spores were introduced to the observation hives. To obtain newly emerged bees, brood combs from three colonies were placed overnight in wire emerging cages inside an incubator (33 °C, 70% RH). The next morning, the newly emerged bees were transferred to a plastic container and *Nosema* spore diagnosis was performed as before in a subsample of them to ensure that the bees were *Nosema* spore free. Emerged bees were initially starved for 2 h and then force-fed 5 µL of a 50% sucrose syrup that contained *N. ceranae* spores at a concentration of 1×10^4 spores/µL, using a micropipette (Eppendorf, Mississauga, ON, CA). The suspension of spores used for survivorship experiments was confirmed to only contain *N. ceranae* spores [40]. Control bees only received 5 µL of the syrup without spores. The experimental bees were identified by marking them with enamel paint of different colors (according to treatment) on their thoraces.

Approximately 1000 marked bees—500 inoculated and 500 non-inoculated with *N. ceranae* spores—were introduced into each observation hive. By co-fostering the bees of both treatments in a common hive, environmental influences were minimized [52] to more accurately measure the effect of *N. ceranae* infection on bee survivorship. After introduction, bee acceptance was determined by visually censing the marked bees as per Unger and Guzman-Novoa [53] 24 h after introducing the bees to the observation hives. Subsequently, censuses were performed every day during early mornings when bees were not foraging, from day two after introduction, until day 35. Bee survivorship was estimated by calculating the proportion of bees of each treatment that were present in each hive every day over the initial number of accepted bees. At the end of the study, the surviving marked bees of each treatment in the observation hives were collected and analyzed for *N. ceranae* spores as a pooled sample, to confirm that infections developed normally in the inoculated bees.

2.7. Statistical Analyses

Percentage data (viability and prevalence) were arcsine-square root transformed, whereas the data on infection intensity were log transformed to normalize them. Data on colony populations and food stores were square root transformed. Transformed data were subjected to analysis of variance (ANOVA) and the means were compared with Tukey tests at a $p < 0.05$ when significance was detected. Additionally, data on infection intensity, prevalence, viability, colony conditions and bee population growth were analyzed with Spearman rank correlations. Data on bee survivorship were analyzed with the Kaplan–Meier method, comparing estimated survival functions. Survival curves were compared with a long-rank test. Differences in colony mortality after the winter were tested with χ^2 tests. All statistical analyzes were performed using the R statistical program [54].

3. Results

3.1. Intensity, Prevalence, and Viability of Nosema ceranae Infections

The intensity of *N. ceranae* infections in the experimental colonies was significantly higher in the spring and summer than in the fall, and there were no differences between infection rates of spring and summer ($F_{2, 512} = 30.45$, $p < 0.0001$; Figure 1). Similarly, a significantly higher percentage of bees had visible *N. ceranae* spores in the spring and summer relative to the fall ($F_{2, 27} = 14.04$, $p < 0.0001$). Additionally, there were significant differences between seasons for the percentage of viable *N. ceranae* spores ($F_{2, 27} = 63.22$, $p < 0.0001$). Spore viability was approximately four-fold higher in the spring than in the summer and fall (Table 1). Furthermore, the intensity of *N. ceranae* infections was significantly correlated with the percentage of *N. ceranae* infected bees and with spore viability (r = 0.32 and 0.41, respectively, $p < 0.01$); the percentage of *N. ceranae* infected bees and spore viability were also correlated ($r = 0.56$, $p < 0.0001$).

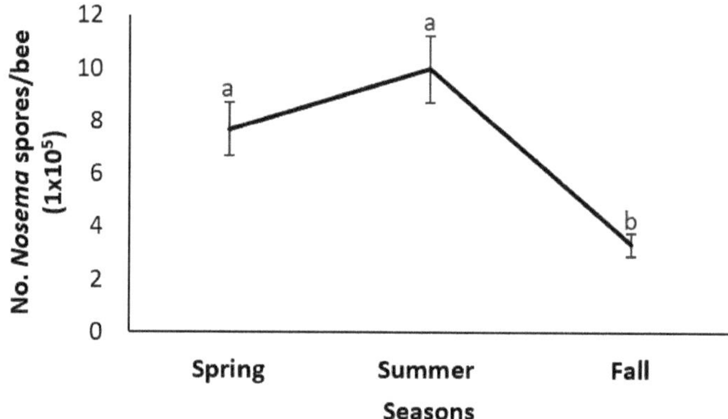

Figure 1. *Nosema ceranae* infection intensity (mean No. spores/bee ± SE) of honey bee colonies during spring, summer and fall ($n = 179$). Different letters indicate significant differences between seasons based on analyses of variance and Tukey tests of arcsine-square root transformed data at a significant level of <0.05. Actual data are presented.

Table 1. Prevalence and viability of *Nosema ceranae* spores in honey bee colonies during spring, summer and fall ($n = 30$).

Season	% Infected Bees ± SE	% Spore Viability ± SE
Spring	73.9 ± 1.8 [a]	59.0 ± 6.0 [a]
Summer	68.0 ± 1.8 [a]	13.7 ± 1.6 [b]
Fall	48.0 ± 5.6 [b]	16.3 ± 1.8 [b]

Different letters indicate significant differences between seasons based on analyses of variance and Tukey tests of arcsine-square root transformed data at a significance level of <0.05. Actual data are presented.

3.2. Nosema ceranae Infections and Colony Conditions

Colonies with low levels of *N. ceranae* infection (<1,000,000 spores/bee) had significantly more bees ($F_{2,169} = 3.36$, $p < 0.05$), brood ($F_{2,154} = 4.09$, $p < 0.05$), pollen stores ($F_{2,148} = 4.51$, $p < 0.05$), and honey stores ($F_{2,154} = 10.53$, $p < 0.0001$) than colonies with high levels of *N. ceranae* infection (>2,000,000 spores/bee). They also differed with colonies having threshold levels of *N. ceranae* infection (>1,000,000 <2,000,000 spores/bee) for brood and honey stores, which were higher in the colonies with low levels of *N. ceranae* infection (Table 2). Furthermore, the intensity of *N. ceranae* infections negatively and significantly correlated with colony growth, bee populations, brood, and stores of pollen and honey (Table 3), suggesting a detrimental influence of the microsporidian on these colony variables.

Table 2. Population and food store parameters (mean ± SE) of honey bee colonies categorized by intensity of *Nosema ceranae* infection (low, threshold, high) in the summer ($n = 172$).

Infection Category	No. Spores Per Bee [1]	No. Frames with Bees	No. Frames with Brood	No. Frames with Pollen	No. Frames with Honey
Low	<1.0	19.7 ± 1.0 [a]	8.4 ± 0.3 [a]	2.1 ± 0.1 [a]	14.1 ± 1.2 [a]
Threshold	1.0–2.0	17.9 ± 1.3 [a,b]	6.9 ± 0.6 [b]	1.7 ± 0.2 [a,b]	9.0 ± 1.3 [b]
High	>2.0	14.3 ± 2.6 [b]	6.7 ± 1.4 [b]	1.4 ± 0.3 [b]	4.6 ± 1.6 [c]

Different letters indicate significant differences between infection categories based on analyses of variance and Tukey tests of square root transformed data at a significant level of <0.05. Actual data are presented. [1] millions.

Table 3. Correlations between infection intensity of *Nosema ceranae* (No. spores/bee), colony growth, populations, and stored food of honey bee colonies ($n = 151$).

Variables Correlated	Correlation	p
No. spores–Frames w/bees	−0.33	<0.0001
No. spores–Frames w/brood	−0.31	<0.0001
No. spores–Frames w/pollen	−0.24	0.0024
No. spores–Frames w/honey	−0.52	<0.0001
No. spores–Colony growth	−0.18	<0.05
Frames w/bees–Frames w/brood	0.56	<0.0001
Frames w/bees–Frames w/pollen	0.30	0.0002
Frames w/bees–Frames w/honey	0.74	<0.0001
Frames w/bees–Colony growth	0.96	<0.0001
Frames w/brood–Frames w/pollen	0.02	0.7678
Frames w/brood–Frames w/honey	0.47	<0.0001
Frames w/brood–Colony growth	0.38	<0.0001
Frames w/pollen–Frames w/honey	0.44	<0.0001
Frames w/pollen–Colony growth	0.23	<0.01
Frames w/honey–Colony growth	0.79	<0.01

The percent survivorship of overwintered colonies was lower for colonies that had high infections of *N. ceranae* during the summer (43.8%) compared to those that had threshold and low infection levels of the parasite (55.6% and 64.2%, respectively). However, these differences as well as those for colony populations and food stores were not significant ($p > 0.05$). Similarly, there was a trend in *N. ceranae* infection levels of overwintered colonies, as colonies that were classified as having high infection rates of the microsporidian in the summer had about twice the number of spores/bee in the following spring, compared to colonies that had been classified as having threshold and low infection rates of the parasite; but again, those differences were not significant ($p > 0.05$; Table 4).

Table 4. Populations, food store parameters (mean ± SE), and percent survival of honey bee colonies categorized by the intensity of *Nosema ceranae* infections (low, threshold, high) in the summer, that were alive the following spring ($n = 104$).

Variables	Low	Threshold	High
No. spores/bee [1]	1.88 ± 0.25	1.75 ± 0.61	3.21 ± 1.58
No. Frames w/bees	6.3 ± 0.5	6.3 ± 0.9	5.5 ± 1.3
No. Frames w/brood	3.4 ± 0.3	4.1 ± 0.6	3.0 ± 0.5
No. Frames w/pollen	1.9 ± 0.2	2.1 ± 0.2	1.9 ± 0.9
No. Frames w/honey	8.3 ± 0.4	8.2 ± 0.7	9.7 ± 0.7
% Colony survival	64.2	55.6	43.8

[1] millions.

3.3. Survivorship of Bees in Colonies

Worker bees inoculated with *N. ceranae* spores had significantly lower survivorship rates than control bees (Long-rank test, $\chi^2 = 120$, $p < 0.01$; Figure 2). Additionally, at 35 days of age, only about 3% of the *N. ceranae* inoculated bees remained in the colonies vs. 25% of the control group, difference that was significant ($\chi^2 = 244.6$; $p < 0.0001$). The surviving bees were collected and analyzed at the end of the study, showing rates of infection of >16,000,000 spores/bee in the inoculated group of bees but <1,000,000 spores/bee in the control group. These results demonstrate that *N. ceranae* is pathogenic to honey bees in their natural environment and can impact the population growth of colonies during the seasons of the year of more bee activity.

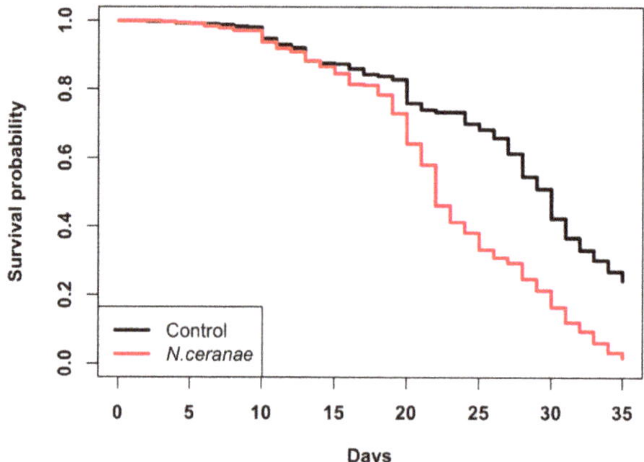

Figure 2. Probability of survival of adult worker honey bees inoculated with 50,000 *Nosema ceranae* spores/bee (red) or not inoculated (black) when newly emerged. Experimental bees were marked, introduced into colonies and daily censed. Survival functions were estimated with the Kaplan-Meier method.

4. Discussion

This study evidenced significant seasonal associations for patterns of *N. ceranae* viability, infection prevalence, and infection intensity in Canadian honey bee colonies. The study also showed that the intensity of *N. ceranae* infections is negatively related to honey bee populations, food stores and bee survivorship.

Seasonal patterns of *N. ceranae* infections were evident. Spore counts of the parasite were significantly higher in the spring and summer than in the fall. Similarly, the percentage of *N. ceranae* infected bees was also lowest in the fall, but did not differ between spring and summer. Additionally, the highest rate of *N. ceranae* spore viability was found in the spring, while the lowest rates were observed in the summer and fall. Furthermore, spore counts and rates of spore viability and infected bees were positively and significantly correlated. Contrary to these results, Jack et al. [55] found a non-significant correlation between prevalence and intensity of *N. ceranae* infections within colonies ($r = 0.29$, $p > 0.05$). The lack of significance of the correlation between *N. ceranae* prevalence and infection intensity in the study by Jack et al. [55] could have been due, at least partially, to the use of composite samples of bees of different ages collected from different locations in the experimental colonies, when in their study, *N. ceranae* prevalence and infection intensity differed significantly between age cohorts and sample locations. In this study, the data used for the correlation analyses were obtained from samples that in all cases were collected from the hive entrance (old bees). A previous study using bees collected from the same location in hives, also found a significant correlation between intensity and prevalence of *N. ceranae* infections [43]. Therefore, it is important to standardize methods as to the location of the hive where samples of bees are collected for diagnosis and research purposes, to make results of different studies comparable.

The epidemiological results of this study indicate that spring and summer are the most favorable seasons for the spread and development of *N. ceranae* infections. After the summer, conditions for the transmission and multiplication of *N. ceranae* decrease, as evidenced by the low rates of infection intensity, prevalence, and, above all, spore viability found in the fall. Conversely, lack of seasonality for *N. ceranae* infections has been reported from Europe and Hawaii [12,22,32], but the pattern of infection intensity and prevalence found in this study was similar to that historically known for *N. apis*, with a peak starting in mid-spring and declining in mid-summer and fall [56]. Similar results have been recently found for *N. ceranae* in Eastern and Western Canada [14,42] and in other parts of North America [13,34].

The seasonal pattern of *N. ceranae* viability, prevalence, and infection intensity is likely influenced by several factors, which include the dynamics of honey bee populations within colonies during different seasons of the year in temperate and cold climates. During late winter and early spring, the queen increases her egg-laying rate, which progressively results in higher numbers of adult bees emerging in a colony. Consequently, an increasing number of newly emerged bees rapidly acquire spores of the microsporidian by horizontal transmission via trophalaxis with older bees that overwintered and are a source of spores for new bees [57,58], or by cleaning comb cells soiled with feces contaminated with spores of the fungus [4]. This rapid transmission of microsporidian spores causes infection levels of nosema disease to spike in the spring, which remain high until the summer, because the queen continues to maintain a high egg-laying rate until the middle of that season, and many young bees continue to emerge and become infected.

While patterns of infection intensity and prevalence have been previously reported for *N. ceranae*, to the best of our knowledge, this is the first study reporting seasonal viability patterns for *N. ceranae* spores. Despite the limited knowledge on viability rates of *N. ceranae* spores in different seasons, the results of this study show that its seasonal variability coincides with the reproductive cycle of honey bee colonies. Therefore, these results suggest that the viability of *N. ceranae* spores is a relevant factor for the establishment, development, and spread of the disease. It is possible that *N. ceranae* has evolved mechanisms that allow the parasite to have high viability and infectivity rates during the seasons of maximum bee reproduction (spring and summer), which allow the microsporidian to multiply and spread more efficiently. However, despite the synchronous patterns of *N. ceranae* infections, prevalence, and spore viability, many questions remain to be answered about the fundamental causes as to why these aspects of *N. ceranae* infections fluctuate through the year; therefore, further studies are warranted. Regardless of the need for additional research, the results of this study have implications for good beekeeping practices and support the notion that the best time to apply treatments for nosema disease control is during early spring, when infection levels of the parasite rapidly grow. Future studies should also be conducted to determine if a fall treatment would be beneficial in reducing winter and spring infections of *N. ceranae*.

The negative association and correlations between the intensity of *N. ceranae* infections and colony conditions found in this study suggest that at or above threshold levels of infection (>1,000,000 spores/bee), the parasite may have a negative impact on bee populations and food stores. In fact, the colonies with threshold and high infection levels grew at a significantly slower rate than the colonies with low rates of infection. Conversely, a former study conducted in Atlantic Canada, did not find an association between *N. ceranae* infections and variation in bee populations and food stores [59]. The study by Williams et al. [59] measured colony conditions in September and only evaluated every other frame in the colonies. Perhaps if the evaluations had been conducted earlier in the summer and all frames had been assessed, the results would have been similar to ours. Additionally, our study was conducted with at least three times more colonies and the colonies were categorized according to *N. ceranae* infection levels. Higher numbers of colonies analyzed by infection levels could have increased the probability of detecting significant differences for colony conditions, particularly for colonies with high *N. ceranae* infection intensity. Similar to our findings, other studies have reported that *N. ceranae* infections may have deleterious associations with bee populations and honey yields [23,60,61], but this is the first study to show an association between degrees of *N. ceranae* infection and colony conditions. However, this, as well as previous reports, is a correlational study that reflects the association between *N. ceranae* infections and colony conditions during a snapshot in time. Therefore, studies that more precisely demonstrate cause and effect at the colony level for longer periods of time are warranted. Additionally, further research is required to more precisely determine treatment thresholds for *N. ceranae* infections during different seasons in North America. There is no consensus on the economic injury level or treatment thresholds for nosema disease caused by *N. ceranae*. Therefore, research is required to better understand injury levels for *N. ceranae* in different regions, with different bee genotypes and with different *N. ceranae* haplotypes, because all these factors can differentially and synergistically result in high or low virulence of *N. ceranae* infections in honey bee colonies. These studies would generate information to establish treatment thresholds for nosema disease that are specific for different regions. Our data support the notion that treatment is justified when infections of *N. ceranae* exceed 1,000,000 spores/bee in the region where the study was conducted.

The length of life of honey bees strongly influences colony growth, population size, and honey yields in honey bee colonies [50,52,62]. Therefore, if *N. ceranae* shortens the lifespan of bees, it could have serious detrimental effects on colony fitness and productivity. Here, we provide evidence that the survivorship rate and lifespan of honey bees in colonies are significantly reduced when workers are challenged with an inoculum of 50,000 *N. ceranae* spores/bee. The reason why *N. ceranae* infections may be associated to reduced bee populations and colony growth, as found in this study, could be explained, at least partially, by the significantly lower survival rate of the infected bees. Other studies have previously shown that *N. ceranae* infections shorten the length of life of infected honey bees. For example, Williams et al. [18], using an inoculum of 30,000 *N. ceranae* spores/bee, found that 95% of the bees in their experimental group died during a 28 to 30-day incubation period, compared to only 25% of the control group. Furthermore, Eiri et al. [63] reported that the longevity of adult bees that were experimentally infected during their larval stage with 40,000 *N. ceranae* spores/bee was reduced by 28% compared to bees that did not receive spores. Similarly, Higes et al. [16] reported that the mortality rate of bees infected with an inoculum of 125,000 spores/bee was 94.1% compared to 5.9% of the bees in the control group. The above studies, however, were conducted in laboratory settings using cage experiments. Ours is the first study to show reduced survivorship in *N. ceranae*-infected bees within the more natural environment of colonies for a period of more than 30 days, which agrees with another study conducted in hives, that showed shorter lifespans for *N. apis*-infected bees [64]. The reduction in longevity of bees infected with *N. ceranae* could be due to multiple deleterious effects of the parasite on bee health. For example, *N. ceranae*-infected workers are energetically deprived, exhibit precocious foraging, and show a decrease in the amount of vitellogenin, a protein that affects longevity in insects [11,19,65,66]. In addition, *N. ceranae* causes irreparable damage to the bees' digestive tract, with epithelial cells of the midgut suffering lysis [67], which reduces the absorption of nutrients, leading to the premature death of the bees, as a consequence of digestive failure and starvation.

High rates of colony losses are experienced during late winter and early spring in temperate and cold climates [68–70]. However, *N. ceranae* infections in this study were not associated with overwintering colony conditions and mortality. Overall, the data shows a trend of more intense infections of the parasite with lower populations and lower survival rates of colonies, but this trend was not significant. It is possible that the parasite may have some influence on the survival rate and populations of honey bee colonies during winter, but apparently it is not a major cause of winter mortality as previously reported from other North American and European studies [8,10,30,31]. Conversely, this study and several others have shown that *Nosema* spp. infections detrimentally affect colony population growth during spring [23,30,60]. Therefore, this study supports the notion of low *N. ceranae* pathogenicity and virulence during winter, and high *N. ceranae* pathogenicity and virulence during spring and summer.

Based on the seasonal patterns of *N. ceranae* infection, on its association with poor colony growth, small populations, and reduced food stores during spring and summer, it would be advisable to implement control measures against the microsporidian during early spring. IPM management practices, including chemical treatments, hive sterilizations, selective breeding, and supplemental feeding, are options to manage and control nosema disease in honey bee colonies [37].

This study generated additional evidence about the epidemiology and potential negative impact of *N. ceranae* on honey bee colonies in North America, which could have detrimental consequences on colony productivity and on the pollination services that honey bees provide. The study also opens questions that warrant further investigation aimed at better understanding the impact of *N. ceranae* infections on honey bee health and productivity, and at establishing seasonal control strategies against the microsporidian.

5. Conclusions

This study provides data on seasonal infection intensity, prevalence, and viability of *N. ceranae* and on the relationship of *N. ceranae* infections with honey bee colony populations, food stores, bee survivorship, and overwinter colony mortality in a North American region. The highest rates of *N. ceranae* infection, prevalence, and spore viability were found in the spring and summer, while the lowest were recorded in the fall. *N. ceranae* infection levels of >1,000,000 spores/bee were significantly associated with reduced bee populations and food stores in colonies. Furthermore, worker bee survivorship was significantly reduced by *N. ceranae* infections, although no association between *N. ceranae* and winter colony mortality was found. It is concluded that *N. ceranae* infections are highest in spring and summer and may be detrimental to honey bee populations and colony productivity. Our results support the notion that treatment is justified when infections of *N. ceranae* exceed 1,000,000 spores/bee.

Author Contributions: Conceptualization, E.G.-N. and L.E.; methodology, B.E., A.D.l.M., and E.G.-N.; validation, B.E., A.D.l.M., B.L., L.E., P.G.K., E.G.-N., and C.A.M.-F.; formal analysis, T.P. and E.G.-N.; investigation, B.E., A.D.l.M., B.L., L.E., P.G.K., C.A.M-F., and N.M.; resources, E.G.-N.; writing—original draft preparation, B.E., A.D.l.M., and B.L.; writing—review and editing, B.E., A.D.l.M., N.M., and E.G.-N; supervision, E.G.-N., L.E., and P.G.K.; project administration, E.G.-N; funding acquisition, E.G.-N. All authors have read and agreed to the published version of the manuscript.

Funding: This research was funded by the Ontario Ministry of Agriculture, Food and Rural Affairs, grant number 200456.

Acknowledgments: We thank the technology transfer team of the Ontario Beekeepers Association for collaborating with evaluations of colonies, as well as participating beekeepers for managing the colonies. Mollah Hamiduzzaman performed the PCR analyses of the samples. Logistic assistance was provided by Abril Soria, Laura Campbell, David Stotesbury, Jessica Gu, Emma Feiler, and Hailey Ashbee.

Conflicts of Interest: The authors declare no conflict of interest.

References

1. Fries, I.; Feng, F.; da Silva, A.; Slemenda, S.B.; Pieniazek, N.J. *Nosema ceranae* n. sp. (Microspora, Nosematidae), morphological and molecular characterization of a microsporidian parasite of the Asian honey bee *Apis cerana* (Hymenoptera, Apidae). *Eur. J. Parasitol.* **1996**, *32*, 356–365. [CrossRef]

2. Higes, M.; Martín, R.; Meana, A. *Nosema ceranae*, a new microsporidian parasite in honeybees in Europe. *J. Invertebr. Pathol.* **2006**, *92*, 93–95. [CrossRef]

3. Huang, W.F.; Jiang, J.H.; Chen, Y.W.; Wang, C.-H. A *Nosema ceranae* isolate from the honeybee *Apis mellifera*. *Apidologie* **2007**, *38*, 30–37. [CrossRef]

4. Fries, I. Protozoa. In *Honey Bee Pests, Predators and Diseases*, 3rd ed.; Morse, R.A., Flottum, K., Eds.; A I Root Company: Medina, OH, USA, 1997; pp. 57–76.

5. Fries, I. Infectivity and multiplication of *Nosema apis* Z. in the ventriculus of the honey bee. *Apidologie* **1988**, *19*, 319–328. [CrossRef]

6. Goblirsch, M. *Nosema ceranae* disease of the honey bee (*Apis mellifera*). *Apidologie* **2018**, *49*, 131–150. [CrossRef]

7. Cilia, G.; Cabbri, R.; Maiorana, G.; Cardaio, I.; Dall'Olio, R.; Nanetti, A. A novel TaqMan® assay for *Nosema ceranae* quantification in honey bee, based on the protein coding gene Hsp70. *Eur. J. Protistol.* **2018**, *63*, 44–50. [CrossRef]

8. Martín-Hernández, R.; Bartolomé, C.; Chejanovsky, N.; Le Conte, Y.; Dalmon, A.; Dussaubat, C.; García-Palencia, P.; Meana, A.; Pinto, M.A.; Soroker, V.; et al. *Nosema ceranae* in *Apis mellifera*: A 12 years postdetection perspective. *Environ. Microbiol.* **2018**, *20*, 1302–1329. [CrossRef] [PubMed]

9. Cilia, G.; Sagona, S.; Giusti, M.; Jarmela dos Santos, P.E.; Nanetti, A.; Felicioli, A. Nosema ceranae infection in honeybee samples from Tuscanian archipelago (Central Italy) investigated by two qPCR methods. *Saudi J. Biol. Sci.* **2019**, *26*, 1553–1556. [CrossRef]

10. Meixner, M.D.; Francis, R.M.; Gajda, A.; Kryger, P.; Andonov, S.; Uzunov, A.; Topolska, G.; Costa, C.; Amiri, E.; Berg, S.; et al. Occurrence of parasites and pathogens in honey bee colonies used in a European genotype-environment interactions experiment. *J. Apic. Res.* **2014**, *53*, 215–229. [CrossRef]

11. Martín-Hernández, R.; Botías, C.; Barrios, L.; Martínez-Salvador, A.; Meana, A.; Mayack, C.; Higes, M. Comparison of the energetic stress associated with experimental *Nosema ceranae* and *Nosema apis* infection of honeybees (*Apis mellifera*). *Parasitol. Res.* **2011**, *109*, 605–612. [CrossRef]

12. Higes, M.; Martín-Hernández, R.; Meana, A. *Nosema ceranae* in Europe: An emergent type C nosemosis. *Apidologie* **2010**, *41*, 375–392. [CrossRef]

13. Traver, B.E.; Williams, M.R.; Fell, R.D. Comparison of within hive sampling and seasonal activity of *Nosema ceranae* in honey bee colonies. *J. Invertebr. Pathol.* **2012**, *109*, 187–193. [CrossRef]

14. Ibrahim, A.; Melathopoulos, A.P.; Pernal, S.F. Integrated management of *Nosema* and detection of antibiotic residues. *Hivelights* **2012**, *25*, 11–17.

15. Sánchez Collado, J.G.; Higes, M.; Barrio, L.; Martín-Hernández, R. Flow cytometry analysis of *Nosema* species to assess spore viability and longevity. *Parasitol. Res.* **2014**, *113*, 1695–1701. [CrossRef]

16. Higes, M.; García-Palencia, P.; Martín-Hernández, R.; Meana, A. Experimental infection of *Apis mellifera* honeybees with *Nosema ceranae* (Microsporidia). *J. Invertebr. Pathol.* **2007**, *94*, 211–217. [CrossRef]

17. Paxton, R.J.; Klee, J.; Korpela, S.; Fries, I. *Nosema ceranae* has infected *Apis mellifera* in Europe since at least 1998 and may be more virulent than *Nosema apis*. *Apidologie* **2007**, *38*, 558–565. [CrossRef]

18. Williams, G.R.; Shutler, D.; Burgher-MacLellan, K.L.; Rogers, R.E.L. Infra-Population and community dynamics of the parasites *Nosema apis* and *Nosema ceranae*, and consequences for honey bee (*Apis mellifera*) hosts. *PLoS ONE* **2014**, *9*, e99465. [CrossRef]

19. Goblirsch, M.; Huang, Z.Y.; Spivak, M. Physiological and behavioral changes in honey bees (*Apis mellifera*) induced by *Nosema ceranae* infection. *PLoS ONE* **2013**, *8*, e58165. [CrossRef]

20. Milbrath, M.O.; van Tran, T.; Huang, W.-F.; Solter, L.F.; Tarpy, D.R.; Lawrence, F.; Huang, Z.Y. Comparative virulence and competition between *Nosema apis* and *Nosema ceranae* in honey bees (*Apis mellifera*). *J. Invertebr. Pathol.* **2015**, *125*, 9–15. [CrossRef]

21. Huang, W.-F.; Solter, L.; Aronstein, K.; Huang, Z. Infectivity and virulence of *Nosema ceranae* and *Nosema apis* in commercially available North American honey bees. *J. Invertebr. Pathol.* **2015**, *124*, 107–113. [CrossRef]

22. Martín-Hernández, R.; Meana, A.; Prieto, L.; Salvador, A.M.; Garrido-Bailón, E.; Higes, M. Outcome of colonization of *Apis mellifera* by *Nosema ceranae*. *Appl. Environ. Microbiol.* **2007**, *73*, 6331–6338. [CrossRef] [PubMed]

23. Higes, M.; Martín-Hernández, R.; Botías, C.; Bailón, E.G.; González-Porto, A.V.; Barrios, L.; del Nozal, M.J.; Bernal, J.L.; Jiménez, J.J.; Palencia, P.G.; et al. How natural infection by *Nosema ceranae* causes honeybee Colony Collapse. *Environ. Microbiol.* **2008**, *10*, 2659–2669. [CrossRef] [PubMed]

24. Higes, M.; Martín-Hernández, R.; Garrido-Bailón, E.; González-Porto, A.V.; García-Palencia, P.; Meana, A.; Nozal, M.J.D.; Mayo, R.; Bernal, J.L. Honeybee colony collapse due to *Nosema ceranae* in professional apiaries. *Environ. Microbiol. Rep.* **2009**, *1*, 110–113. [CrossRef] [PubMed]

25. Dainat, B.; Evans, J.D.; Chen, Y.P.; Gauthier, L.; Neumann, P. Predictive markers of honey bee colony collapse. *PLoS ONE* **2012**, *7*, e32151. [CrossRef] [PubMed]

26. Siede, R.; Berg, S.; Meixner, M. Effects of symptomless infections with *Nosema* sp. on honey bee colonies. In Proceedings of the OIE-Symposium Diagnosis and Control of Bee Diseases, Freiburg, Germany, 26–28 August 2008.

27. Gisder, S.; Hedtke, K.; Möckel, N.; Frielitz, M.-C.; Linde, A.; Genersch, E. Five-year cohort study of *Nosema* spp. in Germany: Does climate shape virulence and assertiveness of *Nosema ceranae*? *Appl. Environ. Microbiol.* **2010**, *76*, 3032–3038. [CrossRef]

28. Stevanovic, J.; Simeunovic, P.; Gajic, B.; Lakic, N.; Radovic, D.; Fries, I.; Stanimirovic, Z. Characteristics of *Nosema ceranae* infection in Serbian honey bee colonies. *Apidologie* **2013**, *44*, 522–536. [CrossRef]

29. Cox-Foster, D.L.; Conlan, S.; Holmes, E.C.; Palacios, G.; Evans, J.D.; Moran, N.A.; Quan, P.-L.; Briese, T.; Hornig, M.; Geiser, D.M.; et al. A Metagenomic survey of microbes in honey bee Colony Collapse Disorder. *Science* **2007**, *318*, 283–287. [CrossRef]

30. Currie, R.W.; Pernal, S.F.; Guzman-Novoa, E. Honey bee colony losses in Canada. *J. Apic. Res.* **2010**, *49*, 104–106. [CrossRef]

31. Guzman-Novoa, E.; Eccles, L.; Calvete, Y.; McGowan, J.; Kelly, P.G.; Correa-Benítez, A. *Varroa destructor* is the main culprit for the death and reduced populations of overwintered honey bee (*Apis mellifera*) colonies in Ontario, Canada. *Apidologie* **2010**, *41*, 443–450. [CrossRef]

32. Martin, S.J.; Hardy, J.; Villalobos, E.; Martín-Hernández, R.; Nikaido, S.; Higes, M. Do the honeybee pathogens *Nosema ceranae* and deformed wing virus act synergistically? *Environ. Microbiol. Rep.* **2013**, *5*, 506–510. [CrossRef]

33. Emsen, B.; Guzman-Novoa, E.; Hamiduzzaman, M.M.; Eccles, L.; Lacey, B.; Ruiz-Pérez, R.A.; Nasr, M. Higher prevalence and levels of *Nosema ceranae* than *Nosema apis* infections in Canadian honey bee colonies. *Parasitol. Res.* **2016**, *115*, 175–181. [CrossRef]

34. Guerrero-Molina, C.; Correa-Benítez, A.; Hamiduzzaman, M.M.; Guzman-Novoa, E. *Nosema ceranae* is an old resident of honey bee (*Apis mellifera*) colonies in Mexico, causing infection levels of one million spores per bee or higher during summer and fall. *J. Invertebr. Pathol.* **2016**, *141*, 38–40. [CrossRef] [PubMed]

35. Invernizzi, C.; Abud, C.; Tomasco, I.H.; Harriet, J.; Ramallo, G.; Campá, J.; Katz, H.; Gardiol, G.; Mendoza, Y. Presence of *Nosema ceranae* in honeybees (*Apis mellifera*) in Uruguay. *J. Invertebr. Pathol.* **2009**, *101*, 150–153. [CrossRef] [PubMed]

36. Guzman-Novoa, E.; Hamiduzzaman, M.M.; Arechavaleta-Velasco, M.E.; Koleoglu, G.; Valizadeh, P.; Correa-Benítez, A. *Nosema ceranae* has parasitized Africanized honey bees in Mexico since at least 2004. *J. Apic. Res.* **2011**, *50*, 167–169. [CrossRef]

37. Holt, H.L.; Grozinger, C.M. Approaches and challenges to managing *Nosema* (Microspora: Nosematidae) parasites in honey bee (Hymenoptera: Apidae) Colonies. *J. Econ. Entomol.* **2016**, *109*, 1487–1503. [CrossRef] [PubMed]

38. Gary, N.E.; Lorenzen, K. Vacuum devices for capturing and partitioning commingled subpopulations of honey bees (Hymenoptera: Apidae). *Ann. Entomol. Soc. Am.* **1990**, *83*, 1152–1154. [CrossRef]

39. Fries, I.; Chauzat, M.; Chen, Y.; Doublet, V.; Genersch, E.; Gisder, S.; Higes, M.; McMahon, D.P.; Martín-Hernández, R.; Natsopoulou, M.; et al. Standard methods for *Nosema* research. *J. Apic. Res.* **2013**, *52*, 1–28. [CrossRef]

40. Hamiduzzaman, M.M.; Guzman-Novoa, E.; Goodwin, P.H. A multiplex PCR assay to diagnose and quantify *Nosema* infections in honey Bees (*Apis mellifera*). *J. Invertebr. Pathol.* **2010**, *105*, 151–155. [CrossRef]

41. World Organization for Animal Health (OIE) Terrestrial Manual 2018 Section 3.2.4. Nosemosis of Honey Bees. Available online: https://www.oie.int/fileadmin/Home/eng/Health_standards/tahm/3.02.04_NOSEMOSIS_FINAL.pdf (accessed on 6 August 2020).

42. Copley, T.R.; Chen, H.; Giovenazzo, P.; Houle, E.; Jabaji, S.H. Prevalence and seasonality of *Nosema* species in Québec honey bees. *Can. Entomol.* **2012**, *144*, 577–588. [CrossRef]

43. McGowan, J.; De la Mora, A.; Goodwin, P.H.; Habash, M.; Hamiduzzaman, M.M.; Kelly, P.G.; Guzman-Novoa, E. Viability and infectivity of fresh and cryopreserved *Nosema ceranae* spores. *J. Microbiol. Meth.* **2016**, *131*, 16–22. [CrossRef]

44. Fenoy, S.; Rueda, C.; Higes, M.; Martín-Hernández, R.; del Aguila, C. High-level resistance of *Nosema ceranae*, a parasite of the honeybee, to temperature and desiccation. *Appl. Environ. Microbiol.* **2009**, *75*, 6886–6889. [CrossRef] [PubMed]

45. Campbell, A.T.; Robertson, L.J.; Smith, H.V. Viability of *Cryptosporidium parvum* Oocysts: Correlation of in vitro excystation with inclusion or exclusion of fluorogenic vital dyes. *Appl. Environ. Microbiol.* **1992**, *58*, 3488–3493. [CrossRef] [PubMed]

46. Delaplane, K.S.; van der Steen, J.; Guzman-Novoa, E. Standard methods for estimating strength parameters of *Apis mellifera* colonies. *J. Apic. Res.* **2013**, *52*, 1–12. [CrossRef]

47. Jaycox, E.R. *Laboratory Diagnosis of Bee Diseases*; Publication H-688; University of Illinois: Urbana, IL, USA, 1977.

48. Ontario Treatment Recommendations for Honey Bee Disease and Mite Control. Available online: http://www.omafra.gov.on.ca/english/food/inspection/bees/2017-treatment.htm (accessed on 10 August 2020).

49. Mortensen, A.N.; Jack, C.J.; McConnell, M.; Teigen, L.; Ellis, J. How to Quantify Nosema Spores Infection Rate in a Honey Bee Colony; Bulletin ENY-167; University of Florida Extension. Available online: https://edis.ifas.ufl.edu/in1123 (accessed on 10 August 2020).

50. Woyke, J. Correlations and interactions between population, length of worker life and honey production by honeybees in a temperate region. *J. Apic. Res.* **1984**, *23*, 148–156. [CrossRef]

51. Szabo, T.I.; Lefkovitch, L.P. Effect of brood production and population size on honey production of honeybee colonies in Alberta, Canada. *Apidologie* **1989**, *20*, 157–163. [CrossRef]
52. Guzman-Novoa, E.; Gary, N.E. Genotypic variability of components of foraging behavior in honey bees (Hymenoptera: Apidae). *J. Econ. Entomol.* **1993**, *86*, 715–721. [CrossRef]
53. Unger, P.; Guzman-Novoa, E. Honey bee (Hymenoptera: Apidae) guarding behaviour may not be a rare task. *Can. Entomol.* **2012**, *144*, 538–541. [CrossRef]
54. R Core Team. *R: A Language and Environment for Statistical Computing*; R Foundation for Statistical Computing: Vienna, Austria. Available online: http://www.R-project.org/ (accessed on 10 August 2020).
55. Jack, C.J.; Lucas, H.M.; Webster, T.C.; Sagili, R.R. Colony level prevalence and intensity of *Nosema ceranae* in Honey bees (*Apis mellifera* L.). *PLoS ONE* **2016**, *11*, e0163522. [CrossRef]
56. Pickard, R.S.; El-Shemy, A.A.M. Seasonal variation in the infection of honeybee colonies with *Nosema apis* Zander. *J. Apic. Res.* **1989**, *28*, 93–100. [CrossRef]
57. Smith, M.L. The Honey bee parasite *Nosema ceranae*: Transmissible via food exchange? *PLoS ONE* **2012**, *7*, e43319. [CrossRef]
58. Huang, W.-F.; Solter, L.F. Comparative development and tissue tropism of *Nosema apis* and *Nosema ceranae*. *J. Invertebr. Pathol.* **2013**, *113*, 35–41. [CrossRef]
59. Williams, G.R.; Shutler, D.; Little, C.M.; Burgher-Maclellan, K.L.; Rogers, R.E.L. The microsporidian *Nosema ceranae*, the antibiotic fumagilin-B®, and western honey bee (*Apis mellifera*) colony strength. *Apidologie* **2011**, *42*, 15–22. [CrossRef]
60. Botías, C.; Martín-Hernández, R.; Barrios, L.; Meana, A.; Higes, M. *Nosema* spp. infection and its negative effects on honey bees (*Apis mellifera* Iberiensis) at the colony level. *Vet. Res.* **2013**, *44*, 25. [CrossRef] [PubMed]
61. Villa, J.D.; Bourgeois, A.L.; Danka, R.G. Negative evidence for effects of genetic origin of bees on *Nosema ceranae*, positive evidence for effects of *Nosema ceranae* on bees. *Apidologie* **2013**, *44*, 511–518. [CrossRef]
62. Becerra-Guzmán, F.; Guzman-Novoa, E.; Correa-Benítez, A.; Zozaya-Rubio, A. Length of life, age at first foraging and foraging life of Africanized and European honey bee (*Apis mellifera*) workers, during conditions of resource abundance. *J. Apic. Res.* **2005**, *44*, 151–156. [CrossRef]
63. Eiri, D.M.; Suwannapong, G.; Endler, M.; Nieh, J.C. *Nosema ceranae* can infect honey bee larvae and reduces subsequent adult longevity. *PLoS ONE* **2015**, *10*, e0126330. [CrossRef]
64. Mattila, H.R.; Otis, G.W. Effects of pollen availability and *Nosema* infection during the spring on division of labor and survival of worker honey bees (Hymenoptera: Apidae). *Environ. Entomol.* **2006**, *35*, 708–717. [CrossRef]
65. Dussaubat, C.; Maisonnasse, A.; Crauser, D.; Beslay, D.; Costagliola, G.; Soubeyrand, S.; Kretzchmar, A.; Le Conte, Y. Flight behavior and pheromone changes associated to *Nosema ceranae* infection of honey bee workers (*Apis mellifera*) in field conditions. *J. Invertebr. Pathol.* **2013**, *113*, 42–51. [CrossRef]
66. Zheng, H.-Q.; Lin, Z.-G.; Huang, S.-K.; Sohr, A.; Wu, L.; Chen, Y.P. Spore loads may not be used alone as a direct indicator of the severity of *Nosema ceranae* infection in honey bees *Apis mellifera* (Hymenoptera:Apidae). *J. Econ. Entomol.* **2014**, *107*, 2037–2044. [CrossRef]
67. Dussaubat, C.; Brunet, J.L.; Higes, M.; Colbourne, J.K.; Lopez, J.; Choi, J.H.; Martín-Hernández, R.; Botías, C.; Cousin, M.; McDonnell, C.; et al. Gut pathology and responses to the microsporidium *Nosema ceranae* in the honey bee *Apis mellifera*. *PLoS ONE* **2012**, *7*, e37017. [CrossRef]
68. Kevan, P.G.; Guzman-Novoa, E.; Skinner, A.; Englesdorp, D.V. Colony Collapse Disorder (CCD) in Canada: Do we have a problem? *Hivelights* **2007**, *20*, 14–16.
69. Williams, G.R.; Tarpy, D.R.; vanEngelsdorp, D.; Chauzat, M.P.; Cox-Foster, D.L.; Delaplane, K.S.; Neumann, P.; Pettis, J.S.; Rogers, R.E.L.; Shutler, D. Colony Collapse Disorder in context. *BioEssays* **2010**, *32*, 845–846. [CrossRef] [PubMed]
70. Guzman-Novoa, E. Colony collapse disorder and other threats to honey bees. In *One Health Case Studies: Addressing Complex. Problems in a Changing World*; Cork, S., Hall, D.C., Liljebjelke, K., Eds.; 5M Publishing Ltd.: Sheffield, UK, 2016; pp. 204–216.

Article

Prevalence of the Microsporidian *Nosema* spp. in Honey Bee Populations (*Apis mellifera*) in Some Ecological Regions of North Asia

Nadezhda V. Ostroverkhova [1,2,*], Olga L. Konusova [1], Aksana N. Kucher [1], Tatyana N. Kireeva [1] and Svetlana A. Rosseykina [1]

[1] Invertebrate Zoology Department, Biology Institute, National Research Tomsk State University, 36 Lenina Avenue, 634050 Tomsk, Russia; olga.konusova@mail.ru (O.L.K.); kucheraksana@gmail.com (A.N.K.); emilia30@mail.ru (T.N.K.); rosseykina75@mail.ru (S.A.R.)

[2] Department of Biology and Genetics, Siberian State Medical University, 2 Moskovsky Trakt, 634055 Tomsk, Russia

* Correspondence: nvostrov@mail.ru; Tel.: +7-3822-529-461

Received: 7 July 2020; Accepted: 10 August 2020; Published: 13 August 2020

Abstract: Two species of microsporidia, *Nosema apis* and *Nosema ceranae*, are obligate intracellular parasites that are widespread in the world and cause the infectious disease (Nosemosis) of the Western honey bee *Apis mellifera*. Information on the prevalence and distribution of *Nosema* species in North Asia conditions is scarce. The main aim of the present study is to determine the prevalence of *Nosema* spp. (Nosemosis) in honey bees inhabiting some inland regions of North Asia (Western and Eastern Siberia, Altai Territory, Russia, and northeastern part of Kazakhstan). The objective of the paper is also to assess the influence of climatic factors on the spread of *N. ceranae*. Eighty apiaries in four ecological regions of North Asia (southern taiga, sub-taiga zone, forest steppe, and mountain taiga forests) were investigated with regard to distribution, prevalence, and diversity of *Nosema* infection in honey bees using duplex-PCR. *Nosema* infected bees were found in 65% apiaries of ecoregions studied, and coinfection was predominant (36.3% of *Nosema*-positive apiaries). Both *N. apis* and *N. ceranae* occur across subarctic and warm summer continental climates, but while *N. apis* predominates in the former, *N. ceranae* is more predominant in the latter. No statistically significant differences in *Nosema* distribution were identified in various climatic zones. In the sub-taiga zone (subarctic climate), low presence of colonies with pure *N. ceranae* and a significantly higher proportion of coinfection apiaries were revealed. Long-term epidemiological study of *Nosema* spp. prevalence in the sub-taiga zone showed a surprising percentage increase of *Nosema*-positive apiaries from 46.2% to 74.1% during 2012–2017. From 2012 to 2015, *N. apis* became a predominant species, but in 2016–2017, the coinfection was mainly detected. In conclusion, the results of this investigation showed that *N. ceranae* is widespread in all study ecoregions of North Asia where it exists in combination with the *N. apis*, but there is no replacement of *N. apis* by *N. ceranae* in the studied bee populations.

Keywords: honey bee; *Apis mellifera*; *Nosema ceranae*; *Nosema apis*; epidemiology; replacement; ecoregions; North Asia

---

1. Introduction

In Western honey bees, two species of Microsporidia causing nosemosis have been described: *Nosema apis* Zander, 1909 [1] and *Nosema ceranae* Fries et al., 1996 [2]. The microsporidian *N. apis*, a relatively benign pathogen, was considered to be the only causative agent of nosemosis in *Apis mellifera* for a long time [3]. Since 2006, a second *Nosema* species, *N. ceranae*, was identified in *A. mellifera* populations worldwide [4–13]. It should be noted that *N. ceranae* had been described as a microsporidian parasite of the Eastern honey bee *Apis cerana* in 1996 [2]. The parasite *N. apis*, responsible for nosemosis type A, is characterized by moderate virulence (mainly nosemosis is accompanied by dysentery), and bee colonies can often cure themselves under favorable environmental conditions [3,6]. *N. ceranae*, responsible for nosemosis type C, has been associated with reduced honey production, weakness, and increased colony mortality not preceded by any visible symptoms [14–16]. *Nosema* species also differ morphologically, genetically, in their ability to adapt to temperature, survival, and effects on the host [17–24]. For *N. ceranae*, a relatively new parasite for the *A. mellifera*, data on the virulence, the prevalence in different climates, as well as the role and effect on the viability and survival of bee colonies are contradictory [14,19,25–30].

Conflicting results from different studies can be attributed to many factors, such as the biological characteristics of honey bees (caste, age of the bees, commercially and traditionally managed bees) [31,32], the genetic diversity of honey bees, bee subspecies and lineages [24,33–35], climatic and environmental differences [7,18,36], beekeeping practices [10], as well as diagnostic methods [37] and research conditions (number of analyzed bees, time and method of sampling, natural population research or experiment) [31,38,39]. *Nosema* species can only be confirmed using molecular methods [39], which can have different levels of resolution, for example, the single-copy *Hsp70* gene method qPCR detects a lower amount of *N. ceranae* copies compared to the multicopy *16S rRNA* gene method [37]. When conducting experimental infection studies, the laboratory results may be affected by specific conditions such as the duration of the experiment, the temperature and humidity, the method of infection, number of bees in the cages, the type of their diet, etc. [31,39].

To understand the mechanisms of *Nosema* disease and the effect of *N. ceranae* infection on the host it is necessary to study the prevalence and distribution of *Nosema* species, primarily *N. ceranae*, in different regions and climatic zones and the long-term dynamics of infection of honey bees with various *Nosema* species.

The prevalence of *Nosema* species in *A. mellifera* populations in Eurasia is currently studied in most European countries [6,25,40–42] and some countries of Southwest Asia as well [29,43–47]. As the sole causative agent of nosemosis, *N. ceranae* was detected in Croatia [48], Central Italy [49,50], Iran [43,47], and Saudi Arabia [45]. In South-European countries, such as Italy [6,49,50] and Greece [25], *N. ceranae* had indeed practically replaced *N. apis* while this was not observed in Northern Europe (Ireland, Sweden, Norway, and Germany) [30,38]. For example, in Sweden, in the spring of 2007, 89.0% of *Nosema*-positive bee colonies had *N. apis* and 10.0% colonies had mixed *N. apis/N. ceranae* infections [38]. On the contrary, in Scotland, 86.2% of *Nosema*-positive bee colonies had mixed infections [40]. In Germany, in 2009, three infection categories were widespread: 48.5% *Nosema*-positive bee colonies were with *N. apis*, 33.8% were infected with *N. ceranae*, and 17.6% had mixed infection [27].

Climate is considered to be one of the main factors in the spread of *Nosema* species [26,30,49]. In warmer climates, *N. ceranae* is more competitive than *N. apis*; in contrast, in cold climates, *N. ceranae* spores appear to be much more vulnerable than the *N. apis* spores [26]. Laboratory data also suggest that the spread of *N. ceranae* across the globe is reduced in colder climates as *N. ceranae* spores are capable of surviving high temperatures and desiccation but are intolerant of cold [17,18,27,36,51]. However, the impact of weather conditions on the distribution of microsporidian *Nosema*, primarily *N. ceranae*, in the field is poorly understood [26].

The different sensitivity to temperatures in the *Nosema* species may be a potential explanation for the wider *N. ceranae* prevalence in warmer (subtropical) climate (Southern Europe, for example, the Mediterranean countries) compared to *N. apis*, which is more prevalent in temperate climates (Northern countries) [26]. Today, bees are threatened, and the cause of the problem is still unknown, which is why it is being described as Colony Collapse Disorder (CCD). Researchers suspect this may be due to a combination of various diseases including *N. ceranae*, environmental pollution, and farming practices, mainly due to large monoculture cropping and toxic phytosanitary products [52–54]. For the first time in Spain, bee colony mortality was clearly attributed to *N. ceranae* infection [7,14–16,55,56]. In 2004–2006, the prevalence of *N. ceranae* in dead bee colonies was about 90% [7]. It is assumed that colony collapse caused by *N. ceranae* is not restricted to Spain but is a global problem, at least it is regarded as a Europe-wide phenomenon. For example, in some Mediterranean countries, such as Greece [25], Israel [29], and Turkey [44,46], bee colony losses are also associated with *N. ceranae* infections. However, in Northern Europe, colony collapse could not be associated with *N. ceranae* [27,57]. These data pointed to climatic factors differentially influencing the prevalence and the virulence of *N. ceranae* in Europe [26–28] and/or differences in *N. ceranae* susceptibility between regionally predominating *A. mellifera* subspecies [24,34,39].

The prevalence of *N. ceranae* in more severe conditions in the continental regions of North Asia has not been adequately studied yet [13,58,59]. The purpose of this study was to determine the prevalence of *Nosema* infection in honey bees inhabiting some ecological regions of North Asia (southern taiga, sub-taiga zone, forest steppe taiga, and mountain taiga forests) and assess climatic factors that influence prevalence.

2. Materials and Methods

2.1. Ecological and Geographical Characteristics of the Region

We explored the *Nosema* infestation of honey bees in several regions of North Asia (Western and Eastern Siberia, Altai Territory, Russia, and northeastern part of Kazakhstan). The study apiaries were located in an area extending from 48° N to 65° N in the meridional direction and from 81° E to 92° E in the zonal direction (Figure 1).

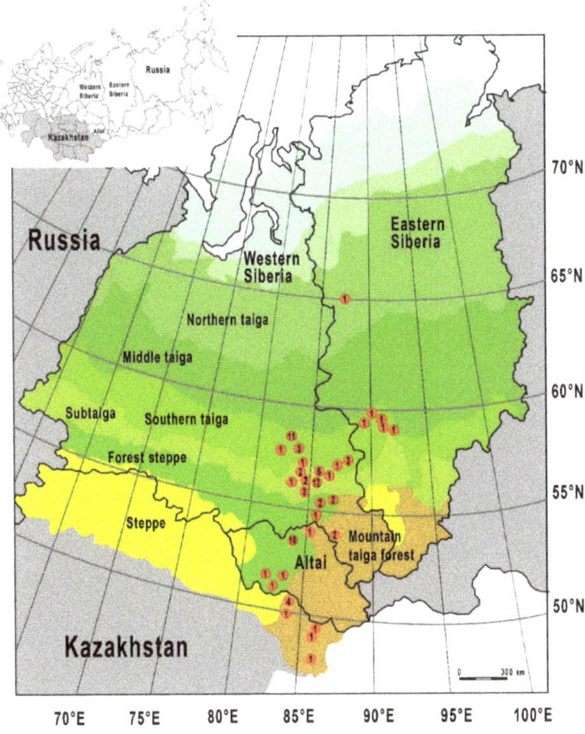

Figure 1. Distribution of monitored apiaries within the study ecoregions of North Asia (southern taiga, sub-taiga zone, forest steppe taiga, and mountain taiga forests). The number of the examined apiaries in the locality is indicated by numbers.

In this region, located at a significant distance from the ocean, the climate is regarded as a continental one (Köppen climate classification). The characteristic features of the continental climate are cold winters and hot summers. Air masses are usually cold in winter, warm in summer, but always relatively dry continental. The average temperature is above 10 °C in the warmest months, and during the coldest month, it is on the average below 0 °C. The annual temperature amplitudes can be 70 °C. The vegetation period in the regions of North Asia is short, wintering of honey bees is long (it usually lasts for 6 months, but sometimes it is up seven months).

In North Asia, various ecological regions are represented (Figure 1). We investigated apiaries located in 4 ecoregions, namely southern taiga, sub-taiga zone, forest steppe taiga, and mountain taiga forests (Table 1). In these regions, beekeeping is well-developed, especially there are lots of apiaries in the sub-taiga zone, forest steppe taiga, and mountain taiga forests. In the southern taiga, there are many apiaries, but they are located in far-away areas. In addition, we also investigated one apiary (near Turukhansk, 65°47'35" N, 87°57'44" E) located in the Northern taiga (the north of Eastern Siberia; Figure 1). Beekeeping is not well-developed in this ecoregion due to a very cold climate, but there exist single apiaries in that area [59].

Selected apiaries were situated in both flatland (southern taiga, sub-taiga zone, forest steppe taiga) and mountainous (mountain taiga forests) parts under different climatic conditions. The southern taiga and sub-taiga are characterized by a subarctic climate, while the forest steppe taiga and mountain taiga forests are distinguished by a warm summer continental climate (Table 1). Although the southern taiga and sub-taiga are characterized by a subarctic climate, there are some differences in the average monthly and average annual temperatures, sum of active temperatures above +10.0 °C (Σt). For the southern taiga zone, significant variability of these indicators is observed. Similarly, for zones with a warm summer continental climate (forest steppe taiga and mountain taiga forests), a significant range of the main climatic indicators is surveyed in the zone of mountain taiga forests (Table 1).

Table 1. Geographic and climatic characteristics of the study ecoregions of North Asia.

Ecological Region	Area Coordinates	Climate (Group D) [#]	Altitude, m	Average Temperature (°C)			Average Annual Precipitation, mm	Frost-Free Period (Days)	Σt, Degree Days [*]
				Annual	In January	In July			
Southern taiga (boreal forest)	from 57°00′00″–59°50′00″ to 82°39′00″–92°08′00″	Subarctic climate	60–130	−0.9–(−2.0)	−19.8–(−21.6)	17.9–18.3	250–500	90–120	1472–1820
Sub-taiga zone	from 56°15′00″–56°48′21″ to 83°58′00″–86°42′00″	Subarctic climates	90–120	−0.6	−19.2	18.1	350–550	100–120	1650–1800
Forest steppe	from 51°00′00″–55°19′59″ to 81°28′00″–85°30′00″	Warm summer continental climate	200–250	2.1–2.6	−16–(−19)	18–20	350–450	120–123	1900–2100
Mountain taiga forests	from 48°34′03″–53°45′00″ to 82°18′25″–87°07′00″	Warm summer continental climate	290–1685	0.3–2.4	−12.6–(−24.0)	17–22	500–900	123–135	2200–2400

[#] Köppen climate classification was used. [*] Degree days are a specialist kind of weather data, calculated from readings of outside air temperature.

2.2. Historical Background

In Siberia and Altai Territory, the honey bee was introduced about 230 years ago. It was a dark forest bee *Apis mellifera mellifera* that well-adapted to the local climate and plant communities as well. The honey bee population is an artificial population whose wintering is controlled by people [60]. The first apiaries were formed in the mountainous regions of Western Siberia and Altai Territory. Subsequently, the development of beekeeping took place in Eastern Siberia, namely in the southern taiga, sub-taiga, and forest steppe.

2.3. Research Algorithm

It should be noted that we have conducted our research according to two lines. The first line of research is the study of *Nosema* infection of honey bees in apiaries in North Asia. A total of 80 distant apiaries located in four ecoregions (southern taiga, sub-taiga zone, forest steppe taiga, and mountain taiga forests) were monitored for *Nosema* infection between spring 2016 and autumn 2017 (for most apiaries, the material was mainly collected in the early summer due to the long cold spring; usually after the first flight of bees). We examined 24 apiaries in the southern taiga, 27 apiaries in sub-taiga, 18 apiaries in forest steppe taiga, and 11 apiaries in the mountain taiga forests (Figure 1). In addition, in the northern taiga, one apiary near Turukhansk was also investigated. Within each apiary, a minimum of three colonies (10% of the total colonies) were randomly selected with regard to *Nosema* detection.

The second line of research is connected with a retrospective analysis (a long-term dynamics) of *Nosema* infestation of apiaries in the sub-taiga zone in 2012–2017. A total of 79 distant apiaries were analyzed. For a retrospective study, we used bee samples collected in sub-taiga apiaries from 2012 to 2017 and stored them in a biobank at a temperature of −20 °C.

It is worth pointing out that the apiaries were visited only once. In each sampling, forager bees were collected at the entrance of each sampled hive. For each bee sample, the beekeeper provided information, including the location, history and other characteristics of the apiary, the origin of bees, the incidence of bee colonies, etc. No hives of the apiaries sampled had a history of external signs referable to nosemosis and no signs of the disease were present at the time of sampling. Mass death of bees after wintering in the study apiaries was not observed. Between 60 to 70 workers from each bee colony were pooled and used for DNA isolation. The presence of nosemosis in the bee colony was examined using a polymerase chain reaction (PCR).

2.4. Experimental Procedures

DNA was extracted from the midgut of bees (a pool of 60–70 individuals was formed) using a DNA purification kit, PureLink™ Mini (Invitrogen, Carlsbad, CA, USA) according to the manufacturer's protocol. After the DNA extraction, the samples were submitted to duplex-PCR [7]. For the diagnosis of two *Nosema* species, we used two types of primers. The primer sequences used to amplify the 321 bp fragment corresponding to the 16S ribosomal gene of *N. apis* were 321APIS-FOR 5′-GGGGGCATGTCTTTGACGTACTATGTA-3′ and 321APIS-REV 5′-GGGGGGGCGTTTAAAATGTG AAACAACTATG-3′. The primer sequences utilized to amplify the 218 bp fragment corresponding to the 16S ribosomal gene of *N. ceranae* were 218MITOC-FOR 5′-CGGCGACGATGTGATAT GAAAATATTAA-3′ and 218MITOC-REV 5′-CCCGGTCATTCTCAAACAAAAAACCG-3′ [7].

PCR was performed using a thermal MyCycler T100 (BioRad, Foster City, CA, USA) in a reaction volume of 20 µL containing 1 µL of template DNA, 1× PCR buffer, 1.5 mM MgCl$_2$, 200 µM of each dNTP, 0.2 µM of each forward and reverse primer and 1U Taq polymerase (Fermentas, Thermo Fisher Scientific, Chelmsford, MA, USA). The routine consisted of an initial denaturation step at 94 °C for 2 min, followed by 35 cycles of 94 °C for 30 s, 58 °C for 30 s and 72 °C for 1 min, and a final extension step at 72 °C for 5 min. PCR products were analyzed on 1.5% (*m/v*) agarose gels. Gels were stained with ethidium bromide and visualized using UV illumination (Gel Doc XR+, BioRad, Foster City, CA,

USA). All analyses were carried out in duplicate and identical results were obtained. For each PCR, positive control (reference *N. apis* and *N. ceranae* DNA extracts as template) was used. Negative control (ddH$_2$O) was also included in each run of PCR amplification to detect possible contamination.

If only *N. ceranae* or only *N. apis* were detected in all examined bees, the apiary infection category was pure *N. ceranae* or pure *N. apis*, respectively. An apiary was considered coinfected if both *Nosema* species were detected in honey bees.

2.5. Meteorological Data

To study the climatic characteristics of ecoregions (Table 1), we used reference material [61–63] including the average indicators of meteorological observations over a long-term period (for 30 years), namely, the average annual temperature and average annual precipitation, the average temperature in January and July, the duration of the frost-free period and the sum of active temperatures (Σt).

To assess long-term temporal trends in the prevalence of two *Nosema* species in the sub-taiga zone and analyze the associations of the prevalence of *Nosema* infection with climatic factors, we used the meteorological data obtained by the Tomsk weather station, 56°29' N, 84°56' E [62]. The following data were employed: average annual temperatures, average daily temperatures, duration of the period with active temperatures, sum of precipitation just for the same period of the year. Reference data were used to calculate two parameters: sums of active temperatures (Σt) and hydrothermal coefficient (HTC).

The sum of active temperatures is one of the main indicators of the territory's thermal resources. Active temperatures designate daily average temperatures above +10.0 °C. The sum of the active temperatures (Σt) is calculated for a period with an average daily air temperature above +10.0 °C.

To characterize the moisture supply (humification conditions) in the area concerned, we used the hydrothermal coefficient suggested by Selyaninov (HTC) [64]. The hydrothermal coefficient (K) is calculated as the ratio of precipitation (R) to the sum of temperatures (Σt) for a period with temperatures above +10.0 °C: K = R*10/Σt. At HTC values 1.1–1.4, moisture supply is regarded as optimal; at values 0.76–1.0 it is insufficient, and at values 1.41–1.5 it is an elevated one.

2.6. Statistical Analysis

The results of the research are presented using numbers and percentages. Statistical significance of qualitative data was determined using the Chi-square test or the Z-test for proportions. To compare apiaries infected and uninfected with *Nosema* between different ecoregions, we used a two-sample proportion Z-test. The Chi-square test was used to compare the incidence of *Nosema* infection in the sub-taiga zone between three sampling years. In the case of a small number of one of the comparison classes, we used the chi-square test with Yates' correction. *Nosema*-positive, *N. apis*-positive, *N. ceranae*-positive, and coinfected apiaries were subgroups to be compared. Statistical analysis was performed using the Statistica program.

3. Results

3.1. Infestation of Apiaries with Nosema Species in Four Ecoregions of North Asia in 2016–2017

Honey bee samples collected in 2016–2017 from 80 apiaries of several ecological regions of North Asia were examined for the presence of *Nosema* infections (Table 2). In all the ecoregions of North Asia studied, two species of Microsporidia were registered in honey bees: *N. apis* and *N. ceranae*. In total, among 80 apiaries which had been analyzed, *Nosema* spp. were detected in 52 honey bee apiaries (65.0%): 15.0% of *Nosema*-positive apiaries were infected with only *N. apis*; 13.75%—only *N. ceranae*; in most apiaries (36.25%), both *Nosema* species were detected. In general, in North Asia, the infestation of apiaries with both types of *Nosema* (coinfection) was higher compared to the infestation of apiaries with only *N. apis* or only *N. ceranae*, but statistically significant differences were not shown (proportion Z-test; Z = 1.31, *p* > 0.05).

Table 2. Prevalence of colonies infected with only *N. apis* (*N. apis*) or only *N. ceranae* (*N. ceranae*) or coinfection (*N. apis* and *N. ceranae*) in apiaries in several ecoregions of North Asia from spring 2016 to autumn 2017.

Ecoregion	Total Number of Analyzed Apiaries	*Nosema* Infection not Detected		Infected Apiaries (Infection Categories)					
				N. apis		*N. ceranae*		*N. apis* and *N. ceranae*	
		N	%	N	%	N	%	N	%
Southern taiga	24	8	33.33	4	16.67	4	16.67	8	33.33
Sub-taiga zone	27	7	25.93	5	18.52	2	7.41	13	48.15
Forest steppe	18	9	50.00	2	11.11	3	16.67	4	22.22
Mountain taiga forests	11	4	36.36	1	9.10	2	18.18	4	36.36
Total	80	28	35.00	12	15.00	11	13.75	29	36.25

The total number of analyzed apiaries along with the numbers (n) and proportions (%) of apiaries within each infection category are presented.

In all study ecoregions of North Asia (southern taiga, sub-taiga, forest steppe, and mountain taiga forests), a significant number of apiaries infected with *Nosema* were detected: ranging from 50.0% of apiaries in the forest steppe zone to 74.1% in the sub-taiga zone (Figure 2).

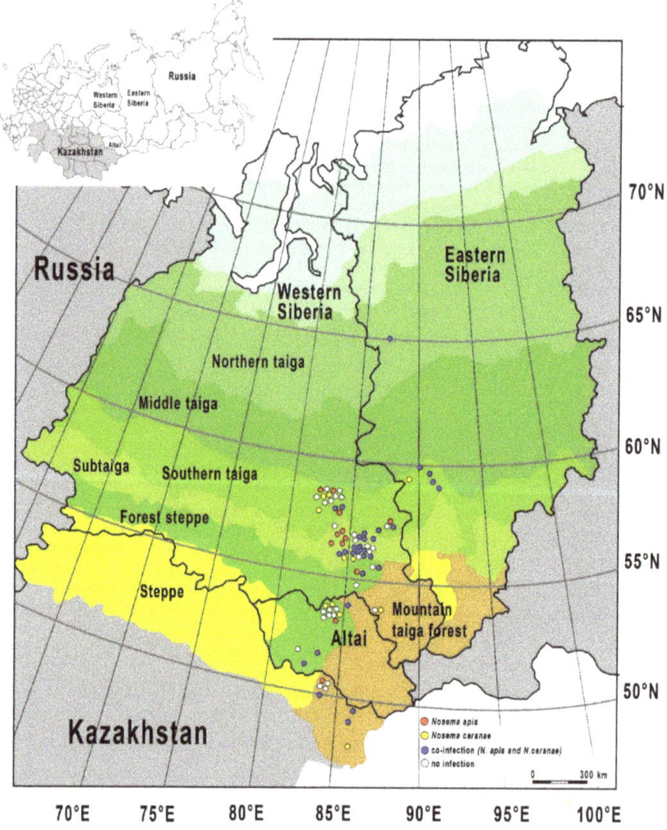

Figure 2. Prevalence of *Nosema* infections in several ecoregions of North Asia (southern taiga, sub-taiga, forest steppe, and mountain taiga forests).

At the same time, the distribution of *Nosema* species differed in the study ecoregions (Table 2). We detected three possible infection categories of apiaries: only *N. apis* infection, only *N. ceranae* infection, and coinfection of two *Nosema* species. coinfection prevailed in apiaries of all the ecoregions and it was detected in 65.0% of *Nosema*-positive apiaries (Figure 2).

The spread of *Nosema* infection in apiaries of various ecoregions was as follows:

(i) In the southern taiga, the number of apiaries infected with either *N. apis* or *N. ceranae* was the same (16.7%).

(ii) In the forest steppe and mountain taiga forests (similar climatic conditions—warm summer continental climate), the spread of *Nosema* species is similar: 9.1% and 11.1% of apiaries with only *N. apis*; 16.7% and 18.2% of apiaries with only *N. ceranae*; 22.2% and 36.4% of apiaries with coinfection (*N. apis* and *N. ceranae*), respectively (proportion Z-test; $Z < 0.39$, $p > 0.05$).

(iii) In the sub-taiga zone, the spread of Nosema infection in apiaries is different from that in the other ecoregions. The smallest number of apiaries with only *N. ceranae* (7.4%) and the highest number of apiaries with mixed infection (48.2%) were revealed. In addition, about a quarter of apiaries (18.5% of *Nosema*-positive apiaries) are infected only with *N. apis* (similar to the southern taiga zone; subarctic climate).

However, no statistically significant differences in the distribution of both common *Nosema* infections and individual *Nosema* infection (pure *N. apis*, pure *N. ceranae*, coinfection) were found between the ecological regions ($Z < 1.76$, $p > 0.05$). Additionally, no statistically significant differences in *Nosema* distribution were identified in various climatic zones (subarctic climate and warm summer continental climate; $Z < 1.52$, $p > 0.05$).

3.2. Dynamics of Infection of Apiaries with Different Nosema Species in Sub-Taiga from 2012 to 2017

Despite the fact that no significant differences were found in the incidence of *Nosema* infection both between ecological regions and climates (probably due to the small number of study apiaries), there exist some trends in research. For example, a difference in *Nosema* prevalence characteristic of the two groups of apiaries with coinfection between the sub-taiga and forest steppe zones approached the level of statistical significance (proportion Z-test; $Z = 1.76$, $p < 0.10$).

To gain a more complete understanding of the spread of *Nosema* infections in the sub-taiga zone, we examined long-term temporal trends in the prevalence of two *Nosema* species from 2012 to 2017 (Table 3).

Table 3. Spreading of *Nosema* species in apiaries of the sub-taiga zone from 2012 to 2017.

Study Period	Total Number of Analyzed Apiaries	*Nosema* Infection not Detected	Infected Apiaries (Infection Categories)		
			N. apis	*N. ceranae*	Coinfection
2012–2013	26	53.85	42.31	3.85	0
2014–2015	26	42.31	42.31	3.85	11.54
2016–2017	27	25.93	18.52	7.41	48.15
Total	79	40.51	34.18	5.06	20.25

In the course of six years from 2012 to 2017, there was a surprising increase in *Nosema* infection in the sub-taiga zone (Table 3). The percentage of *Nosema*-positive apiaries increased from 46.1 to 57.7% over a period of the first four years to 74.1% during the last two years ($\chi^2 = 4.316$, $p = 0.038$). If in 2012–2015, *N. apis* was predominant among the other infectious categories compared to 2016–2017 ($\chi^2 = 10.800$, $p = 0.002$) whereas, in 2016–2017, coinfection (*N. apis* and *N. ceranae*) was mainly detected among *Nosema*-positive groups ($\chi^2 = 5.298$, $p = 0.022$; Table 3).

Three various periods with regard to the frequency of coinfection apiaries were established in the following order: from 2012 to 2013, from 2014 to 2015, and from 2016 to 2017. During the first period (2012–2013), no apiary with coinfection was detected. In the course of the second period (2014–2015), a small number of apiaries with coinfection was found (11.5%). Interestingly, the number of apiaries, either with only *N. apis* (42.3%), or with only *N. ceranae* (3.9%), was the same during these study periods (2012–2013, and 2014–2015). Finally, in the course of the third period (2016–2017), coinfection was identified in about half of the study apiaries (48.2%), which is statistically different from the 2014–2015 period ($\chi^2 = 6.776$, $p = 0.01$). In addition, the number of apiaries where only *N. apis* was diagnosed decreased by half (from 42.3 to 18.5%; $\chi^2 = 3.557$, $p = 0.06$). In 2016–2017, the smallest number of apiaries was detected with only *N. ceranae* (7.4%) but it doubled during the observation period, from 3.9% to 7.4% ($\chi^2 = 0.001$, $p = 0.97$).

We tried to assess the influence of both climatic factors separately (temperature, humidity) and the integrated indicator "hydrothermal coefficient" (HTC) on the spread of *Nosema* infection in the sub-taiga zone in 2012–2017.

An increase in the *Nosema* infection of honey bees and the spread of *N. ceranae* in the sub-taiga zone may be associated with an increase in the duration and heat supply of the active life period of biological objects from 2012 to 2017 (Table 4). On the whole, the indicators of the temperature regime in the sub-taiga zone from 2012 to 2017 reflect a general tendency of warming of a landing air layer in the Russian territory. So, average annual temperature deviations from average long-term values toward an increase by 1.2–3.1 °C were observed in the study area. The biggest deviations were identified in 2013 and from 2015 to 2017. What is more, abnormally high indicators of the vegetation period heat supply were observed in 2012 and in 2015–2017. For example, in 2016, the sum of active temperatures (Σt) comprised 2217 degree days in view of a basic value of 1650–1800 degree days.

Table 4. Temperature regime and humification conditions in the sub-taiga zone according to observations made at the Tomsk weather station, 56°29′ N, 84°56′ E.

Year of Study	Average Annual Temperature (°C)	Σt, Degree Days *	Period with Active Temperatures, Days	Amount of Precipitation for the Period with Active Temperatures, mm	Hydrothermal Coefficient (HTC)
2012	0.6	2064	121	199	0.96
2013	1.7	1587	86	200	1.26
2014	0.8	1590	90	162	1.01
2015	2.5	2107	128	277	1.31
2016	1.7	2217	128	203	0.92
2017	2.2	1912	124	285	1.49

* Degree days are a specialist kind of weather data, calculated from readings of outside air temperature. Σt—the sum of the temperatures for a period with an average daily air temperature above +10.0 °C. Active temperatures are average daily temperatures above +10.0 °C.

We also observe significant variations in such climatic indicators as the period with active temperatures and the hydrothermal coefficient (HTC) in the study area (Table 4). For example, in 2012–2017, the hydrothermal coefficient (HTC) ranged from 0.92 to 1.49. An alternating number of years with an insufficient period of moisture supply (2012, 2014, 2016) was followed by a period of optimal moisture supply. It is interesting to note that the moisture period of 2017 is characterized as an excessive one.

The year 2012 along presents a certain interest because it was characterized by high summer temperatures and a low moisture supply. During the active growing season, 86 cases of the absolute maximum air temperature exceeding in the course of one day were recorded. Most cases were reported in the south of Western Siberia, where abnormally hot weather was observed. For example, in the Tomsk Region (Western Siberia), a significant deviation temperature from the required standard (from 1.3 °C to 7.2 °C) was first observed in June–July over the past 60 years. In Tomsk in particular (56°29′ N, 84°56′ E), the temperature deviation from the average long-term standard came to 5.7 °C in June and 5.3 °C in July. According to the data of the Tomsk weather station, in June, the amount of precipitation was 54% according to the required standard, it comprised only 33% of the standard data. In Tomsk, in July, the HTC value accounted for 0.35, which was indicative of severe drought. That is the reason for the extreme meteorological conditions of 2012 which were quite different from those during a long-term observation period in this area [65].

Under these conditions, we expected a widespread occurrence of Microsporidia, primarily *N. ceranae*, in this region. However, in the 2012 beekeeping season, we did not find a single bee colony infected with *Nosema* spores, and the bees were distinguished by a high flying activity and a high honey productivity as well. Therefore, there is no need for making any assumptions relating to the impact of climatic conditions on the *Nosema* distribution in 2012. It is probably not the temperature, but the good resistance of the examined bee colonies to Microsporidia which is the determining factor.

In 2013–2014, the heat supply to the growing season was very low (86 and 90, respectively), whereas, in 2015–2017, the period with active temperatures was the longest (from 124 to 128 days). If in 2013–2015, in apiaries of the sub-taiga, *N. apis* was the predominant species, then in 2016–2017, coinfection was most common. None of the studied climatic indicators can be associated with the spread of *Nosema*, coinfection in particular, in the sub-taiga zone.

Probably, both biotic and abiotic factors are involved to a greater or lesser extent individually or synergistically. Environmental conditions can exert a direct impact on the parasite or indirectly influence them by altering host physiology, behavior, and immunity [36].

4. Discussion

We have investigated the spread of *Nosema* species in several ecological regions of North Asia (southern taiga, sub-taiga, forest steppe, and mountain taiga forests) and the influence of climatic conditions (subarctic and warm summer continental climates) on the prevalence of *N. ceranae*.

The geographic distribution of *Nosema* spp. in several ecoregions of North Asia can be represented as follows:

(i)   The presence of bee colonies infected with pure *N. ceranae* in all the study ecoregions;

(ii)   No significant differences were identified in the incidence of *Nosema* infection between subarctic and warm summer continental climates. There is only a trend towards a high proportion of apiaries with only *N. apis* infection in the subarctic climate, and vice versa, towards a higher proportion of apiaries with only *N. ceranae* infection in warm summer continental climate;

(iii)   Coinfection detected in most study apiaries of North Asia;

(iv)   A higher proportion of coinfection apiaries and a lower presence of colonies with pure *N. ceranae* in the sub-taiga zone (subarctic climate);

(v)   There is no replacement of *N. apis* by *N. ceranae* in the study honey bee populations of North Asia, but their coexistence is registered.

Our results show the widespread use of *N. ceranae* in all study ecoregions of North Asia, as has been documented in many regions of the world [66]. In addition, *N. ceranae* was found in the apiary in very cold climates in Northern taiga (near Turukhansk). Our epidemiology data clearly show that *N. ceranae* became successfully established and expanded its presence in honey bee populations in both the southern and northern regions of North Asia. It seems likely that changes in the spread of two *Nosema* species should be considered in the context of stable changes to the heat supply regime of the study area. Moreover, the spread of the parasite during a certain period is determined by the weather conditions of the previous period. For example, unfavorable conditions in 2013–2014 could have been the reason for a decrease in bee resistance to diseases and, consequently, for the spread of infection in subsequent years (2015–2017), characterized by an increase in heat supply during the growing season.

The increasing worldwide prevalence of *N. ceranae* in the past decade, and, conversely, a decrease in the *N. apis* prevalence (even the absence of this parasite) in some regions suggests that *N. ceranae* might be displacing *N. apis* [6,26,67]. However, while in Southern Europe, especially in the Mediterranean countries like Spain, Italy, Israel, Greece, and Turkey, *N. ceranae* has been the dominant species for 10 years [6,25,37,49], in Northern Europe (Ireland, Sweden, Norway, and Germany), *N. apis* is still the predominant species [27,30,38]. For example, in Sweden, the majority of bee colonies (89%) were infected only with *N. apis*, and in other bee colonies, coinfection was identified [38]. In Europe, there is a South to North gradient in the distribution of *Nosema* species, which can be determined by the climatic characteristics of the regions [6,27,30]. In European countries with hot summers and moderate winters, *N. ceranae* is predominant and nearly replaced *N. apis* over the past decade in Spain and Italy [5–7,14]. On the contrary, in countries with rather cold and long winters such as Sweden and Germany, *N. apis* is a prevailing species [27,30]. For example, in Northeast Germany, despite a significant increase in

N. ceranae prevalence during the last 12 years, no replacement of *N. apis* by *N. ceranae* took place in the honey bee population. For replacement of *N. apis* by *N. ceranae* at the population level, a simple increase in *N. ceranae* infection prevalence is not sufficient but *N. apis* infection prevalence should have concomitantly decreased during the study period. However, in Northeast Germany, a significant decrease in *N. apis* infection prevalence was only observed in autumn, and no significant change in *N. apis* infection prevalence was found in spring [30].

In this study (for example, sub-taiga zone) and in our earlier studies of various regions of North Asia, for example the Tomsk Region [13], we also showed that starting from a very low level, the prevalence of *N. ceranae* infections significantly increased continuously in the observed honey bee populations over the last six years. However, unlike Northeast Germany, where values for coinfection prevalence of bee colonies ranged between 0.0% and 10.0% during 2005–2016 [30], in North Asia, coinfection is widespread and was found in more than 30% of apiaries. However, a similar distribution pattern of *Nosema* infection has been identified in North Asia and Scotland. So, in Scotland, in the 70.4% of the bee colonies, the presence of both *N. ceranae* and *N. apis* was detected [40]. Interestingly, an increasing gradient of coinfection from North to South was also observed in some countries of the southern hemisphere, for example, in Argentina: coinfection was more prevalent in regions with temperate (77.9%) as compared to those of subtropical climate (22.1%) [68].

The high prevalence of coinfection (*N. ceranae* and *N. apis*) in the studied ecoregions of North Asia may suggest that *N. ceranae* is moving from warm summer continental to subarctic climate regions. Despite the distribution of *N. ceranae* infection in more severe climatic conditions (subarctic climate), we revealed some trends in the prevalence of pure *Nosema* infections in subarctic and warm summer continental climates. In warm summer continental climate, pure *N. ceranae* infection predominates, while in subarctic climate, pure *N. apis* infection is widespread.

As recently demonstrated in laboratory studies, mixed infections (*N. apis* and *N. ceranae*) negatively affected honey bee survival more than single *Nosema* infections [69]. However, contrary to this publication, in Siberia, bee colonies living in nature in very cold climates for a long time were relatively healthy, and *N. ceranae* is not associated with colony depopulation or honey bee collapse.

To determine the possible causes of the observed pattern of *Nosema* infection in honey bees in North Asia, we studied the history of beekeeping in Siberia and Altai Territory, namely, bee diseases and registered cases of mass death of bee colonies. A characteristic feature of the honeybee populations in Siberia is their long-term habitation in an isolated area. Honey bees have adapted to very cold climates. Under these conditions, certain parasite–host relationships, including Microsporidia, were formed.

Previously, nosemosis in honey bees in Siberia and Altai Territory was attributed exclusively to *N. apis*. The first description of *N. ceranae* infection in bees in North Asia using molecular genetic methods refers to 2009: the Tyumen Region, Altai Territory [58], and the Tomsk Region [13,60]. In addition, we identify the *N. ceranae* in the *A. m. mellifera* bee colonies from the long-isolated apiaries in the far-away taiga (Yenisei population of Krasnoyarsk Krai, Eastern Siberia), where new bees have not been imported for more than 60 years [59]. Our results provide evidence that *N. ceranae* infection occurs in bee colonies living in cold climates in Siberia, and this parasite is not associated with a colony depopulation or honey bee collapse. The connection of a long wintering and the nosemosis caused by *N. apis* was noted, but cases of mass bee mortality in study areas of North Asia are rather rare [70,71].

The first case of mass mortality of bee colonies in Siberia from the southern taiga to the forest steppe and mountain forests was described in the 1880s. However, the reasons for such an occurrence remain unknown. The first documented cases of mass bee mortality from nosemosis date back to 1914–1917 [71]. The presence of the causative agent was confirmed both by the clinical picture and by the microcopy of the pathological material. It is interesting to note that in the following years (1980s), the most noticeable losses of bee colonies in Siberia were associated with the rapid spread of varroosis (*Varroa destructor*) [70]. In 2017, in Altai Territory, the mass bee mortality was marked at the end of wintering [72]. The most probable reason for the spread of diseases, including nosemosis, and, possibly, the mass bee mortality is bee hybridization because the southern bee subspecies and hybrids characterized by a decrease in immunity did not withstand a particularly prolonged wintering. It can be assumed that diseases, including nosemosis, are probably a consequence of bee hybridization. The intensive importation of honey bees from the southern regions of Russia had been practiced in the Western Altai since the 1940s. For the southern taiga and sub-taiga this process is typical for the last two decades. Probably, both biotic and abiotic factors are involved to a greater or lesser extent individually or synergistically. Environmental conditions can exert a direct impact on the parasite or indirectly influence them by altering host physiology, behavior, and immunity [36]. In addition to the geographic and climatic factors, the genetic features of the host may affect the coadaptation of the honey bee and *Nosema* parasite, and the *Nosema* spp. prevalence in honey bee populations [24,34,68]. It is assumed that the variation between bee colonies in susceptibility to infection by *N. ceranae* is linked to genetic variability in workers from resistance to tolerance [24,34,39].

Thus, the issues relating to the distribution of *Nosema* spp. and the consequences of infection for bee colonies have not been resolved yet. The virulence of *N. ceranae* could be influenced by climatic conditions [23,26,28] or might actually depend on honeybee race, honeybee genetic diversity [33,34,73–75], and multiple *N. ceranae* genetic variants resulting in different biological consequences [76]. In this regard, it is of considerable interest to study the long-term and seasonal dynamics of *Nosema* infections in bee colonies, the relationship between pathogens and honeybees taking into account their genetic features, and the role of nosemosis pathogens in mass mortality of honeybees. Evidently, extended and detailed research is urgently required in order to elucidate a complete effect of *N. ceranae* infection on *A. mellifera* colonies in various geographical and climatic areas. The protective mechanisms behind this pattern remain to be studied.

5. Conclusions

The results of the research have shown that coinfection of two *Nosema* species, *N. apis* and *N. ceranae*, is widespread in all the ecological regions of North Asia studied. In the sub-taiga, a significant increase in *Nosema* infection has been observed for the last 6 years. There is no replacement of *N. apis* by *N. ceranae* in the studied bee populations. Despite the distribution of *N. ceranae* infection in subarctic climate, we have revealed some trends in the prevalence of pure *Nosema* infections in subarctic climate (pure *N. apis* predominates) and warm summer continental climate (pure *N. ceranae* is widespread). Further long-term research will contribute to understanding the interaction between the two *Nosema* species and the role of various factors, primarily climate, in the spread of these parasites of honey bee populations.

Author Contributions: Conceptualization, N.V.O. and O.L.K.; investigation, T.N.K. and S.A.R.; methodology, N.V.O., O.L.K., A.N.K., T.N.K. and S.A.R.; writing—original draft, N.V.O. and O.L.K.; writing—review and editing, N.V.O. and A.N.K. All authors have read and agreed to the published version of the manuscript.

Funding: This work was supported by the Tomsk State University competitiveness improvement programme.

Acknowledgments: The work would have been impossible without the help and assistance of Pilyukova A.V., a senior teacher of the Tomsk State University.

Conflicts of Interest: The authors declare no conflict of interest.

References

1. Zander, E. Tierische Parasiten als Krankheitserreger bei der Biene. *Münchener Bienenztg.* **1909**, *31*, 196–204.
2. Fries, I.; Feng, F.; da Silva, A.; Slemenda, S.B.; Pieniazek, N.J. *Nosema ceranae* n. sp. (Microspora, Nosematidae), morphological and molecular characterization of a microsporidian parasite of the Asian honey bee *Apis cerana* (Hymenoptera, Apidae). *Eur. J. Protistol.* **1996**, *32*, 356–365. [CrossRef]
3. Fries, I. *Nosema apis*—A parasite in the honey bee colony. *Bee World* **1993**, *74*, 5–19. [CrossRef]
4. Fries, I.; Martin, R.; Meana, A.; García-Palencia, P.; Higes, M. Natural infections of *Nosema ceranae* in European honey bees. *J. Apic. Res.* **2006**, *47*, 230–233. [CrossRef]
5. Higes, M.; Martin, R.; Meana, A. *Nosema ceranae*, a new microsporidian parasite in honeybees in Europe. *J. Invertebr. Pathol.* **2006**, *92*, 93–95. [CrossRef]

6. Klee, J.; Besana, A.M.; Genersch, E.; Gisder, S.; Nanetti, A.; Tam, D.Q.; Chinh, T.X.; Puerta, F.; Ruz, J.M.; Kryger, P.; et al. Widespread dispersal of the microsporidian *Nosema ceranae*, an emergent pathogen of the western honey bee, *Apis mellifera*. *J. Invertebr. Pathol.* **2007**, *96*, 1–10. [CrossRef]

7. Martín-Hernández, R.; Meana, A.; Prieto, L.; Salvador, A.M.; Garrido-Bailon, E.; Higes, M. Outcome of colonization of *Apis mellifera* by *Nosema ceranae*. *Appl. Environ. Microbiol.* **2007**, *73*, 6331–6338. [CrossRef]

8. Chen, Y.; Evans, J.D.; Smith, I.B.; Pettis, J.S. *Nosema ceranae* is a long-present and wide-spread microsporidian infection of the European honey bee (*Apis mellifera*) in the United States. *J. Invertebr. Pathol.* **2008**, *97*, 186–188. [CrossRef]

9. Williams, G.R.; Shafer, A.B.; Rogers, R.E.; Shutler, D.; Stewart, D.T. First detection of *Nosema ceranae*, a microsporidian parasite of European honey bees (*Apis mellifera*), in Canada and central USA. *J. Invertebr. Pathol.* **2008**, *97*, 189–192. [CrossRef]

10. Giersch, T.; Berg, T.; Galea, F.; Hornitzky, M. *Nosema ceranae* infects honey bees (*Apis mellifera*) and contaminates honey in Australia. *Apidologie* **2009**, *40*, 117–123. [CrossRef]

11. Chen, Y.P.; Huang, Z.Y. *Nosema ceranae*, a newly identified pathogen of *Apis mellifera* in the USA and Asia. *Apidologie* **2010**, *41*, 364–374. [CrossRef]

12. Yoshiyama, M.; Kimura, K. Distribution of *Nosema ceranae* in the European honeybee, *Apis mellifera* in Japan. *J. Invertebr. Pathol.* **2011**, *106*, 263–267. [CrossRef] [PubMed]

13. Ostroverkhova, N.V.; Kucher, A.N.; Golubeva, E.P.; Rosseykina, S.A.; Konusova, O.L. Study of *Nosema* spp. in the Tomsk region, Siberia: Co-infection is widespread in honeybee colonies. *Far East. Entomol.* **2019**, *378*, 12–22. [CrossRef]

14. Higes, M.; Martín-Hernández, R.; Botías, C.; Bailón, G.E.; González-Porto, A.V.; Barrios, L.; del Nozal, M.J.; Bernal, J.L.; Jiménez, J.J.; García-Palencia, P.; et al. How natural infection by *Nosema ceranae* causes honeybee colony collapse. *Environ. Microbiol.* **2008**, *10*, 2659–2669. [CrossRef] [PubMed]

15. Higes, M.; Martín-Hernández, R.; Garrido-Bailon, E.; Gonzalez-Porto, A.V.; Garcia-Palencia, P.; Meana, A.; Del Nozal, M.J.; Mayo, R.; Bernal, J.L. Honeybee colony collapse due to *Nosema ceranae* in professional apiaries. *Environ. Microbiol. Rep.* **2009**, *1*, 110–113. [CrossRef] [PubMed]

16. Botías, C.; Martín-Hernández, R.; Barrios, L.; Meana, A.; Higes, M. *Nosema* spp. infection and its negative effects on honey bees (*Apis mellifera iberiensis*) at the colony level. *Vet. Res.* **2013**, *44*, 25. [CrossRef]

17. Higes, M.; García-Palencia, P.; Botias, C.; Meana, A.; Martín-Hernández, R. The differential development of microsporidia infecting worker honey bee (*Apis mellifera*) at increasing incubation temperature. *Environ. Microbiol. Rep.* **2010**, *2*, 745–748. [CrossRef]

18. Martín-Hernández, R.; Meana, A.; Garcia-Palencia, P.; Marin, P.; Botías, C.; Garrido Bailon, E.; Barrios, L.; Higes, M. Effect of temperature on the biotic potential of honeybee microsporidia. *Appl. Environ. Microbiol.* **2009**, *75*, 2554–2557. [CrossRef]

19. Forsgren, E.; Fries, I. Comparative virulence of *Nosema ceranae* and *Nosema apis* in individual European honey bees. *Vet. Parasitol.* **2010**, *170*, 212–217. [CrossRef]

20. Martín-Hernández, R.; Botías, C.; Barrios, L.; Martinez-Salvador, A.; Meana, A.; Mayack, C.; Higes, M. Comparison of the energetic stress associated with experimental *Nosema ceranae* and *Nosema apis* infection of honeybees (*Apis mellifera*). *Parasitol. Res.* **2011**, *109*, 605–612. [CrossRef]

21. Huang, W.F.; Solter, L.F. Comparative development and tissue tropism of *Nosema apis* and *Nosema ceranae*. *J. Invertebr. Pathol.* **2013**, *113*, 35–41. [CrossRef] [PubMed]

22. Chen, Y.P.; Pettis, J.S.; Zhao, Y.; Liu, X.; Tallon, L.J.; Sadzewicz, L.D.; Li, R.; Zheng, H.; Huang, S.; Zhang, X.; et al. Genome sequencing and comparative genomics of honey bee microsporidia, *Nosema apis* reveal novel insights into host-parasite interactions. *BMC Genom.* **2013**, *14*, 451–467. [CrossRef] [PubMed]

23. Van der Zee, R.; Gomez-Moracho, T.; Pisa, L.; Sagastume, S.; Garcia-Palencia, P.; Maside, X.; Bartolomé, C.; Martín-Hernández, R.; Higes, M. Virulence and polar tube protein genetic diversity of *Nosema ceranae* (Microsporidia) field isolates from Northern and Southern Europe in honeybees (*Apis mellifera iberiensis*). *Environ. Microbiol. Rep.* **2014**, *6*, 401–413. [CrossRef] [PubMed]

24. Huang, W.-F.; Solter, L.; Aronstein, K.; Huang, Z. Infectivity and virulence of *Nosema ceranae* and *Nosema apis* in commercially available North American honey bees. *J. Invertebr. Pathol.* **2015**, *124*, 107–113. [CrossRef]

25. Bacandritsos, N.; Granato, A.; Budge, G.; Papanastasiou, I.; Roinioti, E.; Caldon, M.; Falcaro, C.; Gallina, A.; Mutinelli, F. Sudden deaths and colony population decline in Greek honey bee colonies. *J. Invertebr. Pathol.* **2010**, *105*, 335–340. [CrossRef]

26. Fries, I. *Nosema ceranae* in European honey bees (*Apis mellifera*). *J. Invertebr. Pathol.* **2010**, *103*, S73–S79. [CrossRef]

27. Gisder, S.; Hedtke, K.; Möckel, N.; Frielitz, M.C.; Linde, A.; Genersch, E. Five-year cohort study of *Nosema* spp. in Germany: Does climate shape virulence and assertiveness of *Nosema ceranae*? *Appl. Environ. Microbiol.* **2010**, *76*, 3032–3038. [CrossRef]

28. Higes, M.; Martín-Hernández, R.; Martínez-Salvador, A.; Garrido Bailon, E.; González-Porto, A.V.; Meana, A.; Bernal, J.L.; Del Nozal, M.J.; Bernal, J. A preliminary study of the epidemiological factors related to honey bee colony loss in Spain. *Environ. Microbiol. Rep.* **2010**, *2*, 243–250. [CrossRef]

29. Soroker, V.; Hetzroni, A.; Yakobson, B.; David, D.; David, A.; Voet, H.; Slabezki, Y.; Efrat, H.; Levski, S.; Kamer, Y.; et al. Evaluation of colony losses in Israel in relation to the incidence of pathogens and pests. *Apidologie* **2011**, *42*, 192–199. [CrossRef]

30. Gisder, S.; Schüler, V.; Horchler, L.L.; Groth, D.; Genersch, E. Long-term temporal trends of *Nosema* spp. infection prevalence in northeast Germany: Continuous spread of *Nosema ceranae*, an emerging pathogen of honey bees (*Apis mellifera*), but no general replacement of *Nosema apis*. *Front. Cell. Infect. Microbiol.* **2017**, *7*, 301. [CrossRef]

31. Urbieta-Magro, A.; Higes, M.; Meana, A.; Barrios, L.; Martín-Hernández, R. Age and method of inoculation influence the infection of worker honey bees (*Apis mellifera*) by *Nosema ceranae*. *Insects* **2019**, *10*, 417. [CrossRef] [PubMed]

32. Taric, E.; Glavinic, U.; Vejnovic, B.; Stanojkovic, A.; Aleksic, N.; Dimitrijevic, V.; Stanimirovic, Z. Oxidative stress, endoparasite prevalence and social immunity in bee colonies kept traditionally vs. those kept for commercial purposes. *Insects* **2020**, *11*, 266. [CrossRef] [PubMed]

33. Bourgeois, A.L.; Rinderer, T.E.; Sylvester, H.A.; Holloway, B.; Oldroyd, B.P. Patterns of *Apis mellifera* infestation by *Nosema ceranae* support the parasite hypothesis for the evolution of extreme polyandry in eusocial insects. *Apidologie* **2012**, *43*, 539–548. [CrossRef]

34. Fontbonne, R.; Garnery, L.; Vidau, C.; Aufauvre, J.; Texier, C.; Tchamitchian, S.; El Alaoui, H.; Brunet, J.L.; Delbac, F.; Biron, D.G. Comparative susceptibility of three Western honeybee taxa to the microsporidian parasite *Nosema ceranae*. *Infect. Genet. Evol.* **2013**, *17*, 188–194. [CrossRef]

35. Mendoza, Y.; Antúnez, K.; Branchiccela, B.; Anido, M.; Santos, E.; Invernizzi, C. *Nosema ceranae* and RNA viruses in European and Africanized honeybee colonies (*Apis mellifera*) in Uruguay. *Apidologie* **2014**, *45*, 224–234. [CrossRef]

36. Fenoy, S.; Rueda, C.; Higes, M.; Martín-Hernández, R.; del Aguila, C. High-level resistance of *Nosema ceranae*, a parasite of the honeybee, to temperature and desiccation. *Appl. Environ. Microbiol.* **2009**, *75*, 6886–6889. [CrossRef] [PubMed]

37. Cilia, G.; Cabbri, R.; Maiorana, G.; Cardaio, I.; Dall'Olio, R.; Nanetti, A. A novel TaqMan® assay for *Nosema ceranae* quantification in honey bee, based on the protein coding gene Hsp70. *Eur. J. Protistol.* **2018**, *63*, 44–50. [CrossRef] [PubMed]

38. Forsgren, E.; Fries, I. Temporal study of *Nosema* spp. in a cold climate. *Environ. Microbiol. Rep.* **2013**, *5*, 78–82. [CrossRef]

39. Martín-Hernández, R.; Bartolomé, C.; Chejanovsky, N.; Le Conte, Y.; Dalmon, A.; Dussaubat, C.; García-Palencia, P.; Meana, A.; Pinto, M.A.; Soroker, V.; et al. *Nosema ceranae* in *Apis mellifera*: A 12 years post-detection perspective. *Environ. Microbiol.* **2018**, *20*, 1302–1329. [CrossRef]

40. Bollan, K.A.; Hothersall, J.D.; Moffat, C.; Durkacz, J.; Saranzewa, N.; Wright, G.A.; Raine, N.E.; Highet, F.; Connolly, C.N. The microsporidian parasites *Nosema ceranae* and *Nosema apis* are widespread in honeybee (*Apis mellifera*) colonies across Scotland. *Parasitol. Res.* **2013**, *112*, 751–759. [CrossRef]

41. Shumkova, R.; Georgieva, A.; Radoslavov, G.; Sirakova, D.; Dzhebir, G.; Neov, B.; Bouga, M.; Hristov, P. The first report of the prevalence of *Nosema ceranae* in Bulgaria. *PeerJ* **2018**, *6*, e4252. [CrossRef] [PubMed]

42. Matović, K.; Vidanović, D.; Manić, M.; Stojiljković, M.; Radojičić, S.; Debeljak, Z.; Šekler, M.; Ćirić, J. Twenty-five-year study of *Nosema* spp. in honey bees (*Apis mellifera*) in Serbia. *Saudi J. Biol. Sci.* **2020**, *27*, 518–523. [CrossRef] [PubMed]

43. Nabian, S.; Ahmadi, K.; Nazem Shirazi, M.H.; Gerami Sadeghian, A. First detection of *Nosema ceranae*, a microsporidian protozoa of European honeybees (*Apis mellifera*) in Iran. *Iran. J. Parasitol.* **2011**, *6*, 89–95. [PubMed]

44. Tunca, R.I.; Oskay, D.; Gosterit, A.; Tekin, O.K. Does *Nosema ceranae* wipe out *Nosema apis* in Turkey? *Iran. J. Parasitol.* **2016**, *11*, 259–264.

45. Ansari, M.J.; Al-Ghamdi, A.; Adgaba, N.; Khan, K.A.; Alattal, Y. Geographical distribution and molecular detection of *Nosema ceranae* from indigenous honey bees of Saudi Arabia. *Saudi J. Biol. Sci.* **2017**, *24*, 983–991. [CrossRef]

46. Oğuz, B.; Karapinar, E.; Dinçer, E.; Değer, M.S. Molecular detection of *Nosema* spp. and black queen-cell virus in honeybees in Van Province, Turkey. *Turk. J. Vet. Anim. Sci.* **2017**, *41*, 221–227. [CrossRef]

47. Khezri, M.; Moharrami, M.; Modirrousta, H.; Torkaman, M.; Salehi, S.; Rokhzad, B.; Khanbabai, H. Molecular detection of *Nosema ceranae* in the apiaries of Kurdistan province, Iran. *Vet. Res. Forum* **2018**, *9*, 273–278. [CrossRef]

48. Tlak Gajger, I.; Vugrek, O.; Grilec, D.; Petrinec, Z. Prevalence and distribution of *Nosema ceranae* in Croatian honeybee colonies. *Vet. Med.* **2010**, *55*, 457–462. [CrossRef]

49. Papini, R.; Mancianti, F.; Canovai, R.; Cosci, F.; Rocchigiani, G.; Benelli, G.; Canale, A. Prevalence of the microsporidian *Nosema ceranae* in honeybee (*Apis mellifera*) apiaries in Central Italy. *Saudi J. Biol. Sci.* **2017**, *24*, 979–982. [CrossRef]

50. Cilia, G.; Sagona, S.; Giusti, M.; dos Santos, P.E.J.; Nanetti, A.; Felicioli, A. *Nosema ceranae* infection in honeybee samples from Tuscanian Archipelago (Central Italy) investigated by two qPCR methods. *Saudi J. Biol. Sci.* **2019**, *26*, 1553–1556. [CrossRef]

51. Sánchez Collado, J.G.; Higes, M.; Barrio, L.; Martín-Hernández, R. Flow cytometry analysis of *Nosema* species to assess spore viability and longevity. *Parasitol. Res.* **2014**, *113*, 1695–1701. [CrossRef] [PubMed]

52. Chauzat, M.P.; Jacques, A.; Laurent, M.; Bougeard, S.; Hendrikx, P.; Ribière-Chabert, M.; EPILOBEE Consortium. Risk indicators affecting honey bee colony survival in Europe: One year of surveillance. *Apidologie* **2016**, *47*, 348–378. [CrossRef]

53. Seitz, N.; Traynor, K.S.; Steinhauer, N.; Rennich, K.; Wilson, M.E.; Ellis, J.D. A national survey of managed honey bee 2014–2015 annual colony losses in the USA. *J. Apic. Res.* **2016**, *54*, 292–304. [CrossRef]

54. Gomes, T.; Feás, X.; Iglesias, A.; Estevinho, L.M. Study of organic honey from the Northeast of Portugal. *Molecules* **2011**, *16*, 5374–5386. [CrossRef] [PubMed]

55. Cepero, A.; Ravoet, J.; Gómez-Moracho, T.; Bernal, J.L.; del Nozal, M.J.; Bartolomé, C.; Maside, X.; Meana, A.; González-Porto, A.V.; de Graaf, D.C.; et al. Holistic screening of collapsing honey bee colonies in Spain: A case study. *BMC Res. Notes* **2014**, *7*, 649. [CrossRef] [PubMed]

56. Meana, A.; Llorens-Picher, M.; Euba, A.; Bernal, J.L.; Bernal, J.; García-Chao, M.; Dagnac, T.; Castro-Hermida, J.A.; González-Porto, A.V.; Higes, M.; et al. Risk factors associated with honey bee colony loss in apiaries in Galicia, NW Spain. *Span. J. Agric. Res.* **2017**, *15*, e0501. [CrossRef]

57. Genersch, E.; Der Ohe, W.V.; Kaatz, H.; Schroeder, A.; Otten, C.; Büchler, R.; Berg, S.; Ritter, W.; Mühlen, W.; Gisder, S.; et al. The German bee monitoring project: A long term study to understand periodically high winter losses of honey bee colonies. *Apidologie* **2010**, *41*, 332–352. [CrossRef]

58. Tokarev, Y.S.; Zinatullina, Z.Y.; Ignatieva, A.N.; Zhigileva, O.N.; Malysh, J.M.; Sokolova, Y.Y. Detection of two Microsporidia pathogens of the European honey bee *Apis mellifera* (Insecta: Apidae) in Western Siberia. *Acta Parasitol.* **2018**, *63*, 728–732. [CrossRef]

59. Ostroverkhova, N.V. Prevalence of *Nosema ceranae* (Microsporidia) in the *Apis mellifera mellifera* bee colonies from long time isolated apiaries of Siberia. *Far East. Entomol.* **2020**, *407*, 8–20. [CrossRef]

60. Ostroverkhova, N.V.; Konusova, O.L.; Kucher, A.N.; Sharakhov, I.V. A comprehensive characterization of the honeybees in Siberia (Russia). In *Beekeeping and Bee Conservation—Advances in Research*; Dechechi Chambo, E., Ed.; InTech: Rijeka, Croatia, 2016; pp. 1–37. [CrossRef]

61. Shumny, V.K.; Shokin, Y.I.; Kolchanov, N.A. *Biodiversity and Ecosystem Dynamics (Information Technology and Modeling)*; Siberian Branch of the Russian Academy of Sciences: Novosibirsk, Russia, 2006; 648p, Available online: https://znanium.com/catalog/product/924641 (accessed on 28 July 2020).

62. North Eurasian Climate Center. Available online: http://seakc.meteoinfo.ru/ (accessed on 28 July 2020).

63. Federal Service for Hydrometeorology and Environmental Monitoring. All-Russian Research Institute of hydrometeorological information—World Data Center. Available online: http://meteo.ru/ (accessed on 28 July 2020).

64. Selyaninov, G.T. About climate agricultural estimation. *Proc. Agric. Meteorol.* **1928**, *20*, 165–177.

65. Polyakov, D.V.; Barashkova, N.K.; Kuzhevskaya, I.V. Weather and climate description of anomalous summer 2012 in Tomsk Region. *Russ. Meteorol. Hydrol.* **2014**, *39*, 22–28. [CrossRef]

66. Martín-Hernández, R.; Botías, C.; Bailón, E.G.; Martínez-Salvador, A.; Prieto, L.; Meana, A.; Higes, M. Microsporidia infecting *Apis mellifera*: Coexistence or competition. Is *Nosema ceranae* replacing *Nosema apis*? *Environ. Microbiol.* **2012**, *14*, 2127–2138. [CrossRef] [PubMed]

67. Chen, Y.W.; Chung, W.P.; Wang, C.H.; Solter, L.F.; Huang, W.F. *Nosema ceranae* infection intensity highly correlates with temperature. *J. Invertebr. Pathol.* **2012**, *111*, 264–267. [CrossRef]

68. Pacini, A.; Mira, A.; Molineri, A.; Giacobino, A.; Cagnolo, N.B.; Aignasse, A.; Zago, L.; Izaguirre, M.; Merke, J.; Orellano, E.; et al. Distribution and prevalence of *Nosema apis* and *N. ceranae* in temperate and subtropical eco-regions of Argentina. *J. Invertebr. Pathol.* **2016**, *141*, 34–37. [CrossRef] [PubMed]

69. Milbrath, M.O.; van Tran, T.; Huang, W.F.; Solter, L.F.; Tarpy, D.R.; Lawrence, F.; Huang, Z.Y. Comparative virulence and competition between *Nosema apis* and *Nosema ceranae* in honey bees (*Apis mellifera*). *J. Invertebr. Pathol.* **2015**, *125*, 9–15. [CrossRef]

70. Konusova, O.L.; Pogorelov, Y.L.; Ostroverkhova, N.V.; Nechipurenko, A.O.; Vorotov, A.A.; Klimova, E.A.; Prokopiev, A.S. Honey bee and bee-farming in the Tomsk region: Past, present and future. *Tomsk State Univ. J. Biol.* **2009**, *4*, 15–28, (In Russian, English Summary).

71. Sorokin, V.S. *A Short Guide to Caring for Bees in Frame Hives for Siberia*, 2nd ed.; Tomsk Beekeeping Society: Tomsk, USSR, 1924; p. 180.

72. Smolyakova, E.O. The development of beekeeping in the Altai Territory. In *Actual Issues of the Functioning of the Economy in the Altai Territory*; Mishchenko, V.V., Ed.; Altai State University: Barnaul, Russia, 2018; pp. 177–186.

73. Meixner, M.D.; Büchler, R.; Costa, C.; Francis, R.M.; Hatjina, F.; Kryger, P.; Uzunov, A.; Carreck, N.L. Honey bee genotypes and the environment. *J. Apic. Res.* **2014**, *53*, 183–187. [CrossRef]

74. Meixner, M.D.; Francis, R.M.; Gajda, A.; Kryger, P.; Andonov, S.; Uzunov, A.; Topolska, G.; Costa, C.; Amiri, E.; Berg, S.; et al. Occurrence of parasites and pathogens in honey bee colonies used in a European genotype-environment interactions experiment. *J. Apic. Res.* **2014**, *53*, 215–229. [CrossRef]

75. Büchler, R.; Costa, C.; Hatjina, F.; Andonov, S.; Meixner, M.D.; Le Conte, Y.; Uzunov, A.; Berg, S.; Bienkowska, M.; Bouga, M.; et al. The influence of genetic origin and its interaction with environmental effects on the survival of *Apis mellifera* L. colonies in Europe. *J. Apic. Res.* **2014**, *53*, 205–214. [CrossRef]

76. Branchiccela, B.; Arredondo, D.; Higes, M.; Invernizzi, C.; Martín-Hernández, R.; Tomasco, I.; Zunino, P.; Antúnez, K. Characterization of *Nosema ceranae* genetic variants from different geographic origin. *Microb. Ecol.* **2017**, *73*, 978–987. [CrossRef]

veterinary sciences

Article

Variation in the Distribution of *Nosema* Species in Honeybees (*Apis mellifera* Linnaeus) between the Neighboring Countries Estonia and Latvia

Sigmar Naudi [1,*], Juris Šteiselis [2], Margret Jürison [1], Risto Raimets [1], Lea Tummeleht [3], Kristi Praakle [3], Arvi Raie [4] and Reet Karise [1]

[1] Chair of Plant Health, Institute of Agricultural and Environmental Sciences, Estonian University of Life Sciences, Friedrich Reinhold Kreutzwaldi 1a, 51014 Tartu, Estonia; margret.jyrison@emu.ee (M.J.); risto.raimets@emu.ee (R.R.); reet.karise@emu.ee (R.K.)

[2] Latvian Beekeepers Association, Rīgas iela 22, LV-3004 Jelgava, Latvia; juris.steiselis@strops.lv

[3] Institute of Veterinary Medicine & Animal Sciences, Estonian University of Life Sciences, Kreutzwaldi 62, 51006 Tartu, Estonia; lea.tummeleht@emu.ee (L.T.); kristi.praakle@emu.ee (K.P.)

[4] Estonian Ministry of Rural Affairs, Lai tn 39, 15056 Tallinn, Estonia; arvi.raie@agri.ee

[*] Correspondence: sigmar.naudi@emu.ee; Tel.: +372-5688-4696

Citation: Naudi, S.; Šteiselis, J.; Jürison, M.; Raimets, R.; Tummeleht, L.; Praakle, K.; Raie, A.; Karise, R. Variation in the Distribution of *Nosema* Species in Honeybees (*Apis mellifera* Linnaeus) between the Neighboring Countries Estonia and Latvia. *Vet. Sci.* **2021**, *8*, 58. https://doi.org/10.3390/vetsci8040058

Academic Editors: Giovanni Cilia, Antonio Nanetti and Chengming Wang

Received: 18 January 2021
Accepted: 30 March 2021
Published: 1 April 2021

Publisher's Note: MDPI stays neutral with regard to jurisdictional claims in published maps and institutional affiliations.

Abstract: The unicellular spore-forming parasites *Nosema apis* and *Nosema ceranae* are considered to be one of the causes of increased honey bee mortality in recent years. These pathogens attack their honey bee hosts through their gut, causing changes in behavioral stress responses and possibly resulting in decreased honey yield and increased honey bee mortality. The present study aimed to determine the prevalence of *Nosema* spp. (nosemosis) in Estonia and Latvia, as well as the persistence of the disease in previously infected hives. Currently, *N. ceranae* is considered the most virulent species and is predominant worldwide. However, in some regions, usually with colder climates, *N. apis* is still prevalent. To achieve better disease control, it is important to determine the species distribution. For this purpose, we selected 30 apiaries in Estonia and 60 in Latvia that were positive for *Nosema* spp. in the EPILOBEE (2012–2014) study, which was 5 years prior to the present study. The results show that, while both species are present in Estonia and Latvia, *N. apis* is dominant in Estonia (43%), and *N. ceranae* is dominant in Latvia (47%). We also found that the pathogens are very persistent, since 5 years later, only 33% of infected apiaries in Estonia and 20% of infected apiaries in Latvia, we could not detect any pathogens at the time of sampling.

Keywords: *Apis mellifera* L.; *Nosema ceranae*; *Nosema apis*; unicellular; pathogens

1. Introduction

The biggest challenge for beekeepers, from both an economic and ecological perspective, is limiting winter mortality, which means that the honey bee colonies must be healthy. In recent years, winter mortality has been relatively high, which has led to collaboration between scientists and beekeepers to determine the causes [1]. Proposed explanations include increased environmental pressures due to changes in land use [2], the presence of pesticide residues in nectar and pollen [3,4], and changes in the distribution of parasites and pathogens [5]. In northern countries such as Estonia, temperatures have increased and winters are milder, and in the last 10 years, scientists have observed temperature-related changes in insect populations. For example, the typically univoltine Colorado potato beetle can now produce two viable generations in warmer years [6].

Chronic exposure to stressors can cause disorders of the honey bee immune system [7]. One of these stressors is acute *Nosema* infection, which often causes no symptoms, but can decrease honey bee immunity [8], which in turn increases the risk of mortality. Nosemosis is a honeybee disease caused by the unicellular spore-forming fungal parasites *Nosema apis* Zander [9] and *N. ceranae* Fries et al. [10] (division *Microsporidia*) [11]. *N. apis* was thought to be the only cause of nosemosis in the honey bee *Apis mellifera* L. until 2005, when *N. ceranae* was first described in *A. mellifera* L. in Taiwan [12]. The first *N. ceranae* infection was also found in Spain at the same time [13].

N. ceranae has replaced *N. apis* in many countries [14,15]; it has become predominant in Argentina [16], central Italy [17], Croatia [18], Serbia [19], and South-Germany [20]. *N. ceranae* is also dominant in more northern countries, such as Lithuania and Finland [21,22] Essentially, *N. ceranae* has spread throughout the world in a short time.

Nosemosis damages the tissues in the honey bee midgut, is energy-consuming for the host, causes changes in behavioral stress responses, and may reduce the life span of the host [23]. Both *N. ceranae* and *N. apis* affect the epithelial cells of the gut, but it has long been thought that *N. ceranae* can also exist in the hypopharyngeal glands, salivary glands, Malpighian tubules, and body fat [24]. However, recently, it was shown that *N. ceranae* has a specific tropism for the epithelial gut, as this was only tissue invaded by this parasite [25]. Despite their generally similar descriptions, these two species affect honey bees differently *N. apis* outbreaks occur mostly during the springtime and tend to be less severe, while *N. ceranae* outbreaks are detected during the entire period of active colony growth and can cause gradual fading [26]. The hidden course of the disease is also one of the reasons why it is difficult to eliminate. Furthermore, in Europe, no veterinary drugs are registered to control it, because these fungi have developed resistance to antibiotics, which are now ineffective and can leave toxic residues in the hive. The antibiotic fumagillin has been used in beekeeping and has been shown to be effective against nosemosis. However, fumagillin is fairly toxic, can cause chromosomal aberrations, and is carcinogenic to human consumers of honey bee products [27]. Thymol has shown good results in laboratory studies [28,29], and good results against *Nosema* have also been shown for various nutraceutical and immunostimulatory compounds. However, further research is needed in this area [30]. European beekeepers use different measures to control *Nosema* infection. Most of these are hygienic management techniques, but selecting and replacing infected colonies and changing the queen may also help mitigate the infection [31].

The symptoms of nosemosis can vary and can be inconspicuous. The two *Nosema* species are distinguished according to their clinical pattern: *N. apis* causes nosemosis type A, and *N. ceranae* causes nosemosis type C [32]. The latter is more problematic because there is no specific clinical picture. However, the disease might lead to a decrease in honey production and may contribute to mortality [33].

The European EPILOBEE project, which was conducted in 2012–2014, mapped the spread of honey bee viruses and parasites in the member states of the European Union [1]. During the project, 197 samples were taken from hives with clinical symptoms of nosemosis in Estonia in 2012, and 30 of these were positive, and 194 samples were collected from hives in Latvia in 2013, and 60 of these were positive. The persistent nature of the disease, difficulties in self-diagnosis, and variation in the clinical symptoms of the disease caused by the two different species resulted in the need to repeat the survey. Therefore, we aimed to resample previously *Nosema*-positive apiaries to assess the persistence of the disease and determine the distribution of *N. apis* and *N. ceranae* in Estonia and Latvia.

2. Materials and Methods

2.1. The Geographical Location of Apiaries and Honey Bee Sampling

This study was undertaken in Estonia (2017) and Latvia (2018). Apiaries that had previously tested positive for nosemosis (EPILOBEE, 2012–2014), including 30 apiaries (1 apiary = 1 sample) in Estonia and 60 apiaries in Latvia, were resampled. Samples were collected before the major turnover from winter to summer bees (May) to obtain the largest number of spores per bee. In the sampling years, the spring was rather cold, and colonies were just starting their development. For each sample, 60 forage bees were collected from the flying boards of 2–4 neighboring hives using a portable vacuum device. The samples were placed in plastic tubes, cooled immediately for transportation, and frozen at -20 °C until laboratory analyses.

2.2. DNA Extraction and Analysis

Twenty worker honey bees were randomly selected from each sample, and their abdomens were removed using a sterile disposable scalpel. The abdomens were pooled and placed in a 15 × 28.5 cm Universal Extraction Bag (Bioreba). Then, 3 mL of ddH$_2$O was added to facilitate homogenization using a hand homogenizer. Approximately 800 µL of the suspension was collected for DNA extraction.

DNA was isolated using the DNeasy Blood and Tissue Kit (Qiagen, Hilden, Germany) according to the manufacturer's protocol. A multiplex PCR (M-PCR) assay was performed to identify the *Nosema* species using 2 µL of DNA and primers as described by Fries et al. (2013) [26] (Table 1). The PCR program consisted of a 2-min initial denaturation at 95 °C, followed by 35 cycles of denaturation for 30 s at 95 °C, annealing for 30 s at 57 °C, and elongation for 1 min at 72 °C, with a final elongation step at 72 °C for 5 min. The M-PCR products were visualized on a 2% agarose gel.

Table 1. Primers used to detect *Nosema ceranae* and *Nosema apis*.

Name	Primer Sequence	Fragment Size (bp)	Specificity
Mn*Ceranae*-F	5′–CGTTAAAGTGTAGATAAGATGTT–3′	143	*N. ceranae*
Mn*Apis*-F	5′–GCATGTCTTTGACGTACTATG–3′	224	*N. apis*
Muniv-R	5′–GACTTAGTAGCCGTCTCTC–3′		

2.3. Flow Cytometry and Spore Counting

A flow cytometer (BD Accuri C6) was used to determine the number of spores per honey bee, and, in cases of a mixed infection, the total number of spores are presented. An aliquot (50 µL) of the suspension generated for DNA isolation was diluted with 1 mL of ddH$_2$O, and 50 µL of the dilution was used for the spore counting analysis. A disposable sieve (10 µm) was used to remove debris. To determine the number of spores per honey bee, the following formula was used: $x = \frac{3000 \times 1050}{20 \times 50 \times 50} \times$ SA (spores in the sample). The formula is explained as follows: 3000 (3000 µL of water added to facilitate homogenization), 1050 (final mixture from which 50 µL of spore suspension was removed was diluted with 1000 µL of water), 20 (number of bee abdomens used), 50 (50 µL of spore suspension used in the dilution), and 50 (50 µL of the filtered spore suspension was loaded into the flow cytometer for spore counting). The accuracy of the results was compared to the microscopic spore counts from five samples of each infected species. There was no significant difference between the results [KW-H = 0.06, $p = 0.73$].

2.4. Statistics

Data were processed using TIBCO Statistica® 13.3.0. The Kruskal-Wallis test was used to assess the statistical differences in spore number per honey bee in Estonia and Latvia and confirm the accuracy of the flow cytometry results. The chi-square test, which compares the differences in shares of subdivisions, was used to examine the statistical significance of differences between *N. ceranae* and *N. apis* distribution changes in Latvia. Statistical significance was set at $p < 0.05$.

3. Results

3.1. Prevalence of N. apis and N. ceranae

Multiplex PCR showed that both Nosema species were present in Estonia and Latvia, either separately or in co-infections. The species distribution varied between the two countries. *N. apis* was the more prevalent species in Estonia, while *N. ceranae* was more prevalent in Latvia at the time of sampling. Among the 30 sampled apiaries in Estonia, 17% were positive for *N. ceranae*, 43% were positive for *N. apis*, and 7% were co-infected. In the remaining 33% of sampled apiaries, *Nosema* was no longer detected. In Latvia (n = 60), the results were almost the opposite: 47% of the samples were positive for *N. ceranae*, 15% were positive for *N. apis*, and 18% were co-infected. In the remaining 20% of sampled apiaries, *Nosema* was no longer detected in the sampling period (Figure 1).

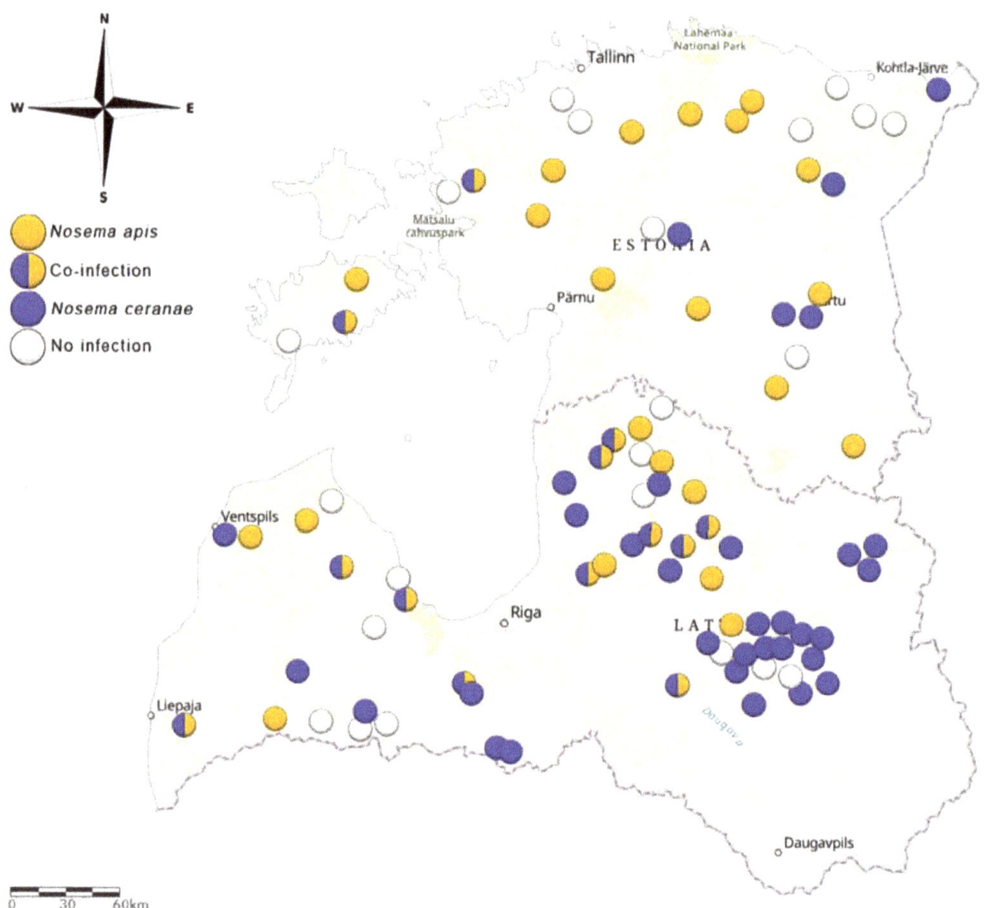

Figure 1. Prevalence of nosemosis causative agents (*Nosema ceranae* and *Nosema apis*) in Estonia in 2017 (May) and Latvia in 2018 (May) (ArcGis online).

3.2. Spore Quantity

The number of spores per honey bee was the highest when both species were present. The median count (spores per honey bee) in co-infections was approximately 12 million in Estonia and approximately 9 million in Latvia. In single species infections the median spore count for *N. apis* was approximately 6 million, while that for *N. ceranae* was approximately 4 million in Estonia. There were no statistically significant differences in spore counts between *Nosema* species or mixed infections (Figure 2). In Latvia, the median spore count for *N. apis* was approximately 1.7 million and that for *N. ceranae* was approximately 2.4 million spores per honey bee (Figure 2), which was significantly lower than that in co-infections.

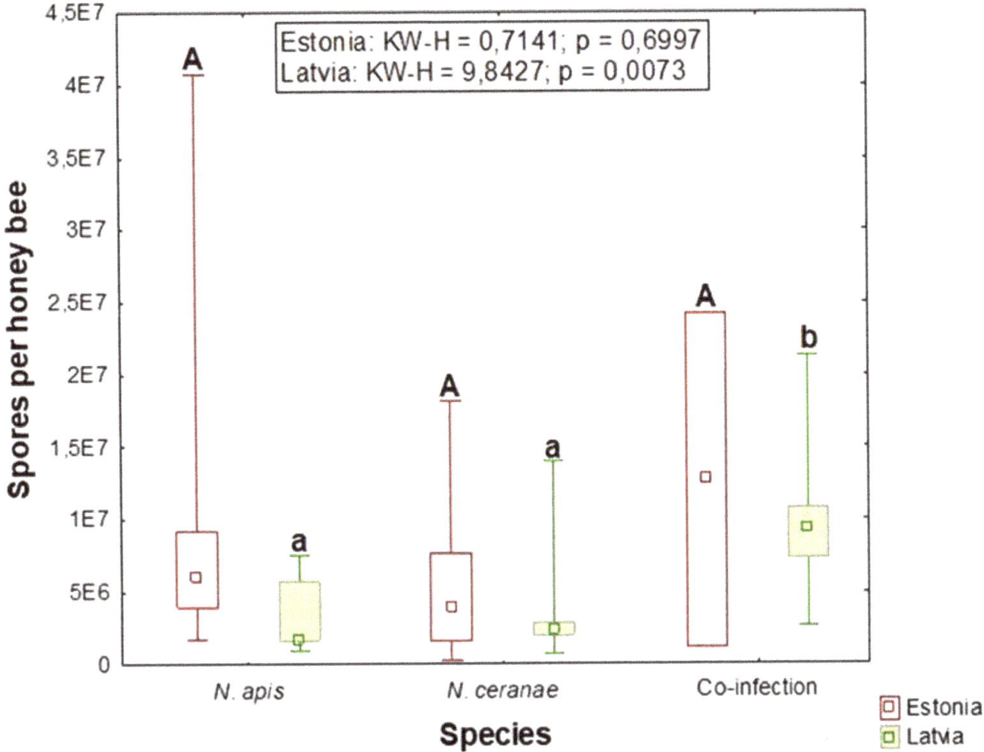

Figure 2. Bar graphs depicting the numbers of *Nosema* spores per honey bee in samples from Estonian and Latvian apiaries, as measured using flow cytometry. Error bars are the minimums and maximums. Columns with different letters indicate significant differences. Statistical differences were calculated by the Kruskal-Wallis test, with significance set at $p < 0.05$.

3.3. Species Distribution Changes in Latvia

A significant change in the distribution of *Nosema* species in Latvia was observed ($\chi^2 = 35.71$, $p < 0.00001$). In 2013, 65% of collected samples were co-infected (unpublished data from Latvian Beekeepers Association), 15% were infected with *N. apis*, and 20% were infected with *N. ceranae* (Figure 3a). However, in 2018, 47% of apiaries were infected with *N. ceranae*, 18% were infected with both species, and 15% were infected with *N. apis*; 20% of the apiaries in 2018 were those in which we could not find any *Nosema* (Figure 3b).

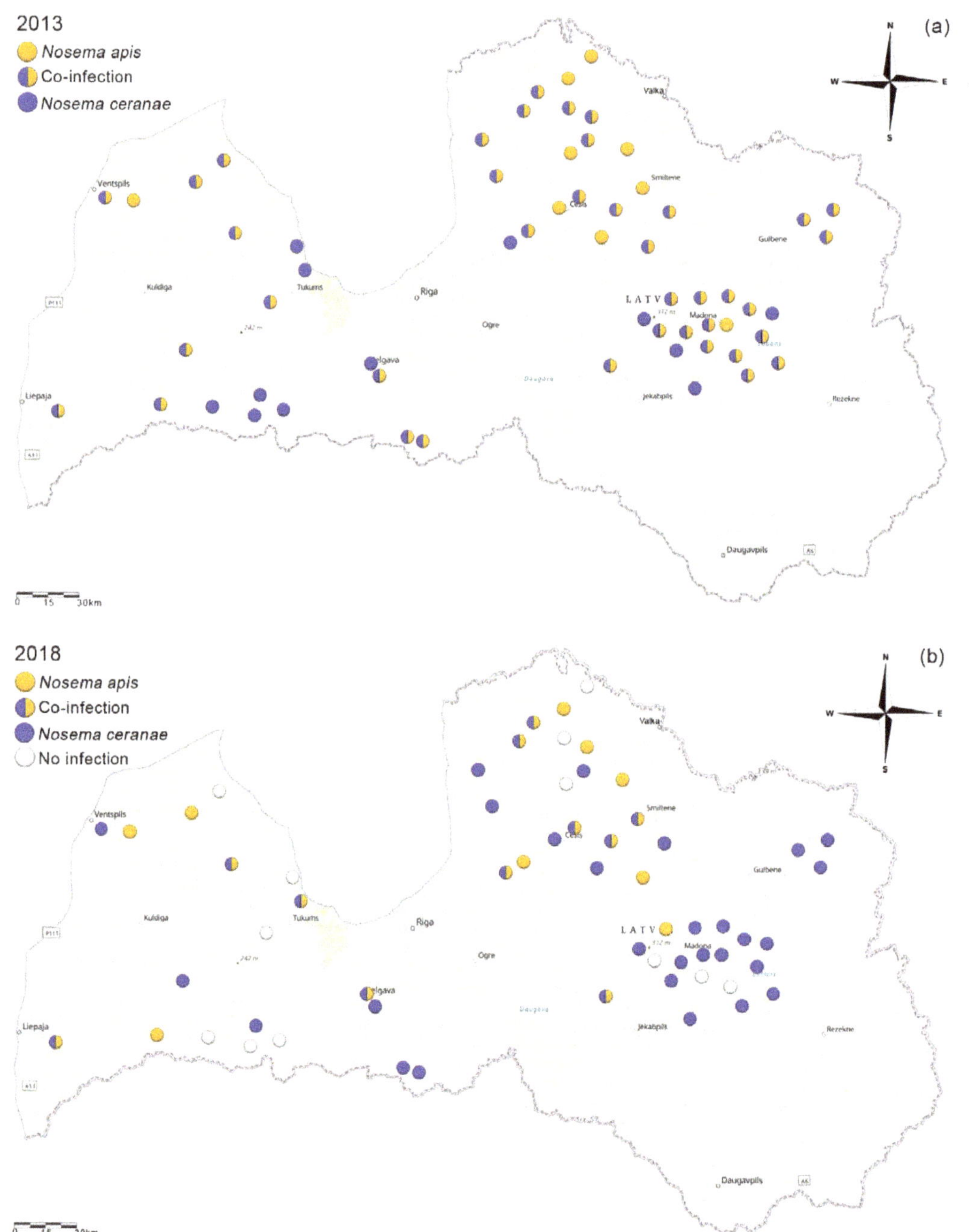

Figure 3. *Nosema* species prevalence in Latvia in 2013 (May) (**a**) and 2018 (May) (**b**) (ArcGis online).

4. Discussion

Our study showed that both *Nosema* species causing nosemosis are present in Estonia and Latvia. The causative agents were found for both single infections and co-infections.

Interestingly, the species composition varied greatly between the two neighboring countries at the time of sampling. *N. apis* was the most abundant species in Estonia, while *N. ceranae* was the most abundant species in Latvia. We observed similar spore counts for single-species infections of both species, whereas co-infected colonies had higher spore counts.

Since its first discovery in *A. mellifera* in Taiwan in 2005 [10], *N. ceranae* has become dominant in *Nosema*-infected apiaries. Over the last few decades, the original pathogenic species, *N. apis*, has been displaced by this more aggressive pathogen. Although for a long time, these two species were thought to be host species specific [13]. It is now clear that these species can exist either individually or together. Still, there is a question of whether a co-infection is more damaging to honey bees than a single infection. In cage studies where the spores of the two pathogens were fed to infection-free honey bees, both *Nosema* spp. were found to be virulent. This was especially true for *N. ceranae* and a mixture of *N. ceranae* and *N. apis*, as both groups of honey bees showed decreased longevity and viability [33,34]. This decrease in survival may result from the reduced defence mechanisms of the honey bees which makes the host more vulnerable to various external factors [35]. In addition, *N. ceranae* tends to be more aggressive during the rapid development phase of honey bee colonies, which in turn contributes to the increased number of spores [15]. In our study, the median number of spores per worker bee ranged from 1.6 to 14 million. A very similar result was shown by Odemer et al. [36]. Although they infected bees artificially and reported cross-infection of their *N. apis* honey bees. This could be explained by the fact that the artificial *Nosema* infection relies on viable spore material. Maybe the spores used for the *N. apis* infection were somehow of lower viability than in a natural setting and expressed therefore lesser spores in the infected honey bees. Additionally, at the end of the paper, they mention the fact that laboratory or semi-laboratory results depend on many factors which may affect the results. It is difficult to draw convincing conclusions about the correlation between high spore counts and colony losses [37] because in field and semi-field studies it is difficult to exclude the possible co-effects of other factors. Another question is whether an infection with these pathogens can lead to colony losses. Exposure to various stressors (pesticides, parasites, etc.) can significantly increase mortality. For example, exposure to various neonicotinoids (e.g., clothianidin and imidacloprid). However, here too, the results diverge. Alaux et al. [38] demonstrated that an interaction between *Nosema* spores and imidacloprid reduced the lifespan of honey bees and neonicotinoid exposure weakened colonies. Odemer et al. [36] demonstrated that by applying neonicotinoid clothianidin in field-relevant sublethal concentrations to free-flying colonies, the neonicotinoid clothianidin did not act synergistically either with *N. apis* or with *N. ceranae*. However, this neonicotinoid is considered to be very toxic to honey bees [39].

Estonia and Latvia are small neighboring countries with similar climates, thus climatic variation is an unlikely cause of the differences in species distribution. It is possible that *N. ceranae* is still increasing its range. This could be investigated by repeating the study after a shorter time and including random hives. A similar increase in the range of *N. ceranae* was also shown by Ostroverkhova et al. [40]; despite large climatic differences in the study region, they were not able to show a climate dependence in the relative spread of these two species. Pacini et al. [16] recorded infections with *N. apis* only in the subtropical regions of Argentina, whereas in the temperate regions, *N. apis* was detected only in co-infected colonies. Conversely, only *N. ceranae* was found in Saudi Arabia [41]. Finally, a study from Mexico showed that *N. apis* was dominant (87%) in an area with a warm climate [42].

Estonia seems to be one of the few countries in the world where *N. apis* is still individually prevalent. In a 4-year study carried out in Lithuania [22] which also looked at the distribution of these pathogens, their proportions were very similar, and neither species was dominant. This result is in contrast to our findings in Estonia and Latvia, where *N. apis* was prevalent in Estonia and *N. ceranae* was prevalent in Latvia. It has reported that *N. ceranae* has become increasingly dominant in Finland since 2006, whereas before that, *N. apis* was dominant [21]. It is possible that the geographical location of Estonia could explain the unique prevalence of *N. apis*, since honey bees cannot cross the Baltic Sea from

Finland, the natural spread of honey bee pathogens could occur only from the south or east. However, *Nosema* can spread with the assistance of people, through the import of infected colonies, small nuclei of colonies or queens. In Estonia, imported queens are widely used. These queens come mainly from Romania and Italy, where *N. ceranae* is a common pathogen [43,44]. However, this does not explain why *N. apis* is more common in Estonia.

As there is no effective cure for *Nosema*, beekeepers need to use uncontaminated equipment in apiaries. The internal temperature of a honey bee colony is always 32–35 °C which is a favorable temperature for the pathogen. A cold climate could somewhat aid in the fight against *Nosema*. This has led beekeepers to believe that cold storage of beekeeping equipment may kill the spores. However, studies have shown contrary and largely variable results. For example, Ozgor and Keskin [45] showed that *N. ceranae* is still infectious after 1 year at −20 °C. They also found that a milder cold temperature (4 °C) was even more conducive to spore survival. Finally, according to Fenoy et al. [46], there was no significant increase in spore mortality within a few hours or after a few months when the spores were exposed to a warm (35 °C) or very warm (60 °C) environment. This indicates that sterilization of beekeeping equipment after every usage in infected colonies is important to avoid spreading the spores from one apiary to another.

Our study samples were collected once in the spring when the age distribution of honey bee colonies exchange for younger bees. However, only one sampling date may create a situation where the *Nosema* prevalence may shift during the season and after overwintering. For future studies, we recommend at least three sampling times (spring, summer, autumn) to investigate the seasonality of the two pathogens in areas with colder climates. Moreover, to get an idea of the species distribution, similar samples should be taken over several years. It is still unclear whether the two pathogen species exhibit seasonality over a longer period. Based on previous studies, *N. apis* infection is present predominantly during the springtime. This is because in spring, only old bees that have overwintered in the hive are present, the queen is just beginning to lay eggs, and the laying intensity is low. Old bees are more susceptible to infection, and when nursing the hive spores spread through the faecal-oral route [47] and reinfect the overwintered worker bees. The latter may also be true for *N. ceranae*. However, several studies have confirmed that it is difficult to find clear links between seasonality and species occurrence. Although spore counts are usually higher in the springtime, they can vary annually and depend on several other factors [19,26,35]. In our experiment, forager bees were collected, which may have affected the study results, because, according to Meana et al. [48], in-hive bees had fewer spore counts than foragers. In the future, in-hive bees should be examined to better describe the extent of the infection in the colony.

5. Conclusions

We conclude that the spread of *N. ceranae* may be lesser in regions with colder climates but further research is needed. To clarify the threat of nosemosis to honey bees in various regions, we need to understand the co-effects of various stressors on infection severity, as well as how to protect honey bees so that their immune system can more effectively fight these internal pathogens. Additionally, from a long-term perspective, nation-wide and pan-European monitoring programs should cover the spread of *Nosema* more accurately and future research should focus on establishing such networks.

Author Contributions: Conceptualization, S.N. and R.K.; investigation, S.N., R.R., M.J., R.K., and J.Š.; writing—original draft preparation, S.N.; writing—review and editing, R.R., L.T., M.J., R.K., J.Š. A.R., and K.P.; visualization, S.N., M.J., and J.Š.; supervision, R.K. and L.T. All authors have read and agreed to the published version of the manuscript.

Funding: This research was supported by institutional research funding (IUT36-2) from the Estonian Ministry of Education, the European Regional Development Fund, and the Estonian Research Council (ETAg) Support for strategic R&D activities (RITA project L180250), the Estonian University of Life Sciences (project P200192PKTE) and the European Union's European Regional Development Fund (Estonian University of Life Sciences ASTRA project "Value-chain based bioeconomy").

Institutional Review Board Statement: Not applicable.

Informed Consent Statement: Not applicable.

Data Availability Statement: Data sharing not applicable.

Acknowledgments: We thank the Estonian and Latvian beekeepers for their cooperation. We also thank Bellis Kullman who assisted us with flow cytometry and Kaarel Mänd for linguistic suggestions.

Conflicts of Interest: The authors declare no conflict of interest.

References

1. Jacques, A.; Laurent, M.; Consortium, E.; Ribière-Chabert, M.; Saussac, M.; Bougeard, S.; Budge, G.E.; Hendrikx, P.; Chauzat, M.-P. A pan-European epidemiological study reveals honey bee colony survival depends on beekeeper education and disease control. *PLoS ONE* **2017**, *12*, e0172591. [CrossRef] [PubMed]
2. Donkersley, P.; Rhodes, G.; Pickup, R.W.; Jones, K.C.; Wilson, K. Honeybee nutrition is linked to landscape composition. *Ecol. Evol.* **2014**, *4*, 4195–4206. [CrossRef] [PubMed]
3. Karise, R.; Raimets, R.; Bartkevics, V.; Pugajeva, I.; Pihlik, P.; Keres, I.; Williams, I.H.; Viinalass, H.; Mänd, M. Are pesticide residues in honey related to oilseed rape treatments? *Chemosphere* **2017**, *188*, 389–396. [CrossRef]
4. Raimets, R.; Bontšutšnaja, A.; Bartkevics, V.; Pugajeva, I.; Kaart, T.; Puusepp, L.; Pihlik, P.; Keres, I.; Viinalass, H.; Mänd, M.; et al. Pesticide residues in beehive matrices are dependent on collection time and matrix type but independent of proportion of foraged oilseed rape and agricultural land in foraging territory. *Chemosphere* **2020**, *238*, 124555. [CrossRef] [PubMed]
5. Traynor, K.S.; Mondet, F.; de Miranda, J.R.; Techer, M.; Kowallik, V.; Oddie, M.A.Y.; Chantawannakul, P.; McAfee, A. Varroa destructor: A Complex Parasite, Crippling Honey Bees Worldwide. *Trends Parasitol.* **2020**, *36*, 592–606. [CrossRef]
6. Hiiesaar, K.; Jõgar, K.; Williams, I.H.; Kruus, E.; Metspalu, L.; Luik, A.; Ploomi, A.; Eremeev, V.; Karise, R.; Mänd, M. Factors affecting development and overwintering of second generation Colorado Potato Beetle (Coleoptera: Chrysomelidae) in Estonia in 2010. *Acta Agric. Scand. Sect. B Soil Plant Sci.* **2013**, *63*, 506–515. [CrossRef]
7. Goulson, D.; Nicholls, E.; Botías, C.; Rotheray, E.L. Bee declines driven by combined stress from parasites, pesticides, and lack of flowers. *Science* **2015**, *347*, 1255957. [CrossRef] [PubMed]
8. Antúnez, K.; Martín-Hernández, R.; Prieto, L.; Meana, A.; Zunino, P.; Higes, M. Immune suppression in the honey bee (*Apis mellifera*) following infection by *Nosema ceranae* (Microsporidia). *Environ. Microbiol.* **2009**, *11*, 2284–2290. [CrossRef]
9. Fries, I. *Nosema Apis*—A Parasite in the Honey Bee Colony. *Bee World* **1993**, *74*, 5–19. [CrossRef]
10. Fries, I.; Feng, F.; da Silva, A.; Slemenda, S.B.; Pieniazek, N.J. *Nosema ceranae* n. sp. (Microspora, Nosematidae), morphological and molecular characterization of a microsporidian parasite of the Asian honey bee *Apis cerana* (Hymenoptera, Apidae). *Eur. J. Protistol.* **1996**, *32*, 356–365. [CrossRef]
11. Forsgren, E.; Fries, I. Comparative virulence of *Nosema ceranae* and *Nosema apis* in individual European honey bees. *Vet. Parasitol.* **2010**, *170*, 212–217. [CrossRef] [PubMed]
12. Huang, W.-F.; Jiang, J.-H.; Chen, Y.-W.; Wang, C.-H. A *Nosema ceranae* isolate from the honeybee *Apis mellifera*. *Apidologie* **2007**, *38*. [CrossRef]
13. Higes, M.; Martín-Hernández, R.; Meana, A.; Higes, M.; Martin, R.; Meana, A. *Nosema ceranae*, a new microsporidian parasite in honeybees in Europe. *J. Invertebr. Pathol.* **2006**, *92*, 93–95. [CrossRef] [PubMed]
14. Klee, J.; Besana, A.M.; Genersch, E.; Gisder, S.; Nanetti, A.; Tam, D.Q.; Chinh, T.X.; Puerta, F.; Ruz, J.M.; Kryger, P.; et al. Widespread dispersal of the microsporidian *Nosema ceranae*, an emergent pathogen of the western honey bee, *Apis mellifera*. *J. Invertebr. Pathol.* **2007**, *96*, 1–10. [CrossRef]
15. Sinpoo, C.; Paxton, R.J.; Disayathanoowat, T.; Krongdang, S.; Chantawannakul, P. Impact of *Nosema ceranae* and *Nosema apis* on individual worker bees of the two host species (*Apis cerana* and *Apis mellifera*) and regulation of host immune response. *J. Insect Physiol.* **2018**, *105*, 1–8. [CrossRef] [PubMed]
16. Pacini, A.; Mira, A.; Molineri, A.; Giacobino, A.; Bulacio Cagnolo, N.; Aignasse, A.; Zago, L.; Izaguirre, M.; Merke, J.; Orellano, E.; et al. Distribution and prevalence of *Nosema apis* and N. ceranae in temperate and subtropical eco-regions of Argentina. *J. Invertebr. Pathol.* **2016**, *141*, 34–37. [CrossRef]
17. Papini, R.; Mancianti, F.; Canovai, R.; Cosci, F.; Rocchigiani, G.; Benelli, G.; Canale, A. Prevalence of the microsporidian *Nosema ceranae* in honeybee (*Apis mellifera*) apiaries in Central Italy. *Saudi J. Biol. Sci.* **2017**, *24*, 979–982. [CrossRef] [PubMed]
18. Tlak Gajger, I.; Vugrek, O.; Grilec, D.; Petrinec, Z. Prevalence and distribution of *Nosema ceranae* in Croatian honeybee colonies. *Vet. Med.* **2010**, *55*, 457–462. [CrossRef]

19. Matović, K.; Vidanović, D.; Manić, M.; Stojiljković, M.; Radojičić, S.; Debeljak, Z.; Šekler, M.; Ćirić, J. Twenty-five-year study of Nosema spp. in honey bees (*Apis mellifera*) in Serbia. *Saudi J. Biol. Sci.* **2020**, *27*, 518–523. [CrossRef] [PubMed]
20. Odemer, R. *Nosema Ceranae*—A New Threat to Honey Bees (*Apis mellifera*)? 2021. Available online: https://osf.io/47df8/ (accessed on 16 March 2021).
21. Paxton, R.J.; Klee, J.; Korpela, S.; Fries, I. *Nosema ceranae* has infected *Apis mellifera* in Europe since at least 1998 and may be more virulent than *Nosema apis*. *Apidologie* **2007**, *38*, 558–565. [CrossRef]
22. Blazyte-Cereskiene, L.; Skrodenyte-Arbaciauskiene, V.; Radžiutė, S.; Nedveckytė, I.; Buda, V. Honey Bee Infection Caused by Nosema spp. in Lithuania. *J. Apic. Sci.* **2016**, *60*. [CrossRef]
23. Holt, H.L.; Aronstein, K.A.; Grozinger, C.M. Chronic parasitization by Nosema microsporidia causes global expression changes in core nutritional, metabolic and behavioral pathways in honey bee workers (*Apis mellifera*). *BMC Genom.* **2013**, *14*, 799. [CrossRef] [PubMed]
24. Chen, Y.P.; Evans, J.D.; Murphy, C.; Gutell, R.; Zuker, M.; Gundensen-Rindal, D.; Pettis, J.S. Morphological, molecular, and phylogenetic characterization of *Nosema ceranae*, a microsporidian parasite isolated from the European honey bee, *Apis mellifera*. *J. Eukaryot. Microbiol.* **2009**, *56*, 142–147. [CrossRef] [PubMed]
25. Higes, M.; García-Palencia, P.; Urbieta, A.; Nanetti, A.; Martín-Hernández, R. *Nosema apis* and *Nosema ceranae* Tissue Tropism in Worker Honey Bees (*Apis mellifera*). *Vet. Pathol.* **2020**, *57*, 132–138. [CrossRef] [PubMed]
26. Emsen, B.; De la Mora, A.; Lacey, B.; Eccles, L.; Kelly, P.G.; Medina-Flores, C.A.; Petukhova, T.; Morfin, N.; Guzman-Novoa, E. Seasonality of *Nosema ceranae* Infections and Their Relationship with Honey Bee Populations, Food Stores, and Survivorship in a North American Region. *Vet. Sci.* **2020**, *7*, 131. [CrossRef]
27. van den Heever, J.P.; Thompson, T.S.; Curtis, J.M.; Ibrahim, A.; Pernal, S.F. Fumagillin: An Overview of Recent Scientific Advances and Their Significance for Apiculture. *J. Agric. Food Chem.* **2014**, *62*, 2728–2737. [CrossRef]
28. Costa, C.; Lodesani, M.; Maistrello, L. Effect of thymol and resveratrol administered with candy or syrup on the development of *Nosema ceranae* and on the longevity of honeybees (*Apis mellifera* L.) in laboratory conditions. *Apidologie* **2009**, *41*, 141–150. [CrossRef]
29. Maistrello, L.; Lodesani, M.; Costa, C.; Leonardi, F.; Marani, G.; Caldon, M.; Mutinelli, F.; Granato, A. Screening of natural compounds for the control of nosema disease in honeybees (*Apis mellifera*). *Apidologie* **2008**, *39*, 436–445. [CrossRef]
30. Borges, D.; Guzman-Novoa, E.; Goodwin, P.H. Control of the microsporidian parasite *Nosema ceranae* in honey bees (*Apis mellifera*) using nutraceutical and immuno-stimulatory compounds. *PLoS ONE* **2020**, *15*, e0227484. [CrossRef]
31. Botías, C.; Anderson, D.L.; Meana, A.; Garrido-Bailón, E.; Martín-Hernández, R.; Higes, M. Further evidence of an oriental origin for *Nosema ceranae* (Microsporidia: Nosematidae). *J. Invertebr. Pathol.* **2012**, *110*, 108–113. [CrossRef]
32. Higes, M.; Martín-Hernández, R.; Meana, A. *Nosema ceranae* in Europe: An emergent type C nosemosis. *Apidologie* **2010**, *41*, 375–392. [CrossRef]
33. Martín-Hernández, R.; Bartolomé, C.; Chejanovsky, N.; Conte, Y.L.; Dalmon, A.; Dussaubat, C.; García-Palencia, P.; Meana, A.; Pinto, M.A.; Soroker, V.; et al. *Nosema ceranae* in *Apis mellifera*: A 12 years postdetection perspective. *Environ. Microbiol.* **2018**, *20*, 1302–1329. [CrossRef]
34. Natsopoulou, M.E.; Doublet, V.; Paxton, R.J. European isolates of the Microsporidia *Nosema apis* and *Nosema ceranae* have similar virulence in laboratory tests on European worker honey bees. *Apidologie* **2016**, *47*, 57–65. [CrossRef]
35. Gisder, S.; Schüler, V.; Horchler, L.L.; Groth, D.; Genersch, E. Long-Term Temporal Trends of Nosema spp. Infection Prevalence in Northeast Germany: Continuous Spread of *Nosema ceranae*, an Emerging Pathogen of Honey Bees (*Apis mellifera*), but No General Replacement of *Nosema apis*. *Front. Cell. Infect. Microbiol.* **2017**, *7*. [CrossRef]
36. Odemer, R.; Nilles, L.; Linder, N.; Rosenkranz, P. Sublethal effects of clothianidin and Nosema spp. on the longevity and foraging activity of free flying honey bees. *Ecotoxicol. Lond. Engl.* **2018**, *27*, 527–538. [CrossRef]
37. Higes, M.; Martín-Hernández, R.; Botías, C.; Bailón, E.G.; González-Porto, A.V.; Barrios, L.; Nozal, M.J.; del Bernal, J.L.; Jiménez, J.J.; Palencia, P.G.; et al. How natural infection by *Nosema ceranae* causes honeybee colony collapse. *Environ. Microbiol.* **2008**, *10*, 2659–2669. [CrossRef]
38. Alaux, C.; Brunet, J.-L.; Dussaubat, C.; Mondet, F.; Tchamitchan, S.; Cousin, M.; Brillard, J.; Baldy, A.; Belzunces, L.P.; Le Conte, Y. Interactions between Nosema microspores and a neonicotinoid weaken honeybees (*Apis mellifera*). *Environ. Microbiol.* **2010**, *12*, 774–782. [CrossRef]
39. Iwasa, T.; Motoyama, N.; Ambrose, J.T.; Roe, R.M. Mechanism for the differential toxicity of neonicotinoid insecticides in the honey bee, *Apis mellifera*. *Crop Prot.* **2004**, *23*, 371–378. [CrossRef]
40. Ostroverkhova, N.V.; Konusova, O.L.; Kucher, A.N.; Kireeva, T.N.; Rosseykina, S.A. Prevalence of the Microsporidian Nosema spp. in Honey Bee Populations (*Apis mellifera*) in Some Ecological Regions of North Asia. *Vet. Sci.* **2020**, *7*, 111. [CrossRef] [PubMed]
41. Ansari, M.J.; Al-Ghamdi, A.; Nuru, A.; Khan, K.A.; Alattal, Y. Geographical distribution and molecular detection of *Nosema ceranae* from indigenous honey bees of Saudi Arabia. *Saudi J. Biol. Sci.* **2017**, *24*, 983–991. [CrossRef] [PubMed]
42. González, S.A.C.; Valencia, G.L.; Cabrera, C.O.; Gómez, S.D.G.; Torres, K.M.; Blandón, K.O.E.; Velázquez, J.G.G.; Paz, L.E.S.; Muñoz, E.T.; Navarro, F.J.M. Prevalence and geographical distribution of *Nosema apis* and *Nosema ceranae* in apiaries of Northwest Mexico using a duplex real-time PCR with melting-curve analysis. *J. Apic. Res.* **2020**, *59*, 195–203. [CrossRef]

3. Cilia, G.; Sagona, S.; Giusti, M.; Jarmela dos Santos, P.E.; Nanetti, A.; Felicioli, A. *Nosema ceranae* infection in honeybee samples from Tuscanian Archipelago (Central Italy) investigated by two qPCR methods. *Saudi J. Biol. Sci.* **2019**, *26*, 1553–1556. [CrossRef] [PubMed]
4. Mederle, N.; Lobo, M.; Morariu, S.; Morariu, F.; Darabus, G.; Mederle, O.; Matos, O. Microscopic and molecular detection of *nosema ceranae* in honeybee *apis mellifera* L. from Romania. *Rev. Chim.* **2018**, *69*, 3761–3772. [CrossRef]
5. Ozgor, E.; Keskin, N. Determination of spore longevity and viability of *Nosema apis* and *Nosema ceranae* according to storage conditions. *EuroBiotech J.* **2017**, *1*, 217–221. [CrossRef]
6. Fenoy, S.; Rueda, C.; Higes, M.; Martín-Hernández, R.; del Aguila, C. High-Level Resistance of *Nosema ceranae*, a Parasite of the Honeybee, to Temperature and Desiccation. *Appl. Environ. Microbiol.* **2009**, *75*, 6886–6889. [CrossRef]
7. Graystock, P.; Goulson, D.; Hughes, W.O.H. Parasites in bloom: Flowers aid dispersal and transmission of pollinator parasites within and between bee species. *Proc. R. Soc. B Biol. Sci.* **2015**, *282*, 20151371. [CrossRef]
8. Meana, A.; Martín-Hernández, R.; Higes, M. The reliability of spore counts to diagnose *Nosema ceranae* infections in honey bees. *J. Apic. Res.* **2010**, *49*, 212–214. [CrossRef]

veterinary sciences

MDPI

Article

Effects of Synthetic Acaricides and *Nosema ceranae* (Microsporidia: Nosematidae) on Molecules Associated with Chemical Communication and Recognition in Honey Bees

Martín Pablo Porrini [1,*], Paula Melisa Garrido [1], María Laura Umpiérrez [2], Leonardo Pablo Porrini [1], Antonella Cuniolo [1], Belén Davyt [2], Andrés González [2], Martín Javier Eguaras [1] and Carmen Rossini [2]

[1] Centro de Investigación en Abejas Sociales (CIAS), Instituto de Investigaciones en Producción Sanidad y Ambiente (IIPROSAM), Consejo Nacional de Investigaciones Científicas y Técnicas (CONICET), Universidad Nacional de Mar del Plata (UNMdP), Funes 3350, Mar del Plata 7600, Argentina; pmgarrid@mdp.edu.ar (P.M.G.); leoporrini@gmail.com (L.P.P.); antocuniolo@gmail.com (A.C.); meguaras@mdp.edu.ar (M.J.E.)

[2] Laboratorio de Ecología Química, Facultad de Química, Universidad de la República Uruguay, Montevideo 11800, Uruguay; mlumpierr@fq.edu.uy (M.L.U.); bdavyt@gmail.com (B.D.); agonzal@fq.edu.uy (A.G.); crossini@fq.edu.uy (C.R.)

* Correspondence: mporrini@mdp.edu.ar; Tel./Fax: +54-223-4752426 (ext. 223)

Received: 6 October 2020; Accepted: 3 December 2020; Published: 8 December 2020

Abstract: Acaricides and the gut parasite *Nosema ceranae* are commonly present in most productive hives. Those stressors could be affecting key semiochemicals, which act as homeostasis regulators in *Apis mellifera* colonies, such as cuticular hydrocarbons (CHC) involved in social recognition and ethyl oleate (EO) which plays a role as primer pheromone in honey bees. Here we test the effect of amitraz, coumaphos, *tau*-fluvalinate and flumethrin, commonly applied to treat varroosis, on honey bee survival time, rate of food consumption, CHC profiles and EO production on *N. ceranae*-infected and non-infected honey bees. Different sublethal concentrations of amitraz, coumaphos, *tau*-fluvalinate and flumethrin were administered chronically in a syrup-based diet. After treatment, purified hole-body extracts were analyzed by gas chromatography coupled to mass spectrometry. While *N. ceranae* infection was also shown to decrease EO production affecting survival rates, acaricides showed no significant effect on this pheromone. As for the CHC, we found no changes in relation to the health status or consumption of acaricides. This absence of alteration in EO or CHC as response to acaricides ingestion or in combination with *N. ceranae*, suggests that worker honey bees exposed to those highly ubiquitous drugs are hardly differentiated by nest-mates. Having determined a synergic effect on mortality in worker bees exposed to coumaphos and Nosema infection but also, alterations in EO production as a response to *N. ceranae* infection it is an interesting clue to deeper understand the effects of parasite-host-pesticide interaction on colony functioning.

Keywords: *Apis mellifera*; nosemosis; acaricides; primer pheromone; hydrocarbon profiles; survival

---

1. Introduction

Current research efforts looking for causes of honeybee colony losses agree on a common hypothesis: the causes are multifactorial due to combined effects of different stressors acting as drivers of lethal or sublethal effects on colony members. These effects include diseases and xenobiotics as two of most likely causes.

Regarding chemical substances, research has focused mainly on the effects of agricultural pesticides; however, there are a growing number of studies referring to colony disruptions produced by in-hive pesticides. This is the case of acaricides, which are periodically incorporated into apiaries suffering varroosis worldwide, representing a huge source of wax and food contamination [1,2].

Among honey bee diseases, *Nosema* spp. infections have been intensively studied, focusing mainly on *Nosema ceranae* as a relatively new microsporidiosis affecting *Apis mellifera*, which has generated a continuously reviewed body of information [2–6]. Further than causing a large list of effects at individual and colony level, *N. ceranae* (recently proposed to be reclassified as *Vairimorpha ceranae* [7]), was also studied in combination with different chemical stressors, evidencing interactions with agricultural pesticides [5]. However, just a few studies have focused on the interaction of *N. ceranae* with acaricides [8,9].

The combination of chemicals and diseases in honeybee colonies clearly represents a challenge to the homeostasis, an equilibrated status depending on, among other factors, volatile chemical compounds (semiochemicals) which are social key regulators in *Apis mellifera* [10–13]. Cuticular hydrocarbons (CHC) and ethyl oleate (EO) are among those semiochemicals, which are also potentially involved in social immunity [14,15].

In the hive, olfactory discrimination mediated by CHC plays a critical role in self-segregation to differentiate between castes, hive of origin and physiological health statuses [16–20]. Regarding nosemosis, changes in the hydrocarbon profiles of the bees have been reported as a consequence of the infection [21,22], however is still discussed if behavioral changes are associated with these CHC alterations [23].

On the other hand, EO, identified as the major component of a primer pheromone that acts as a worker inhibitory factor in honey bees [24,25], is involved in the nurse-to-forager transition, which can be altered by means of *N. ceranae* infection [15,26].

Although the referred homeostatic mechanisms are essential for social functioning and are feasible to be altered by different stressors, the studies on the impact of diseases and chemicals on honeybee pheromones and recognition cues are scarce. This background gave rise to our hypothesis that nosemosis and acaricides can impact on chemical communication cues. Therefore, in the present work, we addressed a laboratory approach to study the effect of different synthetic acaricides and *Nosema ceranae* on honeybee survival, EO production and CHC profiles.

2. Materials and Methods

2.1. Chemicals

The acaricides coumaphos, amitraz, flumethrin and *tau*-fluvalinate and the internal standards (arachidic acid methyl and tridecane) were obtained from Sigma-Aldrich, Saint Louis, MO, USA; and HPLC-grade solvents from Machery-Nagel, Bethlehem, PA, USA or LiChrosolv were used. Concentrations were chosen based on previous data reporting the presence of these pesticides in pollen and honey. *Tau*-fluvalinate was detected at a concentration of 750 ppb in honey [27] and of 487.2 ppb in pollen [28]: the selected dose was approximately the average between those values (666 ppb). In the case of coumaphos, 2020 ppb were found in honey [29] and 5800 ppb in pollen [30], so the selected dose was approximately the average (3333 ppb). The maximum detection in honey for amitraz was 555 ppb and for flumethrin was 1 ppb [31].

2.2. Experimental Procedure

Experiments were performed using hybrid honey bees of *Apis mellifera mellifera* and *Apis mellifera ligustica* from hives located at the experimental apiary of the Social Bees Research Centre (38°10′06″ S, 57°38′10″ W), Mar del Plata, Argentina. Colonies used to obtain brood combs were selected based on their low abundance of *N. ceranae* spores (<5 × 10⁵ spores/bee estimated in forager bees using the technique described by Fries [32] and their low prevalence of *Varroa destructor* Anderson and Trueman (Acari: Varroidae) mites, according to the season. The mite prevalence was evaluated by the natural mite fall method [33] and the rates of phoretic mite infestation [34]. Furthermore, neither of the colonies used to obtain the imagoes presented any visible clinical symptoms of other diseases (i.e., American foulbrood, chalkbrood or viruses).

To avoid the presence of long lasting acaricide residues, commonly found in new commercial beeswax [35], plastic foundations covered with a thin layer of virgin wax were used. These foundations were placed in the selected colonies three months before building the assays began and serve as brood combs. For every assay, frames of sealed brood from one hive were transported in thermic boxes and maintained under incubator conditions until imagoes emerged (34 °C; 60% RH). Newly emerged honey bees were placed in groups of 200–300 in wooden cages ($11 \times 9 \times 6$ cm^3) with a plastic mesh until they were randomly assigned to disposable treatment cages. For each these included in the experimental design, groups of 2-day-old honey bees (N = 50 imagoes/group) were caged in transparent and ventilated plastic flasks (900 cm^3) with inputs for gravity feeding devices and a removable side door (Figure 1), specifically designed following the criteria for good rearing of adult honey bees [36]. The honey bees were maintained in the flasks and placed in an incubator (28 °C; 30% RH) during the assays. In all experiments, each group consisted of five experimental replicates (flasks) unless otherwise noted.

Figure 1. (**A**) Langstroth combs with virgin beeswax over plastic foundation as initial matrix. (**B**) Confinement devices made using plastic flasks, plastic mesh for ventilation, perforated cap for feeding devices and removable lateral door. The dimensions of each cage were 150×70 mm^2 (height × diameter). (**C**) Individual inoculation devices (Porrini et al., 2013).

Honey bees were fed ad libitum with syrup (sucrose-water solution; 2:1 w:v) and beebread (stored pollen) collected from combs of the same hive of origin and not treated with acaricides during the previous 9 months. Besides, water and one of the different treatment diets (diluted in the sucrose syrup) or control syrup were also provided throughout the assays. Except for pollen (replaced weekly), all diet items were replaced daily.

Three experiments were sequentially run to assess the effect of infection status and acaricide ingestion on the recorded variables.

2.2.1. Experiment I—Effect of *Nosema ceranae* Infection and Ethanol

This assay was designed to study the effect of *N. ceranae* infection on EO production and CHC profiles, but also to test the effect of ethanol (3.2%) on those two variables since other experiments with acaricides were performed adding that solvent into the diet.

To achieve the infection, two-day-old honey bees were individually fed according to a technique improved to avoid trophallaxis and possible contamination between confined individuals and to homogenize the stress during inoculation procedures [37]. The infective dose consisted of 10 µL of 50% sugar solution including freshly extracted spores of *N. ceranae* at a concentration of 1×10^5 spores/µL. Spores were obtained from heavily infected forager bees and were concentrated and cleaned from other midgut contents using a triangulation technique [38]. The infection with *N. ceranae* was confirmed by PCR analysis at the end of the assays [39]. After artificial inoculation, honey bees were confined in plastic flasks as described in Section 2.2. Four experimental groups were established according to infection and diet administered for 13 days: Infected honey bees with ethanolic diet (INF+EtOH); infected honey bees without ethanolic diet (INF); non-infected honey bees with ethanolic diet (CTRL+EtOH) and non-infected honey bees without ethanolic diet (CTRL).

2.2.2. Experiment II—Effect of Acaricides

The experiment was designed to simulate realistic conditions of exposure to acaricides to analyze its effect on the recorded variables (Section 2.1). To prepare the diets, acaricidal compounds were dissolved in EtOH 96% and then added to sugar syrup. As previously explained, honey bees were fed for 13 days with solutions of coumaphos (3333 ppb), amitraz (555 ppb), flumethrin (1 ppb) or *tau*-fluvalinate (666 ppb). Therefore, five experimental groups were established: control honey bees under ethanolic diet (CTRL); honey bees fed on coumaphos (COUM); honey bees fed on amitraz (AMI); honey bees fed on flumethrin (FLUM) and honey bees fed on *tau*-fluvalinate (FLUV).

2.2.3. Experiment III—Effect of the Combination of *Nosema ceranae* Infection and Coumaphos

The aim of this assay was to study the effect of *N. ceranae* infection combined with an orally administered concentration of coumaphos on EO production and CHC. To perform *N. ceranae* infection, workers were confined and treated following the protocol detailed in Experiment I (Section 2.2.1). After inoculation, two groups of honey bees (infected and non-infected) were supplied with water, beebread and one of three different syrup diets ad libitum for 13 days. The diets consisted of sugar syrup including EtOH 3.2% or coumaphos (3333 ppb) diluted in EtOH 3.2%. Then, four experimental groups were established: Infected honey bees that received syrup diet (INF); infected honey bees on syrup with coumaphos (INF+Coum); non-infected honey bees on syrup (CTRL) and non-infected honey bees on syrup with coumaphos (CTRL+Coum).

2.3. Recorded Variables

During the assays, different parameters such as survival and substance consumption were recorded, but also, different analyses were performed on sacrificed honey bees, such as parasite development and pheromones and CHC quantifications.

2.3.1. Survival and Diet Consumption

Honey bees without movement response after mechanical stimulation were recorded as dead and removed from containers on a daily basis. In addition, gravity feeders were weighted every day to estimate consumed amounts. Diet evaporation was checked in feeding devices and inside flasks without honey bees to correct the consumed volumes.

2.3.2. *Nosema ceranae* Development

In Experiments I and III, 13 days post-infection, 15–20 honey bees per replicate were analyzed to individually quantify the parasite development. The digestive tract of the honey bees was dissected, keeping only the midguts, which were stored at −20 °C until quantification. The number of spores was individually estimated with a hemocytometer under a light microscope [40].

2.3.3. EO and CHC Quantification

In all assays, after a period of 13 days, pools of five honey bees/replicates were sacrificed and stored for EO and CHC quantification. The extraction protocol was adapted from a previous report [15]. Briefly, after sacrifice, whole-body extracts were prepared in hexane (1.6 mL) with the addition of 200 μL of two internal standard solutions (arachidic acid methyl ester at 5 ng/μL, and tridecane at 50 ng/μL). Subsequently, samples were crushed (2 min, RT) and centrifuged (20 min, 4 °C, 2000 rpm). The supernatants were applied onto SPE-Silica columns (Alltech, Lexington, KY, USA, Clean TM 1000 mg/8 mL) and eluted with mixtures of hexane and ethyl ether. The first fraction (3 mL of hexane-ethyl ether; 98.5:1.5) contained the CHC. The fraction containing fatty acid esters (including EO) was eluted with hexane-thyl ether (94:6), was left to evaporate under a liquid nitrogen stream, and 1 μL analyzed by gas chromatography coupled to mass spectrometry (GCMS). GCMS analyses were done using a QP-2010 Plus Shimadzu acquiring mass spectra from m/z 28 to 350 in the scan mode (70 eV) on an RTX-5ms column (Restek, Bellefonte, PA, USA; 30 m × 0.25 mm i.d., 0.25 μm film thickness), operated with a constant carrier (He) flow of 1 mL/min. The temperature of the GC oven was programmed from an initial temperature of 40 °C (1 min), then heated to 300 °C at 5 °C/min, and held for 1 min. The injector temperature was 250 °C and the interphase temperature was 300 °C in GCMS analysis. Chemical characterization was performed by comparison of the mass spectra and arithmetic retention indexes calculated from data of a solution of hydrocarbons injected in the same conditions [41] to those reported in the NIST2008 [42] and SHIM2205 [41] databases and in the literature as well as by comparisons with synthetic standard in the case of EO. Quantification was done based on the relative areas of compounds compared to the areas of the corresponding internal standards.

3. Statistical Analyses

Statistical analyses were run on MINITAB 17.3.1 and Past (PAleontological Statistics, Oslo, Norway). In each experiment survival curves were compared among treatments by log-rank and Holm–Sidak method. Food consumption was also compared for every assay using ANOVA or Kruskal–Wallis test. Infection levels with *Nosema* spores were also compared by Mann–Whitney tests. To compare levels of EO and CHC between treatments, data was assessed for normality using Anderson–Darling procedure before subjecting it to an ANOVA/GLM and Tukey HSD tests, using as factors the dietary treatment administered ("DIET") and the condition of "infected" or "not-infected" (HEALTH STATUS) where corresponded. In the case of CHC profiles principal component analyses (PCA) were also run on scaled and centered data. The significance level used for every assay was 0.05.

4. Results

4.1. Experiment I—Effect of Nosema ceranae Infection on EO Production and CHC Profile

4.1.1. *Nosema ceranae* Development

At the end of the experiment, no spores were found in midguts from non-infected honey bees. Midguts extracted from infected honey bees fed with or without ethanol reached a mean $\pm$ SE of $1.5 \times 10^6 \pm 0.4 \times 10^6$ spores/bee (INF+EtOH), and $2.0 \times 10^6 \pm 0.5 \times 10^6$ spores/bee (INF). No significant differences were found between spore counts (Mann–Whitney rank sum test; U statistic = 79; T = 119; $p = 0.44$).

4.1.2. Survival and Diet Consumption

Significant differences were found between survival curves (Figure 2) after 13 days of experiment (log-rank test; statistic: 103.491; df = 3; $p < 0.001$). Infected honey bees with or without EtOH included in their diets, showed lower survival than non-infected honey bees (pairwise multiple comparison procedures Holm–Sidak method; $p = 0.05$). Overall cumulative mortality reached 40 ± 7% (pooled data, mean ± SD) for infected honey bees and a 7 ± 2% for non-infected ones. The average survival (estimated by the test) for each treatment was 12.79 days for CTRL, 11.45 days for INF, 12.77 days for CTRL+EtOH and 11.46 days for INF+EtOH. Consumption rates were not different between treatments, with average consumption (grouping all treatments) of 31 ± 5 mg/bee/d (ANOVA, df = 3, F = 1.478, $p = 0.258$).

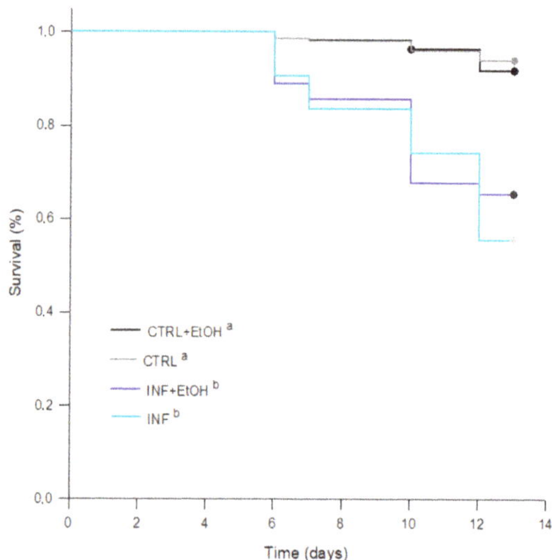

Figure 2. Survival curves for bees under treatments in Experiment I. Significant differences were found between survival curves after 13 days of experiment (log-rank test; statistic: 103.491; df = 3; $p < 0.001$). Significant differences between treatments are indicated with different letters (pairwise multiple comparison procedures Holm–Sidak method; $p = 0.05$). The average survival (estimated by the test) for each treatment was 12.79 days for CTRL, 11.45 days for INF, 12.77 days for CTRL+EtOH and 11.46 days for INF+EtOH.

4.1.3. Ethyl Oleate

Nosema-infected honey bees showed a decrease in EO compared to the level detected in non-infected bees (10 ± 6 and 41 ± 3 ng/bee, respectively; Figure 3A). This reduction, was registered even if honey bees had ingested ethanol in the diet: infected individuals fed on ethanol-supplemented diet showed an average production ± SD of 330 ± 91 ng/bee compared to non-infected honey bees on the same diet (851 ± 96 ng/bee); Figure 3B; ANOVA-GLM: $F_{1,16} = 47.99$ for diet, $p < 0.001$; $F_{1,16} = 11.43$ for health status, $p = 0.005$ and $F_{1,16} = 9.06$, for the interaction diet/health status, $p = 0.01$).

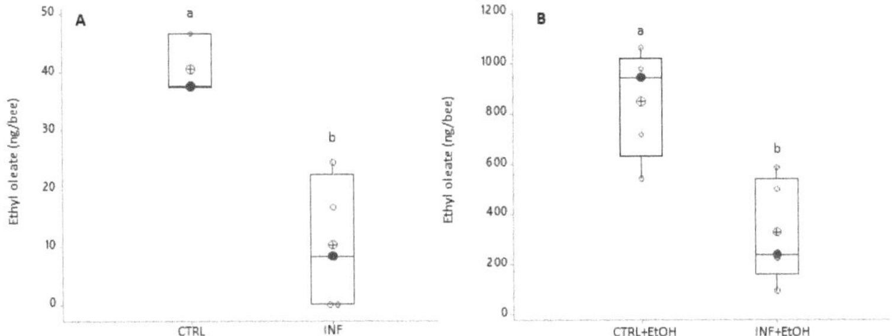

Figure 3. Ethyl oleate variation as a function of infection status and ethanol consumption (black circles are the median values; crossed circles are the mean values and open circles are individual data). X-axis labels are as follows: (**A**) Treatments without ethanolic diet: not-infected honey bees (CTRL); infected honey bees (INF); (**B**) Treatments with ethanolic diet: not-infected honey bees (CTRL+EtOH) and infected honey bees (INF+EtOH). Different letters indicate significant differences: differences between honey bees with different diets and health status were significant (ANOVA-GLM: $F_{1,16}$ = 47.99 for diet, p < 0.001; $F_{1,16}$ = 11.43 for health status, p = 0.005 and $F_{1,16}$ = 9.06 for the interaction diet*health status, p = 0.01).

4.1.4. Cuticular Hydrocarbons

The PCA (var-covar matrix) on the 39 quantified CHC did not find any obvious grouping between honey bees with different health statuses (not-infected vs. *Nosema* infected) or diets (syrup vs. syrup+ethanol). The PCA on all data from the four categories showed that the data were well explained by three components (Component 1:72%; Component 2:11%, Component 3:6% Joliffe cut-off 2.64) and the CHC that better explained the variance in the data (higher loadings) were mainly the same as in all other experiments. No differences were found either on individual CHC when analyzing them by ANOVA (GLMs, all p values > 0.05) or when analyzing them by chemical class (alkanes, alkenes, branched alkanes and alkadienes, Figure 4A–D, Figure S1, Table S1). In addition, different models built by discriminant analyses did not pass validation (permutation tests, p > 0.05), indicating an absence of a pattern related to CHC from honey bees under different treatments.

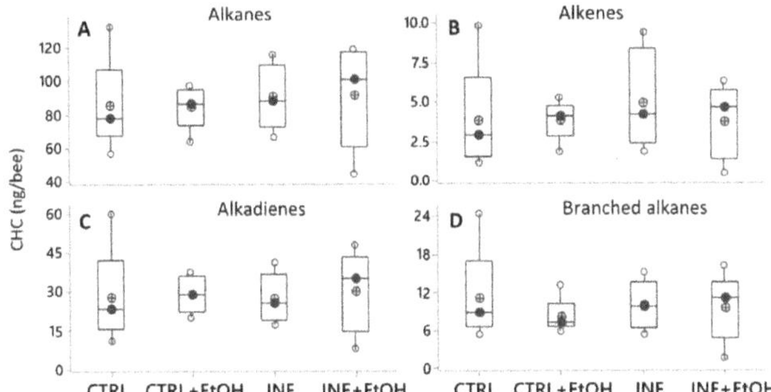

Figure 4. Cuticular hydrocarbons (CHC) profiles shown as compound families (**A–D**), sowing variation as a function of diets. Black circles are the median values; crossed circles are the mean values and open circles are individual data. No differences were found for any of the compound families related to diet (with or without ethanol) or health (with or without *Nosema* infection; ANOVA (GLM), p > 0.05 in all cases. X-axis labels are as follows: not-infected honey bees without ethanolic diet (CTRL); infected honey bees without ethanolic diet (INF); not-infected honey bees with ethanolic diet (CTRL+EtOH) and infected honey bees fed on ethanolic diet (INF+EtOH).

4.2. Experiment II–Effect of Acaricides on EO Production and CHC Profile

4.2.1. Survival and Diet Consumption

After 13 days of diet consumption, honey bees under flumethrin or fluvalinate treatment (exhibiting average survival estimated by the test of 12.83 days and 12.61 days, respectively), did not suffer mortality different from bees under the control diet (12.71 days on avg.). However, bees under coumaphos treatment (12.57 days of survival on avg.) suffered a significant reduction in survival (log-rank test; $p = 0.047$). Furthermore, honey bees treated with amitraz showed a higher probability of survival compared to the control (13.00 day survival on avg.; log-rank test; $p = 0.01$) and the other treatments (Figure 5).

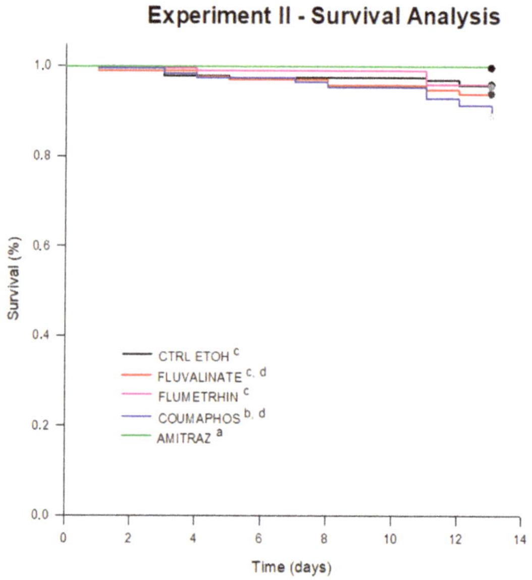

Figure 5. Survival curves for bees under treatments in Experiment II. Significant differences were found between survival curves after 13 days of experiment (log-rank test; statistic: 16,957; df = 4; $p = 0.002$). Significant differences between treatments are indicated with different letters (pairwise multiple comparison procedures Holm–Sidak method; $p < 0.05$). The average survival (estimated by the test) for each treatment was 12.71 days for CTRL (control diet), 12.83 days for FLUM (flumethrin), 12.61 days for FLUV (fluvalinate), 13.00 days for bees under AMI (amitraz) and 12.57 days of survival for bees under COUM (coumaphos treatment).

Similar food consumption was found between acaricide-treated bees and the control (21 ± 1 mg/bee/d on avg.; ANOVA, df = 5, F = 0.516, $p = 0.761$).

Ethyloleate—When honey bees were fed with diets including acaricides the level of ethyloleate did not vary compared to the level of honey bees fed with no acaricides (Figure 6, ANOVA-GLM: $F_{4,16} = 1.09$, $p = 0.404$; Kruskal–Wallis test, H = 2.45, df = 4, $p = 0.65$).

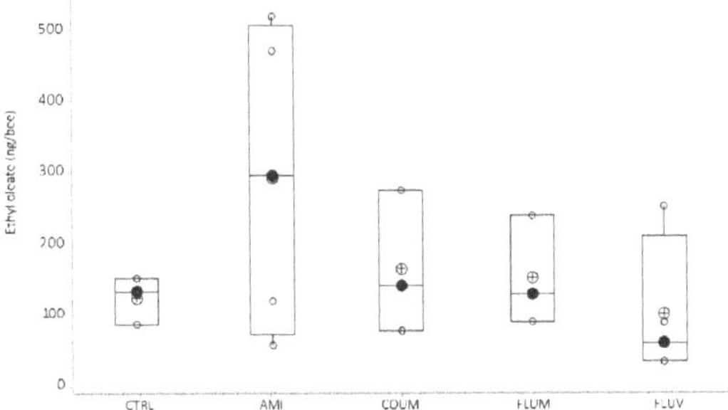

Figure 6. Ethyl oleate variation as a function of acaricide treatment (black circles are the median values; crossed circles are the mean values and open circles are individual data). Differences between honey bees with different diets are not significant ($p > 0.05$, Mann–Whitney tests, see text for further details). X-axis labels are as follows: control honey bees under ethanolic syrup diet (CTRL+EtOH); honey bees feed on amitraz in ethanolic syrup diet (Amit+EtOH); honey bees fed on coumaphos in ethanol syrup diet (Coum+EtOH); honey bees feed on flumethrin in ethanol syrup diet (Flum+EtOH) and honey bees on fluvalinate in ethanol diet (Fluv+EtOH).

4.2.2. Cuticular Hydrocarbons

After a PCA (var-covar matrix) on the 39 quantified CHC, any obvious groupings were found between honey bees fed with different diets (control syrup, amitraz, coumaphos, flumethrin and fluvalinate). The PCA on all data from the six categories showed that the data was well explained by two components (Component 1:91%; Component 2:3%, Joliffe cut-off 15.63) and the CHC that better explained the variance in the data (higher loadings) were again the same as in all other experiments (the alkanes n-heneicosane, n-tricosane, n-pentacosane, n-heptacosane and n-nonacosane, and the alkenes, 9-hentriacontene and triacontene). Besides, different models built by discriminant analyses did not pass validation (permutation tests, $p > 0.05$), indicating an absence of patterns related to CHC from honey bees under different treatments. No differences were either found in individual CHC when analyzing them by ANOVA (GLMs, $p > 0.05$ in all cases) or when analyzing them by chemical class (alkanes, alkenes, branched alkanes and alkadienes (Figure 7A–D, Figure S1, Table S2).

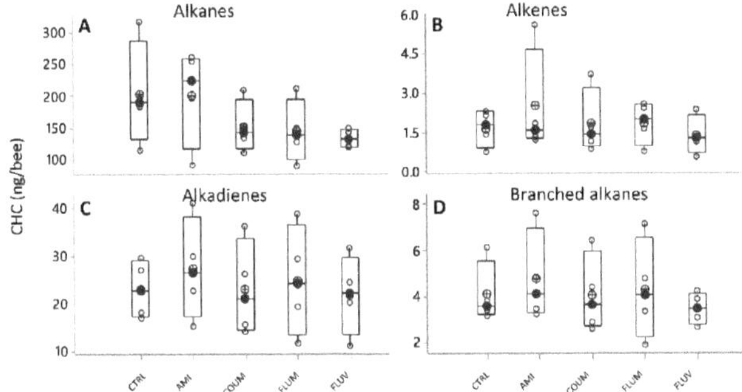

Figure 7. Cuticular hydrocarbons (CHC) profiles shown as compound families (**A–D**), showing variation as a function of diet (black circles are the median values; crossed circles are the mean values and open circles are individual data). No significant differences were found (ANOVA, GLMs, $p > 0.05$ in all cases). X-axis labels are as follows: control honey bees under ethanolic syrup diet (CTRL+EtOH); honey bees feed on amitraz in ethanolic syrup diet (Amit++EtOH); honey bees fed on coumaphos in ethanol syrup diet (Coum+EtOH); honey bees fed on flumethrin in ethanol syrup diet (Flum+EtOH) and honey bees on fluvalinate in ethanol diet (Fluv+EtOH).

4.3. Experiment III—Effect of Nosema ceranae Infection and Coumaphos on EO Production and CHC Profile

4.3.1. *Nosema ceranae* Development

No spores were found in non-infected honey bees (CTRL and CTRL+Coum). Infected honey bees from different dietary treatments (INF and INF+Coum) showed no statistical differences among them (Mann–Whitney rank sum test; U statistic = 71; p = 0.63) and reached a mean ± SE of $1.4 \times 10^6 \pm 0.5 \times 10^6$ spores/bee (INF), and $2.1 \times 10^6 \pm 1.1 \times 10^6$ spores/bee (INF+Coum).

4.3.2. Survival and Diet Consumption

Significant differences were found between survival curves for every treatment (log-rank test; statistic: 92.85; df = 3; $p < 0.001$; pairwise multiple comparison of survival curves Holm–Sidak method; overall significance level = 0.05). The average survival times, estimated by the test in days for every treatment were: 11.46 days for infected honey bees fed on syrup diet (INF); 11.19 days for infected honey bees fed on syrup diet and coumaphos (INF+Coum); 12.77 days for non-infected honey bees fed on syrup diet (CTRL) and 12.39 days for non-infected honey bees fed on syrup diet and coumaphos (CTRL+Coum). Infected honey bees showed significantly lower survival than non-infected ones, independently of the inclusion of coumaphos in the diet (Figure 8). In absence of infection, coumaphos included in diet caused a significant reduction in the survival time compared to the control treatment. Furthermore, a combined effect of "*Nosema* infection" and "coumaphos ingestion" was detected in "survival" since INF+Coum treatment showed a significantly lower survival rate than other treatments.

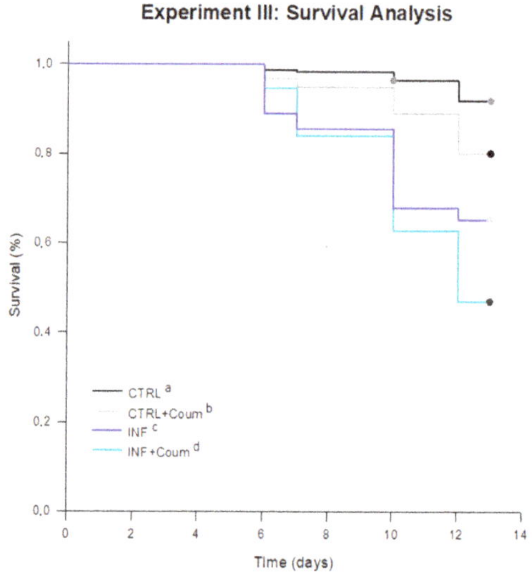

Figure 8. Survival curves for bees under treatments in Experiment II. Significant differences were found between survival curves for every treatment (log-rank test; statistic: 92.85; df = 3; $p < 0.001$). The average survival times, estimated by the test in days for every treatment were: 12.77 days for non-infected honey bees fed on syrup diet (CTRL), 12.39 days for non-infected honey bees fed on syrup diet and coumaphos (CTRL+Coum), 11.46 days for infected honey bees fed on syrup diet (INF); 11.19 days for infected honey bees fed on syrup diet and coumaphos (INF+Coum). Significant differences between treatments are indicated with different letters (pairwise multiple comparison procedures Holm–Sidak method; $p < 0.05$).

Consumption rates were mainly similar between treatments with an average value (±SD) of 37 ± 15 mg/bee/day. The only treatment showing larger consumption rate than the others was the one combining coumaphos and *Nosema* infection, from day 8 post-infection onwards (Kruskal–Wallis one way analysis of variance on ranks, H = 12,462; df = 5; p = 0.029).

4.3.3. Ethyloleate

In this experiment the ANOVA (GLM) on ethyloleate levels showed a significant variation as a function of the treatment ($F_{3,18}$ = 3.45, p < 0.04), with infected bees exhibiting the lower EO levels (Figure 9, post-GLM comparisons, Fisher pairwise comparisons).

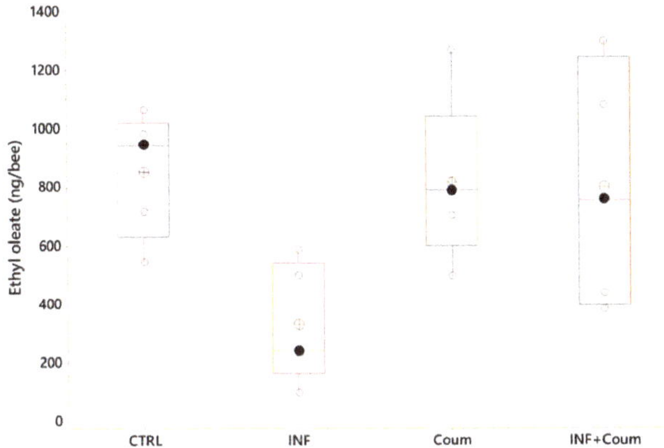

Figure 9. Ethyl oleate variation as a function of acaricide consumption and health status (black circles are the median values; crossed circles are the mean values and open circles are individual data). *X*-axes labels are as follows: non-infected honey bees without ethanolic diet (CTRL); non-infected honey bees on ethanolic syrup diet (CTRL+EtOH); non-infected honey bees on ethanolic and coumaphos syrup diet (CTRL+EtOH+Coum); infected honey bees without ethanolic diet (INF); infected honey bees on ethanolic syrup diet (INF+EtOH) and infected honey bees on ethanolic and coumaphos syrup diet (INF+EtOH+Coum). Differences between honey bees with different diets are significant ($F_{2,25}$ = 19.03, p < 0.001) but not with different health status ($F_{1,25}$ = 3.54, p = 0.075).

4.3.4. Cuticular Hydrocarbons

A PCA (var-covar matrix) on the 39 quantified CHCs showed that the data was well explained by three components (Component 1: 75%; Component 2:11%, Component 3:6%; Joliffe cut-off 33,528) and the CHCs that better explained the variance in the data (higher loadings) were like the ones previously found: mainly linear CHC (n-tricosane, n-pentacosane, n-heptacosane, n-nonacosane and n-hentriacontane) and the alkenes 9-heptacosene, 9-nonacosene, 9-ticosane and 9-hentriacontene. The PCA distribution showed non-grouping patterns among honey bees on different treatments; moreover, all of the models built by discriminant analyses did not pass validation (permutation tests, p > 0.05), indicating an absence of pattern related to CHC from honey bees under different treatments. ANOVAs (GLM) on CHC compound families (alkanes, alkenes, alkadienes and branched alkanes) showed non-significant effect correlated to the treatments for the four compound families (p > 0.05, Figure 10, Figure S1, Table S3). No significant effects were found either for the 39 individual CHC computed (GLMs, p > 0.05 in all cases) related to the treatments.

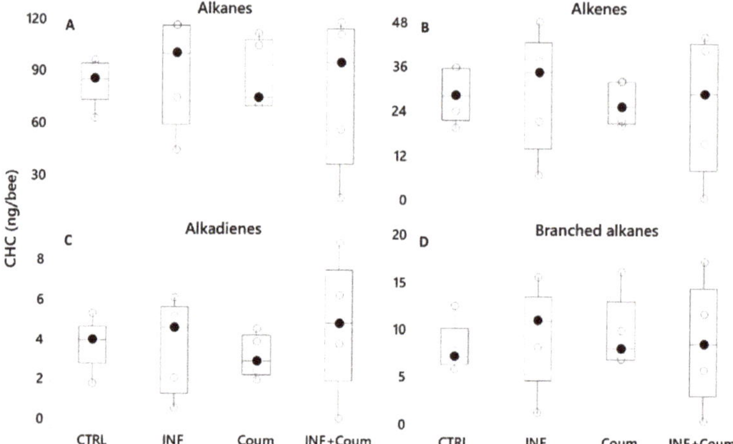

Figure 10. Cuticular hydrocarbons (CHC) profiles shown as compound families (**A–D**), showing variation as a function of diet and health status (black circles are the median values; crossed circles are the mean values and open circles are individual data). No significant differences were found ($p > 0.05$ in all cases). X-axes labels are as follows: non-infected honey bees without ethanolic diet (CTRL); non-infected honey bees on ethanolic syrup diet (CTRL+EtOH); non-infected honey bees on ethanolic and coumaphos syrup diet (CTRL+EtOH+Coum); infected honey bees without ethanolic diet (INF); infected honey bees on ethanolic syrup diet (INF+EtOH) and infected honey bees on ethanolic and coumaphos syrup diet (INF+EtOH+Coum).

5. Discussion

After performing different assays, to study the effect of *Nosema* disease and chemical stressors on pheromone production, cuticular fingerprint and survival, we can highlight two major results: a reduction in EO production in infected workers and no alterations in CHC as response to acaricides ingestion or under a combination of coumaphos ingestion and *N. ceranae* infection.

5.1. Effect of Nosema ceranae Infection

In line with previous findings, our experimentation showed that *Nosema*-infected honey bees exhibited lower survival than non-infected ones, confirming the negative impact of the disease caused by *Nosema ceranae*, even when administering a diet based on fresh pollen under laboratory conditions [43–45]. Furthermore, the infection itself did not alter syrup consumption when comparing daily food intake, similarly to a previous report [46]. Moreover, in the cited work, an increase in *Nosema* spp. infection was observed when feeding honey bees with 5% ethanol in syrup, different levels to our finding in which *Nosema* spores reached similar counts in control and ethanol-fed honey bees, indicating that the consumption of ethanol at 3.2% does not affect *Nosema* development, at least, combined with a fresh pollen diet. The result on consequences of ethanol ingestion in infected bees is a relevant topic since it is commonly used to dissolve some substances studied as therapeutical control of *N. ceranae* (i.e., propolis, herbal extracts).

The infection itself decreased EO production in both control and ethanol-fed honey bees, diminishing approximately 3 to 4 times the amount measured, specifically from 41 ng/bee to 10 ng/bee without ethanol incorporated in the diet, and from 851 ng/bee to 330 ng/bee with an ethanolic diet. The production of EO is energetically costly; its low levels could be explained in part by the negative effects of *N. ceranae* on energy metabolism and the disruption of the epithelial cells of the midgut [47–51]. However, this decrease in EO levels is contrary to previous reports by Dussaubat et al. [15,26], where non-infected worker honey bees exhibited lower EO levels than infected ones (approximately 100 ng/bee vs. 600 ng/bee, respectively). Perhaps, this difference might arise from methodological differences; Dussaubat et al. [15] administered a higher initial inoculum (200,000 spores/bee) to 1-day-old honey bees and found a maximum level of 25 million spores. In our study, 3-day-old honey bees were infected with 100,000 spores/bee, and a maximum of 6–7 million spores were observed. However, since the degree of parasite development was found to be correlated with EO production [15], the relatively low development of the parasite in our assay may explain overall lower levels of EO, but hardly explain that infected honey bees exhibited lower EO titers than healthy (not infected) ones. Additionally, Dussaubat et al. [15] administered a mixed inoculum including *N. ceranae* and *Nosema apis* spores while, in our assay, *N. ceranae* was the only species inoculated. Possibly, this is another cause of differences in the EO amounts found, mainly because of the different physiological effects reported for each *Nosema* species on worker honey bees [52,53].

The methodological differences between our work and the cited ones, surely cannot by themselves completely explain the dichotomy in the results obtained about EO production, therefore, further laboratory and field experiments should assess the change of EO expression along the ontogeny of the workers, under different initial doses of spores in the inoculum. Furthermore, differences in pheromone production resulting from either *N. ceranae*, *N. apis* or both, infections, could reveal interesting results, analogous to the differences shown for cuticular hydrocarbon profiles, differentially affected by infection with those microsporidian species [22].

It has been hypothesized that immune stimulation by pathogens could result in cuticular hydrocarbon changes that may indicate to other honey bees that these individuals are suffering impairment to their health statuses. This has been confirmed for different immune response elicitors (non-replicative pathogen molecules) causing an immune challenge that resulted in hydrocarbon change [54]. Since we found no changes in the individual CHC, the CHC family (Figure 3) or the CHC overall profiles of honey bees in relation with its health status or consumption of ethanol, our data would refute this hypothesis, at least for bees who reach the infection level reported here. Nevertheless, two previous studies have shown plasticity on CHC profiles correlated with health status.

As mentioned, Murray et al. [22] demonstrated differences in the CHC profiles of young (of undefined age) honey bees infected by *N. ceranae* compared to non-infected ones, although no behavioral modification towards *Nosema*-infested nest-mates was found in the same study (this last being a current matter of discussion after the results presented by Biganski et al. [23]). Furthermore, McDonnell et al. [21] measured CHC from honey bees originated from three different hives, showing that, when comparing the overall chemical profile of two of them, infected honey bees showed differences with non-infected ones after 10 days, counting from artificial infection date. In our opinion, these partially contrasting results with previous reports might be attributed to sensitive differences in experimental procedures between assays. The most relevant of those differences, were the ones related to the age of the bees and the homogeneity of the cohort studied, but also, to some differences in the extraction methods performed to obtain the samples for the analyses of the CHC profiles. Future studies should cautiously consider the age and cohort studied, achieving analysis of the CHC profiles in marked honey bees (*Nosema* infected and non-infected), released into the hive and recovered after different periods, to understand how the parasite can affect communication cues under realistic conditions.

5.2. Effect of Acaricides

Exposure to pesticides, including both the ones used for agricultural pests and those administered in the hive, can substantially affect physiological pathways in honeybee workers. Specifically, the sublethal effects of many acaricides, including those caused by *tau*-fluvalinate and coumaphos on physiological functions [55], behavior [56], reproductive castes [57] and immune responses [9,58,59], have been previously reported. However, scarce antecedents are available studying the effects of chemical stressors on honey bee pheromones. In the present work, although some toxicity was found in honey bees consuming coumaphos in their diet, none of the four acaricides tested showed effects on EO titers or CHC profiles. As far as we know, the absence of effect on worker chemical communication and recognition cues, after a chronic exposure to sublethal doses of acaricides, constitutes a new report. Moreover, the absence of effect on EO production was also reported for the pesticide imidacloprid, a neonicotinoid commonly used in integrated pest management in crops [15] and frequently found in hives [28]. The cited report, coupled with our results regarding acaricides, conceivably suggests a general pattern which needs to be corroborated.

It cannot be ruled out that the concentrations tested here (chosen from available data about contaminated honey and pollen stored in hives [29,30] may not have been enough to trigger some effects on the measured variables. Therefore, further experiments designed to emulate a post-treatment scenario, including the study of the combined effect of two or more acaricides at higher sublethal doses could yield different results [60]. A further consideration comes from the quality of the administered diet, since it has been demonstrated that pollen-based diets reduce the sensitivity of workers to pesticides exposure [61]; therefore, providing honey bees with high-quality nutrition, as we did, may have attenuated the effect of pesticides.

5.3. Effect of Nosema ceranae Infection and Coumaphos

It has been demonstrated that there is a possibility of joint negative effects of *Nosema* spp. and chemical stressors such as imidacloprid, fipronil or thiacloprid, impacting directly on honey bee survival, social immunity or sometimes promoting parasite development [8,43,44,62–65]. Previous results revealed no evidence of combined effects of *Nosema ceranae* and acaricides on honeybee survival and in a variety of immune related genes [9]. In agreeing with that, we found no evidence of synergistic effects of both coumaphos and *Nosema* infection EO production or CHC profiles. However, compared with survival data obtained by Garrido et al. [9] until day 9 of consumption, we found a decrease in survival rate caused by the coumaphos diet from day 10 onwards, evidencing a possible cumulative effect of the drug in the worker honey bee organism. Furthermore, the combination of both stressors (coumaphos and *Nosema* disease) significantly decreased the survival rate, suggesting a synergy between both factors.

The accumulation of lipophilic acaricides on beeswax is well known [35,66–68]. During our assays, honey bees only received acaricide exposure as imagoes while pre-imaginal stages were developed in combs with a presumed absence of acaricide residuals (see M & M section). Therefore, it will be interesting to analyze the effect of exposure to sublethal doses even during larval development in future designs.

Since in-field studies demonstrated a complex relationship between acaricides and colony homeostasis as well as different effects of acaricides on the progression of nosemosis disease [69–71], it is relevant to delve deep into the subjacent causes involved in that complexity by means of realistic field or semi-field assays.

6. Conclusions

In the laboratory assays, we found a reduction in EO production in infected young workers, even with low or medium *N. ceranae* spore development, which could possibly influence the nurse-to-forager transition in a hive. Beyond differences in our results with those ones found in previous works, reporting significant alterations in EO production as a response to *N. ceranae* infection can contribute to deeply understanding the effects of the disease on colony functioning. Since a high prevalence of the disease could be found in a hive, reaching percentages over 50% of individuals infected [72,73], any alteration derived from the infection at individual level could be magnified at colony level especially, in sensitive periods during the year or under stressing situations.

On the other hand, not finding evidence of alteration in CHC as response to acaricides ingestion or in combination with *N. ceranae*, is an innovative result that could suggest the impossibility of workers to recognize nest-mates exposed to sublethal doses of those highly ubiquitous drugs.

Finally, we encourage performing of toxicological-parasitological studies under field-realistic conditions since those are necessary to reveal the influence of *Nosema*-infected honey bees on intra-colony interactions, to better understand the relationship between stressors and social immunity, but also to deepen knowledge about how chemical substances can influence the parasitism in a population.

Supplementary Materials: The following are available online at http://www.mdpi.com/2306-7381/7/4/199/s1, Figure S1: PCA graphs for experiments I (A), II (B) and III (C), Table S1: CHC from honeybees in Experiment I, Table S2: CHC from honeybees in Experiment II, Table S3: CHC from honeybees in Experiment III.

Author Contributions: M.P.P. and P.M.G. conceived and designed the experiments; M.P.P., P.M.G., L.P.P., A.C., M.L.U., B.D., A.G. and C.R. performed the experiments and analyzed the data; P.M.G., M.P.P. and C.R. performed the statistical analysis; M.P.P., C.R., P.M.G. and M.J.E. wrote the article or revised it critically. All authors have read and agreed to the published version of the manuscript.

Funding: This research was funded by the National Research Council (CONICET), UNMdP (National University of Mar del Plata); Financial Project PICT 394/2014 and PICT 2044/2016, MinCyT—Inter-American Development Bank; the Programa Regional MERCOSUR *EDUCATIVO* (Grant MRC-C-2011-1-7); the Agencia Nacional de Investigación e Innovación (ANII, Grant FSA-INNOVAGRO-1-2013-12956) and the Comisión Sectorial de Investigación Científica (CSIC, Universidad de la República, Grant from CSIC-Grupos program).

Acknowledgments: We would like to thank the reviewer's suggestions which really improved the presentation of the manuscript.

Conflicts of Interest: The authors declare no conflict of interest.

References

1. Drummond, F.; Ballman, E.S.; Eitzer, B.D.; Du Clos, B.; Dill, J. Exposure of Honey Bee (*Apis mellifera* L.) Colonies to Pesticides in Pollen, A Statewide Assessment in Maine. *Environ. Èntomol.* **2018**, *47*, 378–387. [CrossRef] [PubMed]

2. Goblirsch, M. *Nosema ceranae* disease of the honey bee (*Apis mellifera*). *Apidologie* **2017**, *49*, 131–150. [CrossRef]

3. Higes, M.; Meana, A.; Bartolomé, C.; Botías, C.; Martín-Hernández, R. *Nosema ceranae* (Microsporidia), a controversial 21st century honey bee pathogen. *Environ. Microbiol. Rep.* **2013**, *5*, 17–29. [CrossRef]

4. Soklič, M.; Gregorc, A. Comparison of the two microsporidia that infect honey bees—A review. *Agriculture* **2016**, *13*, 49–56. [CrossRef]

5. Hernández, R.M.; Bartolomé, C.; Chejanovsky, N.; Le Conte, Y.; Dalmon, A.; Dussaubat, C.; García-Palencia, P.; Meana, A.; Pinto, M.A.; Soroker, V.; et al. *Nosema ceranae* in *Apis mellifera*: A 12 years postdetection perspective. *Environ. Microbiol.* **2018**, *20*, 1302–1329. [CrossRef] [PubMed]

6. Burnham, A.J. Scientific Advances in Controlling *Nosema ceranae* (Microsporidia) Infections in Honey Bees (*Apis mellifera*). *Front. Veter. Sci.* **2019**, *6*, 79. [CrossRef]

7. Tokarev, Y.S.; Huang, W.-F.; Solter, L.F.; Malysh, J.M.; Becnel, J.J.; Vossbrinck, C.R. A formal redefinition of the genera Nosema and Vairimorpha (Microsporidia: Nosematidae) and reassignment of species based on molecular phylogenetics. *J. Invertebr. Pathol.* **2020**, *169*, 107279. [CrossRef]

8. Pettis, J.S.; Lichtenberg, E.M.; Andree, M.; Stitzinger, J.; Rose, R.; Vanengelsdorp, D. Crop Pollination Exposes Honey Bees to Pesticides Which Alters Their Susceptibility to the Gut Pathogen *Nosema ceranae*. *PLoS ONE* **2013**, *8*, e70182. [CrossRef]

9. Garrido, P.M.; Porrini, M.P.; Antúnez, K.; Branchiccela, B.; Martínez-Noël, G.M.; Zunino, P.; Salerno, G.; Eguaras, M.; Ieno, E.N. Sublethal effects of acaricides and *Nosema ceranae* infection on immune related gene expression in honeybees. *Veter. Res.* **2016**, *47*, 51. [CrossRef]

10. Alaux, C.; Maisonnasse, A.; Le Conte, Y. Pheromones in a Superorganism. *Anxiety* **2010**, *83*, 401–423. [CrossRef]

11. Ali, M.F.; Morgan, E.D. Chemical Communication in Insect Communities: A Guide To Insect Pheromones With Special Emphasis On Social Insects. *Biol. Rev.* **1990**, *65*, 227–247. [CrossRef]

12. Ayasse, M.; Paxton, R.J.; Tengö, J. Mating behavior and chemical communication in the order hymenoptera. *Annu. Rev. Èntomol.* **2001**, *46*, 31–78. [CrossRef] [PubMed]

13. Slessor, K.N.; Winston, M.L.; Le Conte, Y. Pheromone Communication in the Honeybee (*Apis mellifera* L.). *J. Chem. Ecol.* **2005**, *31*, 2731–2745. [CrossRef] [PubMed]

14. Greene, M. Cuticular hydrocarbon cues in the formation and maintenance of insect social groups. In *Insect Hydrocarbons*; Cambridge University Press: Cambridge, UK, 2010; pp. 244–253.

15. Dussaubat, C.; Maisonnasse, A.; Alaux, C.; Tchamitchan, S.; Brunet, J.-L.; Plettner, E.; Belzunces, L.P.; Le Conte, Y. *Nosema* spp. Infection Alters Pheromone Production in Honey Bees (*Apis mellifera*). *J. Chem. Ecol.* **2010**, *36*, 522–525. [CrossRef]

16. Salvy, M.; Martin, C.; Bagnères, A.-G.; Provost, É.; Roux, M.; Le Conte, Y.; Clément, J. Modifications of the cuticular hydrocarbon profile of Apis mellifera worker bees in the presence of the ectoparasitic mite *Varroa jacobsoni* in brood cells. *Parasitology* **2001**, *122*, 145–159. [CrossRef] [PubMed]

17. Kather, R.; Drijfhout, F.P.; Martin, S.J. Task Group Differences in Cuticular Lipids in the Honey Bee *Apis mellifera*. *J. Chem. Ecol.* **2011**, *37*, 205–212. [CrossRef]

18. Rademacher, E.; Harz, M. Oxalic acid for the control of varroosis in honey bee colonies—A review. *Apidologie* **2006**, *37*, 98–120. [CrossRef]

19. Rosenkranz, P.; Aumeier, P.; Ziegelmann, B. Biology and control of Varroa destructor. *J. Invertebr. Pathol.* **2010**, *103*, S96–S119. [CrossRef]

20. Scholl, J.; Naug, D. Olfactory discrimination of age-specific hydrocarbons generates behavioral segregation in a honeybee colony. *Behav. Ecol. Sociobiol.* **2011**, *65*, 1967–1973. [CrossRef]

21. McDonnell, C.M.; Alaux, C.; Parrinello, H.; Desvignes, J.P.; Crauser, D.; Durbesson, E.; Beslay, D.; Le Conte, Y. Ecto- and endoparasite induce similar chemical and brain neurogenomic responses in the honey bee (*Apis mellifera*). *BMC Ecol.* **2013**, *13*, 25. [CrossRef]

22. Murray, Z.L.; A Keyzers, R.; Barbieri, R.F.; Digby, A.P.; Lester, P.J. Two pathogens change cuticular hydrocarbon profiles but neither elicit a social behavioural change in infected honey bees, *Apis mellifera* (Apidae: Hymenoptera). *Austral Èntomol.* **2016**, *55*, 147–153. [CrossRef]

23. Biganski, S.; Kurze, C.; Müller, M.Y.; Moritz, R.F.A. Social response of healthy honeybees towards *Nosema ceranae*-infected workers: Care or kill? *Apidologie* **2018**, *49*, 325–334. [CrossRef]

24. Leoncini, I.; Crauser, D.; Robinson, G.E.; Le Conte, Y. Worker-worker inhibition of honey bee behavioural development independent of queen and brood. *Insectes Sociaux* **2004**, *51*, 392–394. [CrossRef]

25. Pankiw, T. Brood Pheromone Modulation of Pollen Forager Turnaround Time in the Honey Bee (*Apis mellifera* L.). *J. Insect Behav.* **2007**, *20*, 173–180. [CrossRef]

26. Dussaubat, C.; Maisonnasse, A.; Crauser, D.; Beslay, D.; Costagliola, G.; Soubeyrand, S.; Kretzchmar, A.; Le Conte, Y. Flight behavior and pheromone changes associated to *Nosema ceranae* infection of honey bee workers (*Apis mellifera*) in field conditions. *J. Invertebr. Pathol.* **2013**, *113*, 42–51. [CrossRef]

27. Atienza, J.; Jiménez, J.; Bernal, J.; Martín, M. Supercritical fluid extraction of fluvalinate residues in honey. Determination by high-performance liquid chromatography. *J. Chromatogr. A* **1993**, *655*, 95–99. [CrossRef]

28. Chauzat, M.-P.; Carpentier, P.; Martel, A.-C.; Bougeard, S.; Cougoule, N.; Porta, P.; Lachaize, J.; Madec, F.; Aubert, M.; Faucon, J.-P. Influence of pesticide residues on honey bee (Hymenoptera: Apidae) colony health in France. *Environ. Èntomol.* **2009**, *38*, 514–523. [CrossRef]

29. Martel, A.-C.; Zeggane, S.; Aurières, C.; Drajnudel, P.; Faucon, J.-P.; Aubert, M. Acaricide residues in honey and wax after treatment of honey bee colonies with Apivar® or Asuntol®50. *Apidologie* **2007**, *38*, 534–544. [CrossRef]

30. Mullin, C.A.; Frazier, M.; Frazier, J.L.; Ashcraft, S.; Simonds, R.; Vanengelsdorp, D.; Pettis, J.S. High Levels of Miticides and Agrochemicals in North American Apiaries: Implications for Honey Bee Health. *PLoS ONE* **2010**, *5*, e9754. [CrossRef]

31. Johnson, R.M.; Ellis, M.D.; Mullin, C.A.; Frazier, M. Pesticides and honey bee toxicity—USA. *Apidologie* **2010**, *41*, 312–331. [CrossRef]

32. Fries, I. *Contribution to the Study of Nosema Disease (Nosema apis Z.) in Honey Bee (Apis mellifera L.) Colonies*; Swedish University of Agricultural Sciences: Upsala, Sweden, 1998.

33. Branco, M.R.; Kidd, N.A.; Pickard, R.S. A comparative evaluation of sampling methods for *Varroa destructor* (Acari: Varroidae) population estimation. *Apidologie* **2006**, *37*, 452–461. [CrossRef]

34. Fries, I.; Aarhus, A.; Hansen, H.; Korpela, S. Comparison of diagnostic methods for detection of low infestation levels of *Varroa jacobsoni* in honey-bee (*Apis mellifera*) colonies. *Exp. Appl. Acarol.* **1991**, *10*, 279–287. [CrossRef]

35. Bonvehí, J.S.; Orantes-Bermejo, J. Acaricides and their residues in Spanish commercial beeswax. *Pest Manag. Sci.* **2010**, *66*, 1230–1235. [CrossRef] [PubMed]

36. Williams, G.R.; Alaux, C.; Costa, C.; Csáki, T.; Doublet, V.; Eisenhardt, D.; Fries, I.; Kuhn, R.; McMahon, D.P.; Medrzycki, P.; et al. Standard methods for maintaining adult *Apis mellifera* in cages under in vitro laboratory conditions. *J. Apic. Res.* **2013**, *52*, 1–36. [CrossRef]

37. Porrini, M.P.; Garrido, P.M.; Eguaras, M.J. Individual feeding of honey bees: Modification of the Rinderer technique. *J. Apic. Res.* **2013**, *52*, 194–195. [CrossRef]

38. Cole, R. The application of the "triangulation" method to the purification of Nosema spores from insect tissues. *J. Invertebr. Pathol.* **1970**, *15*, 193–195. [CrossRef]

39. Martín-Hernández, R.; Meana, A.; Prieto, L.; Salvador, A.M.; Garrido-Bailón, E.; Higes, M. Outcome of Colonization of *Apis mellifera* by *Nosema ceranae*. *Appl. Environ. Microbiol.* **2007**, *73*, 6331–6338. [CrossRef]

40. Cantwell, G. Standard methods for counting Nosema spores. *Am. Bee J.* **1970**, *110*, 222–223.

41. Adams, R. *Identification of Essential Oil Components by Gas Chromatography/Mass Spectrometry*; Allured Pub. Corp.: Carol Stream, IL, USA, 2007.

42. Linstrom, P.; Mallard, W. *NIST Chemistry WebBook, NIST Standard Reference Database Number 69*; National Institute of Standards and Technology: Gaithersburg, MD, USA, 2018. [CrossRef]

43. Alaux, C.; Brunet, J.-L.; Dussaubat, C.; Mondet, F.; Tchamitchian, S.; Cousin, M.; Brillard, J.; Baldy, A.; Belzunces, L.P.; Le Conte, Y. Interactions between Nosema microspores and a neonicotinoid weaken honeybees (Apis mellifera). *Environ. Microbiol.* **2010**, *12*, 774–782. [CrossRef]

44. Vidau, C.; Diogon, M.; Aufauvre, J.; Fontbonne, R.; Viguès, B.; Brunet, J.-L.; Texier, C.; Biron, D.G.; Blot, N.; El Alaoui, H.; et al. Exposure to Sublethal Doses of Fipronil and Thiacloprid Highly Increases Mortality of Honeybees Previously Infected by *Nosema ceranae*. *PLoS ONE* **2011**, *6*, e21550. [CrossRef]

45. Porrini, M.P.; Sarlo, E.G.; Medici, S.K.; Garrido, P.M.; Porrini, D.P.; Damiani, N.; Eguaras, M.J. *Nosema ceranae* development in *Apis mellifera*: Influence of diet and infective inoculum. *J. Apic. Res.* **2011**, *50*, 35–41. [CrossRef]

46. Ptaszyńska, A.; Borsuk, G.; Mułenko, W.; Olszewski, K. Impact of ethanol on *Nosema* spp. infected bees. *Med. Weter.* **2013**, *69*, 736–741.

47. Dussaubat, C.; Sagastume, S.; Gómez-Moracho, T.; Botías, C.; García-Palencia, P.; Martín-Hernández, R.; Le Conte, Y.; Higes, M. Comparative study of *Nosema ceranae* (Microsporidia) isolates from two different geographic origins. *Veter. Microbiol.* **2013**, *162*, 670–678. [CrossRef] [PubMed]

48. Paxton, R.J.; Klee, J.; Korpela, S.; Fries, I. *Nosema ceranae* has infected *Apis mellifera* in Europe since at least 1998 and may be more virulent than *Nosema apis*. *Apidologie* **2007**, *38*, 558–565. [CrossRef]

49. Mayack, C.; Naug, D. Parasitic infection leads to decline in hemolymph sugar levels in honeybee foragers. *J. Insect Physiol.* **2010**, *56*, 1572–1575. [CrossRef] [PubMed]

50. Martín-Hernández, R.; Botías, C.; Barrios, L.; Martínez-Salvador, A.; Meana, A.; Mayack, C.; Higes, M. Comparison of the energetic stress associated with experimental *Nosema ceranae* and *Nosema apis* infection of honeybees (*Apis mellifera*). *Parasitol. Res.* **2011**, *109*, 605–612. [CrossRef]

51. Li, W.; Chen, Y.; Cook, S.C. Chronic *Nosema ceranae* infection inflicts comprehensive and persistent immunosuppression and accelerated lipid loss in host *Apis mellifera* honey bees. *Int. J. Parasitol.* **2018**, *48*, 433–444. [CrossRef]

52. Antúnez, K.; Martín-Hernández, R.; Prieto, L.; Meana, A.; Zunino, P.; Higes, M. Immune suppression in the honey bee (*Apis mellifera*) following infection by *Nosema ceranae* (Microsporidia). *Environ. Microbiol.* **2009**, *11*, 2284–2290. [CrossRef]

53. Sinpoo, C.; Paxton, R.J.; Disayathanoowat, T.; Krongdang, S.; Chantawannakul, P. Impact of Nosema ceranae and *Nosema apis* on individual worker bees of the two host species (*Apis cerana* and *Apis mellifera*) and regulation of host immune response. *J. Insect Physiol.* **2018**, *105*, 1–8. [CrossRef]

54. Richard, F.-J.; Aubert, A.; Grozinger, C.M. Modulation of social interactions by immune stimulation in honey bee, Apis mellifera, workers. *BMC Biol.* **2008**, *6*, 50. [CrossRef]

55. Van Buren, N.W.M.; Mariën, A.G.H.; Oudejans, R.C.H.M.; Velthuis, H.H.W. Perizin, an acaricide to combat the mite *Varroa jacobsoni*: Its distribution in and influence on the honeybee Apis mellifera. *Physiol. Èntomol.* **1992**, *17*, 288–296. [CrossRef]

56. Teeters, B.S.; Johnson, R.M.; Ellis, M.D.; Siegfried, B.D. Using video-tracking to assess sublethal effects of pesticides on honey bees (*Apis mellifera* L.). *Environ. Toxicol. Chem.* **2012**, *31*, 1349–1354. [CrossRef] [PubMed]

57. Haarmann, T.; Spivak, M.; Weaver, D.; Weaver, B.; Glenn, T. Effects of fluvalinate and coumaphos on queen honey bees (Hymenoptera: Apidae) in two commercial queen rearing operations. *J. Econ. Èntomol.* **2002**, *95*, 28–35. [CrossRef]

58. Boncristiani, H.; Underwood, R.; Schwarz, R.; Evans, J.D.; Pettis, J.; Vanengelsdorp, D. Direct effect of acaricides on pathogen loads and gene expression levels in honey bees *Apis mellifera*. *J. Insect Physiol.* **2012**, *58*, 613–620. [CrossRef]

59. Garrido, P.M.; Antúnez, K.; Martin, M.; Porrini, M.P.; Zunino, P.; Eguaras, M.J. Immune-related gene expression in nurse honey bees (*Apis mellifera*) exposed to synthetic acaricides. *J. Insect Physiol.* **2013**, *59*, 113–119. [CrossRef]

60. Dahlgren, L.; Johnson, R.M.; Siegfried, B.D.; Ellis, M.D. Comparative Toxicity of Acaricides to Honey Bee (Hymenoptera: Apidae) Workers and Queens. *J. Econ. Èntomol.* **2012**, *105*, 1895–1902. [CrossRef]

61. Zhu, W.; Schmehl, D.R.; Mullin, C.A.; Frazier, J.L. Four Common Pesticides, Their Mixtures and a Formulation Solvent in the Hive Environment Have High Oral Toxicity to Honey Bee Larvae. *PLoS ONE* **2014**, *9*, e77547. [CrossRef]

62. Aufauvre, J.; Biron, D.G.; Vidau, C.; Fontbonne, R.; Roudel, M.; Diogon, M.; Viguès, B.; Belzunces, L.P.; Delbac, F.; Blot, N. Parasite-insecticide interactions: A case study of *Nosema ceranae* and fipronil synergy on honeybee. *Sci. Rep.* **2012**, *2*, 326. [CrossRef]

63. Aufauvre, J.; Misme-Aucouturier, B.; Viguès, B.; Texier, C.; Delbac, F.; Blot, N. Transcriptome Analyses of the Honeybee Response to *Nosema ceranae* and Insecticides. *PLoS ONE* **2014**, *9*, e91686. [CrossRef]

64. Doublet, V.; Labarussias, M.; De Miranda, J.R.; Moritz, R.F.A.; Paxton, R.J. Bees under stress: Sublethal doses of a neonicotinoid pesticide and pathogens interact to elevate honey bee mortality across the life cycle. *Environ. Microbiol.* **2015**, *17*, 969–983. [CrossRef] [PubMed]

65. Gregorc, A.; Silva-Zacarin, E.C.; Carvalho, S.M.; Kramberger, D.; Teixeira, É.W.; Malaspina, O. Effects of *Nosema ceranae* and thiametoxam in Apis mellifera: A comparative study in Africanized and Carniolan honey bees. *Chemosphere* **2016**, *147*, 328–336. [CrossRef] [PubMed]

66. Lodesani, M.; Costa, C.; Serra, G.; Colombo, R.; Sabatini, A.G. Acaricide residues in beeswax after conversion to organic beekeeping methods. *Apidologie* **2008**, *39*, 324–333. [CrossRef]

67. Medici, S.; Maggi, M.D.; Sarlo, E.G.; Ruffinengo, S.; Marioli, J.M.; Eguaras, M. The presence of synthetic acaricides in beeswax and its influence on the development of resistance in *Varroa destructor*. *J. Apic. Res.* **2015**, *54*, 267–274. [CrossRef]

68. Boi, M.; Serra, G.; Colombo, R.; Lodesani, M.; Massi, S.; Costa, C. A 10 year survey of acaricide residues in beeswax analysed in Italy. *Pest Manag. Sci.* **2015**, *72*, 1366–1372. [CrossRef] [PubMed]

69. Botías, C.; Barrios, L.; Nanetti, A.; Meana, A.; Martín-Hernández, R.; Garrido-Bailón, E.; Higes, M. *Nosema* spp. parasitization decreases the effectiveness of acaricide strips (Apivar®) in treating varroosis of honey bee (Apis mellifera iberiensis) colonies. *Environ. Microbiol. Rep.* **2011**, *4*, 57–65. [CrossRef]

70. Johnson, R.M.; Dahlgren, L.; Siegfried, B.D.; Ellis, M.D. Acaricide, Fungicide and Drug Interactions in Honey Bees (Apis mellifera). *PLoS ONE* **2013**, *8*, e54092. [CrossRef]

71. Cilia, G.; Garrido, C.; Bonetto, M.; Tesoriero, D.; Nanetti, A. Effect of Api-Bioxal® and ApiHerb® Treatments against Nosema ceranae Infection in Apis mellifera Investigated by Two qPCR Methods. *Veter. Sci.* **2020**, *7*, 125. [CrossRef]

72. Higes, M.; Martín-Hernández, R.; Botías, C.; Bailón, E.G.; González-Porto, A.V.; Barrios, L.; Del Nozal, M.J.; Bernal, J.L.; Jiménez, J.J.; Palencia, P.G.; et al. How natural infection by *Nosema ceranae* causes honeybee colony collapse. *Environ. Microbiol.* **2008**, *10*, 2659–2669. [CrossRef]

73. Porrini, M.P.; Garrido, P.M.; Porrini, D.P.; Eguaras, M.E. *Nosema ceranae* dynamics in productive honey bee hives. In preparation.

Publisher's Note: MDPI stays neutral with regard to jurisdictional claims in published maps and institutional affiliations.

Article

Effect of Api-Bioxal® and ApiHerb® Treatments against *Nosema ceranae* Infection in *Apis mellifera* Investigated by Two qPCR Methods

Giovanni Cilia [1,2], **Claudia Garrido** [3,*], **Martina Bonetto** [2], **Donato Tesoriero** [2] and **Antonio Nanetti** [2]

1   Department of Veterinary Sciences, University of Pisa, Viale delle Piagge 2, 56124 Pisa, Italy; giovanni.cilia@vet.unipi.it
2   CREA Research Centre for Agriculture and Environment, Via di Saliceto 80, 40128 Bologna, Italy; martina.bonetto@outlook.it (M.B.); donato.tesoriero@crea.gov.it (D.T.); antonio.nanetti@crea.gov.it (A.N.)
3   BeeSafe-Bee Health Consulting for Veterinary Medicine and Agriculture, 59071 Hamm, Germany
*   Correspondence: claudia.garrido@bee-safe.eu

Received: 31 July 2020; Accepted: 2 September 2020; Published: 4 September 2020

Abstract: *Nosema ceranae* is a worldwide distributed midgut parasite of western honey bees, leading to dwindling colonies and their collapse. As a treatment, only fumagillin is available, causing issues like resistance and hampered bee physiology. This study aimed to evaluate ApiHerb® and Api-Bioxal® as treatments against *N. ceranae*. The efficacy was tested using two qPCR methods based on the *16S rRNA* and *Hsp70* genes. In addition, these methods were compared for their aptitude for the quantification of the infection. For this, 19 colonies were selected based on the presence of *N. ceranae* infections. The colonies were divided into three groups: treated with ApiHerb, Api-Bioxal with previous queen caging and an untreated control. All colonies were sampled pre- and post-treatment. The bees were analyzed individually and in duplicate with both qPCR methods. All bees in the pre-treatment tested positive for *N. ceranae*. Both treatments reduced the abundance of *N. ceranae*, but ApiHerb also decreased the prevalence of infected bees. Analysis with the *16S rRNA* method resulted in several orders of magnitude more copies than analysis with the *Hsp70* method. We conclude that both products are suitable candidates for *N. ceranae* treatment. From our analysis, the qPCR method based on the *Hsp70* gene results as more apt for the exact quantification of *N. ceranae* as is needed for the development of veterinary medicinal products.

Keywords: microsporidia; *Nosema ceranae*; *Hsp70* gene; *16S rRNA* gene; oxalic acid; garlic

---

1. Introduction

Nosemosis Type C is a worldwide occurring disease of western honey bees (*Apis mellifera*) caused by *Nosema ceranae* [1]. This microsporidium was identified in the Asian honey bee *Apis cerana* [2], which is generally considered the original host [3]. In recent decades, the prevalence of this parasite highly increased also in western honey bees, causing the colonies to decline and collapse [4,5].

N. ceranae is an intracellular obligate parasite and infects the epithelial cells of the ventriculum [2,6] with high tropism for this organ [7]. Nosemosis Type C shows symptoms both at the individual and colony level, including lifespan reduction, lethargic behavior and poor honey and pollen harvest [5,8,9]. Moreover, in some cases, *N. ceranae* infections tend to be asymptomatic, with features difficult to spot in the field [2,10,11].

Besides *A. cerana* and *A. mellifera*, the microsporidium was reported in several other hymenopteran species, and was also found in regurgitated pellets of bee-eaters [12–16]. It was also found in small hive beetles (*Aethina tumida*) [17], although the possibility of transmission to hymenopterans remains unclear [18].

In addition, the pathogen is transmitted during common flower visits to other bee species [19,20]. This gives finding efficient treatments an added, ecological importance: the emergent disease could pose a threat to wild pollinators and to the stability of pollination services. To avoid further dispersal, an efficient control in managed bee populations is crucial. So far, the only effective treatment against nosemosis is fumagillin, a mycotoxin derived from *Aspergillus fumigatus* [21,22]. However, fumagillin treatment could contribute to the development of resistant *N. ceranae* strains and stable residues in honey. Further, being toxic to mammals, treatments with this compound may create issues for food safety. Fumagillin treatments may also affect bee physiology and promote parasite development [23]. Finally, this compound is not legally available in all countries.

Therefore, finding treatments alternative to fumagillin is of high importance for honey bee health as well as for avoiding ecological issues. Recently, some formulations with natural compounds have been evaluated for the treatment of Nosemosis Type C in honey bee colonies [24–26]. ApiHerb®, a commercial dietary supplement, showed effects against *N. ceranae* infections [27,28]. Similarly, oxalic acid, an organic acid used for treatments against the parasitic mite *Varroa destructor* [29–31], was found to be efficient against *N. ceranae* both in the laboratory and the field [32,33]. In addition, oxalic acid treatments are usually done in broodless conditions. In summer, this is achieved by caging the queen [34], but brood interruptions consequent to natural requeening have been shown to be beneficial against *N. ceranae* infections [35].

The aim of this study was to comparatively evaluate the effect of the dietary supplement ApiHerb®, and Api-Bioxal®, a registered veterinary drug against *Varroa destructor* based on oxalic acid dihydrate. We assessed the efficacy of these two treatments on *N. ceranae* infections under field conditions.

To measure the efficacy of treatments, exact quantification of the *Nosema* infection is crucial. Therefore, we compared two available qPCR methods, respectively based on the *16S rRNA* and *Hsp70* genes. The aim of this comparison was to gain insight into the aptitude of these methods for the quantification of the infection.

2. Materials and Methods

2.1. Experimental Design

The experiment was made in autumn 2017, in an apiary of CREA—Research Centre for Agriculture and Environment located in Bologna, Italy (44°31'27.1" N 11°21'03.6" E). The apiary consisted of approximately forty *Apis mellifera ligustica* colonies housed in ten-frame Dadant-Blatt hives.

Nineteen of those colonies were selected for the experiment based on the presence of *N. ceranae* infection, which was detected in a preliminary screening made on pooled samples of 25 bees collected from the external combs of each colony, and analyzed with the *Hsp70* qPCR method (see Section 2.2). Those colonies were then randomly divided into the three treatment groups AB (N = 7), AH (N = 6) and C (N = 6).

All treatments were made with sugar water (1:1, *w/w*) that was added with the underreported formulations following the label instructions and then trickled once or three times with a syringe onto the combs at the dose of 50 mL, as described on the label. In the colonies of Group AB, the queens were kept in queen-excluder cages (Var-Control, API-MO.BRU, Padua, Italy) for 21 days (15/9–6/10) to prevent egg laying. At the end of the period, the queens were released and the broodless colonies received an Api-Bioxal (Chemicals Laif SpA, Padua, Italy) solution. The colonies of Group AH were unmanipulated, broodright and with a laying queen. They were treated three times with an ApiHerb (Chemicals Laif SpA, Padua, Italy) solution, at one-week intervals (6, 13 and 20/10). The above treatments correspond to the posology indicated by the manufacturer. The colonies of Group C served as negative controls and were left unmanipulated and untreated.

Twenty-five adult honey bees were sampled pre-treatment (15/9) and post-treatment (27/10) from the external combs of each colony and stored at −20 °C until analysis (Figure 1).

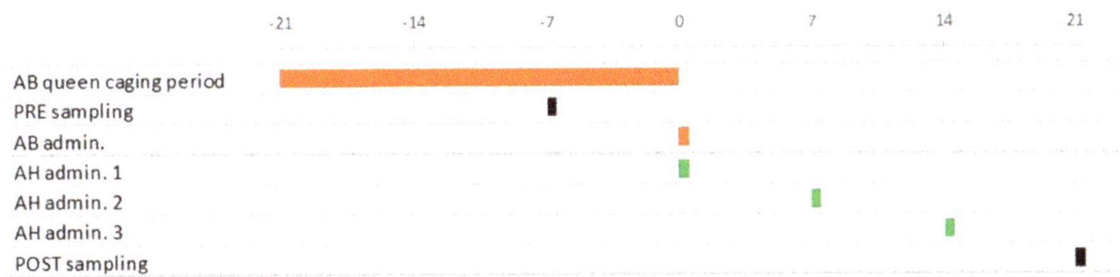

Figure 1. Summary of the working schedule. The horizontal axis represents the timeline, with day 0 corresponding to the 6 October 2017. AB (orange) and AH (green) indicate respectively the Api-Bioxal- and ApiHerb-treated groups.

2.2. DNA Extraction and qPCR Analysis

Each sampled honey bee was analyzed individually after careful dissection. The digestive tract from the ventriculum to the rectum was removed with tweezers and homogenized in 1 mL DNAse-free water with Tissue Lyser II (Qiagen, Hilden, Germany) for 3 min at 30 Hz.

The total DNA was extracted from each homogenate with a Quick DNA Microprep Plus Kit (Zymo Research, Irvine, CA, USA) following the manufacturer's instructions for solid tissue processing.

Two aliquots from each DNA extract were taken and analyzed separately in duplicate by qPCR with primers and probes specific for *N. ceranae*, and respectively designed on sequences of the *16S rRNA* [36] and *Hsp70* [37] genes (Table 1). For each target gene, a total reaction volume of 15 μL was prepared using 2x QuantiTect Probe PCR Master Mix (Qiagen, Hilden, Germany), forward and reverse primers (2 μM), forward and reverse probes (500 nM) and 3 μL DNA extract.

Purified *N. ceranae*-specific amplicons were individually incorporated into a cloning vector using the TA Cloning™ Kit with the pCR™2.1 Vector (Invitrogen, Carlsbad, CA, USA), following the manufacturer's instructions. Recombinant plasmid DNA was purified using the Plasmid Mini Prep Kit (BIO-RAD, Hercules, CA, USA). The copy number of plasmid DNA was calculated based on the molar concentration and molecular mass of the recombinant plasmid consisting of the plasmid vector and the PCR insert. Each template of recombinant plasmids containing the *N. ceranae*-specific DNA fragment was diluted to 10^0 to 10^9 copies. The standard curve was generated by amplifying the serially diluted plasmidas in a duplex qPCR assay. The real-time PCR assay was performed on a Rotorgene Corbett 6000 (Corbett Research, Sydney, Australia) following the amplification and quantification protocols for either gene sequence [36,37]. All the analyses were conducted with two technical replicates for each target gene.

Table 1. Primer sequences used for the qPCR analysis with the *Hsp70* and *16S rRNA* genes and TaqMan Probes.

Gene	Primers	Sequence (5'–3')	Reference
16S rRNA	Forward	AAGAGTGAGACCTATCAGCTAGTTG	[36]
	Reverse	CCGTCTCTCAGGCTCCTTCTC	
	TaqMan Probe	ACCGTTACCCGTCACAGCCTTGTT	
Hsp70	Forward	GGGATTACAAGTGCTTAGAGTGATT	[37]
	Reverse	TGTCAAGCCCATAAGCAAGTG	
	TaqMan Probe	TGAGCCTACTGCGGC	

2.3. Statistical Analysis

The *N. ceranae* abundance was determined at the individual bee level (N = 950) by averaging the two technical replicates of each PCR method. Initially, the individual data were tested against the independent categorical variable "colony" to detect intra-apiary differences. In a second step, the individual bees were also considered technical replicates needed to assess the abundance at the colony level. A new database was then created with the average number of copies detected in the colonies with either method at both sampling points. This database was used to calculate the corresponding prevalence data (i.e., the proportion of positive individuals in the samples (N = 25)) and the relative pre–post variation (RV $= \frac{POST - PRE}{PRE}$) of both abundance and prevalence, that were expressed as percentages. The variables above were tested against the "treatment" as a categorical factor.

Extensive violations to the assumptions of normality and homogeneity of variance were found respectively with the Shapiro–Wilks and Levene's tests. Significant violation of normality remained even after the application of the angular transformation of proportions ($x_1 = \sin^{-1} \sqrt{x}$) and the log transformation of the other continuous variables ($x_1 = \log x + 1$), which made it not possible to analyze the data with a parametric approach.

The effect of categorical factors was therefore tested non-parametrically, by a Kruskal–Wallis one-way ANOVA for independent samples and, when needed, with a post hoc test for multiple comparisons with Bonferroni's correction.

The number of *N. ceranae* copies detected in the same individual honey bee with the two qPCR methods was compared with a two-tailed paired-samples *T* test. Due to the large sample size, checking for the normality assumption was not considered stringent in this case.

The association between continuous variables was evaluated with a two-tailed Pearson's correlation. Besides, the abundance values obtained with both methods for each sampled honey bee were divided (*16S rRNA/Hsp70*) and the ratio was used as the dependent variable in a linear regression model with the *Hsp70* abundance as the explanatory independent variable. Cases generating a divide-by-zero error were excluded, which resulted in the analysis being conducted on a subsample of the original dataset (N = 875).

Frequencies were compared with the Pearson's χ^2 test for independence.

3. Results

3.1. N. ceranae Infection in Test Colonies

All the forager bees collected pre-treatment generated a PCR signal for *N. ceranae* with both methods, resulting in 100% prevalence in all colonies.

The *N. ceranae* abundance measured with the *16S rRNA* method averaged 1.50 +/− 0.34 s.e. (SD = 7.48, 95% CI = 0.82, 2.17) × 10^{12} copies per bee. When the same samples were analyzed with the *Hsp70* method, a lower average abundance value of 3.31 +/− 0.66 s.e. (SD = 14.39, 95% CI = 2.01, 4.60) × 10^9 *Nosema* copies per bee was measured (t(474) = 4.354, p = 0.000) (Table 2).

In both cases, the *Nosema* abundance at the individual bee level was influenced by the colony (*16S rRNA*: H(18) = 468.26, p = 0.000; *Hsp70*: H(18) = 456.21, p = 0.000), but not by the treatment group (*16S rRNA*: (H(2) = 0.031, p = 0.985; *Hsp70*: H(2) = 0.277, p = 0.870).

Table 2. Mean abundance of *N. ceranae* copies detected with the two qPCR methods in the samples collected pre-treatment from the experimental colonies. As data covered various orders of magnitude, for ease of reading, logarithmic representation was adopted. Standard error of means (s.e.) and standard deviations (SD) are shown also.

Colony	Group	N	*16S rRNA* Mean +/− s.e.	SD	*Hsp70* Mean +/− s.e.	SD
3	AB	25	13.45 +/− 12.55	13.25	10.80 +/− 9.45	10.15
4	AB	25	6.73 +/− 5.00	5.70	5.72 +/− 4.24	4.94
5	AB	25	4.73 +/− 3.15	3.85	3.86 +/− 2.23	2.93
8	AB	25	4.75 +/− 3.09	3.79	3.86 +/− 2.22	2.92
10	AB	25	7.75 +/− 6.26	6.96	5.92 +/− 4.22	4.92
13	AB	25	7.09 +/− 5.85	6.55	5.85 +/− 4.30	4.99
14	AB	25	4.54 +/− 3.06	3.76	3.84 +/− 2.39	3.09
F1	AH	25	7.13 +/− 5.67	6.36	5.84 +/− 4.39	5.09
F3	AH	25	7.17 +/− 5.88	6.58	5.79 +/− 4.29	4.99
16	AH	25	5.51 +/− 3.93	4.63	4.30 +/− 3.23	3.93
22	AH	25	6.77 +/− 5.15	5.85	5.58 +/− 4.24	4.94
24	AH	25	5.20 +/− 3.84	4.53	3.93 +/− 2.40	3.10
25	AH	25	4.14 +/− 3.31	4.01	3.70 +/− 2.40	3.10
F4	C	25	5.68 +/− 4.07	4.77	4.68 +/− 3.35	4.05
F7	C	25	5.43 +/− 3.90	4.60	4.66 +/− 3.29	3.99
F9	C	25	4.90 +/− 3.00	3.70	3.91 +/− 2.43	3.12
17	C	25	5.57 +/− 3.92	4.62	4.78 +/− 3.07	3.77
19	C	25	6.94 +/− 5.02	5.72	5.75 +/− 4.26	4.96
26	C	25	7.64 +/− 6.12	6.82	6.86 +/− 5.40	6.10

3.2. Treatment Effect

The *N. ceranae* abundance in the post-treatment samples averaged 2.60 +/− 0.51 s.e. (SD = 11.11, 95% CI = 1.60, 3.60) $\times 10^7$ and 4.82 +/− 0.86 s.e. (SD = 18.71, 95% CI = 3.13, 6.51) $\times 10^5$ copies per bee, respectively with the *16S rRNA* and *Hsp70* methods. Further, in this case, the second method resulted in lower abundance values (t(474) = 5.082, *p* = 0.000).

The number of *N. ceranae* copies detected with either method was influenced by both colony (*16S rRNA*: H(18) = 459.02, *p* = 0.000; *Hsp70*: (H(18) = 458.69, *p* = 0.000) and treatment group (*16S rRNA*: H(2) = 14.20, *p* = 0.001; *Hsp70* (H(2) = 13.66, *p* = 0.001) (Table 3). In detail, a multiple comparison test showed differences between AH and C (with either method: *p* = 0.001) and AB and C (*16S rRNA*: *p* = 0.064; *Hsp70*: *p* = 0.049), but not between AH and AB (*16S rRNA*: *p* = 0.497; *Hsp70*: *p* = 0.341).

Table 3. Post-treatment abundance detected at the colony level (for details, see Table 2). As a treatment effect could be detected, the group means are shown also. Some AH colonies resulted as negative (i.e., zero *N. ceranae* copies/bee), which precluded the transformation into logarithms. Those cases are indicated with an asterisk and represent a non-significant (*16S rRNA* (χ^2 (1, N = 19) = 2.287, *p* = 0.319)) and, respectively, significant (*Hsp70* (χ^2 (1, N = 19) = 7.719, *p* = 0.021)) proportion of colonies that became negative.

Colony	Group	N	*16S rRNA* Mean +/− s.e.	SD	*Hsp70* Mean +/− s.e.	SD
3	AB	25	4.45 +/− 3.36	4.06	3.76 +/− 2.08	2.78
4	AB	25	3.54 +/− 2.68	3.38	2.77 +/− 1.01	1.71
5	AB	25	2.87 +/− 1.42	2.12	2.31 +/− 1.08	1.78
8	AB	25	2.86 +/− 1.45	2.15	2.24 +/− 1.11	1.81
10	AB	25	3.55 +/− 2.45	3.15	2.86 +/− 1.23	1.93
13	AB	25	3.53 +/− 2.50	3.20	2.86 +/− 1.35	2.04
14	AB	25	2.73 +/− 1.36	2.06	2.35 +/− 0.92	1.62
Group AB		175	**3.76 +/− 3.57**	**4.00**	**3.08 +/− 2.89**	**3.31**

Table 3. *Cont.*

			16S rRNA		*Hsp70*	
Colony	Group	N	Mean +/− s.e.	SD	Mean +/− s.e.	SD
F1	AH	25	2.85 +/− 1.11	1.81	2.51 +/− 1.64	2.34
F3	AH	25	2.86 +/− 1.07	1.77	2.47 +/− 1.39	2.09
16	AH	25	0.34 +/− 0.11	0.81	− *	− *
22	AH	25	2.87 +/− 1.14	1.84	2.25 +/− 1.16	1.86
24	AH	25	0.56 +/− 0.30	1.00	− *	− *
25	AH	25	− *	− *	− *	− *
Group AH		**150**	**2.56 +/− 2.21**	**2.60**	**2.12 +/− 1.80**	**2.19**
F4	C	25	5.56 +/− 4.37	5.07	4.93 +/− 3.06	3.76
F7	C	25	5.58 +/− 4.25	4.95	4.93 +/− 3.07	3.77
F9	C	25	4.90 +/− 3.45	4.15	4.00 +/− 2.18	2.88
17	C	25	5.46 +/− 4.07	4.77	4.89 +/− 3.03	3.73
19	C	25	6.84 +/− 5.28	5.98	5.81 +/− 4.06	4.76
26	C	25	8.69 +/− 7.33	8.03	6.92 +/− 5.49	6.19
Group C		**150**	**7.92 +/− 7.91**	**8.30**	**6.18 +/− 6.13**	**6.52**

The treatments also influenced the relative variation of prevalence (*16S rRNA*: H(2) = 7.25, $p = 0.027$; *Hsp70*: (H(2) = 7.31, $p = 0.026$). In the AB and C groups, the prevalence remained at 100%, i.e., the pre-treatment level. On the other hand, in AH colonies, it decreased by 45.33% +/− 20.39 s.e. (SD = 49.94, 95% CI = −97.74, +7.07) and 50.00% +/− 22.36 s.e. (SD = 54.77, 95% CI = −107.48, +7.48) when determined with *16S rRNA* and *Hsp70*, respectively (Table 4).

Table 4. Post hoc test for the multiple comparison of the percent prevalence variations calculated in the treatment groups. The Bonferroni-corrected *p*-values referring to the *16S rRNA* and *Hsp70* methods are shown respectively above and below the diagonal.

	AB	AH	C
AB	-	0.051	1.000
AH	0.049	-	0.064
C	1.000	0.062	-

The percent variation of abundance was also influenced by the treatments (*16S rRNA*: (H(2) = 13.90, $p = 0.001$; *Hsp70*: (H(2) = 14.92, $p = 0.001$) (Table 5).

Table 5. Treatment effect on *N. ceranae* abundance.

	16S rRNA			*Hsp70*		
Group	Mean +/− s.e.	SD	95% CI	Mean +/− s.e.	SD	95% CI
AB	−99.38 +/− 0.28	0.75	−100.07, −98.69	−98.75 +/− 0.56	0.56	−100.12, −97.37
AH	−100.00 +/− 0.00	0.00	−100.00, −99.99	−99.98 +/− 0.01	0.03	−100.00, −99.94
C	+164.80 +/− 170.50	417.64	−273.49, +603.09	+40.52 +/− 12.98	31.80	−7.14, +73.89

The controls differed from both Groups AH (*16S rRNA*: $p = 0.001$; *Hsp70*: $p = 0.000$) and AB (*16S rRNA*: $p = 0.056$; *Hsp70*: $p = 0.080$), but no difference could be detected between the treated groups (*16S rRNA*: $p = 0.4141$; *Hsp70*: $p = 0.224$).

3.3. Comparability of the 16S rRNA and Hsp70 Methods

A paired-samples *T* test conducted on the pre- and post-treatment bee samples resulted in a higher number of copies detected with the *16S rRNA* (7.48 +/− 1.73 s.e., SD = 53.44 × 10^{11}) compared to the

Hsp70 method (1.65 +/− 0.33 s.e., SD = 10.31 × 10^9) (t(949)= −4.314, *p*= 0.000). A positive correlation could be calculated between the two series of analytical data (r(950) = +0.859, *p* = 0.000).

In a linear regression model, the *16S rRNA/Hsp70* ratio was considered the dependent variable against the *Hsp70* abundance as the independent predictor (Table 6).

Table 6. Characterization of the regression model relating the abundance ratio *16S rRNA/Hsp70* (dependent variable) to the *Hsp70* abundance (independent variable).

Model	F(1873) = 1627.70, *p* = 0.000, Adj. R^2 = 0.650
Intercept	12.82 +/− 1.80 s.e. (95% CI = 9.28, 16.35), t(873) = 7.119, *p* = 0.000
Slope	6.68 +/− 0.16 s.e. (95% CI = 6.36, 7.00) × 10^{-9}, t(873) = 40.345, *p* = 0.000

Despite that the model and both regression parameters could be confirmed (*p* = 0.000), the analysis of the standardized regression residuals showed a pattern of progressive spread as the independent variable increased (Figure 2).

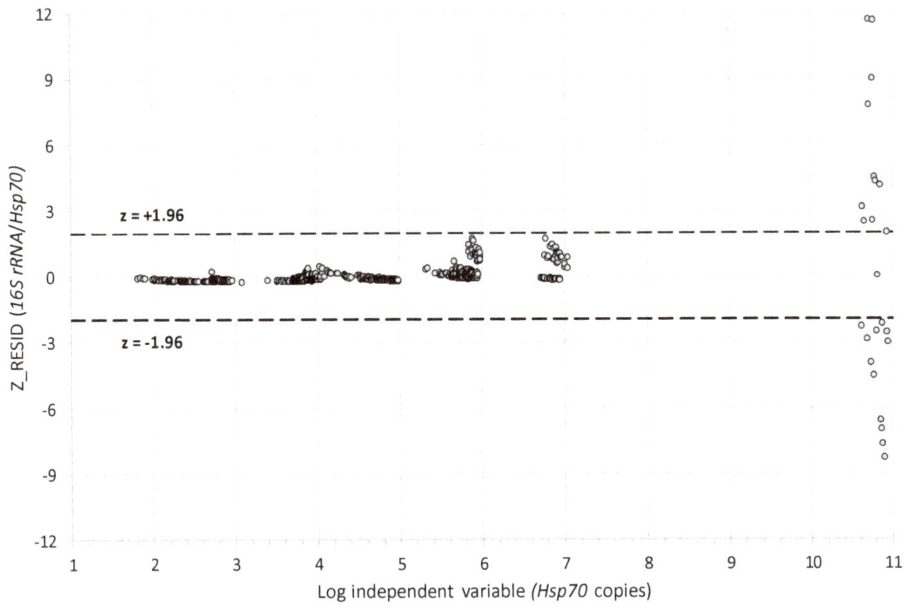

Figure 2. Scatterplot showing the relationship between the error variance of the *16S rRNA/Hsp70* ratio and the abundance measured with the *Hsp70* method (N = 875). The interval of the 95% normally distributed data (z = +/− 1.96) is shown as a reference.

4. Discussion

This trial was conducted to address the largely understudied problem of *N. ceranae* infection management in contexts where the use of antibiotics is ruled out, raising the consequent demand of naturally based treatments.

The colonies included in the experiment were apparently healthy, of regular development for the considered area and season, and did not show signs of dwindling. Despite the good general colony conditions, all the old workers that were sampled pre-treatment from the external combs were found infected. The genetic copies found in the bee intestines with the two q-PCR techniques varied deeply in number between the colonies. The detected high prevalence, compared to previous knowledge on the disease obtained in a similar southern European environment [5], makes it likely that the experimental colonies were approaching the collapse threshold. This may be particularly true for the colony 3AB, which showed conspicuous *N. ceranae* abundance in all sampled bees. It is worth mentioning that the high average abundance detected with both methods makes this colony a putative strong outlier. However, as that was a normal, naturally infected colony of the apiary, its removal from the dataset was considered arbitrary and, for the information content coming from this highly infected and nevertheless asymptomatic case, unjustified also.

The products under consideration were ApiHerb, a patented herbal feeding supplement for honey bees, and Api-Bioxal, a formulation containing dihydrate oxalic acid as the active ingredient. The latter is registered in various countries to control varroa infestations. Previous trials confirmed an influence of either product on the midgut microbiota of honey bees fed in laboratory conditions [38].

The quantitative ApiHerb composition is patent-protected and, therefore, it has not been disclosed. However, the label reports, in decreasing order, dextrose, garlic (*Allium sativum*) and cinnamon (*Cinnamomum zeylanicum*) as the three most abundant ingredients, plus a range of vitamins and flavors. Excluding dextrose as a possible active ingredient and considering that non-confidential information from the manufacturer reports garlic and cinnamon to be respectively contained within the ranges 25–50 and 1–5% [personal communication], it is likely that garlic is the major ingredient responsible for the ApiHerb effect detected on the intestinal microflora of the honey bees.

Garlic preparations and components are reported to have antifungal properties [39–42]. In in vitro tests, aqueous extracts of this plant resulted in decreased spore vitality in *N. bombycis*, the agent of the pebrine disease of silkworm, and, once administered in vivo to *Bombyx mori* larvae, in a reduced prevalence of infected individuals [43]. When administered in the laboratory as ethanolic extracts to artificially *N. ceranae*-infected honey bees, the treatment did not result in significant infection inhibition, as a possible consequence of the low allicin stability in the solvent that was used [44]. However, the administration in sugar water after solubilization into ethanol of allyl sulphide, another garlic compound, significantly reduced the abundance of *N. ceranae* spores in artificially infected bees [45].

Previous trials highlighted the ability of ApiHerb to inhibit *N. ceranae* infections in *A. mellifera*. In a laboratory assay, caged bees were fed ApiHerb for 24 h. Afterwards, they were infected and fed sugar water for ten days, until dissection. In comparison to untreated controls, they showed significantly less *N. ceranae* spores [46]. Field tests resulted in a significant decrease in both number of spores (Italy and Mexico) and prevalence of infected house and forager bees (Spain) compared to untreated controls [47,48].

In the present trial, after four ApiHerb administrations, the post-treatment abundance was various orders of magnitude lower than the pre-treatment level, corresponding to a variation approaching −100%. This drastic and slightly variable effect must be compared with the high and nevertheless very variable increased abundance recorded in the untreated controls in the pre-/post-treatment period. The experiment allowed us to measure also the proportion of infected individuals on the same sample bees. In ApiHerb-treated colonies, the prevalence of infected bees decreased by approximately 45–50%, with variations depending on the q-PCR method that was used, resulting in some negative colonies at the post-treatment check.

The potential of natural queen replacement and the consequent brood hiatus in the control of *N. ceranae* infections was pointed out previously [35], although the dynamics of the healing effect could not be fully clarified. In an independent experiment, a 0.25 M oxalic acid solution was administered twice by trickling to naturally infected free-flying colonies, and by feeding to artificially infected honey bees reared in the laboratory, with an effect on prevalence and abundance, respectively [29]. In this trial, the Api-Bioxal treatment corresponded to a single administration of a 0.47 M oxalic acid solution, that was trickled as a single dose to colonies that were artificially broodless after caging the queen. In a way, this treatment combined two techniques—artificial brood interruption and oxalic acid administration—that showed an effect against the infection in separate experiments. Therefore, it is likely that the recorded effect is the result of the combination between two independent actions.

In the pre-/post-treatment interval, the abundance decreased by one or more orders of magnitude in all Api-Bioxal-treated colonies, corresponding to an approximate average variation of −99%. Although this difference did not diverge significantly from the one recorded in the ApiHerb-treated colonies, it must be noted that the Api-Bioxal treatment did not result in a reduced prevalence also. This contrasts with the above-mentioned experiment, where repeated administrations may have resulted in prolonged coverage and a consequent effect on the prevalence of infected individuals.

Except for the season of application, the Api-Bioxal treatment of this trial simulates the routine varroa control technique used by the beekeepers treating their colonies with oxalic acid after a period of queen caging [34], therefore possibly acting as a double-effect treatment.

The two assays in this study are based on different qPCR targets, *16S rRNA* and *Hsp70*, and characterized by high repeatability [36,37,49,50]. However, there were systematic differences between the two methods. The assay based on the *16S rRNA* gene measured a higher number of copies than those detected with the *Hsp70* gene. The ratio between the measurements obtained with the two methods is not constant. The significant, positive slope resulting from the regression analysis showed that the results of the *16S rRNA* analysis diverge significantly with an increasing number of copies detected with the *Hsp70* method. The residue analysis shown in Figure 2 indicates an uneven distribution of the error variance throughout the number of copies resulting from the analysis using the *Hsp70* gene and, therefore, problematic use of such a model in the straightforward conversion of the abundance values between the two methods.

Both methods have the same linear dynamic range (LDR) and efficiency [36,37]. The diverging results, therefore, arise not because of sensitivity issues, but the different properties of the compared methods. *16S rRNA* is a multi-copy gene and it is present in microsporidia with a variable number of sequences [51]. The fluctuations in the number of copies of the *16S rRNA* gene in the genome of *N. ceranae* may ultimately affect the quantification [52]. In addition, studies on *N. bombi* showed that genes coding for rRNA were intrinsically polymorphic and of elusive nature [53]. These results suggest that high caution is necessary for using these microsporidian sequences, especially when quantification is needed.

On the other hand, *Hsp70* is a single-coding gene in a highly conserved region of the *N. ceranae* genome, which results in accurate quantification [54,55]. In the development of veterinary medicines against Nosemosis Type C, a precise evaluation of both abundance and prevalence reduction is required, which makes the use of quantification methods based on single-copy gene sequences essential. The method based on the *Hsp70* gene, therefore, seems more appropriate for the evaluation of efficacy against *N. ceranae*.

5. Conclusions

Despite the wide-spread use of fumagillin in many places, we demonstrate that it is possible to efficiently control an emergent pathogen like *N. ceranae* with sustainable methods. The products used in this study, however, were not developed for the treatment of *N. ceranae*. For an extension of use, further research is necessary. In a first step, a scientifically based optimization of parameters like the concentration of active substances, treatment period, and posology is still needed. Studies are needed to elucidate the mode of action against *N. ceranae* of oxalic acid as well as the compounds in ApiHerb.

Given the increasing importance of Nosemosis Type C worldwide, safe veterinary medicinal products to control this disease become urgent. In this context, standardizing the protocols with a guideline for the development of these products is an evident requirement. Until now, such a guideline only exists for the registration of treatments against the parasitic mite *Varroa destructor* [56]. With honey bee health becoming an increasingly serious issue for food security, extending these standards to further honey bee diseases appears an indispensable adaptation. Future research focusing on developing integrated treatments will provide us with a more holistic view of honey bee health.

Author Contributions: Conceptualization, G.C., and A.N.; methodology, G.C., A.N., and D.T.; investigation, G.C., D.T., and M.B.; data curation, G.C., A.N., D.T., M.B., and C.G.; writing—original draft preparation, C.G., G.C., and A.N.; writing—review and editing, G.C., A.N., D.T., M.B., and C.G.; supervision, A.N. All authors have read and agreed to the published version of the manuscript.

Funding: This research received no external funding.

Acknowledgments: We thank Chemicals Laif Spa for kindly providing the products used in this study.

Conflicts of Interest: The authors declare no conflict of interest.

References

1. Fries, I.; Chauzat, M.-P.; Chen, Y.-P.; Doublet, V.; Genersch, E.; Gisder, S.; Higes, M.; McMahon, D.P.; Martín-Hernández, R.; Natsopoulou, M.; et al. Standard methods for *Nosema* research. *J. Apic. Res.* **2013**, *52*, 1–28. [CrossRef]

2. Fries, I.; Feng, F.; da Silva, A.; Slemenda, S.B.; Pieniazek, N.J. *Nosema ceranae* n. sp. (Microspora, Nosematidae), morphological and molecular characterization of a microsporidian parasite of the Asian honey bee *Apis cerana* (Hymenoptera, Apidae). *Eur. J. Protistol.* **1996**, *32*, 356–365. [CrossRef]

3. Botías, C.; Anderson, D.L.; Meana, A.; Garrido-Bailón, E.; Martín-Hernández, R.; Higes, M. Further evidence of an oriental origin for *Nosema ceranae* (Microsporidia: Nosematidae). *J. Invertebr. Pathol.* **2012**, *110*, 108–113. [CrossRef] [PubMed]

4. Cox-Foster, D.L.; Conlan, S.; Holmes, E.C.; Palacios, G.; Evans, J.D.; Moran, N.A.; Quan, P.-L.L.; Briese, T.; Hornig, M.; Geiser, D.M.; et al. A metagenomic survey of microbes in honey bee colony collapse disorder. *Science* **2007**, *318*, 283–287. [CrossRef]

5. Higes, M.; Martín-Hernández, R.; Botías, C.; Bailón, E.G.; González-Porto, A.V.; Barrios, L.; Del Nozal, M.J.; Bernal, J.L.; Jiménez, J.J.; Palencia, P.G.; et al. How natural infection by *Nosema ceranae* causes honeybee colony collapse. *Environ. Microbiol.* **2008**, *10*, 2659–2669. [CrossRef]

6. Higes, M.; García-Palencia, P.; Martín-Hernández, R.; Meana, A. Experimental infection of *Apis mellifera* honeybees with *Nosema ceranae* (Microsporidia). *J. Invertebr. Pathol.* **2007**, *94*, 211–217. [CrossRef]

7. Higes, M.; García-Palencia, P.; Urbieta, A.; Nanetti, A.; Martín-Hernández, R. *Nosema apis* and *Nosema ceranae* tissue tropism in worker honey bees (*Apis mellifera*). *Vet. Pathol.* **2019**. [CrossRef]

8. Eiri, D.M.; Suwannapong, G.; Endler, M.; Nieh, J.C. *Nosema ceranae* can infect honey bee larvae and reduces subsequent adult longevity. *PLoS ONE* **2015**, *10*, e0126330. [CrossRef]

9. Higes, M.; Martín-Hernández, R.; Garrido-Bailón, E.; González-Porto, A.V.; García-Palencia, P.; Meana, A.A.; del Nozal, M.J.; Mayo, R.; Bernal, J.L. Honeybee colony collapse due to *Nosema ceranae* in professional apiaries. *Environ. Microbiol. Rep.* **2009**, *1*, 110–113. [CrossRef]

10. Giersch, T.; Berg, T.; Galea, F.; Hornitzky, M. *Nosema ceranae* infects honey bees (*Apis mellifera*) and contaminates honey in Australia. *Apidologie* **2009**, *40*, 117–123. [CrossRef]

11. Higes, M.; Martín, R.; Meana, A. *Nosema ceranae*, a new microsporidian parasite in honeybees in Europe. *J. Invertebr. Pathol.* **2006**, *92*, 93–95. [CrossRef] [PubMed]

12. Martín-Hernández, R.; Bartolomé, C.; Chejanovsky, N.; Le Conte, Y.; Dalmon, A.; Dussaubat, C.; García-Palencia, P.; Meana, A.; Pinto, M.A.; Soroker, V.; et al. *Nosema ceranae* in *Apis mellifera*: A 12 years postdetection perspective. *Environ. Microbiol.* **2018**, *20*, 1302–1329. [CrossRef] [PubMed]

13. Ravoet, J.; De Smet, L.; Meeus, I.; Smagghe, G.; Wenseleers, T.; De Graaf, D.C. Widespread occurrence of honey bee pathogens in solitary bees. *J. Invertebr. Pathol.* **2014**, *122*, 55–58. [CrossRef] [PubMed]

14. Plischuk, S.; Lange, C.E. *Bombus brasiliensis* Lepeletier (Hymenoptera, Apidae) infected with *Nosema ceranae* (Microsporidia). *Rev. Bras. Entomol.* **2016**, *60*, 347–351. [CrossRef]

15. Plischuk, S.; Martín-Hernández, R.; Prieto, L.; Lucía, M.; Botías, C.; Meana, A.; Abrahamovich, A.H.; Lange, C.; Higes, M. South American native bumblebees (Hymenoptera: Apidae) infected by *Nosema ceranae* (Microsporidia), an emerging pathogen of honeybees (*Apis mellifera*). *Environ. Microbiol. Rep.* **2009**, *1*, 131–135. [CrossRef]

16. Porrini, M.P.; Porrini, L.P.; Garrido, P.M.; de Melo e Silva Neto, C.; Porrini, D.P.; Muller, F.; Nuñez, L.A.; Alvarez, L.; Iriarte, P.F.; Eguaras, M.J. *Nosema ceranae* in South American native stingless bees and social wasp. *Microb. Ecol.* **2017**, *74*, 761–764. [CrossRef]

17. Cilia, G.; Cardaio, I.; dos Santos, P.E.J.; Ellis, J.D.; Nanetti, A. The first detection of *Nosema ceranae* (Microsporidia) in the small hive beetle, *Aethina tumida* Murray (Coleoptera: Nitidulidae). *Apidologie* **2018**, *49*, 619–624. [CrossRef]

18. Higes, M.; Martín-Hernández, R.; Garrido-Bailón, E.; Botías, C.; García-Palencia, P.; Meana, A. Regurgitated pellets of *Merops apiaster* as fomites of infective *Nosema ceranae* (Microsporidia) spores. *Environ. Microbiol.* **2008**, *10*, 1374–1379. [CrossRef]

19. Graystock, P.; Goulson, D.; Hughes, W.O.H. Parasites in bloom: Flowers aid dispersal and transmission of pollinator parasites within and between bee species. *Proc. Biol. Sci.* **2015**, *282*. [CrossRef]

20. Purkiss, T.; Lach, L. Pathogen spillover from *Apis mellifera* to a stingless bee. *Proc. R. Soc. B Biol. Sci.* **2019**, *286*, 20191071. [CrossRef]

21. Williams, G.R.; Sampson, M.A.; Shutler, D.; Rogers, R.E.L. Does fumagillin control the recently detected invasive parasite *Nosema ceranae* in western honey bees (*Apis mellifera*)? *J. Invertebr. Pathol.* **2008**, *99*, 342–344. [CrossRef] [PubMed]

22. Higes, M.; del Nozal, M.J.; Alvaro, A.; Barrios, L.; Meana, A.; Martín-Hernández, R.; Bernal, J.L.; Bernal, J. The stability and effectiveness of fumagillin in controlling *Nosema ceranae* (Microsporidia) infection in honey bees (*Apis mellifera*) under laboratory and field conditions. *Apidologie* **2011**, *42*, 364–377. [CrossRef]

23. Huang, W.-F.; Solter, L.F.; Yau, P.M.; Imai, B.S. *Nosema ceranae* escapes fumagillin control in honey bees. *PLoS Pathog.* **2013**, *9*, e1003185. [CrossRef] [PubMed]

24. Botías, C.; Martín-Hernández, R.; Meana, A.; Higes, M. Screening alternative therapies to control Nosemosis type C in honey bee (*Apis mellifera iberiensis*) colonies. *Res. Vet. Sci.* **2013**, *95*, 1041–1045. [CrossRef]

25. Maistrello, L.; Lodesani, M.; Costa, C.; Leonardi, F.; Marani, G.; Caldon, M.; Mutinelli, F.; Granato, A. Screening of natural compounds for the control of *Nosema* disease in honeybees (*Apis mellifera*). *Apidologie* **2008**, *39*, 436–445. [CrossRef]

26. Bravo, J.; Carbonell, V.; Sepúlveda, B.; Delporte, C.; Valdovinos, C.E.; Martín-Hernández, R.; Higes, M. Antifungal activity of the essential oil obtained from *Cryptocarya alba* against infection in honey bees by *Nosema ceranae*. *J. Invertebr. Pathol.* **2017**, *149*, 141–147. [CrossRef]

27. Michalczyk, M.; Sokół, R.; Koziatek, S. Evaluation of the effectiveness of selected treatments of *Nosema* spp. infection by the hemocytometric method and duplex PCR. *Acta Vet.* **2016**, *66*, 115–124. [CrossRef]

28. Michalczyk, M.; Sokół, R. Estimation of the influence of selected products on co-infection with *N. apis*/*N. ceranae* in *Apis mellifera* using real-time PCR. *Invertebr. Reprod. Dev.* **2018**, *62*, 92–97. [CrossRef]

29. Charriére, J.-D.; Imdorf, A. Oxalic acid treatment by trickling against *Varroa destructor*: Recommendations for use in central Europe and under temperate climate conditions. *Bee World* **2002**, *83*, 51–60. [CrossRef]

30. Milani, N. Activity of oxalic and citric acids on the mite Varroa destructor in laboratory assays. *Apidologie* **2001**, *32*, 127–138. [CrossRef]

31. Nanetti, A.; Büchler, R.; Charriere, J.D.; Friesd, I.; Helland, S.; Imdorf, A.; Korpela, S.; Kristiansen, P. Oxalic acid treatments for varroa control (review). *Apiacta* **2003**, *38*, 81–87.

32. Nanetti, A.; Rodriguez-García, C.; Meana, A.; Martín-Hernández, R.; Higes, M. Effect of oxalic acid on *Nosema ceranae* infection. *Res. Vet. Sci.* **2015**, *102*, 167–172. [CrossRef] [PubMed]

33. Porrini, M.P.; Garrido, P.M.; Silva, J.; Cuniolo, A.; Román, S.; Iaconis, D.; Eguaras, M. Ácido oxálico: Potencial antiparasitario frente a *Nosema ceranae* por administración oral y exposición total. In Proceedings of the I Workshop Latinoamericano en Sanidad Apicola, Mar de Plata, Argentina, 25–26 October 2018; p. 64.

34. Büchler, R.; Uzunov, A.; Kovačić, M.; Prešern, J.; Pietropaoli, M.; Hatjina, F.; Pavlov, B.; Charistos, L.; Formato, G.; Galarza, E.; et al. Summer brood interruption as integrated management strategy for effective Varroa control in Europe. *J. Apic. Res.* **2020**, 1–10. [CrossRef]

35. Botías, C.; Martín-Hernández, R.; Días, J.; García-Palencia, P.; Matabuena, M.; Juarranz, Á.; Barrios, L.; Meana, A.; Nanetti, A.; Higes, M. The effect of induced queen replacement on *Nosema* spp. infection in honey bee (*Apis mellifera iberiensis*) colonies. *Environ. Microbiol.* **2012**, *14*, 845–859. [CrossRef]

36. Bourgeois, A.L.; Rinderer, T.E.; Beaman, L.D.; Danka, R.G. Genetic detection and quantification of *Nosema apis* and *N. ceranae* in the honey bee. *J. Invertebr. Pathol.* **2010**, *103*, 53–58. [CrossRef]

37. Cilia, G.; Cabbri, R.; Maiorana, G.; Cardaio, I.; Dall'Olio, R.; Nanetti, A. A novel TaqMan® assay for *Nosema ceranae* quantification in honey bee, based on the protein coding gene Hsp70. *Eur. J. Protistol.* **2018**, *63*, 44–50. [CrossRef]

38. Cilia, G.; Fratini, F.; Tafi, E.; Turchi, B.; Mancini, S.; Sagona, S.; Nanetti, A.; Cerri, D.; Felicioli, A. Microbial profile of the *Ventriculum* of honey bee (*Apis mellifera ligustica* Spinola, 1806) fed with veterinary drugs, dietary supplements and non-protein amino acids. *Vet. Sci.* **2020**, *7*, 76. [CrossRef]

39. Pai, S.T.; Platt, M.W. Antifungal effects of *Allium sativum* (garlic) extract against the *Aspergillus* species involved in otomycosis. *Lett. Appl. Microbiol.* **1995**, *20*, 14–18. [CrossRef]

40. Li, W.-R.; Shi, Q.-S.; Liang, Q.; Huang, X.-M.; Chen, Y.-B. Antifungal effect and mechanism of garlic oil on *Penicillium funiculosum*. *Appl. Microbiol. Biotechnol.* **2014**, *98*, 8337–8346. [CrossRef]

41. Asehraou, A.; Mohieddine, S.; Faid, M.; Serhrouchni, M. Use of antifungal principles from garlic for the inhibition of yeasts and moulds in fermenting green olives. *Grasas Aceites* **1997**, *48*, 68–73. [CrossRef]

42. Singh, U.P.; Pandey, V.N.; Wagner, K.G.; Singh, K.P. Antifungal activity of ajoene, a constituent of garlic (*Allium sativum*). *Can. J. Bot.* **1990**, *68*, 1354–1356. [CrossRef]

43. Devi, A.G.; Sharma, S.D.; Balavenkatasubbaiah, M.; Chandrasekharan, K.; Nayaka, A.R.N.; Kumar, J.J. Identification of botanicals for the suppression of pebrine disease of the silkworm, *Bombyx mori* L. *Uttar Pradesh J. Zool.* **2010**, 173–179.

44. Porrini, M.P.; Fernández, N.J.; Garrido, P.M.; Gende, L.B.; Medici, S.K.; Eguaras, M.J. In vivo evaluation of antiparasitic activity of plant extracts on *Nosema ceranae* (Microsporidia). *Apidologie* **2011**, *42*, 700–707. [CrossRef]

45. Borges, D.; Guzman-Novoa, E.; Goodwin, P.H. Control of the microsporidian parasite *Nosema ceranae* in honey bees (*Apis mellifera*) using nutraceutical and immuno-stimulatory compounds. *PLoS ONE* **2020**, *15*, e0227484. [CrossRef] [PubMed]

46. Bessi, E.; Nanetti, A. Evaluation of tree different strategies to *Nosema* control. In Proceedings of the 39th Apimondia International Apicultural Congress, Dublin, Ireland, 21–26 August 2005.

47. Sóstenes Rodríguez Dehaibes, R.; Luna-Olivares, G.; Chávez-Hernández, E. Resultados preliminares de ApiHerb para el control de *Nosema* spp. en México. *Vida Apic.* **2018**, *207*, 38–39.

48. Nanetti, A.; Martín-Hernández, R.; Gómez-Moracho, T.; Cabbri, R.; Higes, M. Api-Herb en el control orgánico de la Nosemosis tipo C (Nosema ceranae, Microsporidia). In Proceedings of the III Congreso Ibérico de Apicultura, Mirandela, Portugal, 13–15 April 2015; Vilas-Boas, M., Dias, L.G., Moreira, L.M., Eds.; Instituto Politécnico de Bragança: Bragança, Portugal, 2014; p. 36.

49. Cilia, G.; Sagona, S.; Giusti, M.; Jarmela dos Santos, P.E.; Nanetti, A.; Felicioli, A. *Nosema ceranae* infection in honeybee samples from Tuscanian Archipelago (Central Italy) investigated by two qPCR methods. *Saudi J. Biol. Sci.* **2018**. [CrossRef] [PubMed]

50. Martín-Hernández, R.; Meana, A.; Prieto, L.; Salvador, A.M.; Garrido-Bailón, E.; Higes, M. Outcome of colonization of *Apis mellifera* by *Nosema ceranae*. *Appl. Environ. Microbiol.* **2007**, *73*, 6331–6338. [CrossRef]

51. Sagastume, S.; Martín-Hernández, R.; Higes, M.; Henriques-Gil, N. Genotype diversity in the honey bee parasite *Nosema ceranae*: Multi-strain isolates, cryptic sex or both? *BMC Evol. Biol.* **2016**, *16*, 1–11. [CrossRef]

52. Sagastume, S.; Del Águila, C.; Martín-Hernández, R.; Higes, M.; Henriques-Gil, N. Polymorphism and recombination for rDNA in the putatively asexual microsporidian *Nosema ceranae*, a pathogen of honeybees. *Environ. Microbiol.* **2011**, *13*, 84–95. [CrossRef]

53. O'Mahony, E.M.; Tay, W.T.; Paxton, R.J. Multiple rRNA variants in a single spore of the microsporidian *Nosema bombi*. *J. Eukaryot. Microbiol.* **2007**, *54*, 103–109. [CrossRef]

54. Gomez-Moracho, T.; Maside, X.; Martin-Hernandez, R.; Higes, M.; Bartolomé, C. High levels of genetic diversity in *Nosema ceranae* within *Apis mellifera* colonies. *Parasitology* **2014**, *141*, 475–481. [CrossRef] [PubMed]

55. Wang, Y.; Ru, Y.; Liu, W.; Wang, D.; Zhou, J.; Jiang, Y.; Shi, S.; Qin, L. Morphological characterization and HSP70-, IGS-based phylogenetic analysis of two microsporidian parasites isolated from *Antheraea pernyi*. *Parasitol. Res.* **2017**, *116*, 971–977. [CrossRef] [PubMed]

56. European Medicines Agency. *Guideline on Veterinary Medicinal Products Controlling Varroa Destructor Parasitosis in Bees*; European Medicines Agency: London, UK, 2011; p. 9. Available online: https://www.ema.europa.eu/en/documents/scientific-guideline/guideline-veterinary-medicinal-products-controlling-varroa-destructor-parasitosis-bees_en.pdf (accessed on 27 July 2020).

Review

Veterinary Diagnostic Approach of Common Virus Diseases in Adult Honeybees

Julia Dittes [1,2,*], Heike Aupperle-Lellbach [3], Marc O. Schäfer [4], Christoph K. W. Mülling [2] and Ilka U. Emmerich [5]

1   Centre for Applied Training and Learning, Faculty of Veterinary Medicine, Leipzig University, An den Tierkliniken 19, 04103 Leipzig, Germany
2   Institute of Veterinary Anatomy, Histology and Embryology, Faculty of Veterinary Medicine, Leipzig University, An den Tierkliniken 43, 04103 Leipzig, Germany; c.muelling@vetmed.uni-leipzig.de
3   LABOKLIN GmbH & CO.KG, Labor für klinische Diagnostik, Steubenstraße 4, 97688 Bad Kissingen, Germany; aupperle@laboklin.com
4   Friedrich-Loeffler-Institut, Federal Research Institute for Animal Health, Südufer 10, 17493 Greifswald, Insel Riems, Germany; Marc.Schaefer@fli.de
5   VETIDATA, Institute of Pharmacology, Pharmacy and Toxicology, Faculty of Veterinary Medicine, Leipzig University, An den Tierkliniken 39, 04103 Leipzig, Germany; emmerich@vetmed.uni-leipzig.de
*   Correspondence: julia.dittes@vetmed.uni-leipzig.de; Tel.: +49-341-973-8247

Received: 20 September 2020; Accepted: 19 October 2020; Published: 21 October 2020

Abstract: Veterinarians are educated in prevention, diagnosis and treatment of diseases in various vertebrate species. As they are familiar with multifactorial health problems in single animals as well as in herd health management, their knowledge and skills can be beneficial for the beekeepers and honeybee health. However, in education and in practice, honeybees are not a common species for most veterinarians and the typical veterinary diagnostic methods such as blood sampling or auscultation are not applicable to the superorganism honeybee. Honeybee colonies may be affected by various biotic and abiotic factors. Among the infectious agents, RNA-viruses build the largest group, causing covert and overt infections in honeybee colonies which may lead to colony losses. Veterinarians could and should play a more substantial role in managing honeybee health—not limited to cases of notifiable diseases and official hygiene controls. This review discusses the veterinary diagnostic approach to adult bee examination with a special focus on diagnosis and differential diagnosis of the common virus diseases Acute Bee Paralysis Virus (ABPV)-Kashmir Bee Virus (KBV)-Israeli Acute Paralysis Virus (IAPV)-Complex, Chronic Bee Paralysis Virus (CBPV) and Deformed Wing Virus (DWV), as well as coinfections like *Varroa* spp. and *Nosema* spp.

Keywords: honeybee veterinary medicine; acute bee paralysis; chronic bee paralysis; deformed wing virus; *varroa* infestation; nosemosis

---

1. Introduction

In contrast to human medicine, one of the most prominent features of veterinary medicine is that a veterinarian has to deal with a large variety of species. Whereas dogs, cats and horses are common patients in a veterinary practice, the honeybee (*Apis mellifera* L.) is only rarely in contact with veterinarians as a patient. Official veterinarians and diagnostic laboratories are responsible for the analysis of honey samples and for animal disease control in case of, e.g., American foulbrood, but the general disease prevention and control are done by the beekeepers. The majority of the establishments for veterinary education in Europe accredited by ESEVT (European System of Evaluation of Veterinary Training) offer elective or no training in honeybee veterinary medicine. Only in 24 of 68 surveyed schools honeybee veterinary medicine is part of the curriculum as a separate subject [1]. This is a very limited consideration in the curriculum and veterinary practice for a food-producing animal with such a tremendous importance for global agriculture [2].

However, in recent years, the honeybee has aroused increasing interest among veterinarians. While only 9 veterinarians held a specialization about honeybees in Germany in 2014 [3], five years later, there were 17 bee veterinarians or nearly twice as many experts in this field [4]. Honeybee diseases are on the rise [5] and at the same time, a decline of honeybee colonies and beekeepers in Europe can be seen [6]. More and more, veterinarians publish specialist literature to deal with honeybee diseases [7,8]. At the supranational level, the European Parliament in 2008 called for the Commission "to incorporate into its veterinary policy, research into, and actions to tackle bee disease" [9], and in April 2011, a European Union Reference Laboratory for Bee Health was designated to coordinate diagnostic methods, disease monitoring and expert training [10,11].

In honeybee colonies, a large number of individuals together form a superorganism. Both this unit as well as the individual bee have to be investigated carefully. General principles of herd health management and hygiene concepts are well-known to veterinarians and thus, can be applied to the honeybee. Nevertheless, dealing with diseases in honeybees is different from the normal approach in veterinary medicine. Usual examination techniques, known from mammals, are not applicable to insect species, although the general approach and procedures are similar. Food control is not possible as the feed is provided by the environment and not by the beekeeper or the veterinarian [7]. The honeybee is a food-producing animal and assuring the quality of honey is an additional objective for veterinarians and beekeepers.

A variety of biotic and abiotic factors have an impact on honeybee colonies. Belsky et al. presented a broad overview of these stressors: habitat and climate changes, weather, the density of apiaries and food resources, as well as transportation of colonies, are external aspects influencing the honeybee [12]. Equally, the bees depend on intrinsic factors such as genetics and queen longevity [13]. The intensive agriculture with monocultures decreases the plant diversity and thus, the food supply for the honeybees. Between mass flowering of, e.g., rapeseed (*Brassica napus*) and sunflower (*Helianthus annuus*), pollen harvest severely declines [12] and limits adequate nutrition. Such an insufficient protein diet weakens the bees in defending against pathogens [14]. Various bacteria, microsporidia, viruses and pests cause bee diseases, among which viruses have become more relevant during recent years [5].

This review presents basic information on selected virus diseases and the process of colony examination and diagnostics from the veterinary perspective, with a focus on adult honeybees.

2. Virus Diseases in Honeybees and Contributing Factors

Viruses, mostly positive single-strand RNA viruses, are the largest class of honeybee infecting pathogens [13]. Over 20 bee viruses have been identified to date, including the Acute bee paralysis virus (ABPV), Kashmir bee virus (KBV), Israeli acute paralysis virus (IAPV), forming the ABPV-KBV-IAPV complex, Black queen cell virus (BQCV), Chronic bee paralysis virus (CBPV), Deformed wing virus (DWV) and Sac brood virus (SBV). Detailed reviews about the most important ones can be found in the literature [15–18]. Information on symptoms and transmission routes of ABPV-KBV-IAPV, CBPV and DWV, which is relevant to the veterinary diagnostic approach, will be presented in this review. Table 1 provides a summary of taxonomy, symptoms, affected castes and main transmission routes.

Table 1. Overview of selected honeybee viruses (taxonomy, symptoms, affected casts of bees and transmission routes), modified after Vidal-Naquet [19].

Virus	Taxonomy	Symptoms in a Colony	Affected Castes	Transmission (Main Routes)
Chronic bee paralysis virus (CBPV)	unclassified (RNA)	hairless black syndrome and paralysis syndrome, high mortality after a few days	w, d, q	c, o
Acute bee paralysis virus (ABPV)	*Dicistroviridae* (RNA)	paralysis and high mortality after 1–2 days (experimentally)	w, d	vec, ver (v, to)
Kashmir bee virus (KBV)	*Dicistroviridae* (RNA)	mortality without other symptoms	w, d	vec, ver
Israeli acute paralysis virus (IAPV)	*Dicistroviridae* (RNA)	paralysis and death	w, d	vec, ver
Deformed Wing Virus (DWV)	Picorna-like virus (RNA)	crippled bees with deformed wings and shortened abdomens	w, d, q	vec, ver (v, to)

Legend: RNA = Ribonucleic acid, w = worker, d = drone, q = queen, c = direct contact, o = oral–fecal, vec = vector-borne, ver = vertical, v = venereal, to = transovarial.

2.1. Acute Bee Paralysis Virus—Kashmir Bee Virus—Israeli Acute Paralysis Virus—Complex

ABPV, KBV and IAPV are three similar icosahedral viruses of the family *Dicistroviridae*. During the last years, this complex has been reported in association with colony collapse disorder (CCD), a phenomenon of severe colony weakening without visible worker bee mortality [7,20,21]. ABPV and IAPV cause symptoms such as trembling, paralysis, inability to fly, darkening and loss of hair from thorax and abdomen, affecting mostly individual bees and not the whole colony [22]. Hou and Chejanovsky describe experimental symptoms of IAPV such as disorientation, shivering wings, crawling and progressive paralysis until death in infection experiments [20]. ABPV and KBV are a reason for a sharp decline in the adult bee population [22].

Transmission of these viruses is possible vertically as well as horizontally. Furthermore, the *Varroa* mite plays an important role as a mechanical vector [15]. It transfers the virus while feeding from hemolymph of the bees and increases the viral load in the colony [15].

2.2. Chronic Bee Paralysis Virus

CBPV is an unclassified RNA virus that causes two different syndromes in honeybees. Type A, the paralytic form, is characterized by trembling, disorientation and ataxia. Type B affected bees show black, hairless and greasy shining abdomens. Both can occur in the colony at the same time and lead to massive worker bee losses [23].

Comprehensive information on this virus can be found in our case report presenting an overt infection with CBPV in two colonies [24].

2.3. Deformed Wing Virus

DWV is a picorna-like virus, that is often involved in winter losses of honeybee colonies. Two main variants, A and B, can be differentiated [25–28]. In most cases, it can be detected in colonies as a covert infection (see definition of covert under Section 3 in this review) without causing clinical symptoms [15]. Transmission routes are vertical (via eggs and sperm) or horizontal through larval food, trophallaxis or cannibalism of pupae during hygienic behavior [18]. However, *Varroa destructor* plays the key role in transmission as well as virulence and pathology of the virus [15]. Obvious symptoms are seen in the colony only if DWV replicates in the mite before being transferred to the bees. Clinical signs are deformed wings, bloated, discolored, shortened abdomens, hypoplastic glands and pupal death [29]. Infected bees are not viable and die within less than 67 h after emergence [15].

3. Veterinary Diagnostic Approach

As mentioned above, honeybee health is often influenced by many different factors [7]. Without clinical findings, the appearance of a pathogen in a colony does not constitute a disease. For that reason, the terms overt and covert infections were introduced by de Miranda and Genersch to describe honeybee virus diseases [30]. The descriptive terms overt and covert are widely used in insect virology. Overt infections are characterized by obvious clinical findings related to the virus infection and a high virus production rate. Acute and chronic forms are differentiated. In covert infections, low titers of virus particles are present in the absence of clinical symptoms. Vertical transmission allows virus persistence over several generations and competent virus particles can turn into overt infections due to various influencing factors. Persistent infections with low-level virus production can be distinguished from latent infections without virus production [30].

From the veterinary perspective, a holistic diagnostic work-up of medical issues in honeybee colonies is important, because the environment, the colony, the pathogens and every bee are each just a link in the chain leading to occurrences of infections [7]. Figure 1 shows a detailed plan for diagnostics and management in general. Starting from the environmental observations, followed by an examination of the hive, the colony and the bees, samples are taken, and relevant laboratory diagnostics carried out and further illnesses investigated. The resulting problem list leads to a prognosis and a management plan. The main goal is the healthy colony formed by fit individuals.

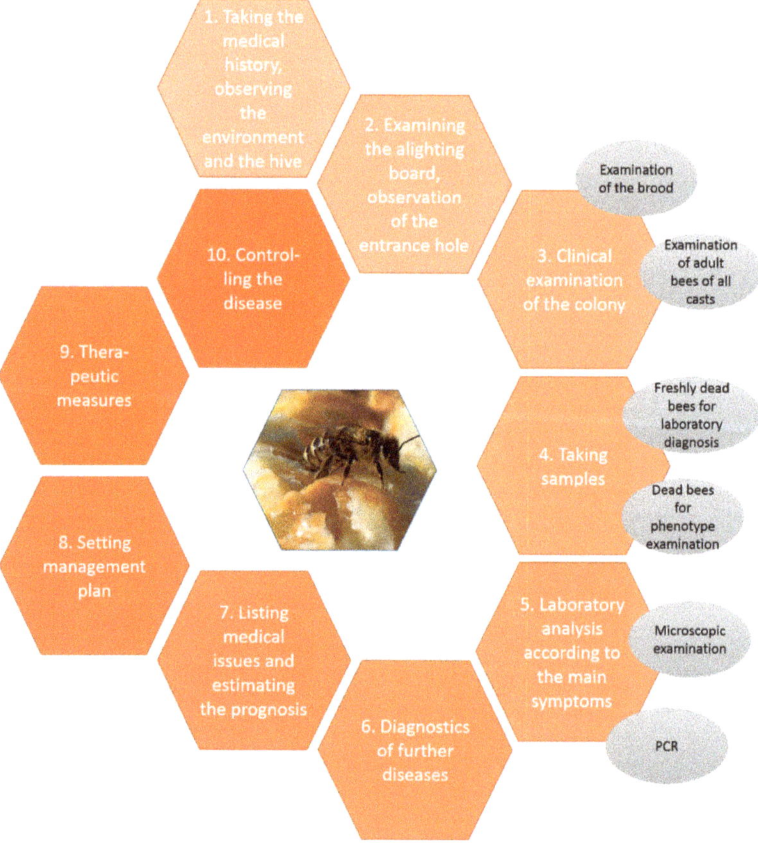

Figure 1. Diagnostic approach in bee diseases, © Julia Dittes.

3.1. Medical History, Appeareance of the Hive and Environment

Like in any other species, clinical examination starts with a detailed anamnesis. Gathering information from the beekeeper gives a first overview about the acute problem, which often is superficial and skewed by incorrectly interpreted data [31]. Asking detailed questions can help: When did the beekeeper first observe symptoms or changes in the colonies? How frequently are the hives inspected during the season? How is *Varroa* infestation monitored and controlled? Is the beekeeper migrating the colonies? Maybe the beekeeper can also offer information about weather, crops in the region around the apiary or the density of apiaries in the vicinity—information needed for epidemiological evaluation.

The location and the environment around the beehive have an immense relevance to the occurrence of health problems and should therefore be carefully observed. The weather, the food range, the landmarks and further factors influence the bees in a similar way as a stable or field does to cattle, for example. A complete hazard analysis is shown by Formato and Smulders [10]. Vitally important are accessible water, identifiable food sources, climate and wind [31]. The hive model, its material and the hive's general condition should be noticed. Fecal spots as well as traces of predators and pests may be seen during external examination.

Figure 2A,B show two apiaries in different locations. The size of the colony and the offered space in the hive should be evaluated [7,31].

(A)   (B)

Figure 2. (**A**) Apiary in April 2019: colonies with two brood chambers in wooden Zander hives and polystyrene Segeberger Classic hives, standing in pairs of two. © Julia Dittes. (**B**) Apiary with Zander hives with one or two brood chambers and honey supers on top, standing next to each other. © Heike Aupperle-Lellbach.

3.2. Examination of the Alighting Board and Observation of the Entrance Hole

First pathologic findings can be identified during observation of the alighting board and entrance hole. The mass of bees flying in and out—depending on temperature, weather, time of day and state of colony development—provides a first impression about the strength of a colony. Dead bees as well as bees and larvae with alterations of the phenotype, carried out of the hive, may be found in front of the entrance. Veterinarian and beekeeper should take their time to observe the entrance hole carefully (Figure 3A,B). The way the bees fly, the manner of landing and the behavior of guarding bees offer a lot of information, because unusual flying behavior may be a neurologic symptom of virus diseases. Furthermore, the ground in front of the hive should be investigated for waste, feces, dead bees and larval mummies [7]. These observations, together with an assessment of the environment and hive from the outside, enable a first evaluation of the colony condition [31].

(A)   (B)

Figure 3. (**A**) Observation of the entrance hole enables a first evaluation of colony strength and activity according to season, weather and environment. © Julia Dittes. (**B**) Alighting board of a Chronic Bee Paralysis Virus-infected honeybee colony with some black hairless individuals (yellow circles). © Silvia Heisch.

3.3. Clinical Examination of the Superorganism "Honeybee Colony" and Observations of Living Bees

The "internal" clinical examination of the bee colony and of individual bees are the next steps. It is important to have a look at the adult bees of all castes (workers, drones and the queen), as well as the brood in different stages and the hive material. Honeybees are social insects that can survive within the colony as a superorganism only. The colony is functional if each single individual performs its tasks. In the following, we will focus on the clinical findings in adult bees.

Clinical examination of the bees and the frames should be performed together with the beekeeper. This way, it is possible to analyze the beekeeper´s operations while handling the bees [7]. General features to assess after opening the hive are the colony behavior, the odor and the colony strength [7,31]. Normally, a honeybee colony smells like wax. If vinegar or foul smell is perceived, there is likely a disease. The intercomb-spaces, where bees are visible, can be counted to give a first idea of colony size. This might be confirmed later by estimating the number of bees, brood cells and food, e.g., using the Liebefeld method [32,33]. The bees are carefully examined to find body deformities, wing abnormalities like V-wings or K-wings, changes in color and size, phoretic *Varroa* mites as well as behavioral, neurological, social or digestive symptoms. Signs of CBPV disease are hairless, black bees and/or neurological symptoms such as trembling, circling or paralysis [23,24]. Abnormal wing position and wing form are shown in case of CBPV or DWV. Phoretic *Varroa* indicate an intense infestation rate with the mite. Phenotype changes, e.g., crippled bees, high mortality and paralysis symptoms, can also be a sign of intoxication. Table 2 lists different phenotype changes in honeybees, which can be seen in a sample of dead bees as well, and their possible causes.

While examining the bees, the frames are assessed: the color and brightness of the wax correspond to the age of the combs. Old wax, having been used in several brood cycles, is much darker than newly produced wax. The frames themselves can be clean or show fecal spots in case of digestive problems [34]. Pollen and honey stores are evaluated to get an idea about the alimentation of the colony.

It is advisable to have a look at how easily the frames can be removed from the hive by the beekeeper. If they are not easy to remove, propolis foraging activity or insufficient surveillance by the beekeeper could be the reason [7].

For all symptoms, it is determined whether they affect a small number of individuals or numerous bees, which would indicate a severe problem. Furthermore, it is relevant which castes of bees show the symptoms.

During the colony investigation, the queen is sought. It is observed whether she shows any abnormalities and whether the symptoms are equal to those of the workers. The queen does not always have the same virus load and symptoms as her colony [35].

Table 2. Phenotype changes in honeybees and their possible causes.

Phenotype Feature	Possible Causes (Selection)
shortened abdomens	DWV, Varroosis, CBPV
hairless, black abdomens	CBPV, black robbers for alimentary, genetic or mechanical reasons
crippled wings, legs, antennae	Varroosis, DWV, intoxication, CBPV
bloated abdomens and diarrhea (pressure on abdomen light-brown fluid)	Nosemosis, *Malpighamoeba mellificae*, CBPV
extended proboscis	virus diseases, intoxication, unspecific

Legend: DWV = Deformed wing virus, CBPV = Chronic bee paralysis virus.

3.4. Taking Samples for Laboratory Analysis

To investigate the phenotype of bees and pathological changes in more detail and to perform a laboratory analysis, bee samples have to be taken. Most suitable are living symptomatic bees, euthanized. However, in case of increased forager mortality, freshly dead bees from the soil in front of the hive could be appropriate as well, but it has to be considered that there can be false-negative results in virus diagnostics because RNA is unstable in the environment. Different diagnostic methods require a specified number of bees, a fact that should be considered when taking the samples. To assess the size of a sample, the following reference value can be considered: 100 mL are equivalent to approximately 330 honeybees and 31 g. The required number of bees is caught and euthanized by freezing (about 15 min, −20 °C) [8]. Other ways of euthanizing are using 96% ethanol, carbon dioxide or sampling after asphyxiation with sulfur. The sample with dead bees is sent chilled in an air-permeable case to the laboratory. In general, it is advisable to contact the responsible laboratory for information on size and condition of the sample.

3.4.1. Examination of Dead Bees

Dead bees are examined, and their phenotype is described. They can be sorted according to their size and symptoms: Bees can be smaller than normal and show shortened abdomens (Figure 4A) when infected with several viruses, e.g., Deformed Wing Virus (DWV). More indicative for DWV, associated with a severe *Varroa* infestation load, are crippled wings in freshly hatched honeybees (Figure 4B). Williams et al. describe a ranking in six categories, according to the severity of wing abnormities [36]. Hairless black abdomens are seen in case of a CBPV infection [24] or as a result of genetics, alimentation within honeydew flow period or for mechanical reasons. Robbery and fighting as well as maturing in foragers may result in breaking hairs and black abdomens, which, however, only affects individuals, whereas genetics, alimentation and CBPV affect the whole colony. If the abdomens of several bees seem to be bloated, pressure on the abdomen may lead to a light-brown fluid leaking from the gut. In such cases, the bee´s gastro-intestinal tract can be pulled out for further investigation. Often, an extended proboscis is seen (Figure 4A), which is a more unspecific sign and not necessarily evidence for an intoxication. A summary of phenotype changes is listed in Table 2 above.

(**A**)   (**B**)

Figure 4. (**A**) sample of dead bees of a honeybee colony infected with CBPV in 2019: inhomogeneous size of worker bees, shortened abdomen (orange arrows), extended proboscis (blue arrow); © Julia Dittes. (**B**) freshly hatched honeybee infected by DWV; © Heike Aupperle-Lellbach.

3.4.2. Examination of Debris

A plastic drawer is put under the hive to sample debris. Examination of debris provides significant information about the health, development status and strength of a honeybee colony. Building materials such as wax scales, cell lids and drops of propolis, as well as cell components like pollen, sugar and melicitose crystals or drops of diluted food [37], can be found (Figure 5). The respective amounts of these components are indicators for the strength of the colony.

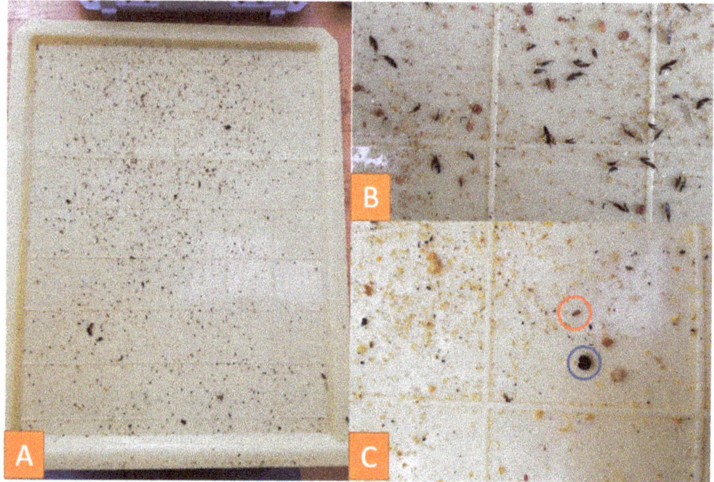

Figure 5. Photographs of a plastic drawer after three days under the hive of an *Apis mellifera carnica* colony in September with debris containing a lot of *Varroa* mites (**A**), separated legs and wings of bees (**B**), pollen (blue cycle) and wax cylinders (red cycle) (**C**). © Julia Dittes.

Furthermore, traces of predators can be seen: Parts of bees are a sign of wasps or mice in the hive (Figure 5B). Mice leave 3–8 mm long feces on the drawer. Smaller, dark brown feces belong to the greater wax moth, *Galleria mellonella*, which can also be detected by its creamy-white to grey larvae [7].

Wood chips or straw from the feeding trough or condensation water can be found, too.

Debris examination is additionally used for *Varroa* infestation control. The mite can be seen with the naked eye. Their number, counted after the drawer was under the hive for a defined time, can be used to quantify the infestation [7].

The validity of the information gathered from debris depends on the weather, the season and the time the drawer was under the hive [37]. In relation to these conditions, it constitutes an important part of examination.

3.5. Laboratory Diagnosis

3.5.1. PCR to Detect Viral Diseases

This step has to be adjusted to the clinical observations in each specified case. Honeybee viruses are mainly detected by Real-Time Reverse Transcription (RT)-Polymerase Chain Reaction (PCR). In the following, the methods are described for CBPV, ABPV and DWV in the way they are performed at the Friedrich-Loeffler-Institute, the Federal Research Institute for Animal Health. A sample of 50 symptomatic bees is an appropriate sample size to submit to the laboratory.

From each bee sample, ten bees are homogenized using the gentleMACS^{TM} Dissociator. Total RNA is purified from 150 µL of clarified bee homogenate using the RNeasy Mini Kit (Qiagen, Venlo, NL) according to the manufacturer's instructions. For each of the viruses, a one-step real-time RT-PCR is subsequently performed in duplicate using the AgPathID^{TM} One-Step RT-PCR Kit (Applied Biosystems^{TM,}, Waltham, MA, USA) in a 96-well reaction plate with 2.5 µL RNA in a final volume of 12.5 µL.

For CBPV detection, it contained 320 nM of forward and reverse primer (qCBPV 9: 5′- CGC AAG TAC GCC TTG ATA AAG AAC -3′; qCBPV 10: 5′- ACT ACT AGA AAC TCG TCG CTT CG -3′), and 200 nM of the CBPV probe (CBPV 2 probe: 5′- FAM- TCA AGA ACG AGA CCA CCG CCA AGT TC -BHQ1 -3′) [38].

For ABPV detection, it comprised 800 nM of forward and reverse primer (ABPV1: 5′- CAT ATT GGC GAG CCA CTA TG -3′; ABPVRn: 5′- CTA CCA GGT TCA AAG AAA ATT TC -3′), and 90.4 nM of the ABPV probe (ABPVnTaq: 5′- FAM- ATA GTT AAA ACA GCT TTT CAC ACT GG -BHQ1 -3′) [39].

For DWV-A detection, it contained 350 nM of forward and reverse primer (F-DWV_4250: 5′- GCG GCT AAG ATT GTA AAT TG -3′; R-DWV_4321: 5′- GTG ACT AGC ATA ACC ATG ATT A -3′), and 100 nM of the DWV-A probe (Pr-DWV_4293: 5′- FAM- CCT TGA CCA GTA GAC ACA GCA TC -BHQ1 -3′) [40]. For DWV-B detection, it contained 1.2 µM of forward and reverse primer (F-VDV1_4218: 5′ GGT CTG AAG CGA AAA TAG -3′; R-VDV1_4290: 5′- CTA GCA TAT CCA TGA TTA TAA AC -3′), and 400 nM of the DWV-B probe (Pr-VDV1_4266: 5′- FAM- CCT TGT CCA GTA GAT ACA GCA TCA CA -BHQ1 -3′) [40].

The thermal cycling conditions are 10 min at 45 °C (RT = reverse transcription), 10 min at 95 °C (RT inactivation, initial denaturation, activation of DNA Polymerase), followed by 41 amplification cycles at 94 °C for 15 s and 60 °C for 45 s. The results are expressed as the mean of the two replicates for each reaction. Figure 6 shows an amplification diagram for evaluation. The assay is combined with an internal control assay in a duplex real-time RT-PCR—with the exception of DWV-B detection. This is carried out as a simplex approach, since another primer–probe combination interferes with the PCR approach and hinders or prevents the amplification of the virus fragment. DWV-B therefore is detected in parallel to the closely related DWV-A virus, which also helps in differentiating DWV disease. The performed internal controls are, on the one hand, a universal internal control system based on IC2-RNA [41], and, on the other hand, the detection of β-actin in the extracted samples [42].

Both internal controls monitor that the RNA extraction was successful in all samples (and extraction controls), as well as confirm their transcription into cDNA and the amplification of those during real-time RT-PCR. Furthermore, this allows to see whether a uniform effective RNA extraction took place.

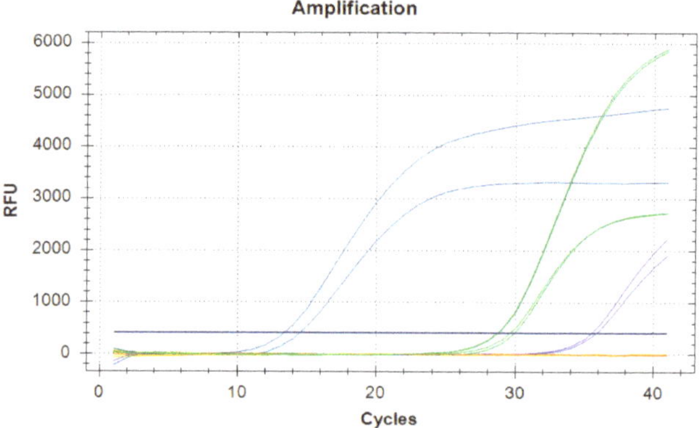

Figure 6. Amplification of a sample of a honeybee colony investigated for CBPV and ABPV in duplicate, result: ABPV = Ct 35.23, CBPV = Ct 14.33; green = positive controls, blue = CBPV, violet = ABPV, where the sample lines cross the blue straight line, Ct is determined, © Julia Dittes.

3.5.2. Monitoring the *Varroa* infestation

The *Varroa* mite is one of the main stressors, probably globally "the greatest threat" [7] for honeybee colonies. If there is no control, colonies with a high mite load are weakening until they ultimately collapse. Additionally, *Varroa destructor* serves as a mechanical and biological vector for various honeybee viruses and suppresses the immune response of honeybees [43]. Therefore, it should be monitored carefully.

To evaluate the mite load in a colony, different methods are described in the literature. It is possible to examine the debris, the brood and adult bees. Drone brood can be investigated by removing the lids of the cells and studying the larvae and combs individually (Figure 7A). In case of a larger number of samples, the brood cells can be washed out into a sieve system where mites and brood are gathered separated from each other [44].

Figure 7. (**A**) *Varroa* infestation control via examination of drone brood: red-eyed larva with *Varroa* mite, © Julia Dittes. (**B**) DWV-infected adult honeybee with phoretic *Varroa* mites on thorax, © Ilka Emmerich.

Adult bees play a role in assessing the mite load (Figure 7B). Examination of the natural mite fall in the debris is a common method to determine the mite load. The natural mite fall is the number of mites per day falling naturally from the bees down to the bottom board. A drawer is placed under the hive over a period of 2 to 5 days (Figure 8A). It can be lined with a sticky paper towel. The drawer is examined, and the mites are counted (Figure 8B). The oval, reddish-brown *Varroa* mites are 1.2 to

1.7 mm [7] and can easily be detected in the debris (Figure 8C). Finally, the number of mites per day has to be calculated.

Figure 8. *Varroa* infestation control via debris examination: (**A**) A drawer with a sticky paper is put under the hive, (**B**) counting the number of *Varroa* mites, (**C**) close-up of *Varroa mites* on the drawer. © Jens Emmerich.

Bak et al. compared the flotation and the powdered sugar shake method in 2009 with similar results for both [45]. About 300 young adult bees (is equivalent to 100 mL) are sampled from a frame with uncapped brood for either flotation or powdered sugar method. For flotation, the bees are shaken in alcohol solution for five minutes, after freezing or killing them in alcohol. The mites are removed from the bees and can be counted (Figure 9).

Figure 9. *Varroa* infestation control with the alcohol washing method, (**A**) shaking the honeybee sample in the ethanol solution, (**B**) straining the liquid in a sieve system, (**C**) mites in the smaller sieve can be counted. © Heike Aupperle-Lellbach.

The second method leaves the bees alive. Dusted with icing sugar, they are shaken gently two times for two minutes. The grooming behavior of the bees is stimulated and the mites' feet do not stick any longer, thus, the mites are dislodged from the bees (Figure 10). Either the sugar is sieved out or dissolved in water, so that the separated mites can be counted [45,46]. If the mite infestation rate is under 5%, the colony is slightly affected, if the value is above 10%, an immediate treatment should be applied [7].

Figure 10. *Varroa* infestation control via iced sugar shaking method, (**A**) iced sugar-dusted bees in a shaking bucket, (**B**) dusted bees back in the hive after the procedure, (**C**) dusted mites on a paper towel. © Ilka Emmerich.

3.5.3. Detection of Nosemosis

Fecal spots at or in honeybee hives can be a sign for nosemosis, but they do not have to be [34]. Dysentery can be caused by different pathogens such as CBPV, *Malpighamoeba mellificae* and *Nosema*-species, or be a symptom of stress. In 21 examined honeybee colonies with fecal spots, only 43% could be proven to be *Nosema*-positive [34].

Nosema-species can occur in a honeybee colony as a covert infection without harming it. But, it can serve as an additional and weakening stressor to the honeybees. In some cases, it can be a precursor of other diseases by opening infection routes due to epithelial damage in the midgut [47].

For investigation of nosemosis, abdomens of 20 freshly dead bees are separated in a mortar (Figure 11A). They are crushed with the pestle while adding 5 mL of aqua purificata (Figure 11B). Then, water is added to form a solution of 1 mL per bee. A drop of that sample is put onto a microscope slide and covered with a coverslip without trapping air bubbles (Figure 11C). The slide is examined under the microscope at 400× magnification [48]. This solution is suitable for detection of both *Nosema* and *Malpighamoeba mellificae*.

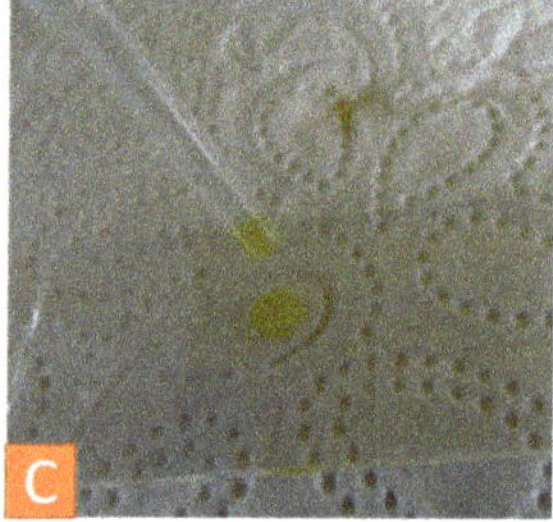

Figure 11. Detection of Nosemosis: (**A**) abdomens of 20 honeybees in a mortar, (**B**) crushing the abdomens with aqua purificata, (**C**) drop of the sample on a microscopic slide. © Julia Dittes.

Nosema spores are oval-shaped and about 4 to 7 µm in length and 2 to 4 µm in width. The two species, *N. apis* and *N. ceranae*, cannot be distinguished by this method [7]. For further differentiation, a PCR would be required [49].

An estimation of the number of spores can be carried out using the scheme described by Ritter in 1996 [50]. The spores seen per visual field are counted and classified into three categories: Less than 20 spores per visual field signify a slight infestation, between 20 and 100 spores per visual field indicate a moderate infestation and more than 100 spores constitute a severe infestation [50]. The three grades are presented in Figure 12. For a more specific result, a hemacytometer should be used [48].

3.5.4. Intoxications

If a massive number of dead bees is found in front of a hive or crippled bees are seen in the colonies, the beekeeper will often suspect a bee poisoning incident with a bee damage as a result of exposure to toxic plant protection products. However, there are some facts which have to be considered. Are all colonies in an apiary affected by increased mortality or is it only one or two? Is the apiary near fields making an application of a pesticide possible? And, are there other symptoms occurring?

In Germany, the Institute for Bee Protection in the Federal Research Centre for Cultivated Plants, Julius-Kühn-Institut (JKI), in cooperation with the national plant protection services examines samples of honeybees with suspected poisoning and plants. The bee sample undergoes a pollen analysis with pollen from the pollen basket (corbicula) or bee hairs to narrow down the incident to plants, which the bees visited. A *Nosema*-detection is performed, because *Nosema*-infected bees are more susceptible to intoxications. Finally, a bioassay with larvae of *Aedes aegyptii* L. is done to detect the presence of bee toxic insecticides. In 2018, 141 samples were sent in with the suspicion of bee intoxication, but only 61 samples were suitable for investigation. In 19 of these 61 incidents, bee toxic insecticides have been detected [51]. Further information can be found on the JKI website [52].

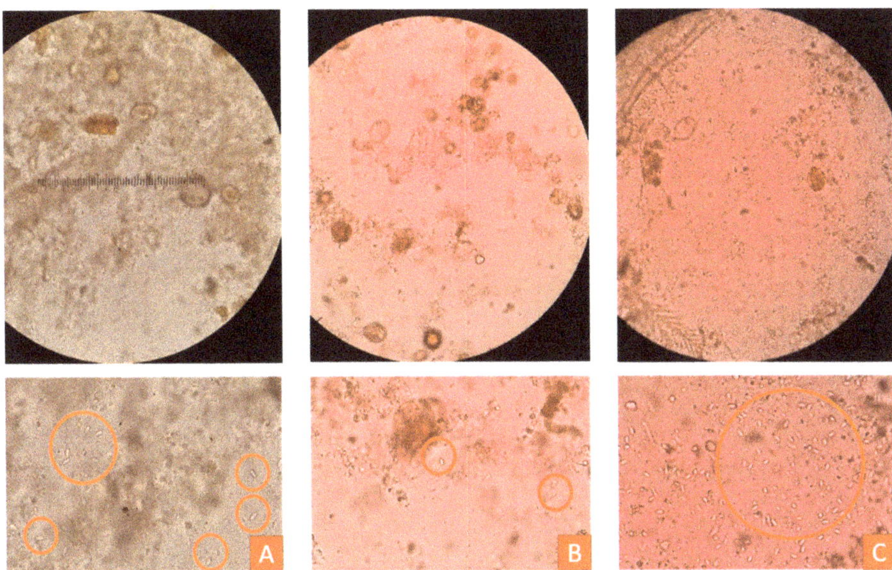

Figure 12. Microscopic investigation of bee samples of two honeybee colonies to estimate nosemosis infestation (400×), orange circles = Nosema spores (oval-shaped, 4–7 µm in length, 2–4 µm in width), upper row = visual fields, lower row = close-ups. (**A**) Sample of a colony with a moderate infestation (between 20 and 100 spores per visual field), (**B**) sample of a colony with a slight infestation (less than 20 spores per visual field), (**C**) sample of a colony with a severe infestation (more than 100 spores per visual field), © Julia Dittes.

The numbers clearly illustrate the relevance of sending in appropriate bee material. An appropriate sample for this investigation should consist of 1000 bees (is equivalent to 100 g). The sample should be taken under witness during the first 24 h after seeing symptoms and immediately sent chilled to the investigating institution [53]. The basic veterinary knowledge about correct sampling can be valuable here for taking an appropriate bee sample and, if possible, a suspicious plant sample as well as taking and storing retention samples in case of subsequent questions. A detailed photographic documentation of the incident is advisable.

3.6. List of Medical Issues, Diagnoses and Prognosis

After examinations, the next step is to list medical issues and establish a final diagnosis. According to the list of medical issues, a prognosis is derived, and a treatment plan created. An example in case of an overt CBPV infection can be found in Dittes et al. [24]. Investigations show that asymptomatic healthy colonies may have higher virus loads than diseased colonies [54,55]. Therefore, the laboratory results must precisely be linked to the information gathered by anamnesis, clinical findings and observations to evaluate the relevance of pathogens. The results form the basis for the choice of therapeutic measures.

If a notifiable disease is diagnosed, the relevant authorities have to be informed. There are notifiable diseases to the World Organization for Animal Health (OiE) and to the veterinary authorities in each country, depending on local regulations. The Terrestrial Animal Health Code lists infections with *Paenibacillus larvae* and *Melissococcus plutonius* and infestations with *Acarapis woodi*, *Varroa* spp., *Aethina tumida* and *Tropilaelaps* spp. in Article 1.3.8 [56]. In Germany, infections with *Paenibacillus larvae* and infestations with

Aethina tumida and *Tropilaelaps* spp. are notifiable. Currently, virus diseases form a part to the notifiable diseases only in Romania [7].

3.7. Outcome Control

An infected honeybee colony should be monitored closely during disease management to recognize problems and success as soon as possible and intervene in case of complications. It is advisable to use a logbook to record detailed information on every colony during disease management as well as during the whole season. Veterinarians are familiar with the management and interpretation of patient records as well as drawing conclusions out of them. A detailed logbook is important to gain basic information to choose and apply appropriate therapy concepts for disease control. The main point to prevent DWV infection and disease, as well as further *Varroa*-associated diseases, is to control the *Varroa* mite infestation. Further studies have to show recommendable therapy measures to control other viral diseases e.g., CBPV in honeybee colonies with overt infections.

4. Conclusions

A holistic approach to bee disease diagnosis is important to establish a correct and comprehensive diagnosis and save colonies. Disease outbreaks in honeybees are often tied to more complex interactions than in other species. Especially, the biological form of life in a superorganism has to be considered during the entire course of examination and management. Furthermore, the diagnostic techniques and possible ways of treatment are more limited than in the "normal" veterinary patient. There are no vaccinations and less veterinary medical products available for honeybees. The lifetime of an individual bee and the dependence on the season have to be considered.

A detailed and careful investigation is the base of diagnosis, and treatment decisions are to be made focusing on the colony as a whole rather than the individual animal. Depending on the colony location, the number of colonies in an apiary and beekeeping strategies as well as further influencing factors, diagnostics and the management plan have to be adapted to the individual requirements.

Veterinarians can be a valuable asset to the beekeeper because of their ability to prevent, diagnose and treat diseases in various species and in populations. In the future, veterinarians should recognize the important role and also the opportunities they can have within the honeybee sector. Taking the chance offered, the veterinarians may be part of education of beekeepers, part of sanitary audits and part of bee health overall in the veterinary practice, in addition to their official function in Veterinary Authorities.

Author Contributions: Conceptualization, C.K.W.M., H.A.-L. and I.U.E.; methodology, J.D., I.U.E. and M.O.S.; validation, J.D. and I.U.E.; formal analysis, J.D.; investigation, J.D., I.U.E. and M.O.S.; resources, C.K.W.M.; writing—original draft preparation, J.D.; writing—review and editing, I.U.E., H.A.-L., C.K.W.M. and M.O.S.; visualization, J.D., I.U.E. and H.A.-L.; supervision, C.K.W.M., I.U.E. and H.A.-L.; project administration, C.K.W.M. All authors have read and agreed to the published version of the manuscript.

Funding: This research received no external funding.

Conflicts of Interest: The authors declare no conflict of interest, although Heike Aupperle-Lellbach and Marc Schäfer are working in laboratories that regularly offer the laboratory diagnostics. The company (LABOKLIN GmbH & Co KG) had no role in the design, execution, interpretation or writing of the study.

References

1. Iatridou, D.; Pohl, L.; Tlak Gajger, I.; De Briyne, N.; Bravo, A.; Saunders, J. Mapping the teaching of honeybee veterinary medicine in the European Union and European Free Trade Area. *Vet. Rec. Open* **2019**, *6*. [CrossRef]
2. Available online: https://deutscherimkerbund.de/163-Bienen_Bestaeubung_Zahlen_die_zaehlen (accessed on 13 September 2020).
3. Bundestierärztekammer, E.I.A. Statistik 2014: Tierärzteschaft in der Bundesrepublik Deutschland. *DTÄB* **2015**, *5*, 670–683.
4. Bundestierärztekammer, E.I.A. Statistik 2019: Tierärzteschaft in der Bundesrepublik Deutschland. *DTÄB* **2020**, *68*, 860–870.

5. Ritter, W. Bee diseases are a worldwide problem. *Forum OiE* **2014**, *2*, 5.

6. Potts, S.G.; Roberts, S.P.M.; Dean, R.; Marris, G.; Brown, M.A.; Jones, R.; Neumann, P.; Settele, J. Declines of managed honey bees and beekeepers in Europe. *J. Apic. Res.* **2010**, *49*, 15–22. [CrossRef]

7. Vidal-Naquet, N. *Honeybee Veterinary Medicine: Apis mellifera L.*, 1st ed.; 5m Publisher: Sheffield, UK, 2015; pp. 71–75.

8. Aupperle, H.; Genersch, E. *Diagnostic Colour Atlas of Bee Pathology*, 1st ed.; Publisher Verlag Laboklin: Bad Kissingen, Germany, 2016; p. 173.

9. Available online: https://www.europarl.europa.eu/sides/getDoc.do?reference=P6-TA-2008-0567&type=TA&language=EN&redirect (accessed on 12 September 2020).

10. Formato, G.; Smulders, F.J.M. Risk management in primary apicultural production. Part 1: Bee health and disease prevention and associated best practices. *Vet. Q.* **2011**, *31*, 29–47. [CrossRef]

11. Available online: https://ec.europa.eu/food/animals/live_animals/bees/health_en (accessed on 12 September 2020).

12. Belsky, J.; Joshi, N.K. Impact of Biotic and Abiotic Stressors on Managed and Feral Bees. *Insects* **2019**, *10*, 233. [CrossRef]

13. McMenamin, A.J.; Brutscher, L.M.; Glenny, W.; Flenniken, M.J. Abiotic and biotic factors affecting the replication and pathogenicity of bee viruses. *Curr. Opin. Insect Sci.* **2016**, *16*, 14–21. [CrossRef]

14. DeGrandi-Hoffman, G.; Chen, Y.; Huang, E.; Huang, M.H. The effect of diet on protein concentration, hypopharyngeal gland development and virus load in worker honey bees (*Apis mellifera L.*). *J. Insect Pathol.* **2010**, *56*, 1184–1191. [CrossRef]

15. Chen, Y.P.; Siede, R. Honey bee viruses. *Adv. Virus Res.* **2007**, *70*, 33–80.

16. Genersch, E.; Aubert, M. Emerging and re-emerging viruses of the honey bee (*Apis mellifera L.*). *Vet. Res.* **2010**, *41*, 54. [CrossRef] [PubMed]

17. Brutscher, L.M.; Daughenbaugh, K.F.; Flenniken, M.L. Antiviral defense mechanisms in honey bees. *Curr. Opin. Insect Sci.* **2015**, *10*, 71–82. [CrossRef] [PubMed]

18. Gisder, S.; Genersch, E. Special Issue: Honey Bee Viruses. *Viruses* **2015**, *7*, 5603–5608. [CrossRef]

19. Vidal-Naquet, N. Les maladies de l'abeille domestique d'élevage *Apis mellifera L. Bull. Acad. Vét. Fr.* **2012**, *165*, 307–316. [CrossRef]

20. Hou, C.; Chejanovsky, N. Acute paralysis viruses of the honey bee. *Virol. Sin.* **2014**, *29*, 324–326. [CrossRef]

21. Chen, Y.P.; Pettis, J.S.; Corona, M.; Chen, W.P.; Li, C.J. Israeli Acute Paralysis Virus: Epidemiology, Pathogenesis and Implications for Honey Bee Health. *PLoS Pathog.* **2014**, *10*, e1004261. [CrossRef]

22. De Miranda, J.R.; Cordoni, G.; Budge, G. The Acute bee paralysis virus—Kashmir bee virus—Israeli acute paralysis virus complex. *J. Invertebr. Pathol.* **2010**, *103*, 30–47. [CrossRef]

23. Ribière, M.; Olivier, V.; Blanchard, P. Chronic bee paralysis: A disease and a virus like no other? *J. Invert. Pathol.* **2010**, *103*, 120–131. [CrossRef]

24. Dittes, J.; Schäfer, M.O.; Aupperle-Lellbach, H.; Mülling, C.K.W.; Emmerich, I.U. Overt infection with Chronic Bee Paralysis Virus (CBPV) in two honey bee colonies. *Vet. Sci.* **2020**, *7*, 142. [CrossRef]

25. Campbell, E.M.; Budge, G.E.; Watkins, M.; Bowman, A.S. Transcriptome analysis of the synganglion from the honey bee mite, *Varroa destructor* and RNAi knockdown of neural peptide targets. *Insect. Biochem. Molec.* **2016**, *70*, 116–126. [CrossRef]

26. Gisder, S.; Möckel, N.; Eisenhardt, D.; Genersch, E. In vivo evolution of viral virulence: Switching of deformed wing virus between hosts results in virulence changes and sequence shifts. *Environ. Microbiol.* **2018**, *20*, 4612–4628. [CrossRef] [PubMed]

27. Martin, S.J.; Highfield, A.C.; Brettell, L.; Villalobos, E.M.; Budge, G.E.; Powell, M.; Nikaido, S.; Schroeder, D.C. Global Honey Bee Viral Landscape Altered by a Parasitic Mite. *Science* **2012**, *336*, 1304–1306. [CrossRef] [PubMed]

28. Moore, J.; Jironkin, A.; Chandler, D.; Burroughs, N.; Evans, D.J.; Ryabov, E.V. Recombinants between *Deformed wing virus* and *Varroa destructor virus-1* may prevail in *Varroa destructor*-infested honeybee colonies. *J. Gen. Virol.* **2011**, *92*, 156–161. [CrossRef]

29. Koziy, R.V.; Wood, S.C.; Kozii, I.V.; van Rensburg, C.J.; Moshynkyy, I.; Dvylyuk, I.; Simko, E. Deformed Wing Virus Infection in Honey Bees (*Apis mellifera L.*). *Vet. Pathol.* **2019**, *56*, 636–641. [CrossRef] [PubMed]

30. De Miranda, J.R.; Genersch, E. Deformed wing virus. *J. Invertebr. Pathol.* **2010**, *103*, 48–61. [CrossRef]

31. Roy, C. La sémiologie an apiculture. Journées Nationales Gtv—Nantes. 2011. Available online: https://www.researchgate.net/publication/343513678_La_semiologie_en_apiculture_-_JOURNEES_NATIONALES_GTV_-_NANTES_2011 (accessed on 20 September 2020).

32. Imdorf, A.; Buehlmann, G.; Gerig, L.; Klichenmann, V.; Wille, H. Überprüfung der Schätzmethode zur Ermittlung der Brutfläche und der Anzahl Arbeiterinnen in freifliegenden Bienenvölkern. *Apidologie* **1987**, *18*, 137–146. [CrossRef]

33. Deutsches Bienenjournal: Bienenvolk Schätzen: Volksstärken Genau Erfassen (20. Oktober 2016). Available online: https://www.bienenjournal.de/imkerpraxis/ratgeber/volksstaerken-erfassen-quiz/ (accessed on 24 May 2020).

34. Horchler, L.; Gisder, S.; Boecking, O.; Genersch, E. Diagnostic value of faecal spots on and in honey bee (*Apis mellifera*) hives. *Berl. Münch. Tierärztl. Wochenschr.* **2018**, *132*, 41–48.

35. Kevill, J.L.; Lee, K.; Goblirsch, M.; McDermott, E.; Tarpy, D.R.; Spivak, M.; Schroeder, D.C. The Pathogen Profile of a Honey Bee Queen Does Not Reflect That of Her Workers. *Insects* **2020**, *11*, 382. [CrossRef]

36. Williams, G.R.; Rogers, R.E.L.; Kalkstein, A.L.; Taylor, B.A.; Shutler, D.; Ostiguy, N. Deformed wing virus in western honey bees (*Apis mellifera*) from Atlantic Canada and the first description of an overtly-infected emerging queen. *J. Invertebr. Pathol.* **2009**, *101*, 77–79. [CrossRef] [PubMed]

37. Hüsken, W.S. Gemülldiagnostik, Netstal, Presentation. 2019. Available online: http://www.imkerverband-sgap.ch/up/files/Vorderland/Gemuelldiagnose_Hysken17.pdf (accessed on 20 September 2020).

38. Blanchard, P.; Ribière, M.; Celle, O.; Lallemand, P.; Schurr, F.; Olivier, V.; Iscache, A.L.; Faucon, J.P. Evaluation of a real-time two-step RT-PCR assay for quantitation of Chronic bee paralysis virus (CBPV) genome in experimentally-infected bee tissues and in life stages of a symptomatic colony. *J. Virol. Methods* **2007**, *141*, 7–13. [CrossRef] [PubMed]

39. Ciglenečki, U.J.; Toplak, I. Development of a real-time RT-PCR assay with TaqMan probe for specific detection of acute bee paralysis virus. *J. Virol. Meth.* **2012**, *184*, 63–68. [CrossRef] [PubMed]

40. Schurr, F.; Tison, A.; Militano, L.; Cheviron, N.; Sircoulomb, F.; Rivière, M.P.; Ribière-Chabert, M.; Thiéry, R.; Dubois, E. Validation of quantitative real-time RT-PCR assays for the detection of six honeybee viruses. *J. Virol. Meth.* **2019**, *270*, 70–78. [CrossRef] [PubMed]

41. Hoffmann, B.; Depner, K.; Schirrmeier, H.; Beer, M. A universal heterologous internal control system for duplex real-time RT-PCR assays used in a detection system for pestiviruses. *J. Virol. Meth.* **2006**, *136*, 200–209. [CrossRef] [PubMed]

42. Chen, Y.P.; Higgins, J.A.; Feldlaufer, M.F. Quantitative Real-Time Reverse Transcription-PCR Analysis of Deformed Wing Virus Infection in the Honeybee (*Apis mellifera* L.). *Appl. Environ. Microbiol.* **2005**, *71*, 436–441. [CrossRef]

43. Yang, X.; Cox-Foster, D.L. Impact of an ectoparasite of the immunity and pathology of an invertebrate: Evidence for host immunosuppression and viral amplification. *Proc. Natl. Acad. Sci. USA* **2005**, *102*, 7470–7475. [CrossRef]

44. World Organisation for Animal Health (OIE). *Manual of Diagnostic Tests and Vaccines for Terrestrial Animals*; Office International des Epizooties: Paris, France, 2008; pp. 1092–1106.

45. Bak, B.; Wilde, J.; Siuda, M.; Kobylinska, M. Comparison of two methods of monitoring honeybee infestation with *Varroa destructor* mite. *Anim. Sci.* **2009**, *46*, 33–38.

46. Brunnemann, G.; Poker, V.; Büchler, R. Bienenprobe mit Puderzucker—Die neue bienenschonende *Varroa*-Befallsmessung. *ADIZ* **2011**, *8*, 7–9.

47. Toplak, I.; Ciglenečki, U.J.; Aronstein, K.; Gregorc, A. Chronic Bee Paralysis Virus and Nosema ceranae Experimental Co-Infection of Winter Honey Bee Workers. *Viruses* **2013**, *5*, 2282–2297. [CrossRef]

48. World Organisation for Animal Health (OIE). *Manual of Diagnostic Tests and Vaccines for Terrestrial Animals*; OIE: Paris, France, 2018.

49. Gisder, S.; Genersch, E. Molecular differentiation of *Nosema apis* and *Nosema ceranae* based on species-specific sequence differences in protein coding gene. *J. Invertebr. Pathol.* **2013**, *113*, 1–6. [CrossRef]

50. Ritter, W. *Diagnostik und Bekämpfung der Bienenkrankheiten*, 1st ed.; Gustav Fischer Verlag: Jena, Germany, 1996; p. 51.

51. Available online: https://bienenuntersuchung.julius-kuehn.de/index.php?menuid=93 (accessed on 5 September 2020).

52. Available online: https://www.julius-kuehn.de/en/ (accessed on 5 September 2020).

53. Available online: https://bienenuntersuchung.julius-kuehn.de/index.php?menuid=39 (accessed on 8 September 2020).

Vet. Sci. **2020**, *7*, 159

54. Berényi, O.; Bakonyi, T.; Derakhshifar, I.; Köglberger, H.; Nowotny, N. Occurrence of Six Honeybee Viruses in Diseased Austrian Apiaries. *Appl. Environ. Microbiol.* **2006**, *72*, 2414–2420. [CrossRef]

55. Gauthier, L.; Tentcheva, D.; Tournaire, M.; Dainat, B.; Cousserans, F.; Colin, M.E.; Bergoin, M. Viral load estimation in asymptomatic honey bee colonies using the quantitative RT-PCR technique. *Apidologie* **2007**, *38*, 426–435. [CrossRef]

56. World Organisation for Animal Health (OIE). *Terrestrial Animal Health Code 2019*; OIE: Paris, France, 2019.

Publisher's Note: MDPI stays neutral with regard to jurisdictional claims in published maps and institutional affiliations.

Article

Honey Bee Virus Transmission via Hive Products

Dominik Schittny [1], Orlando Yañez [1,*] and Peter Neumann [1,2]

[1]   Institute of Bee Health, Vetsuisse Faculty, University of Bern, 3097 Bern, Switzerland;
      dominik.schittny@ana.unibe.ch (D.S.); peter.neumann@vetsuisse.unibe.ch (P.N.)
[2]   Agroscope, Swiss Bee Research Center, 3097 Bern, Switzerland
*   Correspondence: orlando.yanez@vetsuisse.unibe.ch

Received: 27 May 2020; Accepted: 10 July 2020; Published: 21 July 2020

Abstract: The global trade of honey bee hive products has raised concern about pathogen transmission. However, the efficacy of hive products as virus vehicles is poorly understood. Here, we investigated the transmission capacity of hive products for Deformed wing virus genotype A (DWV-A) in a fully-crossed hoarding cage experiment and estimated the transmission risk by screening commercial products. Western honey bee workers were provided with honey, pollen and wax either contaminated with high ($\sim 2 \times 10^9$), medium ($\sim 1.7 \times 10^8$), low ($\sim 8 \times 10^6$) or zero (control) DWV-A genome copies. For 10 days, mortality was monitored. Then, virus titers were quantified in bee heads and 38 commercial products using RT-qPCR. For honey and pollen, a positive association between DWV-A concentration and mortality was observed. High concentrations always resulted in infections, medium ones in 47% of cases and low ones in 20% of cases. No significant difference was observed between the tested products. In commercial honey and pollen, 7.7×10^2–1.8×10^5 and 1.4×10^3–1.3×10^4 DWV-A copies per gram were found, respectively. The results show that DWV-A transmission via hive products is feasible. The risk of introducing novel viruses and/or strains should be considered in trade regulations by including virus analyses for health certificates of hive products

Keywords: honey bee; virus; DWV-A; hive products; honey; pollen; wax

---

1. Introduction

The international trade of honey bee products increased over the past decades, thereby enhancing chances for the spread of bee diseases [1,2]. Indeed, many honey bee pathogens have already been discovered in traded hive products [2]. To protect the health of humans, animals and plants, most countries joined the Terrestrial Animal Health Code (26th edition) released by the World Organization for Animal Health (OIE) in 2017, in which the trade of animal products is also regulated. However, amongst the honey bee pathogens, viruses are not covered by the terrestrial code due to a lack of specific criteria (OIE 2016), even though they are frequently associated with honey bee products and may potentially cause harmful effects [3–5]. This seems surprising because it seems most likely that viruses are also spreading as a side effect of the worldwide trade of bee products. Regarding the transmission of viruses, the international movement of live honey bees arguably plays the main role in the global dispersal of bee viruses [6], facilitating the transmission between colonies. Imported bee packages containing viral agents may act as a source of infection for other colonies in the surrounding area (horizontal transmission). The viral particles can be horizontally transmitted in different ways, such as when an infected bee drifts from its own colony to another [7], contact between bees during robbing or while foraging in common food sources [8], and also by human activity, when contaminated material is shared between colonies and apiaries [8]. The international trade of queens, in addition, allows the introduction of viral agents inside the recipient colonies. It has been shown that queens can hold many viruses at the same time and are able to transmit them vertically to their offspring [3,4,9–11]. Viruses have also been detected in honey bee products such as honey, pollen and royal jelly freshly extracted from colonies [3–5], as well as pollen pellets recently brought to colonies [5]. Moreover, the infectivity of viruses carried by frames containing honey and pollen (bee bread) has also been shown when colonies became infected after receiving such contaminated frames [5]. Still, the role of honey bee hive products for the transmission of viruses in the trade scenario has not been explored in detail.

In addition to the risk of introducing novel viruses, novel strains of already established ones may pose additional threats. Due to genetic recombination, significant changes in the viral genome may occur resulting from the insertion of gene fragments from another viral strain during coinfection of a host cell [12]. Indeed, recombination is increasingly recognized as a major driver of virus evolution [13]. While the high mutation rates observed in RNA viruses often generate deleterious mutations [14], recombination events purge those deleterious mutations [15] and can often result in adaptations for the virus, such as expanding the host range, evasion of host immunity and changes in virulence [16,17].

Although the emergence of recombinants from deformed wing virus (DWV) genotypes were associated with increasing virulence to western honey bees, *Apis mellifera* [18–24], the potential of genetic recombination is not fully understood in the context of honey bee viruses.

Ubiquitous DWV is amongst the most harmful pathogens of honey bees [6,25–28]. It can cause clear clinical symptoms, such as crippled wings and a reduced host lifespan [27,29,30], and is a known key driver of honey bee colony losses [31,32]. DWV is a positive sense single-stranded RNA virus (family Iflaviridae; genus Iflavirus) [26,33] and is a recent global epidemic in honey bees [32]. The latter is probably driven by the ectoparasitic mite, *Varroa destructor*, because it is a very efficient vector of DWV. It generates a disease epidemic within the honey bee colony, which then dwindles until it dies [26,34,35]. *V. destructor* have also reduced the genetic diversity of DWV [36], promoting the spread and global distribution of DWV genotype A (DWV-A). The emergence of new genotypes, such as DWV genotype B (DWV-B, also known as Varroa destructor virus-1), has raised concern about their differences in virulence [22,37–39]. DWV is infecting honey bees, wild bees and probably other arthropods [5,40] (reviewed in [41]). A better knowledge of the transmission of this particular honey bee virus is therefore of importance, due to the considerable concern for both apiculture and nature conservation efforts.

In this study, we investigate whether or not honey bee hive products can, in principle, act as matrices for DWV-A transmission and how efficient they are. To achieve this goal, fully-crossed laboratory hoarding cage experiments were conducted and complemented with a survey of DWV-A titers in commercial bee products. Since royal jelly and propolis are known to have antimicrobial properties [42], our study focused on honey, pollen and wax.

2. Materials and Methods

2.1. Study Set Up

In Bern, Switzerland, twelve queenright local honey bee colonies, *A. mellifera*, were screened for DWV-A infections. Adult workers ($N = 30$) were collected from middle frames of each colony in March and April 2015, pooled and tested for DWV-A using RT-qPCR [43]. The three colonies with the lowest infection levels were chosen and tested again in June prior to the experiment, when they had a mean of 1.6×10^3, 3.9×10^3 and 6.1×10^3 virus copies per bee, respectively.

One sealed worker brood frame was taken from each experimental colony and placed in an incubator at 34 °C and 70% relative humidity (RH) until adult emergence [44]. After 48 h, freshly emerged workers were randomly distributed between the experimental hoarding cages. The fully-crossed hoarding cage experiments [44] were conducted from June to August 2015 and designed to test whether or not honey bee products spiked with DWV-A are able to induce an infection in honey bees. To see if the efficacy in DWV-A transmission is different among the tested honey bee products, three treatments with different initial amounts of DWV-A (high, medium and low) were used for each product.

Each series of experiments, corresponding to each honey bee product, consisted of four treatments: three treatments where the honey bee product had been spiked with three different concentrations of DWV-A ($5 \times 10^9/5 \times 10^8/5 \times 10^6$ copies per ml for honey; $1 \times 10^9/1 \times 10^7/1 \times 10^6$ copies per g for pollen; $2.5 \times 10^8/2.5 \times 10^6/2.5 \times 10^5$ copies per cm^2 for wax; Table 1) and a non-spiked treatment with UV-sterilized products as control. All treatments consisted of five repetitions. Each cage was equipped with a honey solution feeder (3 mL syringe), a pollen paste feeder (modified centrifugation tube) and a piece of wax foundation (4 cm^2). Then, 30 newly emerged workers were introduced into each cage and kept for 10 days at 30 °C and 70% RH in the incubator. Dead individuals were removed from the cages daily, recorded and stored at −20 °C. Honey and pollen consumption was also controlled daily and feeders refilled if required. After two days, the spiked honey and pollen products have been consumed in all cases, resulting in an average consumption of 1×10^8 and 3.3×10^7 copies per bee for high, 1×10^7 and 3.3×10^5 copies per bee for medium and 1×10^5 and 3.3×10^4 copies per bee for low DWV treatment, respectively. Then the feeders were replaced with sterilized food. For the wax product, the spiked piece of wax was available for the entire 10 days. At day 10, all remaining bees were stored at −20 °C.

Table 1. Initial Deformed wing virus genotype A (DWV-A) genome copy numbers in the different treatments. The number of DWV-A copies per contaminated honey, pollen and wax are given. For comparison, the average number of DWV-A genome copies consumed per bee individual (virus copies per cage divided by number of bees in cage) is also listed.

Treatment	Hive Product		
Group	Honey	Pollen	Wax
High (per mL/g/cm^2)	5.0×10^9	1.0×10^9	2.5×10^8
High (per bee)	1.0×10^8	3.3×10^7	3.3×10^7
Medium (per mL/g/cm^2)	5.0×10^8	1.0×10^7	2.5×10^6
Medium (per bee)	1.0×10^7	3.3×10^5	3.3×10^5
Low (per mL/g/cm^2)	5.0×10^6	1.0×10^6	2.5×10^5
Low (per bee)	1.0×10^5	3.3×10^4	3.3×10^4

2.2. Cage Experiment

2.2.1. Bee Product Preparation

Only honey and pollen that tested negative for DWV-A by RT-qPCR [43] was used for the experiments. Additionally, these were irradiated with UV light for 120 min. During the UV-treatment, the honey (25 g) was mixed by slowly rotating the honey containing tube each 30 min. A 50% (*w/w*) honey solution was prepared using the UV light-treated honey and Milli-Q water (Millipore Corporation, Billerica, MA, USA). The solution was mixed and stored at −20 °C until used in the experiments.

The pollen grains were crushed to a powder using a stone mortar. The grained pollen was spread on a sheet of paper and irradiated with UV light for 120 min. Every 30 min, the pollen powder was mixed using a sterilized spatula. The pollen paste was prepared with the following proportions: 40% UV-light treated pollen, 50% powder sugar and 10% MilliQ water. The pollen paste was then wrapped into aluminum foil and frozen at −20 °C until usage.

The wax was provided as small pieces of organic wax foundation, which was cut in square pieces with an edge length of 20 mm. The pieces of foundation were displayed on a sheet of paper and each side was irradiated with UV light for 30 min. After irradiation, the wax pieces were stored at −20 °C.

2.2.2. Propagation of DWV-A

DWV-A was propagated using standard methods [45]. Red-eyed worker pupae were microinjected with 2×10^7 virus copies in 2 µL PBS solution (Phosphate Buffered Saline; pH 7.4) between the 2nd and 3rd integuments using a 50 µL micro syringe (Hamilton Microliter™ Syringes, Reno, Nevada, USA) and 30-gauge disposable needles. Pupae were incubated at 30 °C and collected after 6 days. Each pupa was macerated individually in 500 μL PBS and homogenized with 100 μL chloroform using strong vortex. After centrifugation at 13,000 rpm for 10 min, the supernatant was collected and stored at −20 °C. DWV-A was quantified using RT-qPCR [43] and diluted to 1×10^7 copies per μL to make a stock solution for the spiking of the bee products.

2.2.3. Spiking Bee Products with DWV-A

In a pilot study, the average amount of honey and pollen consumed by a single bee per day was estimated. On average, one bee consumed 20+/−3.1 mg of honey solution and 16+/−3.6 mg of pollen paste per day. The DWV-A concentration in the bee products was estimated to apply the required number of virus copies within the first two days of the experiment for honey and pollen.

The honey solution was spiked according to the different DWV-A concentrations, keeping the ratio of honey and water 1:1 (*w/w*). In the case of pollen, the DWV-A solution replaced the water in the recipe for the pollen paste (10%). A homogenous spiking of wax was not possible because the virus RNA would degrade at temperatures where the wax would melt. Therefore, the virus solution was applied on the surface of the pieces of wax foundation and allowed to rest overnight for the water to evaporate. Since bees do not consume wax, the desired virus amount was applied on the piece of wax foundation, which was left in the cage for the entire duration of the experiment.

2.2.4. Cage Construction

The cages were clear polystyrene cups with a diameter of 63 mm and an inner volume of 75 cm^3 (RIWISA AG Kunststoffwerke, Hägglingen, Switzerland). The cups were turned upside down with the lid acting as the bottom of the cage. Each cage was fitted with three holes for ventilation (large hole in lid covered with a mesh fabric permeable to air) as well as for holding the pollen (2 mL micro-centrifuge tubes) and honey (2 mL plastic syringes) feeders.

2.2.5. Detection of DWV-A

To test for overt virus infection, only bee heads ($N = 10$ per cage) were considered for analyses [46]. The heads were removed using a scalpel sterilized using EtOH and flaming after each cut. Heads from the same treatment were homogenized individually in 100 μL TN-Buffer (10 mM Tris HCl, 10 mM NaCl) using a metal bead (5 mm diameter) and an electronic crushing shaker machine (Retsch Mixer Mill MM 300, Haan, Germany). A NucleoSpin® RNA II kit (Macherey-Nagel, Oensingen, Switzerland) was used for RNA extraction following the manufacturer's recommendations and using 50 μL of pooled bee head homogenates from the same cage. The purified RNA was then eluated using 60 μL of RNase free water (Macherey-Nagel, Oensingen, Switzerland).

Reverse transcription was performed using standard protocols [32]. The concentration of RNA was measured using a spectrophotometer (Witec NanoDrop® ND 1000 Spectrophotometer, Sursee, Switzerland). Then, 1 μg of RNA and 0.75 μL of 100 mM hexamer primer (Microsynth AG, Balgach, Switzerland) were heated at 70 °C for 5 min and then cooled down to 4 °C. To obtain a final volume of 25 μL, a master mix, consisting of 5 μL M-MLV Reaction Buffer (Promega, Fitchburg, Wisconsin, USA), 1.25 μL 2.5 mM dNTP Mix (Bioline, London, UK) and 1 μL 200 u/μL M-MLV reverse transcriptase (Promega, Fitchburg, Wisconsin, USA), was added and heated to 37 °C for 60 min, before cooling down to 4 °C. A total of 10 fold dilutions were used for the quantification assays.

Each sample was tested for DWV-A using quantitative PCR (qPCR) [32]. The qPCR was conducted using the KAPA SYBR® Fast Universal qPCR kit (KAPA Biosystems, Wilmington, Massachusetts, USA). Briefly, a 12 μL total volume reaction consist of 6 μL of 2x reaction buffer, 0.24 μL forward and reverse primers each (10 μM, Table 2), 2.52 μL water and 3 μL of cDNA. The qPCR was run using an Illumina® Eco Real-Time PCR System (Illumina, San Diego, CA, USA. The amplification conditions were initiated by heating to 95 °C for three minutes in order to activate the polymerase. Then, during each of 40 repeating cycles, the samples were heated to 95 °C for three seconds and cooled down to 57

°C for 30 s. For melting, curve analysis samples were heated to 95 °C for 15 s, cooled down to 55 °C for another 15 s and heated from 55 °C to 95 °C while the strand dissociation was recorded.

Table 2. PCR primers used for the relative virus quantification of DWV-A.

Target	Sequence (5′–3′)	[bp]	Ref
DWV-A	TTC ATT AAA GCC ACC TGG AAC ATC	136	[47]
	TTT CCT CAT TAA GTG TGT CGT TGA		
β-actin	CGT TGT CCC GAG GCT CTT T	66	[48]
	TGT CTC ATG AAT ACC GCA AGC		

Standard curves prepared from DWV-A and *A. mellifera* β-Actin gene were used for virus quantification and normalization, respectively. The standard curve dilutions (10^{-2} to 10^{-5} ng/reaction) were prepared from purified PCR products (Table 2). Two kinds of negative controls were applied (1: RNA-extraction control without bee sample to check for possible contamination in the reagents; 2: PCR negative control, using water instead of cDNA template).

2.3. Survey of DWV-A in Commercial Honey and Pollen Products

Honey ($N = 34$) and pollen ($N = 5$) products were acquired from a variety of local Swiss grocery stores. The honeys originated from all continents except Antarctica and the pollen originated from Spain. For extracting RNA from honey and pollen, 120 mg of these bee products and 200 μL of TN-Buffer were mixed thoroughly by using a shaker (Retsch Mixer Mill MM 300, Haan, Germany) and a metal bead (2 mm diameter) that was put inside the sample tube. After shaking, the sample was centrifuged for 5 min at 14,000 rpm. 50 μL of the supernatant were then used for the RNA extraction following the NucleoSpin® RNA II kit (Macherey-Nagel, Oensingen, Switzerland) protocol. Reverse transcription as well as qPCR [32] were conducted as described above.

2.4. Statistical Analyses

The statistical analyses were performed using the NCSS statistical software version 10. Since the Kolmogorov-Smirnov test rejected normality in all the used data sets (Test value = 0.289 for virus copies data, 0.305 for delta DWV-A data, 0.214 for cage infection data and 0.246 for mortality data), non-parametric tests were used. The sample size was the number of cages per treatment ($N = 5$) because considering each bee individually would result in pseudo replications [39]. For all tests a critical *p*-value of 0.05 was used.

Using the log rank test with a Bonferroni correction, we checked each bee product experiment for significant differences in mortality between treatments of different DWV-A concentrations. Bees from all five cages underlying the same treatment were integrated for the survival analyses. There were 150 bees per treatment and the survival was recorded for 10 days.

The Kruskal-Wallis multiple-comparison z-value test with Bonferroni correction was used to test for differences in DWV-A titers between high, medium and low DWV-A treatments as well as control treatments with sterilized products. As before, all bees from the same treatment were integrated to one population.

To compare the DWV-A transmission efficacy between different honey bee products a Δ DWV-A level was calculated. The average number of DWV-A copies taken up by the bees during the experiment was subtracted from the detected DWV-A copies in the heads. Thus the Δ-DWV-A value was a measure for DWV-A replication within the bees. For the analysis, all bees receiving virus copies via the same honey bee product were integrated. In addition, here the Kruskal-Wallis multiple-comparison z-value with Bonferroni correction was used.

3. Results

3.1. Cage Experiment

3.1.1. Survival

With the exception of one outlier, death rates of less than 15% mortality during the ten-day experiments were observed in all control cages. In the case of the honey assays, the mortality was significantly higher at the high DWV-A treatment compared to all lower DWV-A concentrations and the control (Log Rank test, $p < 0.0001$ per each pair-wise comparison). In the pollen assays, the control cages had significantly lower death rates than those of high ($p < 0.0001$) and medium ($p = 0.0005$) DWV-A treatment. Concerning wax, no significant differences in mortality between different treatments were found (Figure 1).

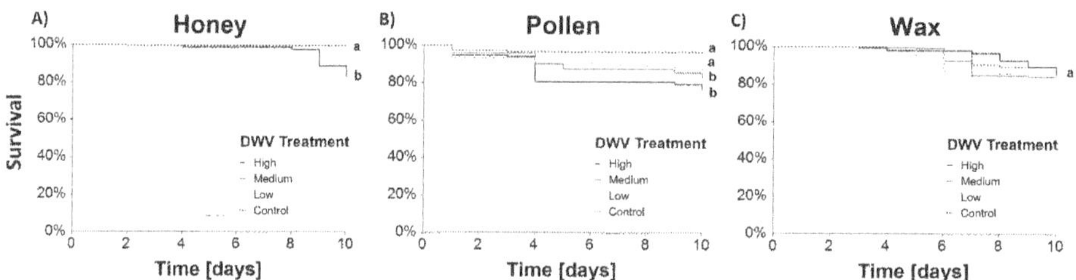

Figure 1. Kaplan-Meier survival plots of caged honey bee workers over the experimental period for each bee product ((**A**) honey, (**B**) pollen and (**C**) wax) with high, medium and low initial DWV-A concentrations. Survival for each treatment was pooled from 5 cages with 30 bees each. Significant differences are indicated by different letters (a,b).

3.1.2. DWV-A Infection Levels

While the Kruskal-Wallis test indicated significantly higher DWV-A titers in the heads of bees from high DWV-A treatments compared to bees from control treatments in honey ($z = 2.83$), pollen ($z = 3.64$) and wax ($z = 2.89$), the high DWV-A treatments also showed significantly higher DWV-A titers compared to the low DWV-A treatments in the case of honey ($z = 2.78$) and pollen ($z = 2.83$) treatments. The critical significance level of the z-value, concerning Bonferroni correction, was 2.64 with the *p*-value set at 0.05. As in the survival curves, the correlation between initial DWV-A concentration and mortality or virus titers, respectively, was the strongest in pollen and the weakest in wax (Figure 2).

According to the detected amount of DWV-A in their heads, a significant bimodal distribution (Kolmogorov-Smirnov normality, test value = 0.289), was found (Figure 3). The group with the lower DWV-A showed between 5×10^3 and 5×10^5 (median = 2.2×10^4) copies per bee, while the group with the higher DWV-A showed between 1×10^9 and 2×10^{11} (median = 2.0×10^{10}) copies per bee. Bees from the higher DWV-A group had significantly more viruses (between one to five orders of magnitude) than the initial fed amount (from 3.3×10^4 to 1×10^8 copies per bee; Kruskal-Wallis test, z-value = 5.65). Therefore, virus replication can be considered to occur in those bees.

Regarding the frequency of infection, the number of bee cages that showed an infection as defined above was different depending on the treatment. High DWV-A treatment resulted in an infection of all cages independent of the bee product. At the medium DWV-A treatment, there were two out of five infected cages in honey and pollen each, while there were three in wax. The largest difference was seen at low virus concentrations. There was one infected cage in the case of honey, two in the case of wax while no cages were infected in the case of pollen.

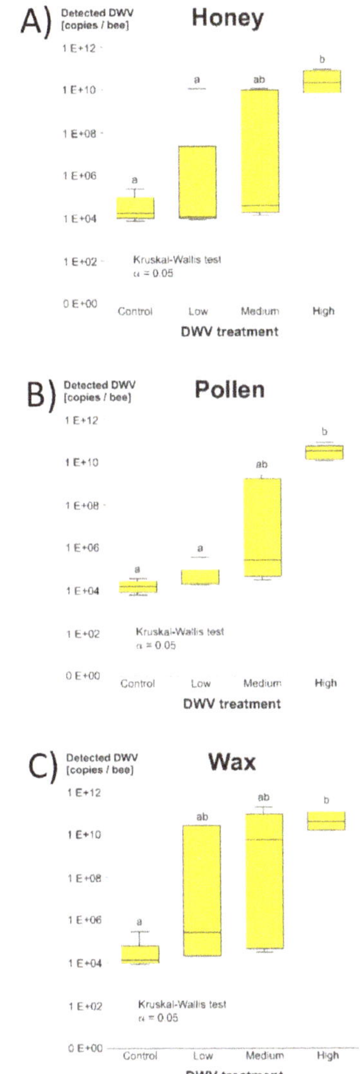

Figure 2. DWV-A copies detected per bee head in the different treatments ((**A**) honey, (**B**) pollen and (**C**) wax). Medians, upper and lower quartiles and maximum and minimum are shown. Significant differences are indicated by different letters (a,b).

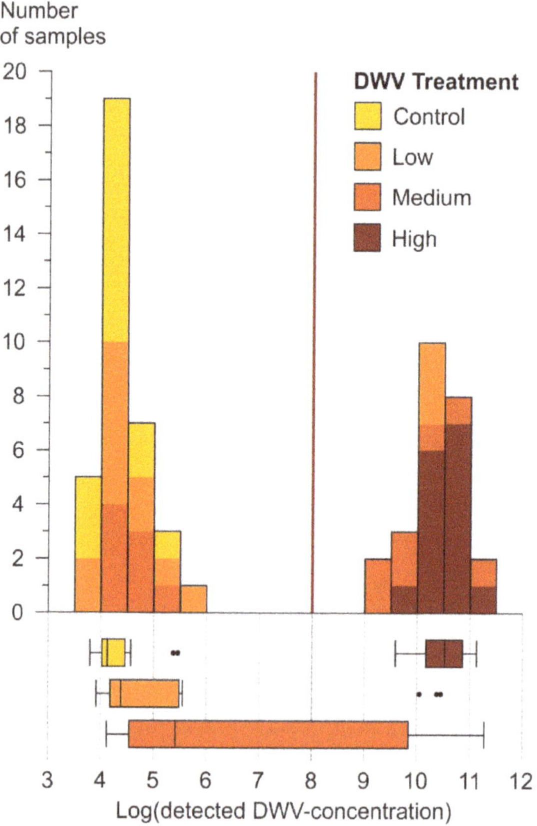

Figure 3. Frequency distribution of DWV-A copy numbers per bee head, all tested hive products pooled together. The highest amount of initially provided DWV-A during the experiment was 1.0 × 10^8 copies per bee (red line). There is a significant bimodal distribution (Kolmogorov-Smirnov normality, test value = 0.289) with one group showing high DWV-A titers and another group showing low DWV-A titers. The virus titers of the two groups are significantly different (Kruskal-Wallis test, z-value = 7.8444 with p-value set at 0.05) and there was significantly more virus detected in the high virus titer group than was initially provided during the experiment (z-value = 5.6537 with p-value set at 0.05). This implies that bees with a high DWV-A titer got infected and that virus replication took place. The box plots at the lower part of the figure show differences between the detected virus titers of different treatments.

The comparison between the different bee products showed no significant differences (z = 0.38 for honey and pollen, 1.88 for honey and wax and 1.50 for pollen and wax) in virus titers between the different honey bee products (Figure 4).

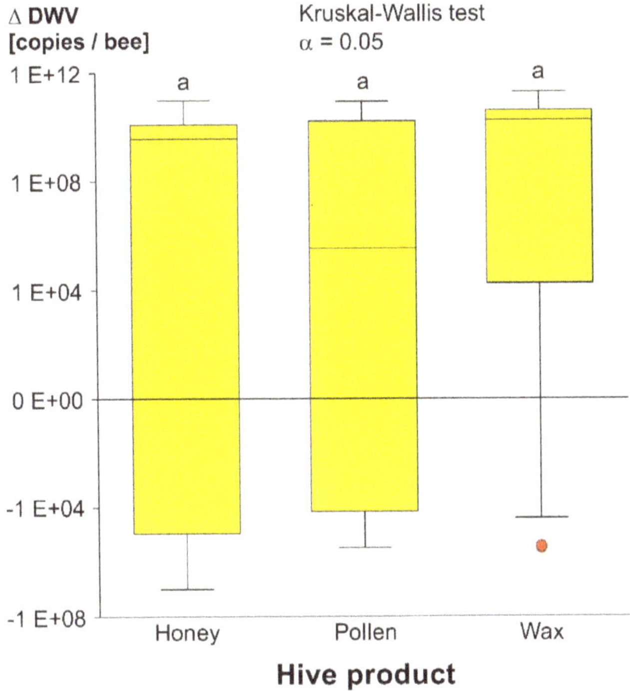

Figure 4. Quantification of replicative DWV-A in bees exposed to DWV-A-spiked honey, pollen and wax. Δ-DWV-A represents the initial amount of DWV-A provided in the products subtracted from the detected DWV-A amount in the bee heads, as a measure for DWV-A replication. For each box, data from high, medium and low initial DWV-A were pooled. There were no significant differences among the bee products.

3.2. Survey of DWV-A in Commercial Honey and Pollen

In all honey and pollen samples, the detected amount of DWV-A was low. In honey, between 7.6×10^2 and 1.8×10^5 virus copies with a median of 1.2×10^4 virus copies per gram were detected. In pollen, between 1.4×10^3 and 1.2×10^4 virus copies with a median of 3.5×10^3 virus copies per gram were detected (Figure 5).

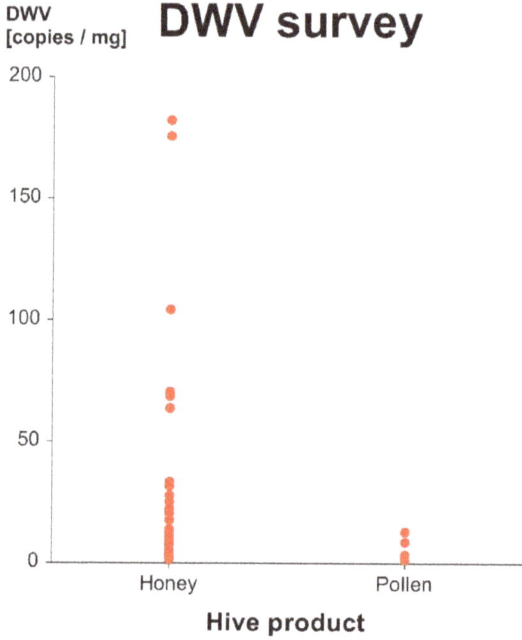

Figure 5. Survey results for commercial honey and pollen. DWV-A copies per mg in 34 commercial honey and 5 pollen samples are shown. Even though a few honey samples showed higher DWV-A titers, no significant difference between the two products was found (Kruskal-Wallis multiple-comparison z-value test, z-value = 1.9324, $p > 0.05$).

4. Discussion

Our results clearly show that DWV-A transmission via hive products is feasible. The data also show that mortality increases when honey bees are fed with higher titers of DWV-A via honey and pollen, but not via wax. DWV-A infection was detected in all cages from the high treatment (fed with at least 3.3×10^7 copies per bee), irrespective of the tested bee product. Only very small amounts of DWV-A were detected in the commercial bee products.

4.1. DWV-A Transmission Experiment

4.1.1. Survival and DWV-A Titers in Caged Bees

The survival plot shows that only the high DWV-A treatment affects mortality in the honey treatments. Moreover, this treatment resulted in high DWV-A infections in all cages. This suggests an association between mortality and DWV-A infection by oral consumption of high DWV-A titers (1.0×10^8 virus copies per bee) via honey. In contrast, mortality was not different from the controls at the medium and low treatments, despite the occurrence of high DWV-A titers in some of those cages. Similar to honey, the high DWV-A treatment affects mortality in the pollen assays. Moreover, mortality in the medium treatment was also higher than in controls. On the other hand, no significant difference in mortality was found between the low treatment and the control, which is consistent with the low DWV-A titers detected in the bee heads from the pollen low treatment.

In the case of wax assays, no differences in mortality were found among different DWV-A treatments and controls. One possible reason could be that wax does not serve as a food resource, so the oral pathway is somehow different in comparison to honey and pollen.

4.1.2. Data Structure of Detected DWV-A in Bee Heads

Looking at the detected DWV-A titers in the bee heads, there was one pattern that could be observed across all bee products and treatments, including the controls. All bees could be divided into either a high (median = 2.0×10^{10} copies per bee) or a low (median = 2.2×10^4 copies per bee) DWV-A titer group (Figure 3). There was a distinct gap between 5×10^5 and 1×10^9 DWV-A copies per bee, with not a single sample in this range. The highest DWV-A amount fed in the experiment was 1×10^8 copies per bee considering that each bee had fed the same (average) amount of spiked bee product. If that assumption is true, all bees from the group with high detected virus titers ($>1 \times 10^9$ copies/bee) had at least one order of magnitude more virus copies in their heads than the maximal amount of virus that they had taken up during the experiment. Therefore, all bees with a DWV-A amount of at least 1×10^9 copies can be considered as having experienced DWV-A replication and thus an overt infection. In contrast, there were samples in the group of low detected DWV-A titers that showed considerably less DWV-A copies than have been fed, indicating that infection does not always occur. Since up to 5×10^5 DWV-A copies per bee were found even in bees from the negative control, these virus titers are likely to represent DWV-A cover infection levels.

4.1.3. Product Comparison

The honey treatments were spiked with more virus copies compared to pollen and wax, because the different physical properties of the hive products only enabled a certain maximum of treatment solution to be absorbed. Hence, a direct comparison of the tested hive products for viral titers was not feasible. Instead, we compared the DWV-A transmission efficacy between the products using a Δ-DWV-A level as a measure for virus replication within the bees. A significant difference of transmission efficacy between honey, pollen and wax was not found. Even though the obtained results from different bee products are mostly similar, there might be a trend that wax transmits DWV-A easier than honey and pollen. This can be seen in Figure 4, where the median of wax is the highest. That implies that there was more virus replication in the wax treatments, compared to the two other bee products. Indeed, the high levels of the detected virus in the low DWV-A treatment of wax were not significantly different from the high DWV-A treatment, which stands in contrast to the observations of the other products. This might be a clue for a lower DWV-A threshold for an infection via wax. Another indicator can be found when looking at the number of individual cages in which DWV-A replication occurred. In the case of wax, replication took place in two out of five cages in low DWV-A treatment and in three out of five cages in medium DWV-A treatment (number of infected cages: honey low = 1; pollen low = 0; honey medium = 2; pollen medium = 2). However, the impression that wax might be the most efficient matrix of all tested products could also have a methodical reason: in contrast to the honey and pollen assays, where bees consumed the spiked products during 48 h, the spiked pieces of wax were in contact with the bees for 10 days. The amount of virus was similar in the wax compared to the pollen and honey assays, but the virus was highly concentrated on the surface due to an inhomogeneous DWV-A distribution. Apart from that, a different transmission route, such as topical transmission [49], may produce different effects in the host parasite interactions since wax does not serve as food for honey bees.

4.2. Survey of DWV-A in Commercial Honey and Pollen

Detected amounts of DWV-A in commercially available honey and pollen were relatively low (1.8×10^5 and 1.2×10^4 copies per gram, the highest values, respectively). However, Mazzei and colleagues [50] found up to 3.0×10^6 virus copies per gram in pollen samples freshly collected by honey bee workers. This amount is 230-fold higher compared to our highest value in pollen. One reason for this difference could be the quick degradation of DWV-A particles at room temperature [51]. In the same line, Graystock and colleagues [52] only detected DWV-A in 2 out of 25 samples of pollen provided as food for bumblebee colonies.

Transmission Risk Under Realistic Conditions

Overall, the transmission risk of DWV-A via bee products under realistic conditions is not very high. Nevertheless, it is possible and should be considered. In the case of pollen, the cage experiment shows that a transmission and infection, defined by detection of high DWV-A titers, can only take place if the supplied DWV-A concentration lies between 1×10^6 and 1×10^7 copies per gram. However, in all samples tested in the survey of commercial pollen none reached a concentration in that range. Therefore, the risk of DWV-A to produce infection when transmitted via pollen is considered minimal.

By contrast, honey appears more efficient as a potential matrix for DWV-A transmission. With one out of five cages infected at a concentration of 5×10^6 virus copies per milliliter, a transmission is possible even at lower concentrations. Since we had no test point at lower concentrations, the minimal infectious dose of DWV-A via honey was not clearly defined. However, there were 3 out of the 33 honey samples that showed concentrations of more than 1×10^5 virus copies per milliliter in the commercial honey survey. Thus, those three samples were very close to the concentration that induced infection in one out of five cages. Based on these data, the risk of DWV-A transmission via honey under realistic conditions is 3/33 × 5 which is around 2% with the transmission threshold set at 1×10^5 DWV-A copies per milliliter. It is known that viral particles exposed to environmental conditions (e.g., humidity and heat) may quickly deteriorate [51,53,54]. This variable has not been thoroughly tested in this study. Nevertheless, it has been shown that viruses contained in frames with honey and bee bread remain infective even after been stored by six months at room temperature [5]. Since honey and bee bread stored in frames may represent a different kind of matrices than commercial processed honey and pollen products, data regarding the infectability of DWV in such commercial processed hive products stored for longer periods of time are needed. Until then, even though the estimated infection risk of 2% appears to be rather small, it is only a question of time before an infection occurs, especially if bees consume large amounts of imported honey. Moritz and Erler [55] observed a positive correlation between the amount of honey produced per colony and the amount of honey imports. Thus, they are considering the possibility of relabeling imported honey to locally produced honey. This illegal relabeling might be disguised by feeding the imported honey to local bee colonies. In that case, local bees would consume huge amounts of imported and possibly contaminated honey, which would increase the chance of infection dramatically.

For wax as a matrix, we have no data for realistic DWV-A concentrations of commercial products. That makes it impossible to evaluate a realistic transmission risk. However, wax is often heated during commercial processing and this may inactivate viruses, thereby decreasing the role of wax foundation as a virus carrier in the global trade scenario. By contrast, wax showed some transmission potential in the cage experiments and should thus also be considered as a potentially effective DWV-A vehicle, e.g., as a result of the common apicultural practice of exchanging non-processed combs between colonies.

As DWV is ubiquitous in most continents, a major concern should be the introduction of novel, potentially more virulent strains. Genetic recombination between different strains may indeed result in new viral variants with enhanced virulence. For example, distribution of DWV-B (also known as VDV-1) has expanded over the last decade [56] and recombination events with DWV-A appear to be frequent [18–23]. Moreover, the selection of a virulent recombinant variant between DWV-A and DWV-B has been already reported [23]. However, due to the potent role of *V. destructor* as a virus vector [35], it appears as if a recombinant DWV variant with lower virulence has been selected to optimize its transmission by *V. destructor* [57]. Therefore, more studies are necessary to assess the real impact of DWV genetic recombination on the emergence of potentially more virulent strains.

Risk is composed of the probability of occurrence and the damage caused if that undesired event actually occurs [58]. Therefore, the potential of hive products for virus transmission reported here should be considered, even though the probability of occurrence is probably rather low. Indeed, the potential damage resulting from novel viral strains could obviously be severe. Therefore, the potential of hive products for virus transmission should be considered in daily beekeeping practice. For example, it appears very risky to feed cheap honey to colonies, which is unfortunately a common practice of some beekeepers. A sustainable way to limit the risk effectively would be to include virus analyses in health certificates for import and export of hive products.

5. Conclusions

Our data clearly show that transmission of DWV-A via hive products is feasible. Since there are currently no regulations aiming to limit the spread of viruses due to global trade, the results are calling for respective mitigation measures, i.e., health certificates for hive products.

Author Contributions: Conceptualization, O.Y. and P.N.; methodology, D.S. and O.Y.; formal analysis, D.S.; resources, O.Y. and P.N.; data curation, D.S.; writing—original draft preparation, D.S.; writing—review and editing, D.S., O.Y. and P.N.; supervision, O.Y. and P.N.; project administration, P.N.; funding acquisition, P.N. All authors have read and agreed to the published version of the manuscript.

Funding: This research was funded by the Swiss Federal Food Safety and Veterinary Office (FSVO; project number 1.15.01) and the Vinetum Foundation.

Acknowledgments: Appreciation is given to Kaspar Roth for technical support.

Conflicts of Interest: The authors declare no conflict of interest.

References

1. Matheson, A. Managing risks in world trade in bees and bee products. *Apiacta* **2000**, *35*, 1–12.
2. Mutinelli, F. The spread of pathogens through trade in honey bees and their products (including queen bees and semen): Overview and recent developments. *Rev. Sci. Tech.* **2011**, *30*, 257–271. [CrossRef] [PubMed]
3. Shen, M.; Cui, L.; Ostiguy, N.; Cox-Foster, D. Intricate transmission routes and interactions between picorna-like viruses (Kashmir bee virus and sacbrood virus) with the honeybee host and the parasitic varroa mite. *J. Gen. Virol.* **2005**, *86*, 2281–2289. [CrossRef]
4. Chen, Y.; Evans, J.; Feldlaufer, M. Horizontal and vertical transmission of viruses in the honey bee, *Apis mellifera. J. Invertebr. Pathol.* **2006**, *92*, 152–159. [CrossRef] [PubMed]
5. Singh, R.; Levitt, A.L.; Rajotte, E.G.; Holmes, E.C.; Ostiguy, N.; Lipkin, W.I.; dePamphilis, C.W.; Toth, A.L.; Cox-Foster, D.L. RNA viruses in hymenopteran pollinators: Evidence of inter-taxa virus transmission via pollen and potential impact on non-*Apis* hymenopteran species. *PLoS ONE* **2010**, *5*, e14357. [CrossRef] [PubMed]
6. Beaurepaire, A.; Piot, N.; Doublet, V.; Antunez, K.; Campbell, E.; Chantawannakul, P.; Chejanovsky, N.; Gajda, A.; Heerman, M.; Panziera, D.; et al. Diversity and Global Distribution of Viruses of the Western Honey Bee, *Apis mellifera. Insects* **2020**, *11*, 239. [CrossRef] [PubMed]
7. Retschnig, G.; Kellermann, L.A.; Mehmann, M.M.; Yañez, O.; Winiger, P.; Williams, G.R.; Neumann, P. Black queen cell virus and drifting of honey bee workers (*Apis mellifera*). *J. Apic. Res.* **2019**, *58*, 754–755. [CrossRef]
8. Fries, I.; Camazine, S. Implications of horizontal and vertical pathogen transmission for honey bee epidemiology. *Apidologie* **2001**, *32*, 199–214. [CrossRef]
9. Chen, Y.; Pettis, J.S.; Feldlaufer, M.F. Detection of multiple viruses in queens of the honey bee *Apis mellifera* L. *J. Invertebr. Pathol.* **2005**, *90*, 118–121. [CrossRef]
10. Yue, C.; Schröder, M.; Gisder, S.; Genersch, E. Vertical-transmission routes for deformed wing virus of honeybees (*Apis mellifera*). *J. Gen. Virol.* **2007**, *88*, 2329–2336. [CrossRef]
11. de Miranda, J.R.; Fries, I. Venereal and vertical transmission of deformed wing virus in honeybees (*Apis mellifera* L.). *J. Invertebr. Pathol.* **2008**, *98*, 184–189. [CrossRef] [PubMed]
12. Jetzt, A.E.; Yu, H.; Klarmann, G.J.; Ron, Y.; Preston, B.D.; Dougherty, J.P. High rate of recombination throughout the human immunodeficiency virus type 1 genome. *J. Virol.* **2000**, *74*, 1234–1240. [CrossRef] [PubMed]
13. Stedman, K.M. Viral recombination: Ecology, evolution, and pathogenesis. *Viruses* **2018**, *10*, 358. [CrossRef] [PubMed]
14. Domingo, E.J.J.H.; Holland, J.J. RNA virus mutations and fitness for survival. *Annu. Rev. Microbiol.* **1997**, *51*, 151–178. [CrossRef] [PubMed]

15. Xiao, Y.; Rouzine, I.M.; Bianco, S.; Acevedo, A.; Goldstein, E.F.; Farkov, M.; Brodsky, L.; Andino, R. RNA recombination enhances adaptability and is required for virus spread and virulence. *Cell Host Microbe* **2016**, *19*, 493–503. [CrossRef]

16. Simon-Loriere, E.; Holmes, E.C. Why do RNA viruses recombine? *Nat. Rev. Microbiol.* **2011**, *9*, 617–626. [CrossRef]

17. Pérez-Losada, M.; Arenas, M.; Galán, J.C.; Palero, F.; González-Candelas, F. Recombination in viruses: Mechanisms, methods of study, and evolutionary consequences. *Infect. Genet. Evol.* **2015**, *30*, 296–307. [CrossRef]

18. Moore, J.; Jironkin, A.; Chandler, D.; Burroughs, N.; Evans, D.J.; Ryabov, E.V. Recombinants between Deformed wing virus and Varroa destructor virus-1 may prevail in *Varroa destructor*—Infested honeybee colonies. *J. Gen. Virol.* **2011**, *92*, 156–161. [CrossRef]

19. Zioni, N.; Soroker, V.; Chejanovsky, N. Replication of Varroa destructor virus 1 (VDV-1) and a Varroa destructor virus 1-deformed wing virus recombinant (VDV-1-DWV) in the head of the honey bee. *Virology* **2011**, *417*, 106–112. [CrossRef]

20. Dalmon, A.; Desbiez, C.; Coulon, M.; Thomasson, M.; Le Conte, Y.; Alaux, C.; Vallon, J.; Moury, B. Evidence for positive selection and recombination hotspots in Deformed wing virus (DWV). *Sci. Rep.* **2017**, *7*, 41045. [CrossRef]

21. Cornman, R.S. Relative abundance of deformed wing virus, Varroa destructor virus 1, and their recombinants in honey bees (*Apis mellifera*) assessed by kmer analysis of public RNA-Seq data. *J. Invertebr. Pathol.* **2017**, *149*, 44–50. [CrossRef] [PubMed]

22. Gisder, S.; Möckel, N.; Eisenhardt, D.; Genersch, E. In vivo evolution of viral virulence: Switching of deformed wing virus between hosts results in virulence changes and sequence shifts. *Environ. Microbiol.* **2018**, *20*, 4612–4628. [CrossRef] [PubMed]

23. Ryabov, E.V.; Wood, G.R.; Fannon, J.M.; Moore, J.D.; Bull, J.C.; Chandler, D.; Mead, A.; Burroughs, N.; Evans, D.J. A virulent strain of deformed wing virus (DWV) of honeybees (*Apis mellifera*) prevails after *Varroa destructor*-mediated, or In Vitro, transmission. *PLoS Pathog.* **2014**, *10*, e1004230. [CrossRef] [PubMed]

24. Ryabov, E.V.; Childers, A.K.; Lopez, D.; Grubbs, K.; Posada-Florez, F.; Weaver, D.; Girten, W.; vanEngelsdorp, D.; Chen, Y.; Evans, J.D. Dynamic evolution in the key honey bee pathogen deformed wing virus: Novel insights into virulence and competition using reverse genetics. *PLoS Biol.* **2019**, *17*, 1–27. [CrossRef] [PubMed]

25. Sumpter, D.J.T.; Martin, S.J. The dynamics of virus epidemics in Varroa-infested honey bee colonies. *J. Anim. Ecol.* **2004**, *73*, 51–63. [CrossRef]

26. de Miranda, J.R.; Genersch, E. Deformed wing virus. *J. Invertebr. Pathol.* **2010**, *103*, S48–S61. [CrossRef]

27. Dainat, B.; Evans, J.D.; Chen, Y.P.; Gauthier, L.; Neumann, P. Predictive markers of honey bee colony collapse. *PLoS ONE* **2012**, *7*, e32151. [CrossRef]

28. Martin, S.J.; Brettell, L.E. Deformed wing virus in honey bees and other insects. *Annu. Rev. Virol.* **2019**, *6*, 49–69. [CrossRef]

29. Chen, Y.P.; Siede, R. Honey bee virus. *Adv. Virus Res.* **2007**, *70*, 33–80.

30. Dainat, B.; Neumann, P. Clinical signs of deformed wing virus infection are predictive markers for honey bee colony losses. *J. Invertebr. Pathol.* **2013**, *112*, 278–280. [CrossRef]

31. Dainat, B.; vanEngelsdorp, D.; Neumann, P. Colony collapse disorder in Europe. *Environ. Microbiol. Rep.* **2012**, *4*, 123–125. [CrossRef] [PubMed]

32. Wilfert, L.; Long, G.; Leggett, H.C.; Schmid-Hempel, P.; Butlin, R.; Martin, S.J.; Boots, M. Deformed wing virus is a recent global epidemic in honeybees driven by Varroa mites. *Science* **2016**, *351*, 594–597. [CrossRef] [PubMed]

33. Lanzi, G.; de Miranda, J.R.; Boniotti, M.B.; Cameron, C.E.; Lavazza, A.; Capucci, L.; Camazine, S.M.; Rossi, C. Molecular and biological characterization of deformed wing virus of honeybees (*Apis mellifera* L.). *J. Virol.* **2006**, *80*, 4998–5009. [CrossRef] [PubMed]

34. Bowen-Walker, P.L.; Martin, S.J.; Gunn, A. The transmission of deformed wing virus between honeybees (*Apis mellifera* L.) by the ectoparasitic mite *Varroa jacobsoni* Oud. *J. Invertebr. Pathol.* **1999**, *73*, 101–106. [CrossRef]

35. Neumann, P.; Yañez, O.; Fries, I.; de Miranda, J.R. Varroa invasion and virus adaptation. *Trends Parasitol.* **2012**, *28*, 353–354. [CrossRef]

36. Martin, S.J.; Highfield, A.C.; Brettell, L.; Villalobos, E.M.; Budge, G.E.; Powell, M.; Nikaido, S.; Schroeder, D.C. Global honey bee viral landscape altered by a parasitic mite. *Science* **2012**, *336*, 1304–1306. [CrossRef]

37. McMahon, D.P.; Natsopoulou, M.E.; Doublet, V.; Fürst, M.A.; Weging, S.; Brown, M.J.F.; Gogol-Döring, A.; Paxton, R.J. Elevated virulence of an emerging viral genotype as a driver of honeybee loss. *Proc. R. Soc. B Biol. Sci.* **2016**, *283*, 20160811. [CrossRef]

38. Natsopoulou, M.E.; McMahon, D.P.; Doublet, V.; Frey, E.; Rosenkranz, P.; Paxton, R.J. The virulent, emerging genotype B of Deformed wing virus is closely linked to overwinter honeybee worker loss. *Sci. Rep.* **2017**, *7*, 5242. [CrossRef]

39. Tehel, A.; Vu, Q.; Bigot, D.; Gogol-Döring, A.; Koch, P.; Jenkins, C.; Doublet, V.; Theodorou, P.; Paxton, R. The Two Prevalent Genotypes of an Emerging Infectious Disease, *Deformed Wing Virus*, Cause Equally Low Pupal Mortality and Equally High Wing Deformities in Host Honey Bees. *Viruses* **2019**, *11*, 114. [CrossRef]

40. Levitt, A.L.; Singh, R.; Cox-Foster, D.L.; Rajotte, E.; Hoover, K.; Ostiguy, N.; Holmes, E.C. Cross-species transmission of honey bee viruses in associated arthropods. *Virus Res.* **2013**, *176*, 232–240. [CrossRef]

41. Yañez, O.; Piot, N.; Dalmon, A.; de Miranda, J.R.; Chantawannakul, P.; Panziera, D.; Amiri, E.; Smagghe, G.; Schroeder, D.C.; Chejanovsky, N. Bee viruses: Routes of infection in Hymenoptera. *Front. Microbiol.* **2020**, *11*, 943. [CrossRef]

42. Fontana, R.; Mendes, M.A.; Souza, B.M.; Donno, K.; César, L.M.M.; Palma, M.S. Jelleines: A family of antimicrobial peptides from the royal jelly of honeybees (*Apis mellifera*). *Peptides* **2004**, *25*, 919–928. [CrossRef] [PubMed]

43. Evans, J.D.; Schwarz, R.S.; Chen, Y.P.; Budge, G.; Cornman, R.S.; de la Rua, P.; de Miranda, J.R.; Foret, S.; Foster, L.; Gauthier, L.; et al. Standard methods for molecular research in *Apis mellifera*. *J. Apic. Res.* **2013**, *52*, 1–54. [CrossRef]

44. Williams, G.R.; Alaux, C.; Costa, C.; Csaki, T.; Doublet, V.; Eisenhardt, D.; Fries, I.; Kuhn, R.; McMahon, D.P.; Medrzycki, P.; et al. Standard methods for maintaining adult *Apis mellifera* in cages under In Vitro laboratory conditions. *J. Apic. Res.* **2013**, *52*, 1–35. [CrossRef]

45. De Miranda, J.R.; Bailey, L.; Ball, B.V.; Blanchard, P.; Budge, G.E.; Chejanovsky, N.; Chen, Y.; Gauthier, L.; Genersch, E.; de Graaf, D.C.; et al. Standard methods for virus research in *Apis mellifera*. *J. Apic. Res.* **2013**, *52*, 1–56. [CrossRef]

46. Yue, C.; Genersch, E. RT-PCR analysis of Deformed wing virus in honey bees (*Apis mellifera*) and mites (*Varroa destructor*). *J. Gen. Virol.* **2005**, *86*, 3419–3424. [CrossRef]

47. Forsgren, E.; De Miranda, J.R.; Isaksson, M.; Wei, S.; Fries, I. Deformed wing virus associated with Tropilaelaps mercedesae infesting European honey bees (Apis mellifera). *Exp. Appl. Acarol.* **2009**, *47*, 87–97. [CrossRef]

48. Gauthier, L.; Ravallec, M.; Tournaire, M.; Cousserans, F.; Bergoin, M.; Dainat, B.; deMiranda, J.R. Viruses associated with ovarian degeneration in Apis mellifera L. queens. *PLoS ONE* **2011**, *6*, e16217. [CrossRef]

49. Bailey, L.; Ball, B.V.; Perry, J.N. Honeybee Paralysis: Its Natural Spread and its Diminished Incidence in England and Wales. *J. Apic. Res.* **1983**, *22*, 191–195. [CrossRef]

50. Mazzei, M.; Carrozza, M.L.; Luisi, E.; Forzan, M.; Giusti, M.; Sagona, S.; Tolari, F.; Felicioli, A. Infectivity of dwv associated to flower pollen: Experimental evidence of a horizontal transmission route. *PLoS ONE* **2014**, *9*, e113448. [CrossRef]

51. Dainat, B.; Evans, J.D.; Chen, Y.P.; Neumann, P. Sampling and RNA quality for diagnosis of honey bee viruses using quantitative PCR. *J. Virol. Methods* **2011**, *174*, 150–152. [CrossRef]

52. Graystock, P.; Yates, K.; Evison, S.E.F.; Darvill, B.; Goulson, D.; Hughes, W.O.H. The Trojan hives: Pollinator pathogens, imported and distributed in bumblebee colonies. *J. Appl. Ecol.* **2013**, *50*, 1207–1215. [CrossRef]

53. Chen, Y.; Evans, J.; Hamilton, M.; Feldlaufer, M. The influence of RNA integrity on the detection of honey bee viruses: Molecular assessment of different sample storage methods. *J. Apic. Res.* **2007**, *46*, 81–87. [CrossRef]

54. Forsgren, E.; Locke, B.; Semberg, E.; Laugen, A.T.; de Miranda, J.R. Sample preservation, transport and processing strategies for honeybee RNA extraction: Influence on RNA yield, quality, target quantification and data normalization. *J. Virol. Methods* **2017**, *246*, 81–89. [CrossRef] [PubMed]

55. Moritz, R.F.A.; Erler, S. Lost colonies found in a data mine: Global honey trade but not pests or pesticides as a major cause of regional honey bee colony declines. *Agric. Ecosyst. Environ.* **2016**, *216*, 44–50. [CrossRef]

56. Ryabov, E.V.; Childers, A.K.; Chen, Y.; Madella, S.; Nessa, A.; Evans, J.D. Recent spread of Varroa destructor virus-1, a honey bee pathogen, in the United States. *Sci. Rep.* **2017**, *7*, 1–10. [CrossRef]

57. Mordecai, G.J.; Wilfert, L.; Martin, S.J.; Jones, I.M.; Schroeder, D.C. Diversity in a honey bee pathogen: First report of a third master variant of the Deformed Wing Virus quasispecies. *ISME J.* **2016**, *10*, 1264–1273. [CrossRef]

58. Winkler, K.; Brehm, H.P.; Haltmeier, J. *Bergsport Sommer—Technik/Taktik/Sicherheit*, 4th ed.; SAC-Verlag: Bern, Switzerland, 2013.

 *veterinary sciences*

Case Report

Overt Infection with Chronic Bee Paralysis Virus (CBPV) in Two Honey Bee Colonies

Julia Dittes [1,2,*], Marc O. Schäfer [3], Heike Aupperle-Lellbach [4], Christoph K. W. Mülling [2] and Ilka U. Emmerich [5]

1 Centre for Applied Training and Learning, Faculty of Veterinary Medicine, Leipzig University, An den Tierkliniken 19, 04103 Leipzig, Germany
2 Institute of Veterinary Anatomy, Histology and Embryology, Faculty of Veterinary Medicine, Leipzig University, An den Tierkliniken 43, 04103 Leipzig, Germany; c.muelling@vetmed.uni-leipzig.de
3 Friedrich-Loeffler-Institut, Federal Research Institute for Animal Health, Südufer 10, 17493 Greifswald, Insel Riems, Germany; Marc.Schaefer@fli.de
4 LABOKLIN GmbH & CO.KG, Labor für klinische Diagnostik, Steubenstraße 4, 97688 Bad Kissingen, Germany; aupperle@laboklin.de
5 VETIDATA, Institute of Pharmacology, Pharmacy and Toxicology, Faculty of Veterinary Medicine, Leipzig University, An den Tierkliniken 39, 04103 Leipzig, Germany; emmerich@vetmed.uni-leipzig.de
* Correspondence: julia.dittes@vetmed.uni-leipzig.de; Tel.: +49-341-973-8247

Received: 31 July 2020; Accepted: 16 September 2020; Published: 22 September 2020

Abstract: Chronic Bee Paralysis Virus (CBPV), a widespread honey bee RNA virus, causes massive worker bee losses, mostly in strong colonies. Two different syndromes, with paralysis, ataxia and flight incapacity on one hand and black hairless individuals with shortened abdomens on the other, can affect a colony simultaneously. This case report presents two *Apis mellifera carnica* colonies with symptoms of paralysis and hairless black syndrome in 2019. Via RT-PCR, a highly positive result for CBPV was detected in both samples. Further problems, such as a *Nosema* infection and *Varroa* infestation, were present in these colonies. Therapy methods were applied to colony 1 comprising queen replacement, shook swarm method and *Varroa* control, whereas colony 2 was asphyxiated after queen loss and colony weakening. After therapy, colony 1 was wintered without symptoms. Beekeeping and sanitary measures can save a CBPV-infected colony, while further complications result in total colony loss.

Keywords: chronic bee paralysis virus; *Varroa* infestation control; nosemosis; hairless black syndrome

---

1. Introduction

The Western honey bee, *Apis mellifera* L., the most important pollinator in agriculture worldwide, is managed with the additional aim of honey production. Different pathogens are threatening these honey bee colonies. According to the collaborative German bee monitoring project with around 100 beekeepers involved, the most important diseases are varroosis, nosemosis and various bee viruses [1]. Nosemosis is a disease caused by the microsporidia *N. apis* and *N. ceranae*. They are obligate intracellular pathogens causing dysentery and diarrhea after parasitizing and damaging the epithelial cells of the midgut. In spring 2018, the honey bee colonies involved in the German monitoring programme were *Nosema*-positive in 44% of cases [2].

Varroa, a reddish-brown crab-shaped mite, is one of the most important reasons for winter losses and the main pathogen posing a risk to honey bee colonies. It weakens the bees by feeding from the bee's fat bodies and harms the brood during their reproduction cycle in the combs [3]. In addition to being harmful itself, *Varroa* serves as a vector for various bee viruses, e.g., for the Deformed Wing Virus. In 2008, Celle et al. detected 1.4×10^4 copies of Chronic Bee Paralysis Virus in *Varroa* mites and recognized the possibility of an infection of different hymenoptera like *Formica rufa* [4]. *Varroa*, then discussed as being a natural reservoir of CBPV, is rather unlikely as a vector, because there is no proven causal relation between Varroa infestation and CBPV in honey bee colonies. CBPV, the first bee virus ever described and isolated in 1963 [5], is a positive-sense, single-stranded RNA virus that is spread all over the world and mostlyaffects strong colonies with two different syndromes, often simultaneously seen in colonies—paralysis, trembling and crawling bees on one hand and hairless, black bees, with shortened abdomens on the other hand—both ending up in massive worker bee losses. Aristotle has already described the symptoms of the hairless black syndrome and called those bees "thieves" [6]. All casts of bees can be infected by the virus, but there seems to be a kind of behavioral protection for the queen bee [7], which might be the reason that often the queen and some workers remain in an affected colony.

The predominant neurologic symptoms, like the eponymous paralysis and ataxia, are caused by the neurotropism of the virus. Although the head comprises just 1/10 of the whole bee mass, half of the CBPV copies in an infected bee are located there, especially in neurons of the mushroom bodies and central complex, regions where movements are controlled [8]. Diseased bees are attacked by their hive mates, who cut off their hairs with their mandibles and cause the hairless black abdomens [9]. Ingesting the hairs of infected bees is one way of infection. Other possible infection routes are infection through feces or trophallaxis [4].

Almost identical symptoms are produced by Acute Bee Paralysis Virus, but bees die in between one and two days after infection. In contrast, chronically paralyzed bees live for several days in cage experiments [10]. In recent years, Chronic Bee Paralysis seems to be an emerging disease. Budge et al. [11] studied CBPV prevalence in England and Wales between 2007 and 2017. The number of reported cases increased from 1 case in 2007 to 45 cases in 2017 [11]. Other countries see rising numbers of CBPV-positive colonies in their monitoring, e.g., the US with 0.7% in 2010 and 16% in 2014 [12] or Italy with 5% in 2009 to 10% in 2010 [13]. The German bee monitoring also mentioned a notably high prevalence of CBPV in the investigated colonies in 2013 (35.8%) and 2014 (20.7%) [2].

This case report shows the course of two colonies in Germany suffering from an overt CBPV infection in 2019.

2. Case Description

2.1. Medical History, Appearance of the Hive and Environment

Both *Apis mellifera carnica* colonies are situated in Leipzig (51°20′36.5244″ N, 12°23′15.9792″ E), a city in the Federal Republic of Germany in a region with oceanic climate with continental influence.

Colony 1 was situated next to two polystyrene hives on a 0.5-m-high metal stand in a wooden Zander hive. The hive consisted of one honey super and two brood chambers with ten frames each. It stood in an institute courtyard in an area with many chestnut, robinia and linden trees in the south of the city (Figure 1A). The entrance holes were oriented to the south.

Colony 2 was part of an apiary with six hives at a location a few hundred metres away from apiary 1. The wooden Zander hives stood in pairs on wooden pedestals or metal racks in the south of a small ancillary building with their entrance holes facing south as well. This hive also consisted of a honey super and two brood chambers with ten frames each (Figure 1B).

Both colonies were managed under the principles of good beekeeping practice and had been in a healthy status during the previous season.

Varroa infestation in both colonies was controlled by a combination of biotechnical methods and miticides. On 8 April 2019, the first frames without foundation and wire were put into the colonies for building drone combs. During the weekly health checks, drone brood was removed from the hive regularly and the building frames put back into the colony.

(**A**)   (**B**)

Figure 1. (**A**) Apiary 1 with colony 1 in a wooden Zander hive with one brood chamber in March 2020 (orange arrow), ©Julia Dittes. (**B**) Apiary 2 with colony 2 in a violet painted wooden Zander hive with two brood chambers and one honey super (second in the row) in June 2019 (orange arrow), ©Julia Dittes.

2.2. Examination of the Alighting Board and Observation of the Entrance Hole

Observation of the entrance holes of both colonies in the beginning of May 2019 showed some isolated bees flying. Compared to the other hives at the apiaries, colony activity was significantly lower. Some dead bees lay on the alighting boards and a large quantity of dead bees was found in front of the hives (Figure 2A), which produced a certain putrid smell. Between those dead bees, some flightless crawling ones were spotted.

(**A**)   (**B**)

Figure 2. (**A**) Dead bees laying on the stone floor tiles and grass in front of the hive of colony 2 (orange circle), ©Julia Dittes. (**B**) Close-up of scattered brown feces found on the hive of colony 1 (orange arrows), ©Julia Dittes.

At the front of the first colony's hive, additionally some scattered dark brown feces were visible (Figure 2B).

2.3. Clinical Examination of the Honey Bee Colonies and Observation of Living Bees

Clinical examination of colony 1 showed an empty honey super with undeveloped wax foundations and many bees with hairless, dark and shortened abdomens as well as some bees looking like ants in the brood chambers (Figure 3). The affected bees were trembling, atactic, with slow or no reaction when exposed to smoke, and held their wings at an unusual angle. Their movements were abnormal and erratic. Workers as well as drones showed clinical signs of paralysis (Supplementary Materials Videos S1 and S2). The nest consisted of five frames, one of them with drone brood, four with worker cells, and two regular drone brood frames. A large number of drones was seen in the colony.

Figure 3. Clinical examination of colony 1: worker bee with hairless, "black" abdomen (blue arrow); worker bee with ant-like phenotype (orange arrow), ©Julia Dittes.

The same clinical findings were observed in colony 2, with a brood nest out of six frames and two drone brood frames in the upper brood chamber. According to the Liebefeld method [14,15], with more than 7000 bees estimated, both colonies had nearly the same number of worker bees in the upper brood chamber, whereas the number of brood cells differed (approximately 14,200 in colony 2; 15,400 worker bee cells in colony 1).

In contrast to colony 1, colony 2 started storing honey (approximately 4.1 kg) in the honey super. During the course of the disease, the dead queen of colony 2 was found in front of the hive on 10 May 2019. Later on, worker bees started laying eggs into the cells of colony 2 (Figure 4).

Figure 4. Close-ups of a drone brood comb of colony 2 with more than one egg or larvae per cell, ©Julia Dittes.

From these unfertilized eggs, only drones developed, and the colony weakened progressively.

Examination of Dead Bees

Samples of dead bees were examined using the methods described in Dittes et al., 2020 [16]. The bees of both colonies showed the same symptoms as previously seen in the colony: Some bees were about 3 to 4 mm smaller than common, most of the bees had an extended proboscis. The abdomens were hairless, black, greasy shining, shortened and in some individuals bloated. Under pressure on the abdomen a light brown fluid leaked from the gut.

The queen of colony 2, showed the same problems, as well as drones of both colonies, so that all casts of bees were affected by the disease.

2.4. Laboratory Diagnosis

2.4.1. PCR-Diagnosis of CBPV and Further Viruses

RT-PCR was performed for both colonies at the end of May 2019 at the Federal Research Institute for Animal Health (FLI, Greifswald–Isle of Riems). After homogenizing ten bees from each sample with the gentleMACS^{TM} Dissociator, total RNA was purified from 150 µL of clarified bee homogenate using the RNeasy Mini Kit (Qiagen, Venlo, NL, USA). A one-step real-time RT-PCR was subsequently performed in duplicate using the AgPathID^{TM} One-Step RT-PCR Kit (Applied Biosystems^{TM}, Waltham, MA, USA) in a 96-well reaction plate [17]. Detailed information of the methods used can be found in Dittes et al., 2020 [16]. The results are expressed as the mean of the two replicates for each reaction.

Both samples were positive for CBPV showing a threshold cycle (Ct) below 20 for each amplification. Furthermore, titres of Deformed Wing Virus (DWV) and Black Queen Cell Virus (BQCV) were detected in the samples of both colonies without causing clinical findings. Acute Bee Paralysis Virus (ABPV) was not or barely detected (Table 1).

Table 1. Virus titres detected in two honey bee colonies in May 2019.

Virus	Titres for Colony 1	Titres for Colony 2
CBPV	++++ (Ct = 13.45)	++++ (Ct = 13.08)
DWV-B	+++ (Ct = 25.31)	+++ (Ct = 23.10)
BQCV	++++ (Ct = 18.27)	+++ (Ct = 27.56)
ABPV	+ (Ct = 34.65)	- (N/A)

Legend: CBPV = Chronic bee paralysis virus, DWV = Deformed wing virus, BQCV = Black Queen Cell Virus, ABPV = Acute bee paralysis virus, Ct = threshold cycle; classification of Ct: ++++, Ct < 20; +++, 20 ≤ Ct < 28; ++, 28 ≤ Ct ≤ 32; +, 32 < Ct ≤ 35; (+), Ct > 35; N/A, no evidence.

2.4.2. Diagnostics of Further Diseases

Monitoring the Varroa Infestation

There are different methods to estimate the mite load [16]. The number of mites in this case was estimated from the natural mite fall, the number of mites per day falling naturally from the bees down to the bottom board. An oil-coated paper towel was placed on a drawer under the wire-mesh floor for 2 to 3 days and the mites, naturally fallen from the bees to the floor, were counted in the debris. Female *Varroa* mites are from 1.2 to 1.7 mm, crab-shaped and reddish brown and can be seen with the naked eye [18]. In Table 2, the control plan and the natural mite fall with a merely low infestation rate are shown for colony 1.

Table 2. *Varroa* screening plan in 2019 by debris examination for the surviving colony 1.

Plastic Drawer		Natural Mite Fall
Into the Hive	**Out of the Hive**	
23 July	25 July	0m/2d → 0m/d
16 August	19 August	0m/3d → 0m/d
11 September	13 September	6m/2d → 3m/d
25 November	28 November	0m/3d → 0m/d

Legend: m = mite, d = day.

Detection of Nosemosis

A microscopic estimation of the number of spores was carried out using the scheme described by Ritter in 1996 [19]. Colony 1 had a moderate infestation (between 20 and 100 spores per visual field), Colony 2 was only slightly affected (less than 20 spores per visual field) (Figure 5). For further differentiation of the species *N. apis* and *N. ceranae*, a PCR would be required, but was not performed in this case.

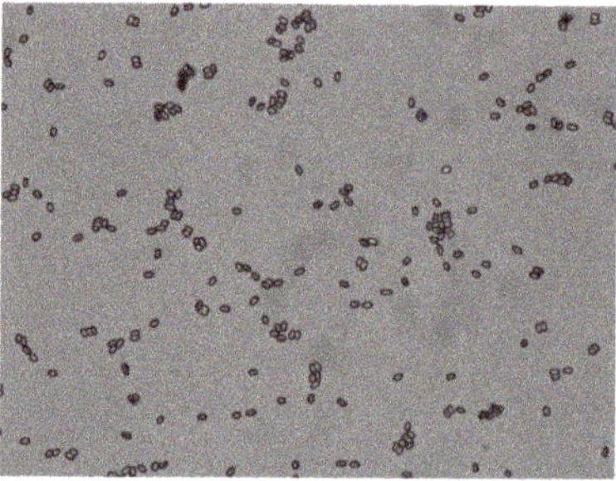

Figure 5. Microscopic image of oval-shaped *Nosema*-spores. (unstained, 400x) Differentiation of the species is not possible under the microscope. ©Heike Aupperle-Lellbach.

2.5. List of Medical Issues and Diagnoses

Table 3 lists the main problems of the colonies. In sum, an overt infection with Chronic Bee Paralysis Virus was diagnosed in both colonies. In addition, a co-infection with *Nosema* was proven with different infestation grades in the colonies.

Table 3. List of medical issues in two honey bee colonies infected with Chronic Bee Paralysis Virus (CBPV) in 2019.

Colony 1	Colony 2
for both colonies:	
- hairless, black bees	
- smaller bees with shortened abdomens	
- atactic, crawling and trembling bees	
- bees with bloated abdomens	
- large number of dead bees in front of the hives	
- positive PCR result for CBPV (Colony 1: ++++ (Ct = 13.45), Colony 2: ++++ (Ct = 13.08))	
moderate infestation of Nosema spores	slight infestation of Nosema spores
	loss of the queen
	laying worker bees and large number of drones
other viruses (without symptoms): DWV-B, BQCV, ABPV (exact Cts in Table 1)	other viruses (without symptoms): DWV-B, BQCV, ABPV (exact Cts in Table 1)

Legend: DWV = Deformed Wing Virus, BQCV = Black Queen Cell Virus, ABPV = Acute Bee Paralysis Virus, Ct = threshold cycle.

Treatment methods were only applied to colony 1. After queen loss, a surplus of drones and laying worker bees, the prognosis was too bad for colony 2.

2.6. Therapeutic Measures and Outcome

Due to the bad prognosis, colony 2 was asphyxiated with sulphur on 17 June 2019.

For therapy of colony 1, measures such as re-queening, shook swarm method and *Varroa* infestation control were applied. Further information on those methods can be found in Dittes et al. 2020 [16]. Table 4 summarizes the therapeutic measures and further development of colony 1.

Table 4. Therapeutic measures and outcome of colony 1 after diagnosis of an overt CBPV infection.

Date	Applied Measures/Methods and Development of Colony 1
29 May 2019	removal of the old queen from the hive, bees start raising a new queen
06 June 2019	destruction of all of the queen cells, insertion of a queen cage with the new queen (Nr. 71, marked green) and some food, gentle acclimatization while workers gnaw through the food
18 June 2019	hive check: many eggs and young brood in the brood combs, queen Nr. 71 with a larger abdomen
19 June 2019	shook swarm method: migration to a new hive body with new pathogen free frames and wax foundations, protection of the queen in a queen cage, bees shaken off in front of the new hive onto a base to enter the new hive by themselves.
21 June 2019	treatment with 5.7% Oxalic acid dihydrate (OXUVAR® 5.7%, Andermatt Biovet GmbH, Lörrach, BW, D)) diluted to a 3.5% solution by spraying; Feeding with sugar syrup (2 kg sugar)
25 June 2019	decreased symptoms of trembling and paralysis or black abdomens
28 June 2019	feeding with sugar syrup (2 kg sugar)
1 July 2019	feeding with sugar syrup (2 kg sugar)
17 July 2019	feeding with sugar syrup (2 kg sugar)
21 August 2019	feeding with sugar syrup (2 kg sugar)
16 September 2019	late summer treatment with formic acid (Ameisensäure 60% ad us. vet., Serumwerk Bernburg AG, Bernburg, ST, D) after the last honey harvest: evaporation of 140 mL of formic acid in the hive using a Liebig Dispenser; observed–mite load (Table 1): about 3 mites per day = medium infestation rate [18]
October 2019–April 2020	overwintering of the colony with only one brood chamber, no application of a winter treatment with oxalic acid because of a low mite infestation rate in November (Table 1)
12 April 2020	successful hibernation; enlargement of the hive with a second brood chamber.
28 April 2020	brood nest out of 9 brood combs during apple blossom

No symptoms of paralysis were shown by the bees during the weekly check-ups in 2020 (Video S3). CBPV was not detected via RT-PCR any longer. The vital colony (Figure 6) started honey production in the following weeks. In total, about 36 kg of honey was harvested from colony 1 in 2020 in an apiary averaging 42 kg of honey per colony.

Figure 6. Development of the rehabilitated colony 1 after wintering/hibernation on 28 April 2020 during apple blossom (brood on 9 combs [Zander 477 × 220 mm], honey super empty), F19: capped drone brood, F20 + F7: brood-comb with eggs, F18 + F6: brood-comb with mostly covered worker brood, F12: build up drone brood frame ©Ilka Emmerich.

3. Discussion

Paralysis of honey bees was systematically studied for the first time by Burnside [20] nearly 100 years ago in 1933. Performing many cage experiments spraying, injecting and feeding healthy bees with extract from sick bees, he observed bees attacking their hive mates [20]. After further infection experiments in 1945, Burnside drew the conclusion that a filtrable virus is the cause of the disease [21]. However, it took another 20 years until Bailey first isolated Chronic Bee Paralysis Virus [5], a virus that causes neurologic symptoms. Clinical findings are trembling, crawling, circling bees and flightless bees without orientation, because the virus affects neurons of higher-order integration centres, optic and antennal lobes, which are involved in locomotion control, learning and orientation behavior [18]. At the same time, phenotype changes in bees are seen: black hairless shiny abdomens and smaller bees with shortened abdomens. In the beginning, CBPV was characterized by two distinct syndromes, the paralysis form and the hairless black syndrome, but both can occur in colonies at the same time, as was observed in the two described ones. All symptoms may lead to a high mortality within a few days, with massive worker bee losses.

Acute Bee Paralysis Virus, which produces the same symptoms but leads to a faster death of affected bees, could be excluded through differential diagnosis via PCR, where it was not or just marginally detected.

In formerly strong colonies just the queen and a few workers remain. Amiri et al. [7] compared the queen's and the worker bee's susceptibility. Infection experiments for both workers and queens resulted in the same symptoms after 6 days and a 100% mortality after 14 days. There seem to be behavioral strategies to protect the queen, e.g., only healthy bees feeding her [7]. The observations in this case confirm that for colony 1. The queen of colony 2 died and was found in front of the hive. However, it is not possible to clearly determine the cause of the queen's death. She could have been affected by CBPV, too, but she also could have been harmed while handling the colony or was to be replaced by a new queen, because of failures detected by the workers. Replacement by supersedure, which is done with old queens, can be nearly ruled out, because it was a young queen raised in 2018, one year prior. Normally, the colony would have reared a new queen. However, instead, laying worker bees started oviposture in a disordered way in the absence of the queen's pheromones. From these unfertilized eggs, only drones developed, and with a declining number of working honey bees, the colony weakened and could not be rehabilitated. Asphyxiating with sulfur is the standard method to end a colony's life, as was done to colony 2.

CBPV often persists as a covert infection in honey bee colonies, detectable via PCR, without causing obvious symptoms, clinical findings and problems within the colony. Different factors favour the change into an overt infection. The interaction between these factors is not understood in detail yet, but periods of bad weather or a lack of nectar and starving are discussed. Taking a look at the weather in May 2019 in Germany, the temperatures were about 1.2 °C colder than normal, especially during the first half [22]. Cold and rainy conditions outside make the bees stay in the hive, intensifying their direct contact and may facilitate the spread of the virus. Direct contact plays a predominant role in transmission during disease outbreak [8].

Infected bees suffer from nibbling attacks of their hive mates, cutting their hairs off. This leads to the black hairless shiny abdomens of the attacked ones and the ingestion of bee hairs by the attackers. Rinderer et al. investigated the effect of hairs removed from the bodies of infected honey bees in infection experiments and saw that bees fed with virus and infected hairs showed a significantly higher mortality than bees fed only with one of both [9].

A high density of bees in the hive also increases the amount of and the contact with feces, another possible way of transmission. The virus load in bee excreta is as high as in the heads of symptomatic bees (10^{10} CBPV RNA copies) [23]. In the described colonies a co-infection with *Nosema* was detected. These microsporidia cause dysentery and diarrhea through damage of the midgut epithelial cells, which results in more feces in the hive, especially under the weather conditions described above. Toplak et al. investigated the effect of a co-infection of CBPV and *Nosema ceranae* in bees under experimental conditions. Bees infected with *Nosema* showed an increased replication ability for CBPV [24]. The damage to the midgut epithelial cells and a suppression of the immune response of the honey bees increase the virulence of the viral pathogens. The co-infection may have had a synergistic effect on the CBPV outbreak in the described cases. While the species were not differentiated via PCR, according to the German bee monitoring, the dominant species was *Nosema ceranae* in more than 96% of the *Nosema*-positive samples in 2018 [1,2], so it is the most likely species here as well.

Furthermore, the *Varroa* mite was discussed to be a vector and natural reservoir for CBPV. Investigations into mites showed a titre of 1.4×10^4 particles per mite, which is quite low. However, minus strand RNA was detected, which is indicative of virus replication. *Varroa* was and is present in the described colonies and also acts as a general weakening factor for honey bee colonies, therefore an intense *Varroa* infestation control has to be done.

There are no specific drugs or therapy measures against viral diseases, so a therapy has to rely on sanitary measures and basics of good bee keeping. Bailey described in 1965 that re-queening was a helpful measure in naturally CBPV-infected colonies [5]. In surviving colony 1, re-queening was the first step to rehabilitation, followed by the shook swarm method. Artificial swarming is often used to decrease the load of a pathogen in the bee hive, e.g., of American foulbrood. The bees got a new hive with new and clean combs. In the same step, a brood-less period was induced, allowing for a *Varroa* treatment by spraying oxalic acid. With the combination of these measures, a total recovery of colony 1 was possible. In colony 2, the intervention came too late to save it.

4. Conclusions

If diagnosed in time, an overt CBPV infection can be treated successfully with bee keeping and sanitary measures. There is a good prognosis for the affected colony. In case of further complications, the infection can result in a total colony loss. A survey about therapy measures and outcome among beekeepers with confirmed CBPV-positive bee colonies during recent years should gather information on successful strategies to be applied to affected colonies in the future.

It remains to be seen how CBPV will spread in the next years, and whether Germany will experience a similar growth to England and Wales. Furthermore, research about the specific factors that lead to an outbreak is required in order to develop strategies for preventing overt infections in colonies.

Supplementary Materials: The following are available online at http://www.mdpi.com/2306-7381/7/3/142/s1, Video S1: trembling and ataxia, hairless, black bees in colony 1_May 2019; Video S2: trembling_paralysis_hairless black bees in colony 2_May 2019; Video S3: normal behaviour of colony 1 in May 2020

Author Contributions: Conceptualization, C.K.W.M., H.A.-L. and I.U.E.; methodology, J.D., I.U.E., M.O.S.; validation, J.D., I.U.E.; formal analysis, J.D.; investigation, J.D., I.U.E. and M.O.S.; resources, C.K.W.M.; writing—original draft preparation, J.D.; writing—review and editing, I.U.E., H.A.-L.; visualization, J.D., I.U.E.; supervision, C.K.W.M., I.U.E., H.A.-L.; project administration, C.K.W.M. All authors have read and agreed to the published version of the manuscript.

Funding: This research received no external funding.

Conflicts of Interest: The authors declare no conflict of interest, although H.A.-L. and M.O.S. are working in laboratories, that regularly offer the laboratory diagnostics. The company (LABOKLIN GmbH & CO.KG) had no role in the design, execution, interpretation, or writing of the study.

References

1. Genersch, E.; von der Ohe, W.; Kaatz, H.; Schroeder, A.; Otten, C.; Büchler, R.; Berg, S.; Ritter, W.; Mühlen, W.; Gisder, S.; et al. The German bee monitoring project: A long term study to understand periodically high winter losses of honey bee colonies. *Apidologie* **2010**, *41*, 332–352. [CrossRef]

2. Rosenkranz, P.; von der Ohe, W.; Schäfer, M.; Genersch, E.; Büchler, R.; Berg, S.; Otten, C. Zwischenbericht 2018 Deutsches Bienenmonitoring–DeBiMo Projektlaufzeit 1.1.2017–31.12.2019, Berichtszeitraum 1.1.2018–31.12.2018. Available online: https://bienenmonitoring.uni-hohenheim.de/88584?tx_ttnews%5Btt_news%5D=44864&cHash=b943160d769aa565de89b191d468bd9e (accessed on 11 June 2020).

3. Ramsey, S.D.; Ochoa, R.; Bauchan, G.; Gulbronson, C.; Mowery, J.D.; Cohen, A.; Lim, D.; Joklik, J.; Cicero, J.M.; Ellis, J.D.; et al. Varroa destructor feeds primarily on honey bee fat body tissue and not hemolymph. *Proc. Natl. Acad. Sci. USA* **2019**, *116*, 1792–1801. [CrossRef] [PubMed]

4. Celle, O.; Blanchard, P.; Olivier, V.; Schurr, F.; Cougoule, N.; Faucon, J.-P.; Ribiere, M. Detection of Chronic bee paralysis virus (CBPV) genome and its replicative RNA form in various hosts and possible ways of spread. *Virus Res.* **2008**, *133*, 280–294. [CrossRef] [PubMed]

5. Bailey, L.; Gibbs, A.J.; Woods, R.D. Two viruses from adult honey bees (*Apis mellifera* Linnaeus). *Virology* **1963**, *21*, 390–395. [CrossRef]

6. Aristotle. *Aristotle's History of Animals in Ten Books*; Cresswell, R., Translator; George Bell & Sons: London, UK, 1902; pp. 262–263.

7. Amiri, E.; Meixner, M.; Büchler, R.; Kryger, P. Chronic bee paralysis virus in honeybee queens: Evaluating susceptibility and infection routes. *Viruses* **2014**, *6*, 1188–1201. [CrossRef] [PubMed]

8. Ribière, M.; Olivier, V.; Blanchard, P. Chronic bee paralysis: A disease and a virus like no other? *J. Invertebr. Pathol.* **2010**, *103*, 120–131. [CrossRef] [PubMed]

9. Rinderer, T.E.; Rothenbuhler, W.C. The fate and effects of hairs removed from honeybees with hairless-black syndrome. *J. Invertebr. Pathol.* **1975**, *26*, 305–308. [CrossRef]

10. Bailey, L. Paralysis of the honey bee, *Apis mellifera* Linnaeus. *J. Invertebr. Pathol.* **1965**, *7*, 132–140. [CrossRef]

11. Budge, G.E.; Simcock, N.K.; Holder, P.J.; Shirley, M.D.F.; Brown, M.A.; Van Weymars, P.S.M.; Evans, D.J.; Rushton, S.P. Chronic bee paralysis as a serious emerging threat to honey bees. *Nat. Commun.* **2020**, *11*, 2164. [CrossRef] [PubMed]

12. Traynor, K.S.; Rennich, K.; Forsgren, E.; Rose, R.; Pettis, J.; Kunkel, G.; Madella, S.; Evans, J.; Lopez, D.; van Engelsdorp, D. Multiyear survey targeting disease incidence in US honey bees. *Apidologie* **2016**, *47*, 325–347. [CrossRef]

13. Porrini, C.; Mutinelli, F.; Bortolotti, L.; Granato, A.; Laurenson, L.; Roberts, K.; Gallina, A.; Silvester, N.; Medrzycki, P.; Renzi, T.; et al. The status of honey bee health in italy: Results from the nationwide bee monitoring network. *PLoS ONE* **2016**, *11*, e0155411. [CrossRef] [PubMed]

14. Imdorf, A.; Buehlmann, G.; Gerig, L.; Klichenmann, V.; Wille, H. Überprüfung der schätzmethode zur ermitllung der brutfläche und der anzahl arbeiterinnen in freifliegenden bienenvölkern. *Apidologie* **1987**, *18*, 137–146. [CrossRef]

15. Deutsches Bienenjournal: Bienenvolk Schätzen: Volksstärken Genau Erfassen (20 Oktober 2016). Available online: https://www.bienenjournal.de/imkerpraxis/ratgeber/volksstaerken-erfassen-quiz/ (accessed on 24 May 2020).

16. Dittes, J.; Schäfer, M.O.; Aupperle-Lellbach, H.; Mülling, C.K.W.; Emmerich, I.U. Veterinary diagnostic approach of common virus diseases in adult honey bees. *Vet. Sci.* **2020**, in press.

17. Blanchard, P.; Ribière, M.; Celle, O.; Lallemand, P.; Schurr, F.; Olivier, V.; Iscache, A.L.; Faucon, J.P. Evaluation of a real-time two-step RT-PCR assay for quantitation of Chronic bee paralysis virus (CBPV) genome in experimentally-infected bee tissues and in life stages of a symptomatic colony. *J. Virol. Methods* **2007**, *141*, 7–13. [CrossRef] [PubMed]

18. Vidal-Naquet, N. *Honeybee Veterinary Medicine: Apis mellifera L.*, 1st ed.; 5m Books Ltd.: Sheffield, UK, 2015; pp. 71–75.

19. Ritter, W. *Diagnostik und Bekämpfung der Bienenkrankheiten*, 1st ed.; Gustav Fischer Verlag: Jena, Germany, 1996; p. 51.

20. Burnside, C.E. Preliminary Observations on "Paralysis" of Honeybees. *J. Econ. Entomol.* **1933**, *26*, 162–168. [CrossRef]

21. Burnside, C.E. The cause of paralysis of bees. *Am. BeeJ.* **1945**, *85*, 354–355.

22. Deutschlandwetter im Mai 2019. Available online: https://www.dwd.de/DE/presse/pressemitteilungen/DE/2019/20190529_deutschlandwetter_mai_news.hthtm (accessed on 29 May 2019).

23. Ribière, M.; Lallemand, P.; Iscache, A.-L.; Schurr, F.; Celle, O.; Blanchard, P.; Olivier, V.; Faucon, J.P. Spread of infectious chronic bee paralysis virus by honeybee (*Apis mellifera* L.) feces. *Appl. Environ. Microbiol.* **2007**, *73*, 771–776. [CrossRef] [PubMed]

24. Toplak, I.; Ciglenečki, U.J.; Aronstein, K.; Gregorc, A. Chronic bee paralysis virus and nosema ceranae experimental co-infection of winter honey bee workers. *Viruses* **2013**, *5*, 2282–2297. [CrossRef] [PubMed]

Article

Silicone Wristbands as Passive Samplers in Honey Bee Hives

Emma J. Bullock [1], Alexis M. Schafsnitz [1], Chloe H. Wang [1], Robert L. Broadrup [1], Anthony Macherone [2,3], Christopher Mayack [4,5] and Helen K. White [1,*]

[1] Department of Chemistry, Haverford College, 370 Lancaster Ave, Haverford, PA 19041, USA; ejbvt15@gmail.com (E.J.B.); amschafsnitz@gmail.com (A.M.S.); chloe.wang18@gmail.com (C.H.W.); rbroadru@haverford.edu (R.L.B.)

[2] Life Science and Chemical Analysis Group, Agilent Technologies, 5301 Stevens Creek Blvd, Santa Clara, CA 95051, USA; anthony_macherone@agilent.com

[3] Department of Biological Chemistry, The Johns Hopkins University School of Medicine, 733 N Broadway, Baltimore, MD 21205, USA

[4] Department of Biology, Swarthmore College, 500 College Ave, Swarthmore, PA 19081, USA; cmayack@sabanciuniv.edu

[5] Molecular Biology, Genetics, and Bioengineering, Faculty of Engineering and Natural Sciences, Sabancı University, 34956 Istanbul, Turkey

* Correspondence: hwhite@haverford.edu or hwhite@alum.mit.edu; Tel.: +1-610-795-6402

Received: 7 June 2020; Accepted: 2 July 2020; Published: 6 July 2020

Abstract: The recent decline of European honey bees (*Apis mellifera*) has prompted a surge in research into their chemical environment, including chemicals produced by bees, as well as chemicals produced by plants and derived from human activity that bees also interact with. This study sought to develop a novel approach to passively sampling honey bee hives using silicone wristbands. Wristbands placed in hives for 24 h captured various compounds, including long-chain hydrocarbons, fatty acids, fatty alcohols, sugars, and sterols with wide ranging octanol–water partition coefficients (K_{ow}) that varied by up to 19 orders of magnitude. Most of the compounds identified from the wristbands are known to be produced by bees or plants. This study indicates that silicone wristbands provide a simple, affordable, and passive method for sampling the chemical environment of honey bees.

Keywords: *Apis mellifera*; bee; silicone band; hive; passive sampler

---

1. Introduction

The dwindling global population of the honey bee pollinator *Apis mellifera*, the main pollinator species used in agriculture, has driven researchers to investigate honey bees and their responses to different environmental and pathological conditions [1,2]. These insects play a vital role in agricultural systems and almost a third of the global food crop depends exclusively on the European *Apis mellifera* species [3]. Since the 1940s, the number of managed honey bee colonies in the United States has declined, with the most dramatic losses occurring in the past decade [4]. Numerous factors have been implicated in the decline in bee health including exposure to pesticides and viruses [5], the spread of *Varroa destructor* mites [6], and parasites such as *Nosema ceranae* [7]; however, none can be identified as a single underlying causal factor.

Due to the vast number of variables and interactions that impact honey bee health, identifying the cause, impact, and risk associated with individual factors is challenging. Targeted exposure studies assessing chemicals and contaminants including pesticides and pathogens have been examined [6]. Recently, an exposomic approach has explored a more comprehensive approach where external chemicals as well as those produced by honey bees are combined into a single analysis [8].

Some compounds produced by bees, known as semiochemicals, are released by individual honey bees as cues or signals in order to relay information to hive mates, allowing honey bees to coordinate their responses to environmental stimuli as a hive [9]. Semiochemicals can also ward off or attract other honey bees [9], can relay the needs of larvae to nurse bees [10,11], and are involved in social recognition [12,13]. Certain chemicals have the potential to relay specific information about the disease state of an individual. For example, ethyl oleate is a chemical produced by honey bees that is significantly increased by infection of the gut fungus *N. ceranae* [13], and cuticular hydrocarbon profiles of honey bees have been observed to change when exposed to bacterial or *N. ceranae* infections [14,15]. It has also been shown that poor nutrition can lead larvae to release E-β-ocimene into the hive [11]. Monitoring chemicals produced by honey bees is of significant interest as it can provide insight into specific infection status, pesticide exposure, as well as nutrition deficiencies, thus providing insight into the overall health of a bee colony.

Semiochemicals and other volatile and semivolatile chemicals (VOCs and SVOCs) found in honey bee hives have previously been sampled by collecting the air in the hive directly via a syringe and then examining the chemical contents of the air sample via gas chromatography–mass spectrometry (GC–MS) [16]; however, this method does not allow the storage of samples prior to analysis [16]. Alternative methods have utilized adsorbent packing materials (such as Super Q, Hayesep Q, Porapak Q, Tenax TA) combined with vacuum/air tube systems [10,16,17]. These methods have limitations as the vacuum/air tube systems are complicated to construct and their presence may cause vibrations that induce a stress response in the bees and bias the chemical profiles measured [16]. Solid-phase microextraction (SPME) fibers have been used as passive samplers, avoiding the problems involved in complicated air flow setups; however, they are somewhat limited as they are designed to target specific chemicals, and are expensive and fragile to use [16–21].

This study aims to apply silicone wristbands, a previously established method [22–26], as a new approach to passively sample a wide range of chemicals present in honey bee hives. Silicone wristbands are cheap, commercially available, and have been used as passive environmental samplers to provide time-averaged concentrations of human exposures to polycyclic aromatic hydrocarbons (PAHs), consumer products, personal care products, pesticides, phthalates, other industrial compounds [22–24], and organophosphate flame retardants [25,26]. Compounds adsorbed by silicone wristbands have been found to remain on the bands for extended periods of time, unlike compounds on SPME fibers [24–27]. Perhaps the most significant advantage of silicone wristbands is that they are able to sample VOCs and SVOCs with octanol-air partition coefficients (K_{oa}) ranging in 10 orders of magnitude [22–24]. The K_{oa} as well as the octanol–water partition coefficients (K_{oaw}) of a chemical compound are values that represent the hydrophobicity of a compound. Specifically, these values describe the ratio of a compound in octanol divided by the concentration of the same compound in air or water for K_{oa} and K_{ow}, respectively. With respect to passive samplers, these values are used to show the range in hydrophobicity of compounds that a sampler can detect, which, as mentioned, is large for silicone wristbands. Their ability to adsorb such a broad range of compounds makes them ideal for sampling chemicals produced by bees and could potentially collect a broader range of compounds than the aforementioned SPME fibers and polymer samplers.

In this study, commercially available silicone wristbands were pre-cleaned and placed in honey bee hives to sample VOCs and SVOCs. This study presents information about the types of chemicals that can be adsorbed and analyzed to provide a reliable chemical profile of honey bee hives from a variety of urban and suburban locations.

2. Materials and Methods

2.1. Materials

All solvents were GC Resolv or Optima grade, obtained from Fisher Scientific (Newark, NJ, USA). Glassware was combusted at 450 °C for 8 h before use.

2.2. Band Preparation

Silicone bands (acquired from 24HourWristbands.com) with a circumference of 20 cm, $\approx$40 cm^2 surface area, and $\approx$4 g of sorbent were prepared according to a modified procedure [22]. Briefly, bands were cleaned in previously combusted (450 °C, 8 h) glass jars with a 1:1 volume mixture of ethyl acetate (EtOAc) to hexanes and shaken at 80 rpm for at least 2.5 h before the removal of the solvent. This procedure was repeated three more times, at which point the bands were removed from the jar and placed on combusted aluminum foil in a high-vacuum pressure oven at 60 °C for 48 h. The bands were stored individually in 40 mL combusted amber glass vials at room temperature prior to use.

2.3. Band Deployment

During July, August, September, and October of 2016, bands were deployed in 10 apiaries in rural, suburban, and urban areas of southeastern Pennsylvania (details provided in Table S1 of the Electronic Supplementary Materials). At each apiary, three hives were sampled across three time points. In order to capture different parts of the environment inhabited by the honey bees, we placed one silicone band under the top of each hive (in-cover) and placed one inside the entrance of each hive. An additional band was placed on top of each hive (outside cover) to serve as an external control (details shown in Figure 1). The bands were retrieved 24 h later and transferred to individual 40 mL amber vials, placed on dry ice, transported back to the lab, and stored at −20 °C.

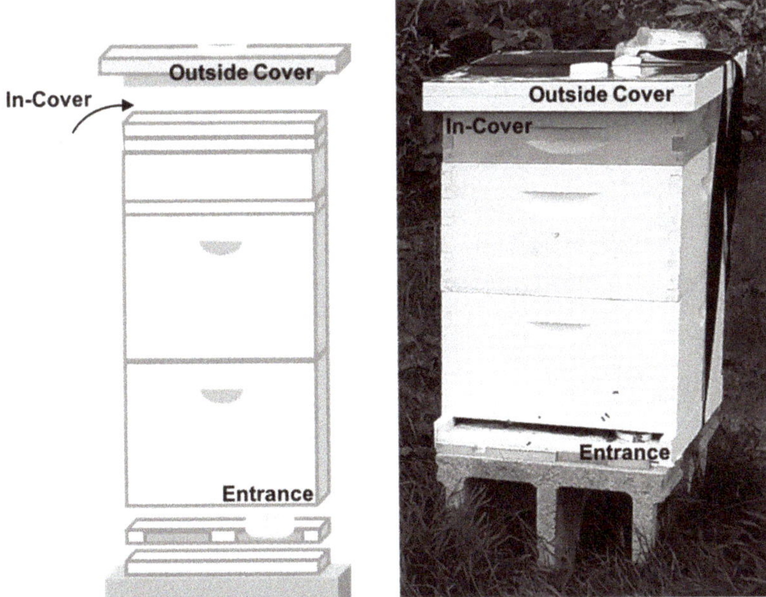

Figure 1. Diagram of a hive (**left**) and photo of a hive in the field (**right**), highlighting the placement of bands. In-cover band for the hive in the field is not shown as it is inside the cover part of the hive. Wristbands shown are 20 cm in circumference and 1 cm width; hives are approximately 1 × 0.5 × 0.4 m.

2.4. Band Extraction

Silicone bands were cleaned and extracted according to a modified procedure [22]. Briefly, each band was rinsed with Milli-Q water to remove any solids, such as dirt, propolis, or stingers, followed by isopropyl alcohol to remove water. The bands were then placed in individual 500 mL jars with 100 mL EtOAc and shaken at 60 rpm for 2 h. The extract was removed and the extraction process was repeated twice more before the extracts were combined, reduced in volume via rotary evaporation to approximately 1 mL, and stored at −20 °C prior to further analysis. The extracts were then solvent-exchanged into hexane, spiked with perdeuterated *n*-hexadecane as an internal standard (average recoveries were 74%), and charged onto a small glass pipette column (0.5 cm × 6 cm) packed with fully activated silica gel (100–200 mesh). The first fraction (F1) was eluted from the column with 4mL of hexane, followed by the second fraction (F2) eluted with 4 mL of an equal mixture of dichloromethane and methanol containing 1% formic acid. Laboratory blanks run alongside samples contained no detectable compounds.

2.5. Gas Chromatographic Analysis of Band Extracts

The F1 extracts were analyzed on a 1D Agilent 7890 series gas chromatograph coupled to a flame ionization detector (FID, Santa Clara, CA, USA). Compounds were separated on a J&W DB-XLB capillary column (30 m, 0.25 mm I.D., 0.25 μm film) with helium carrier gas at a constant flow of 1 mL min^{-1}. The GC oven had an initial temperature of 40 °C (1 min hold) and was ramped at 10 °C min^{-1} until 160 °C (1 min hold), then ramped again at 4 °C min^{-1} until 320 °C (36 min hold). Quantities of *n*-alkanes were calculated using response factors determined from pure standards. Presence of *n*-alkenes in the F1 extracts were confirmed by analyzing select extracts on an Agilent 7890 series gas chromatograph with an Agilent 5975 mass selective detector (MSD) and were noted, though not quantified. All fractions from each in-cover, outside, and entrance band for 10 hives were analyzed via GC–MS with a compound detection limit of 0.1 ng/μL determined from a standard curve using pure standards.

In order to identify unknown compounds, we analyzed F2 extracts via a 1D Agilent 7890 series gas chromatograph coupled to an Agilent 5975 mass selective detector (MSD). Prior to analysis, F2 extracts were derivatized with bis(trimethylsilyl)trifluoroacetamide (BSTFA) in pyridine. Compounds were separated on a J&W DB-XLB capillary column (60 m, 0.25 mm internal diameter (I.D.), 0.25 μm film) with helium carrier gas at a constant flow of 1 mL min^{-1}. The GC oven had an initial temperature of 40 °C (1 min hold) and was ramped at 10 °C min^{-1} until 160 °C (1 min hold), then ramped again at 4 °C min^{-1} until 320 °C (36 min hold). The MS was operated in electron-impact (EI) mode with an ionization energy of 70 eV. Spectra were acquired between *m/z* 40–650 at a scan rate of 1 cycle s^{-1}. Fatty acids and fatty alcohols were identified from mass spectral and retention time characteristics compared to pure standards. All other compounds were tentatively identified from mass spectra and gas chromatographic retention characteristics.

3. Results

3.1. Chemical Profiles of Hive Air Were Dominated by Hydrocarbon Compounds

Bands placed in-cover and at the entrance of hives exhibited similar chemical profiles dominated by *n*-alkanes and *n*-alkenes with odd chain lengths between C_{21}–C_{33} (Figures 2 and 3). The chemical profiles of *n*-alkanes and *n*-alkenes for the entrance bands had more variability between hives (Figure 3), likely due to the fact that entrance bands come into more frequent physical contact with the honey bees than the bands placed in-cover. Bands placed on the outside of hives did not contain compounds at detectable levels (Figure 2), and thus acted as field controls in this study. Differences in the chemical profiles from bands placed in hives at different locations were observed (Figure 3), but were not significant between the location types (rural, suburban, and urban) examined. For the remainder of this study, we chose to focus on the in-cover bands, which have similar chemical profiles to the entrance bands but represent the chemical profile of the hive air with less frequent physical contact, and are practically easier to recover.

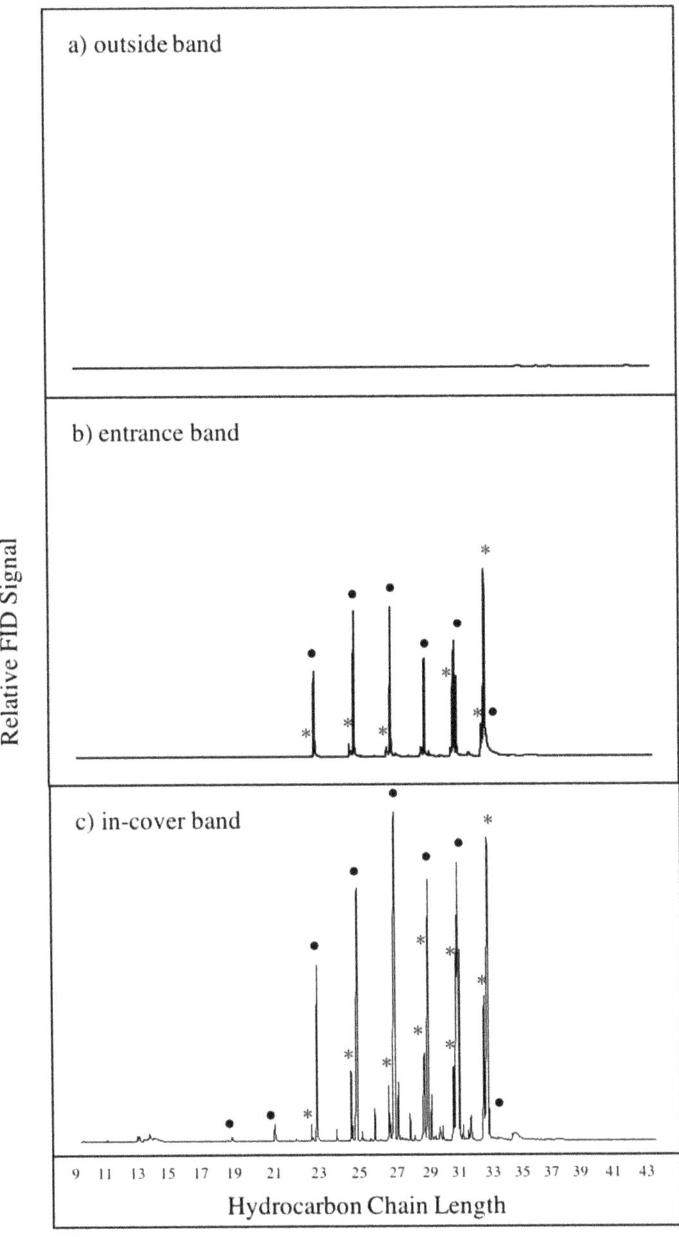

Figure 2. Representative gas chromatograms of compounds extracted from (**a**) an outside band, (**b**) an entrance band, and (**c**) an in-cover band. Alkanes indicated with dots (•) and alkenes indicated with asterisks (*). All bands were deployed in the same hive at Awbury Arboretum on 07/15/2016.

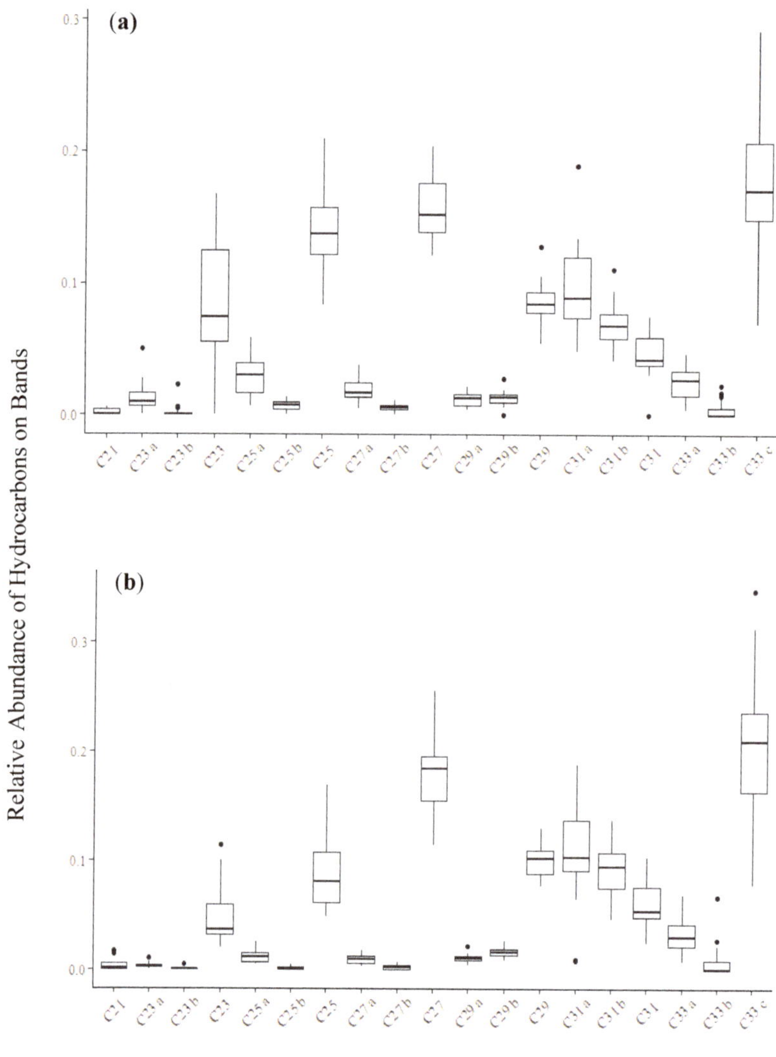

Figure 3. Box and whisker plots illustrating the relative abundance of hydrocarbons for all samples collected from (**a**) the entrance and (**b**) in-cover of hives. Boxes contain the interquartile range—50% of data—with center line indicating the median value of data. Whiskers indicate minimum and maximum values within 1.5 times above the 75th percentile or below the 25th percentile, with dots representing outliers.

3.2. Bands Adsorbed Bee-Associated Chemicals

The predominant compounds detected from the bands (*n*- alkanes and *n*-alkenes with odd chain lengths between C_{21}–C_{33}) are honey bee semiochemicals (Table 1), some of which are known nestmate recognition semiochemicals [28–30]. Four of these compounds—C_{23}- and C_{25}-*n*-alkane, and C_{23}- and C_{25}-*n*-alkene—are also known to be vital for the waggle dance, which relays information about the whereabouts of food [9,31]. Oleic acid, linoleic acid, and α-linolenic acid were also detected in at least half of the samples. These compounds are known to be major constituents of pollen [32] and beeswax [33], and have been detected in worker bees [34,35]. Certain saturated fatty acids and fatty alcohols, previously detected in worker bees [34], drone cocoons [36], or as a constituent of queen retinue pheromone [31], were identified broadly across samples. The alarm pheromone, [Z]-11-eicosenol, which is known to be released by bees, was also observed [31,34]. Glycerol was found in 78% of samples, and is known to participate in the ester biosynthetic pathway in bees [31]. Chrysin is a compound found in honey and was only observed in 13% of samples [37].

Table 1. The percentage of silicone bands placed in honey bee hives that contain honey bee semiochemicals and honey bee-associated compounds organized by compound type.

Compound Type	Chain Lengths	Log K_{ow} Range	% of Bands [a]	Association with Honey Bees
Alkanes	C_{21}–C_{33} odd chain lengths only	10.7–16.6 [b]	44–95	nestmate recognition semiochemical [15,28–30,38] queen tergal gland secretion [38] waggle dance (C_{23} and C_{25} only) [9,31]
Alkenes	C_{23}–C_{33} [b] odd chain lengths only	11.4–16.4 [b]	7–93	nestmate recognition semiochemical [29–31,33] queen tergal gland secretion [38] waggle dance (C_{23} and C_{25} only) [9,31]
Fatty acids	C12:0–C30:0	3.42–13.8	53–97	detected in worker bees (C12:0–C22:0; C26:0–C30:0 even chain lengths) [34,35] detected in *Varroa* destructor (C23:0–C24:0) [35]
Unsaturated fatty acids	C18:1 (oleic acid)	7.64	96	major constituent—beeswax [35] detected in worker bees [34,35]
	C18:2 (linoleic acid)	7.05	81	major constituent—beeswax [31] detected in worker bees [34,35]
	C18:3 (α-linolenic acid)	6.46	48	major constituent—beeswax [31] detected in worker bees [34,35]
Fatty alcohols	C_{16}–C_{32}	6.83–14.10 [b]	36–96	queen retinue pheromone (QRP) (C_{16}) [31] drone cocoon (C_{17}) [36] detected in worker bees (C_{18}–C_{32}) [34]
	C_{19} [c]		40	detected in *Bombus ruderarius* and *Bombus sylvarum* (Hymenoptera, Apidae) [39]
	C_{19} [c,d]		76	detected in *Bombus ruderarius* and *B. sylvarum* (Hymenoptera, Apidae) [39]
	C_{20} [c] [Z]-11-eicosenol		76	alarm pheromone [31,34]
Other	Chrysin	3.52	13	honey, propolis, and beeswax [37]
	Glycerol	-1.76	78	ester biosynthesis in honey bees [31]

[a] Specific percentages are for individual compounds. A range of values corresponds to the range of compounds described. [b] Log K_{ow} values estimated using the Crippen method EPISuite KOWWIN v1.67 estimate (USEPA) (HSDB [40]). [c] Alkenes identified by molecular weight and fragmentation patterns. Exact location of the double bonds is unknown. [d] Alkenes with two double bonds.

3.3. Bands Adsorb Plant-Derived Compounds

Plant-derived compounds were also extracted from bands, as well as the fatty acid C9:0, a common non-selective herbicide (Table 2) [41]. Fatty acids with carbon ranges between C10:0–C22:0 are known to be derived from plants and have been identified in pollen, along with oleic, linoleic, and α-linolenic acid, as previously described [32]. Further, the fatty alcohol 1-tritriacontanol, which has been shown to have a plant origin, was detected [42,43]. Plant-originating allelochemicals were also detected, including benzoic acid and cinnamic acid derivatives [44,45]. In fewer samples, sterols and sugars were observed, likely to originate from pollen and nectar, respectively [46,47]. All compounds identified from bands placed in hives are described in Table S2 of the Supplementary Materials.

Table 2. The percentage of silicone bands placed in honey bee hives that contain plant-associated compounds organized by compound type.

Compound Group	Compounds	Log K_{ow} Range[26]	% of Bands [a]	Associations with Plants
Fatty acids	C10:0–C20:0; C22:0	4.0–9.9	64–97	pollen [32]
	C9:0	3.42	84	nonselective herbicide [41]
Unsaturated fatty acids	C18:1 (oleic acid)	7.64	96	
	C18:2 (linoleic acid)	7.05	81	all major constituents of pollen [32]
	C18:3 (α-linolenic acid)	6.46	48	
Fatty alcohols	C_{33}		13	plant origin [42,43]
Bnzoic and cinnamic acid derivatives	benzoic acid	1.87	79	plant originated allelochemicals [44,45]
	cinnamic acid, p-methoxy	2.68	63	
	cinnamyl cinnamate	3.96	12	
	4-hydroxybenzoic acid	1.58	12	
	hydrocinnamic acid	1.84	10	
	benzyl cinnamate	3.44	9	
	ferulic acid	1.51	7	
	cinnamic acid, 3,4-dihydroxy-	1.15	3	
	benzyl salicylate	4.31 [b]	3	
Sterols	beta-sitosterol (29Δ (5))	9.65 [b]	43	pollen [46]
	stigmasterol (29Δ (5, 22))	9.43	13	pollen [46]
	lanosta-8,24-dien-3-ol, acetate, (2, β)-	11.8 [b]	9	pollen [46]
Sugars	d-mannose	−3.38 [b]	9	nectar [47]
	d-glucose	−2.82	6	nectar [47]
	d-glucopyranose	−2.82	3	nectar [47]
	d-xylose	−2.74 [b]	3	nectar [47]

[a] Specific percentages are for individual compounds. A range of values corresponds to the range of compounds described. [b] Log K_{ow} values estimated using the Crippen method EPISuite KOWWIN v1.67 estimate (USEPA) (HSDB [40]).

4. Discussion

4.1. Implications for Studies of Honey Bee Health

In this approach of using silicone wristbands to passively sample chemicals present in honey bee hives, we found that all but one of the chemicals detected were associated with bees and plants as opposed to human activity or viruses. If these types of compounds were present, our inability to detect them likely arises from them being present at lower abundances, giving a smaller signal that is either masked by the much larger signal of the other abundant compounds present, or that it is outside of our detection limits. Our approach, however, is still of use, as several of the compounds detected on the bands have been implicated in studies concerning honey bee health and nutrition.

Compounds related to honey bee health include those produced by bees such as octadecanoic acid and (9Z)-octadecenoic acid. These two compounds have been observed to be produced more in hives that have large numbers of bees infected with *N. ceranae*, possibly as precursors to short fatty acid chains known to have antibiotic properties [8]. Hexadecanoic acid is a known regulator of fatty acid synthesis in bees and has been observed in lower quantities when a hive is infected [8]. Tricosanoic acid and tetracosanoic acid were both detected on bands, and while not produced by honey bees, are known *Varroa destructor* mite semiochemicals [35]. The alkane tricosane has also been shown to be produced more by bees upon infection by Gram-negative bacteria [14]. Overall, it is known that infection from various pathogens causes honey bees to alter their hydrocarbon profiles [15], which could be examined in future studies by analyzing the chemical profiles of bands placed in hives.

The plant-derived compounds identified in this study could also be used to extrapolate information about hive health, as honey bees obtain all necessary proteins, lipids, and vitamins from pollen [32]. It has been shown that not only the quantity of pollen collected, but also the quality and diversity of compounds available in the pollen, determine hive productivity and longevity [32]. As a result, the detection of numerous compounds from contact with pollen could help researchers investigate the impact of local flora on honey bee hives. These results indicate that deploying silicone bands in honey bee hives may help researchers gain insight into several different aspects of honey bee health, including the regulation of important metabolic pathways, the presence of parasites and pathogens, and the quality of pollen being gathered.

4.2. General Applicability of Bands to Honey Bee Research

The log octanol–water partition coefficients (Log K_{ow}) for compounds in this study range from −3.38 to 16.6, as reported in the Hazardous Substances Data Bank (HSDB) [40]. This exceeds the capacity of most passive samplers currently in use [24], overcoming the major drawback of currently used passive sampling methods in hives that target certain compound groups [16–19]. The diversity in compounds adsorbed by silicone bands thus removes the necessity for multiple passive samplers when the compounds of interest vary significantly in structure and physical and chemical properties. Despite our expectation that this method would primarily target VOCs and SVOCs, non-volatile compounds were consistently seen across samples, arising from direct contact of the bees with the bands. If future implementation of silicone bands in hives chose to sample VOCs and SVOCs only, then an approach to maintain separation between the bees and bands would be needed (e.g., fine wire mesh cage for the bands).

This study also provides a method that allows researchers to quantitatively compare hive chemical profiles without needing to determine compound concentrations in the air of each hive. Previous studies have shown that the consistent adsorption of silicone bands allow samples to be compared without the need for further calculations, provided the dimensions of the silicone sampler and deployment length are the same [25,48]. Silicone bands are more consistent in their compound adsorption ratios than several other passive samplers such as polyurethane foam, urine sampling, and hand wipes [25,26], potentially because the silicone matrix stabilizes compounds until extraction [24].

Unlike SPME fibers, silicone bands can be used for quantitative analysis of compounds, provided that any comparisons are between band samples, as concentrations adsorbed to the bands do not directly represent environmental concentrations [22,24,25]. It has also been shown that concentrations of volatile compounds adsorbed by the silicone matrix remain stable under transport conditions of 30 °C for 7 days and under storage conditions of −20 °C for approximately 6 months [24]. To compare concentrations of chemicals from different compound classes between bands, the specific affinities of the silicone matrix to different compounds would need to be examined. In addition, to connect the chemical compounds detected to the health of the hive, a future study would need to also measure specific infections and diseases, such as *N. ceranae*, the *Varroa* destructor mite, deformed wing virus, and European foulbrood, in addition to chemical compounds of interest. Any future study must be designed to be large enough so that statistical testing between chemicals and health to be made.

4.3. Modifications for Future Studies

One concern with the use of silicone wristbands is the large amount of solvent needed to clean and process the bands; however, in this study, the amount of sample necessary for compounds to be detected via GC–MS and GC–FID was approximately 3×10^{-5} times our original extract per run. This shows that future studies could use fractions of silicone wristbands, rather than the entire band, significantly reducing the amount of solvent needed per sample. Researchers would have to ensure that each section of band being deployed was the same length, width, thickness, and surface area if they wished to compare concentrations of compounds between samples. It would also be necessary to ensure each band is deployed for the exact same duration, as many of the compounds may not reach equilibrium and could exhibit time-dependent concentrations [24].

Since the bands are capable of adsorbing both volatile and non-volatile chemicals, the placement of the bands in the hive will be critical to ensure non-biased quantitative analysis. If researchers are interested in volatile compounds, isolating the bands from the bees using a mesh cage (as previously mentioned) will be essential to obtain non-biased results. If the researcher is interested in non-volatile compounds, such as hydrocarbons or pesticides, they could pin the band flat in front of the entrance of the hive, requiring bees to walk over it when entering and exiting the hive. This approach would make the quantification of chemicals challenging, but it would provide a broad overview of the compounds worker bees are carrying into the hive.

5. Conclusions

Considering the ease with which chemicals in bands can be compared, the minimal disturbance to the hive, and the variety of compounds detectable, using silicone bands to investigate the relationship between chemical compounds and honey bees shows great potential. Further, our results show that bands did not collect detectable compounds from outside of the hive, as no compounds were detected on the outside bands. This contrasts with SPME fibers, which are easily contaminated by background volatiles [16] and are quite expensive. Researchers can use bands as samplers in the open hive environment, as was done in this study, as well as in closed sampling containers. In a closed system, it would be possible to sample the volatile chemicals released by bees or adhered to the surface of bees based on certain castes, age groups, or environmental conditions, without the need for complicated air flow systems or filters. As a result, we believe using silicone band passive samplers provides alternative, flexible, more affordable opportunities to explore the chemical ecology of honey bees and the factors that influence their health, behavior, and survival.

Supplementary Materials: The following are available online at http://www.mdpi.com/2306-7381/7/3/86/s1, Table S1: Locations of hives sampled in this study, Table S2: All compounds identified on bands.

Author Contributions: Conceptualization, R.L.B., A.M., C.M., and H.K.W.; methodology, E.J.B., C.H.W., R.L.B., A.M., C.M., A.M.S., and H.K.W.; formal analysis, E.J.B., C.M., and H.K.W.; data curation, E.J.B., C.M., and H.K.W.; writing—original draft preparation, E.J.B. and H.K.W.; writing—review and editing, all authors; funding acquisition, R.L.B., A.M., C.M., and H.K.W. All authors have read and agreed to the published version of the manuscript.

Funding: This work was supported by funding from an Agilent Applications and Core Technology University Research Grant (no: 3937) to R.L.B., A.M., C.M., and H.K.W. and a Francis Velay Summer Fellow award to E.J.B.

Acknowledgments: The authors acknowledge Julie Brady, Jeff Eckel, Claudia Kent, Don Shump, and Eli St. Amour for access to the hives under their care.

Conflicts of Interest: The authors declare no conflict of interest. A.M is an employee of Agilent Technologies, Inc. and contributed to the conceptualization of this study. The funders had no role in the design of the study; in the collection, analyses, or interpretation of data; in the writing of the manuscript; or in the decision to publish the results.

References

1. Chen, Y.; Schoville, S.; Bonning, B.; Genersch, E. Ecology/Parasites/Parasitoids/Biological control. In *Current Opinion in Insect Science*; Jensen, A., Ed.; Elsevier: Amsterdam, The Netherlands, 2018; Volume 26, pp. 1–154.
2. López-Uribe, M.M.; Simone-Finstrom, M. Special Issue: Honey Bee Research in the US: Current State and Solutions to Beekeeping Problems. *Insects* **2019**, *10*, 22.
3. Klein, A.-M.; Vaissière, B.E.; Cane, J.H.; Steffan-Dewenter, I.; Cunningham, S.A.; Kremen, C.; Tscharntke, T. Importance of pollinators in changing landscapes for world crops. *Proc. R. Soc. B* **2007**, *274*, 303–313. [CrossRef] [PubMed]
4. Ellis, J.D.; Evans, J.D.; Pettis, J. Colony losses, managed colony population decline, and Colony Collapse Disorder in the United States. *J. Apic. Res.* **2010**, *49*, 134–136. [CrossRef]
5. Simon-Delso, N.; San Martin, G.; Bruneau, E.; Minsart., L.A.; Mouret, C.; Hautier, L. Honeybee Colony Disorder in Crop Areas: The Role of Pesticides and Viruses. *PLoS ONE* **2014**, *9*, e103073. [CrossRef] [PubMed]
6. VanEngelsdorp, D.; Evans, J.D.; Saegerman, C.; Mullin, C.; Haubruge, E.; Nguyen, B.K.; Frazier, M.; Frazier, J.; Cox-Foster, D.; Chen, Y.; et al. Colony Collapse Disorder: A Descriptive Study. *PLoS ONE* **2009**, *4*, e6481. [CrossRef]
7. Higes, M.; Meana, A.; Bartolomé, C.; Botías, C.; Martín-Hernández, R.N. *Ceranae* an emergent pathogen for beekeeping. *Environ. Microbiol. Rep.* **2013**, *5*, 17–29. [CrossRef]

8. Broadrup, R.L.; Mayack, C.; Schick, S.J.; Eppley, E.J.; White, H.K.; Macherone, A. Honey bee (*Apis mellifera*) exposomes and dysregulated metabolic pathways associated with *Nosema ceranae* infection. *PLoS ONE* **2019**, *14*, e0213249.
9. Gilley, D.C. Hydrocarbons Emitted by Waggle-Dancing Honey Bees Increase Forager Recruitment by Stimulating Dancing. *PLoS ONE* **2014**, *9*, e105671. [CrossRef]
10. Carroll, M.J.; Duehl, A.J. Collection of volatiles from honeybee larvae and adults enclosed on brood frames. *Apidologie* **2012**, *43*, 715–730. [CrossRef]
11. He, X.J.; Zhang, X.C.; Jiang, W.J.; Barron, A.B.; Zhang, J.H.; Zeng, Z.J. Starving honey bee (*Apis mellifera*) larvae signal pheremonally to worker bees. *Sci. Rep.* **2016**, *6*, 1–9.
12. Slessor, K.N.; Winston, M.L.; Le Conte, Y. Pheromone communication in the honeybee (*Apis mellifera* L.). *J. Chem. Ecol.* **2005**, *31*, 2731–2745. [CrossRef] [PubMed]
13. Duassaubat, C.; Maisonnasse, A.; Alaux, C.; Tchamitchan, S.; Brunet, J.-L.; Plettner, E.; Belzunces, L.P.; Le Conte, Y. Nosema spp. Infection Alters Pheromone Production in Honey Bees (*Apis mellifera*). *J. Chem. Ecol.* **2010**, *36*, 522–525. [CrossRef] [PubMed]
14. Richard, F.; Holt, H.L.; Grozinger, C.M. Effects of immunostimulation on social behavior, chemical communication and genome-wide gene expression in honey bee workers (*Apis mellifera*). *BMC Genom.* **2012**, *13*. [CrossRef] [PubMed]
15. Murray, Z.L.; Keyzers, R.A.; Barbieri, R.F.; Digby, A.P.; Lester, P.J. Two pathogens change cuticular hydrocarbon profiles but neither elicit a social behavioural change in infected honey bees, *Apis mellifera* (Apidae: Hymenoptera). *Austral Entomol.* **2016**, *55*, 147–153. [CrossRef]
16. Torto, B.; Carroll, M.J.; Duehl, A.; Fombong, A.T.; Gozansky, T.K.; Nazzi, F.; Soroker, V.; Teal, P.E.A. Standard methods for chemical ecology research in Apis mellifera. *J. Apic. Res.* **2013**, *52*, 1–34. [CrossRef]
17. Murcia-Morales, M.; Van der Steen, J.J.; Vejsnæs, F.; Díaz-Galiano, F.J.; Flores, J.M.; Fernández-Alba, A.R. APIStrip, a new tool for environmental contaminant sampling through honeybee colonies. *Sci. Total Environ.* **2020**, 138948. [CrossRef]
18. Field, J.A.; Nickerson, G.; James, D.D.; Heider, C. Determination of Essential Oils in Hops by Headspace Solid-Phase Microextraction. *J. Agric. Food Chem.* **1996**, *44*, 1768–1772. [CrossRef]
19. Agelopoulos, N.G.; Pickett, J.A. Headspace Analysis in Chemical Ecology: Effects of Different Sampling Methods on Ratios of Volatile Compounds Present in Headspace Samples. *J. Chem. Ecol.* **1998**, *24*, 1161–1172. [CrossRef]
20. Roberts, D.D.; Pollien, P.; Milo, C. Solid-Phase Microextraction Method Development for Headspace Analysis of Volatile Flavor Compounds. *J. Agric. Food Chem.* **2000**, *48*, 2430–2437. [CrossRef]
21. Gilley, D.C.; Degrandi-Hoffman, G.; Hooper, J.E. Volatile compounds emitted by live European honey bee (*Apis mellifera* L.) queens. *J. Insect Physiol.* **2006**, *52*, 520–527. [CrossRef]
22. O'Connell, S.G.; Kincl, L.D.; Anderson, K.A. Silicone Wristbands as Personal Passive Samplers. *Environ. Sci. Technol.* **2014**, *48*, 3327–3335. [CrossRef] [PubMed]
23. Donald, C.E.; Scott, R.P.; Blaustein, K.L.; Halbleib, M.L.; Sarr, M.; Jepson, P.C.; Anderson, K.A. Silicone wristbands detect individuals' pesticide exposures in West Africa. *R. Soc. Open Sci.* **2016**, *3*, 160433. [CrossRef] [PubMed]
24. Anderson, K.A.; Points, G.L., III; Donald, C.E.; Dixon, H.M.; Scott, R.P.; Wilson, G.; Tidwell, L.G.; Hoffman, P.D.; Herbstman, J.B.; O'Connelll, S.G. Preparation and performance features of wristband samplers and considerations for chemical exposure assessment. *J. Expo. Sci. Environ. Epidemiol.* **2017**, *27*, 551–559. [CrossRef] [PubMed]
25. Hammel, S.C.; Hoffman, K.; Webster, T.F.; Anderson, K.A.; Stapleton, H.M. Measuring Personal Exposure to Organophosphate Flame Retardants Using Silicone Wristbands and Hand Wipes. *Environ. Sci. Technol.* **2016**, *50*, 4483–4491. [CrossRef] [PubMed]
26. Hammel, S.C.; Phillips, A.L.; Hoffman, K.; Stapleton, H.M. Evaluating the Use of Silicone Wristbands to Measure Personal Exposure to Brominated Flame Retardants. *Environ. Sci. Technol.* **2018**, *52*, 11875–11885. [CrossRef] [PubMed]
27. Attari, S.G.; Bahrami, A.; Shahna, F.G.; Heidari, M. Solid-phase microextraction fiber development for sampling and analysis of volatile organohalogen compounds in air. *J. Environ. Health Sci. Eng.* **2014**, *12*. [CrossRef] [PubMed]

28. Dani, F.R.; Jones, G.R.; Corsi, S.; Beard, R.; Pradella, D.; Turillazzi, S. Nestmate Recognition Cues in the Honey Bee: Differential Importance of Cuticular Alkanes and Alkenes. *Chem. Senses* **2005**, *30*, 477–489. [CrossRef]

29. Strachecka, A.; Borsuk, G.; Paleolog, J.; Olszewski, K.; Bajda, M.; Chobotow, J. Body-surface Compounds in Buckfast and Caucasian Honey Bee Workers (*Apis mellifera*). *J. Apic. Sci.* **2014**, *58*, 5–15. [CrossRef]

30. Kather, R.; Drijfhout, F.P.; Martin, S.J. Evidence for colony-specific differences in chemical mimicry in the parasitic mite *Varroa destructor*. *Chemoecology* **2015**, *25*, 215–222. [CrossRef]

31. Trhlin, M.; Rajchard, J. Chemical communication in the honeybee (*Apis mellifera* L.): A review. *Vet. Med.* **2011**, *56*, 265–273. [CrossRef]

32. Mărgăoan, R.; Mărghitaş, L.A.; Dezmirean, D.S.; Dulf, F.V.; Bunea, A.; Socaci, S.A.; Bobiş, O. Predominant and Secondary Pollen Botanical Origins Influence the Carotenoid and Fatty Acid Profile in Fresh Honeybee-Collected Pollen. *J. Agric. Food Chem.* **2014**, *62*, 6306–6316. [CrossRef] [PubMed]

33. Buchwald, R.; Breed, M.D.; Bjostad, L.; Hibbard, B.E.; Greenberg, A.R. The role of fatty acids in the mechanical properties of beeswax. *Apidologie* **2009**, *40*, 585–594. [CrossRef]

34. Teerawanichpan, P.; Robertson, A.J.; Qiu, X. A fatty acyl-CoA reductase highly expressed in the head of honey bee (*Apis mellifera*) involves biosynthesis of a wide range of aliphatic fatty alcohols. *Insect Biochem. Mol. Biol.* **2010**, *40*, 641–649. [CrossRef] [PubMed]

35. Zalewski, K.; Zaobidna, E.; Żółtowska, K. Fatty acid composition of the parasitic mite *Varroa destructor* and its host the worker prepupae of *Apis mellifera*. *Phys. Entom.* **2016**, *41*, 31–37. [CrossRef]

36. Ali, L.; Ahmad, R.; Rehman, N.U.; Khan, A.L.; Hassan, Z.; Rizvi, T.S.; Al-Harrasi, A.; Shinwari, Z.K.; Hussain, J. A New Cyclopropyl-Triterpenoid from *Ochradenus arabicus*. *Helv. Chim. Acta* **2016**, *98*, 1240–1244. [CrossRef]

37. Vit, P.; Soler, C.; Tomás-Barberán, F.A. Profiles of phenolic compounds of *Apis mellifera* and *Melipona* spp. honeys from Venezuela. *Z. Lebensm. Unters. Forsch. A* **1997**, *204*, 43–47. [CrossRef]

38. Okosun, O.O.; Yusuf, A.A.; Crewe, R.M.; Pirk, C.W.W. Effects of age and Reproductive Status on Tergal Gland Secretions in Queenless Honey bee Workers, Apis mellifera scutellata and A. m. capensis. *J. Chem. Ecol.* **2015**, *41*, 896–903. [CrossRef]

39. Terzo, M.; Urbanova, K.; Valerova, I.; Rasmont, P. Intra and interspecific variability of the cephalic labial glands' secretions in male bumblebees: The case of *Bombus (Thoracobombus) ruderarius* and *B. (Thoracobombus) sylvarum* [Hymenoptera, Apidae]. *Apidologie* **2005**, *36*, 85–96. [CrossRef]

40. Hazardous Substances Data Bank (HSDB): A TOXNET Database. Available online: https://toxnet.nlm.nih.gov/newtoxnet/hsdb.htm (accessed on 26 March 2019).

41. Coleman, R.; Penner, D. Organic Acid Enhancement of Pelargonic Acid. *Weed Tech.* **2008**, *22*, 38–41. [CrossRef]

42. Mishra, P.K.; Sing, N.; Ahmad, G.; Dube, A.; Maurya, R. Glycolipids and other constituents from *Desmodium gangeticum* with antileishmanial and immunomodulatory activities. *Bioorg. Med. Chem. Lett.* **2005**, *15*, 4543–4546. [CrossRef]

43. Kamboj, A.; Pooja, A.; Saluja, A.K. Isolation and Characterization of Bioactive Compounds from the Petroleum Ether Extracts of Leaves of *Xanthium Strumarium Linn*. *BioMedRx* **2013**, *1*, 235–238.

44. Yoneyama, K.; Natsume, M. Phenolic Compounds. In *Comprehensive Natural Products II*; Mander, L., Liu, H.-W., Eds.; Elsevier: Kidlington, UK, 2010; Volume 1, pp. 539–558.

45. Widhalm, J.R.; Dudareva, N.A. A Familiar Ring to It: Biosynthesis of Plant Benzoic Acids. *Mol. Plant* **2015**, *8*, 83–97. [CrossRef] [PubMed]

46. Rasmont, P.; Regali, A.; Ings, T.C.; Lognay, G.; Baudart, E.; Marlier, M.; Delcarte, E.; Viville, P.; Marot, C.; Falmagne, P.; et al. Analysis of Pollen and Nectar of *Arbutus unedo* as a Food Source for *Bombus terrestris* (Hymenoptera: Apidae). *J. Econ. Entomol.* **2005**, *98*, 656–663. [CrossRef] [PubMed]

47. Nicolson, S.W. Bee Food: The Chemistry and Nutritional Value of Nectar, Pollen and Mixtures of the Two. *Afr. Zool.* **2011**, *46*, 197–204. [CrossRef]

48. Aerts, R.; Joly, L.; Szternfeld, P.; Tsilikas, K.; De Cremer, K.; Castelain, P.; Aerts, J.M.; Van Orshoven, J.; Somers, B.; Hendrickx, M.; et al. Silicone Wristband Passive Samplers Yield Highly Individualized Peticide Residue Exposure Profiles. *Environ. Sci. Technol.* **2018**, *52*, 298–307. [CrossRef]

Communication

The Honeybee Gut Mycobiota Cluster by Season versus the Microbiota which Cluster by Gut Segment

Jane Ludvigsen, Åsmund Andersen, Linda Hjeljord and Knut Rudi *

Faculty of Chemistry, Biotechnology and Food Science, Norwegian University of Life Sciences, 1430 Ås, Norway; jane.ludvigsen@nmbu.no (J.L.); aan@mycoteam.no (Å.A.); linda.hjeljord@nmbu.no (L.H.)
* Correspondence: knut.rudi@nmbu.no

Abstract: Honeybees represent one of the most important insect species we have, particularly due to their pollinating services. Several emerging fungal and bacterial diseases, however, are currently threatening honeybees without known mechanisms of pathogenicity. Therefore, the aim of the current work was to investigate the seasonal (winter, spring, summer, and autumn) fungal and bacterial distribution through different gut segments (crop, midgut, ileum, and rectum). This was done from two hives in Norway. Our main finding was that bacteria clustered by gut segments, while fungi were clustered by season. This knowledge can therefore be important in studying the epidemiology and potential mechanisms of emerging diseases in honeybees, and also serve as a baseline for understanding honeybee health.

Keywords: honeybee; gut microbiota; gut mycobiota; season

1. Introduction

The importance of the gut microbiota (GM) in regulating honeybee (*Apis mellifera*) health has become increasingly evident in recent years. Regarding nutritional and immunological interactions with the host, the bacterial proportion of the GM is most frequently studied [1–6], but studies investigating the fungal part of the microbiota and how some fungi interact with honeybee pathogens have recently emerged [7]. In humans, the fungi part of the GM has recently been linked to human infant development [8].

The GM of honeybees harbor a specific set of 8–9 bacterial groups, which all seem to have evolved and adapted to a life in the guts of honeybees, and they are only found in honeybees or in closely related bees such as bumble bees [9]. These bacterial groups aid in nutritional breakdown of pollen and nectar, interact with the immune system, and contribute to pathogen defense in the gut [10]. Likewise, specific fungi show interactions with pathogens in honeybees [11,12], but little is currently known about the role of fungi in the GM of honeybees.

The honeybee gastrointestinal tract (GI) can be divided into four main parts: crop, midgut, ileum, and rectum. These four parts have been shown to harbor unique bacterial species for which metabolic properties have been elucidated [13]. Although detailed information about each bacterial species community changes, little information is available about how the bacterial community changes in regard to outer stimuli. Only a few studies have described how the GM composition changes according to season [14,15] and developmental stages of the host [16,17]. The fact that specific bacteria change during the season is an indication that diet contributes to variation in composition. Yun et al. 2018 [13] found that foragers harbor a different set of fungi than that of nurse bees, indicating that diet is a source of variation in the fungal composition as well. They also found that queen bees carry an overload of one type of fungi (*Zygosaccharomyces*), which is different from nurse bees, indicating that the fungal part of the microbiota, as seen for the bacterial part, has adapted to different lifestyles, which is reflected in the overall fungal microbiota [13].

Citation: Ludvigsen, J.; Andersen, Å.; Hjeljord, L.; Rudi, K. The Honeybee Gut Mycobiota Cluster by Season versus the Microbiota which Cluster by Gut Segment. *Vet. Sci.* 2021, *8*, 4. https://doi.org/10.3390/vetsci8010004

Received: 24 November 2020
Accepted: 26 December 2020
Published: 31 December 2020

Publisher's Note: MDPI stays neutral with regard to jurisdictional claims in published maps and institutional affiliations.

All previous studies have investigated the microbiota of the total GI (crop to rectum or midgut to rectum), resulting in a lack of information about the microbiota of different gut parts under these scenarios. Since some honeybee pathogens are gut part specific, information about the variation in the microbiota composition in different gut parts is crucial for understanding the microbiota–pathogen–host dynamics in more detail.

Here, we investigate both the bacterial and fungal parts of the GM (by gut parts)– crop, midgut, ileum, and rectum—throughout an entire season (longitudinal), from March until November in adult honeybee workers. Our results can aid in understanding specific gut part-specific interactions and help in the design of later in-depth functional and metabolic studies.

2. Materials and Methods

2.1. Sampling

The bees were sampled from two neighboring hives located at the Norwegian University of Life Sciences, As, Norway, across seven time points representing before, during and after foraging season. Ten bees where picked randomly from the frame closest to the opening of both hives to represent foragers in April, June, July, and August. In March and November, the bees are not foraging but are clustered together to keep warm, and thus ten bees from both hives were picked from the top of the formed cluster. All bees represent adult bees. All experiments were conducted following Norwegian rules for studies on honeybees [18].

2.2. Gut Dissection

Bees were sampled and put on ice to induce chill-coma before dissection. Then, the gut was removed from the bee after we had sterilized the bee on the outside by washing it in 50% ethanol. By pulling out the stinger, the gut from midgut to rectum was removed. The crop was dissected out separately with sterile dissecting tools. Unfortunately, we did not collect crops from bees sampled in March. The gut was cut into its respective parts under a dissection microscope and on sterile microscopy slides (washed with 70% ethanol and 1:10 chlorine) by cutting at the transition areas between the different gut parts (Figure S1). The dissection was performed in a drop of PCR water (VWR, Radnor, PA, USA) and the Malpighian tubes were cleaned off the midgut and left attached to the ileum part. The different gut parts were collected in tubes prefilled with a bead matrix consisting of 0.2 g of each < 106 μM acid-washed glass beads, 0.425–0.600 mm glass beads, and 2 2.5–3.5 mm acid-washed glass beads (Sigma-Aldrich, Darmstadt, Germany) and stool transport and recovery (S.T.A.R) buffer (Roche, Mannheim, Germany). For midguts and rectum samples, 300 μL of STAR buffer were used and 200 μL were used for crop and ileum. The dataset contained four gut parts (crop, midgut, ileum, and rectum) from 20 bees over seven months (except from March, which only contained midgut, ileum, and rectum) for a total of 540 samples. The samples were frozen at −20 degrees before DNA extraction.

2.3. DNA Extraction

DNA was extracted from a total of 540 samples, with 484 samples yielding sufficient DNA for further processing. Mechanical lysis was performed using FastPrep (MP Biomedicals, Santa Ana, CA, USA) at 1800 rpm for 40 s, two times, with a 5 min cool-down between runs. The samples were then centrifuged for 5 min at 13,000 rpm and 50 μL supernatant was transferred to a 96-well plate for DNA extraction using the MagTM mini kit (LGC, Middlesex, UK) following manufacturers recommendations. The extraction was performed on the KingFisherTM Flex Magnetic Particle Processor, (Thermo ScientificTM, Waltham, MA, USA). The extracted DNA was frozen at −20 degrees before subsequent PCR/qPCR and Illumina sequencing.

2.4. qPCR

Quantification of bacteria and fungi in different gut parts across seasons was performed using qPCR assays targeting the 16S rRNA gene for bacteria and the ITS1 part of the fungal rRNA. Primers targeting the vitellogenin gene of honeybees were used to normalize for possible differences in gut size. Primers used in this study are listed in Table 1. qPCR was performed on LightCycler 480 II (Roche, Mannheim, Germany) using 0.2 µM of forward and reverse primers, 5× HOT FIREPol® EvaGreen qPCRMix Plus (Solis BioDyne, Tartu, Estonia) in 1× concentration, 5 µL gDNA, and adding nuclease-free water (VWR, Radnor, PA, USA) in a total volume of 20 µL. Nuclease-free water (VWR, Radnor, PA, USA) was used as a negative control. PCR conditions for 40 cycles were activated for 15 min at 95 °C, annealing for 30 s at 55 °C and 54 °C for 16S/ITS1 and vitellogenin, respectively, and elongation for 45 s and 30 s for 16S/ITS and vitellogenin, respectively at 72 °C.

Table 1. Primers applied in this work.

Primer	Target	Sequence	Reference
PRK314F	16S rRNA		[19]
PRK806R	16S rRNA		[19]
BITS	ITS1	ACCTGCGGARGGATCA	[20]
B58S3	ITS1	GAGATCCRTTGYTRAAAGTT	[20]
Vitellogenin F	Vitellogenin	GTTGGAGAGCAACATGCAGA	[21]
Vitellogenin R	Vitellogenin	TCGATCCATTCCTTGATGGT	[21]

Relative copy numbers for 16S rRNA and ITS1 genes were calculated based on standard curves generated from Ampure® XP (Beckham coulter, Brea, CA, USA) purified PCR amplified targets, which were quantified using Qubit® dsDNA HS assay kit (Life technologies, Carlsbad, CA, USA), both methods performed according to the manufacturer's recommendations. Standard curves were run using 5× HOT FIREPol® Blend Mastermix Ready to Load (Solis BioDyne, Tartu, Estonia) in 1× concentration, with 0.2 µM of forward and reverse primers and 1 µL of gDNA in a total volume of 25 µL. PCR conditions for 30 cycles were as described above for qPCR with an additional 7 min final elongation step at 72 °C.

2.5. Illumina Sequencing

Illumina sequencing of the 16S rRNA gene and the ITS1 region was performed using the same primers as described above (table primers), and we used the same library preparation methodology as described in [22]. For the initial PCR conditions, we used the same conditions as for preparation of standard curves, but we only ran 25 cycles for 16S rRNA gene compared to 30 cycles for the ITS1. We indexed the ITS primers with 16 forward and 36 reverse indexes (table ITS index primer sequences). Pooling and preparation of the Illumina library was performed following the 16S metagenomic sequencing library preparation protocol (Illumina, San Diego, CA, USA). Quantification of pooled library was performed by ddPCR, BIO-RAD QX 200™ droplet reader (BioRad, Hercules, CA, USA) and diluted to 7 pM and sequenced on MiSeq using v3 reagents (Illumina, San Diego, CA, USA).

2.6. Data Analysis

qPCR copy numbers for all three genes were calculated using SciencePrimer copy-number calculator (Primer 2017), and 16S and ITS were normalized to vitellogenin copy-numbers. Samples in each group that fell outside the 1/3 quartile of the median were removed as outliers.

Illumina fastq files were analyzed as described in [22]. In large, dereplicated, filtered, and generated OTUs in USEARCH v 8, then rarefactioned to 4000 sequences and calculated in QIIME for alpha and beta diversity. Identification of bacterial taxa was done using Greengenes V, and the UNITE (2019) database was used for blast searches for fungal OTUs.

All statistical analyses and plots were conducted in R (v. 3.5.1). Statistical significances were tested using the nonparametric Wilcoxon signed-rank test, with a significance threshold of $p < 0.05$. Dimension reduction of multidimensional data was done using nonmetric multidimensional scaling (NMDS). These analyses were performed using the Phyloseq package (v 1.24.2) in R.

3. Results

The relative abundances of bacterial and fungal taxa with respect to gut segment and season are presented in Figures S2 and S3, in addition to Blast-based taxonomic assignments of fungi (Table S1).

3.1. Higher Bacterial Abundance than Fungal Abundance in All Gut Parts across the Entire Season

Our dataset enabled us to compare bacterial and fungal abundance in different gut parts and track changes in abundance across the season (before, during, and after foraging). There was a higher bacterial abundance in all gut parts compared to fungal abundance, and this increase became more apparent towards the rear end of the gut (Figure 1). In the rectum, the bacterial load was magnitudes higher on average than the fungal load. Our results are consistent with previous findings of higher bacterial load in the hind gut (ileum and rectum) than in the midgut, as we found that the midgut had the lowest bacterial load across all months.

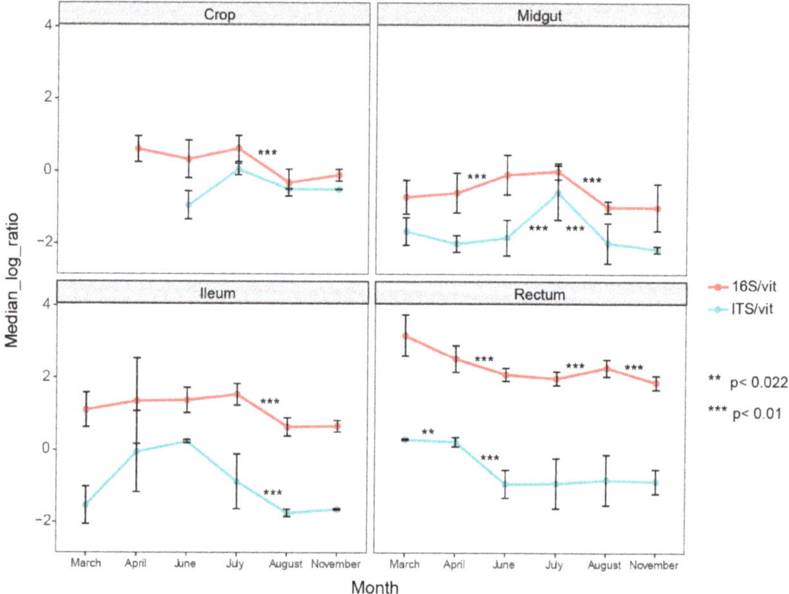

Figure 1. qPCR results showing median relative abundance of bacteria (16S) and fungi (ITS1) across months per gut part. Both 16S and ITS1 gene DNA are normalized against vitellogenin gene DNA. The vitellogenin gene copies were used as a proxy for the weight of the tissue. Error bars represent standard deviations. The asterisks represent statistical significance using the Wilcoxon signed-rank test.

Interestingly, the fungal abundance in the midgut and ileum was substantially influenced by season, as the copy-numbers fluctuated between neighboring months, creating a peak load in these two gut parts in June and July, respectively (Figure 1). The rectum displayed a different trend, with higher fungal load early in the season. The midgut harbored the lowest fungal load compared to the other gut parts (Figure 1).

3.2. Observed Fungal Species Diversity Peaks during Foraging Months

Numbers of observed fungal species displayed an increasing trend from April, which peaked in June (4×) and returned to baseline in August (Figure 2). This peak in fungal diversity was observed in midgut, ileum, and rectum, and the highest number was detected in the midgut. The crop showed a less clear peak in July, but the trend was present there as well. The observed fungal diversity due to these peaks was 4× higher than the number of observed bacterial species.

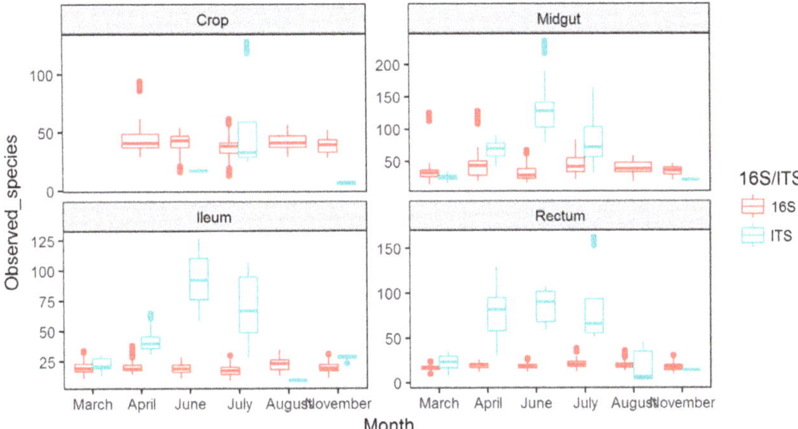

Figure 2. Observed species for bacteria (16S) and fungi (ITS) shown across months for different gut part. The missing data in the figure are due to lack of information for those datapoints. This particularly relates to fungi in crop, as the levels were very low.

For observed bacterial species diversity, the numbers were 2× higher in the crop than in the other gut parts, although a slight increase in diversity was seen in the late foraging months in midgut samples (Figure 2).

3.3. Fungal Communities Cluster by Months and Bacterial Communities Cluster by Gut Parts

NMDS plots revealed that the fungal community is highly influenced by season as the different fungal communities each month cluster apart, i.e., fungal communities were not much different between gut parts and the different gut part communities were similar in regard to different months (Figure 3A). This contrasts with the bacterial communities, which clustered by gut parts, i.e., bacterial communities showed highly gut part-specific communities, which were not strongly influenced by season (Figure 3B).

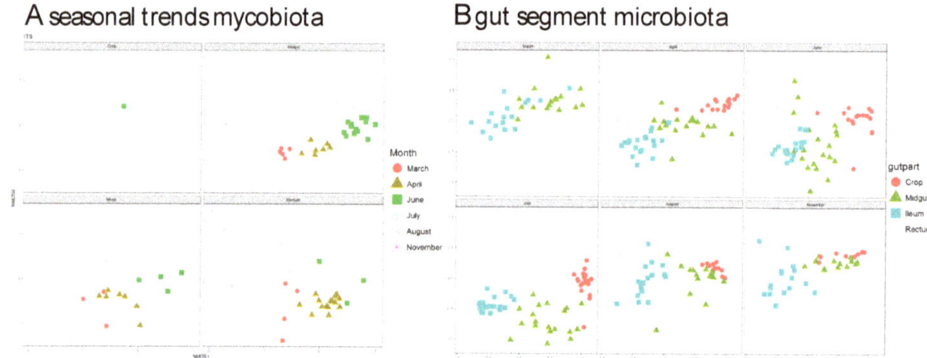

Figure 3. Nonmetric multidimensional scaling (NMDS) plots showing clustering of samples from different gut-parts based on (**A**) mycobiota and microbiota (**B**) composition. The stress values in all cases were < 0.2, suggesting a proper dimension reduction. The reason to present panels connected to gut segment for the mycobiota and season for the microbiota is to highlight the main differences visually.

4. Discussion

Our main finding was that the gut mycobiota changed with season, while the microbiota was mostly affected by the gut segment. This finding can have major implications for understanding the interaction between bacteria and fungi in honeybee health and disease.

Most diseases have a seasonal trend, but the underlying factors determining the seasonality remain unknown [23]. In temperate climates, honeybees stay inside the hive in winter until the weather is warm enough to fly out. This happens around May, and they continue to forage until late August. Our data suggest that the diversity of fungi species is elevated during foraging months, which was demonstrated for all gut parts. Higher species diversity in foragers has previously been shown by Yun et al. 2018 [13], suggesting that only a few fungi are endemic in honeybees and that a large amount of transient fungi might influence honeybees during summer months. Additionally, we saw that during early months there was more yeast-like fungi in midgut, ileum, and rectum. This could be due to bees being inside the hive for a long time in damp conditions. This is congruent with Yun et al. (2018), who found more yeast in nurse bees, which only stay inside the hive as well. The interplay between yeast, fungi, and bacteria could play a role in managing the yeast load in foraging bees, as we saw that the yeast load decreased as more environmental yeast and bacteria were present. Suggestions of this are found in germ-free mice that are highly susceptible to yeast infections [24], and in honeybees, yeast load seems to correlate with Nosema infection, which often peaks early and late in the year [11].

Mortality of honeybees also shows a clear seasonal trend [25]. Therefore, a factor that has not yet been considered in the mortality models is the major change in the fungal population associated with honeybees. Fungal diseases are commonly opportunistic and difficult to trace [26]. The major seasonal trends discovered here highlight the challenge in understanding fungal associated diseases.

In contrast to fungi, bacteria showed a clear association with gut segments through the season. This indicates that bacteria have a crucial role in maintaining honeybee health [27]. Dissection of the different gut parts, as done in this study, can reveal patterns not possible to detect using whole GI tracts. We found in our study that the midgut was more influenced by season than were ileum and rectum parts, which are usually what most studies investigate [14]. The hindgut comprises >90% of the total bacterial load in honeybees and thus will reflect the variation if the whole GI tract is used, and valuable information about which bacteria might be possible transient bacteria will be lost. There were some bacteria in our study that were mostly in the crop and midgut only in foraging months, which thus could be transient bacteria and not part of the endemic honeybee gut microbiota

This type of detailed (both gut part and different foraging months) information can shed light on previous suggestions that these bacteria are a part of the normal flora because they are commonly found in most bees. In addition to obligate pathogens, the absence of health-promoting bacteria can also lead to disease. Such diseases, however, would be much more difficult to detect, as they cannot be linked to specific bacteria [28]. Thus, diseases connected to lack of function could also have a potential role in explaining honeybee diseases, such as the colony collapse disorder (CCD). Thus, the fight towards the obligate pathogens could lead to simultaneous eradication of bacteria that have essential functions, such as vitamin production [29].

A limitation of our study, however, is that we did not consider the microbiota in the mouth part, which could have a substantial influence on both the micro- and mycobiota in the honeybee gastrointestinal tract [30]. Nor did we do specific measurements of diet [31]. Our study is further limited in that we only investigated one location and two hives. Further studies are therefore needed to generalize our findings. Our study also illustrates experimental issues that need to be considered in honeybee studies. Both season and gut segment had a major impact on the gut myco-/microbiota. Without taking the spatiotemporal information into account, misleading conclusions can be drawn related to the association of the honeybee gut myco-/microbiota with health and disease.

5. Conclusions

In conclusion, we have shown major differences connected to gut segment and seasonal associations of the honeybee gut myco-and microbiota. This knowledge can have major implications for honeybee health and disease.

Supplementary Materials: The following are available online at https://www.mdpi.com/2306-7381/8/1/4/s1, Figure S1: Dissection scheme (a) whole gut, (b) the transition between the middle stomach and ileum, and (c) transition between the ileum and the rectum, Figure S2: Relative abundance of the 20 most abundant fungal OTUs across months per gut part, Figure S3: Relative abundance of the bacterial OTUs (with abundance > 1%) across months per gut part, Table S1: Taxonomic assignments of fungal OTUs by Blast.

Author Contributions: Conceptualization, J.L. and K.R.; methodology, Å.A., J.L., and L.H.; writing—review and editing, all authors. All authors have read and agreed to the published version of the manuscript.

Funding: This research received no external funding.

Institutional Review Board Statement: Not applicable.

Informed Consent Statement: Not applicable.

Data Availability Statement: Data is contained within the article or supplementary material.

Acknowledgments: We thank NMBU for providing experimental and financial support.

Conflicts of Interest: The authors declare no conflict of interest.

References

1. Raymann, K.; Bobay, L.M.; Moran, N.A. Antibiotics reduce genetic diversity of core species in the honeybee gut microbiome. *Mol. Ecol.* **2018**, *27*, 2057–2066. [CrossRef] [PubMed]
2. Engel, P.; Kwong, W.K.; McFrederick, Q.; Anderson, K.E.; Barribeau, S.M.; Chandler, J.A.; Cornman, R.S.; Dainat, J.; de Miranda, J.R.; Doublet, V.; et al. The Bee Microbiome: Impact on Bee Health and Model for Evolution and Ecology of Host-Microbe Interactions. *MBio* **2016**, *7*, e02164-15. [CrossRef] [PubMed]
3. Kwong, W.K.; Engel, P.; Koch, H.; Moran, N.A. Genomics and host specialization of honey bee and bumble bee gut symbionts. *Proc. Natl. Acad. Sci. USA* **2014**, *111*, 11509–11514. [CrossRef] [PubMed]
4. Cilia, G.; Fratini, F.; Tafi, E.; Mancini, S.; Turchi, B.; Sagona, S.; Cerri, D.; Felicioli, A.; Nanetti, A. Changes of Western honey bee Apis mellifera ligustica (Spinola, 1806) ventriculus microbial profile related to their in-hive tasks. *J. Apic. Res.* **2020**, 1–5. [CrossRef]
5. du Rand, E.E.; Stutzer, C.; Human, H.; Pirk, C.W.W.; Nicolson, S.W. Antibiotic treatment impairs protein digestion in the honeybee, Apis mellifera. *Apidologie* **2020**, *51*, 94–106. [CrossRef]

6. Ricigliano, V.A.; Anderson, K.E. Probing the Honey Bee Diet-Microbiota-Host Axis Using Pollen Restriction and Organic Acid Feeding. *Insects* **2020**, *11*, 291. [CrossRef]
7. Davis, J.K.; Aguirre, L.A.; Barber, N.A.; Stevenson, P.C.; Adler, L.S. From plant fungi to bee parasites: Mycorrhizae and soil nutrients shape floral chemistry and bee pathogens. *Ecology* **2019**, *100*, e02801. [CrossRef]
8. Schei, K.; Avershina, E.; Oien, T.; Rudi, K.; Follestad, T.; Salamati, S.; Odegard, R.A. Early gut mycobiota and mother-offspring transfer. *Microbiome* **2017**, *5*, 107. [CrossRef]
9. Kwong, W.K.; Moran, N.A. Evolution of host specialization in gut microbes: The bee gut as a model. *Gut Microbes* **2015**, *6*, 214–220 [CrossRef]
10. Raymann, K.; Moran, N.A. The role of the gut microbiome in health and disease of adult honey bee workers. *Curr. Opin. Insect Sci.* **2018**, *26*, 97–104. [CrossRef]
11. Ptaszyńska, A.A.; Paleolog, J.; Borsuk, G. Nosema ceranae Infection Promotes Proliferation of Yeasts in Honey Bee Intestines *PLoS ONE* **2016**, *11*, e0164477. [CrossRef] [PubMed]
12. El Khoury, S.; Rousseau, A.; Lecoeur, A.; Cheaib, B.; Bouslama, S.; Mercier, P.-L.; Demey, V.; Castex, M.; Giovenazzo, P.; Derome, N. Deleterious Interaction Between Honeybees (Apis mellifera) and its Microsporidian Intracellular Parasite Nosema ceranae Was Mitigated by Administrating Either Endogenous or Allochthonous Gut Microbiota Strains. *Front. Ecol. Evol.* **2018**, *6*. [CrossRef]
13. Kešnerová, L.; Mars, R.A.T.; Ellegaard, K.M.; Troilo, M.; Sauer, U.; Engel, P. Disentangling metabolic functions of bacteria in the honey bee gut. *PLoS Biol.* **2017**, *15*, e2003467. [CrossRef] [PubMed]
14. Kešnerová, L.; Emery, O.; Troilo, M.; Liberti, J.; Erkosar, B.; Engel, P. Gut microbiota structure differs between honeybees in winter and summer. *ISME J.* **2020**, *14*, 801–814. [CrossRef]
15. Ludvigsen, J.; Rangberg, A.; Avershina, E.; Sekelja, M.; Kreibich, C.; Amdam, G.; Rudi, K. Shifts in the Midgut/Pyloric Microbiota Composition within a Honey Bee Apiary throughout a Season. *Microbes Environ.* **2015**, *30*, 235–244. [CrossRef] [PubMed]
16. Yun, J.-H.; Jung, M.-J.; Kim, P.S.; Bae, J.-W. Social status shapes the bacterial and fungal gut communities of the honey bee. *Sci. Rep.* **2018**, *8*, 2019. [CrossRef]
17. Kapheim, K.M.; Rao, V.D.; Yeoman, C.J.; Wilson, B.A.; White, B.A.; Goldenfeld, N.; Robinson, G.E. Caste-Specific Differences in Hindgut Microbial Communities of Honey Bees (Apis mellifera). *PLoS ONE* **2015**, *10*, e0123911. [CrossRef]
18. Ministries, N. *National Pollinator Strategy*; Department for Environment, Food & Rural Affairs: London, UK, 2018.
19. Yu, Y.; Lee, C.; Kim, J.; Hwang, S. Group-specific primer and probe sets to detect methanogenic communities using quantitative real-time polymerase chain reaction. *Biotechnol. Bioeng.* **2005**, *89*, 670–679. [CrossRef]
20. Bokulich, N.A.; Mills, D.A. Improved selection of internal transcribed spacer-specific primers enables quantitative, ultra-high-throughput profiling of fungal communities. *Appl. Environ. Microbiol.* **2013**, *79*, 2519–2526. [CrossRef]
21. Amdam, G.V.; Norberg, K.; Fondrk, M.K.; Page, R.E., Jr. Reproductive ground plan may mediate colony-level selection effects on individual foraging behavior in honey bees. *Proc. Natl. Acad. Sci. USA* **2004**, *101*, 11350–11355. [CrossRef]
22. Avershina, E.; Lundgard, K.; Sekelja, M.; Dotterud, C.; Storro, O.; Oien, T.; Johnsen, R.; Rudi, K. Transition from infant- to adult-like gut microbiota. *Environ. Microbiol.* **2016**, *18*, 2226–2236. [CrossRef]
23. Grassly, N.C.; Fraser, C. Seasonal infectious disease epidemiology. *Proc. Biol. Sci.* **2006**, *273*, 2541–2550. [CrossRef] [PubMed]
24. Underhill, D.M.; Iliev, I.D. The mycobiota: Interactions between commensal fungi and the host immune system. *Nat. Rev. Immunol.* **2014**, *14*, 405–416. [CrossRef] [PubMed]
25. Switanek, M.; Crailsheim, K.; Truhetz, H.; Brodschneider, R. Modelling seasonal effects of temperature and precipitation on honey bee winter mortality in a temperate climate. *Sci. Total Environ.* **2017**, *579*, 1581–1587. [CrossRef] [PubMed]
26. Kashyap, D.; Pandey, H.; Jaiswal, K.; Mishra, S. Fungal Diseases of Honey Bees: Current Status and Future Perspective. In *Recent Developments in Fungal Diseases of Laboratory Animals*; Gupta, A., Singh, N.P., Eds.; Springer International Publishing: Cham, Switzerland, 2019; pp. 7–27. [CrossRef]
27. Evans, J.D.; Schwarz, R.S. Bees brought to their knees: Microbes affecting honey bee health. *Trends Microbiol.* **2011**, *19*, 614–620. [CrossRef]
28. Casen, C.; Vebo, H.C.; Sekelja, M.; Hegge, F.T.; Karlsson, M.K.; Ciemniejewska, E.; Dzankovic, S.; Froyland, C.; Nestestog, R.; Engstrand, L.; et al. Deviations in human gut microbiota: A novel diagnostic test for determining dysbiosis in patients with IBS or IBD. *Aliment. Pharmacol. Ther.* **2015**, *42*, 71–83. [CrossRef]
29. Moran, N.A. Symbiosis. *Curr. Biol.* **2006**, *16*, R866–R871. [CrossRef]
30. Dalenberg, H.; Maes, P.; Mott, B.; Anderson, K.E.; Spivak, M. Propolis Envelope Promotes Beneficial Bacteria in the Honey Bee (Apis mellifera) Mouthpart Microbiome. *Insects* **2020**, *11*, 453. [CrossRef]
31. Cilia, G.; Fratini, F.; Tafi, E.; Turchi, B.; Mancini, S.; Sagona, S.; Nanetti, A.; Cerri, D.; Felicioli, A. Microbial Profile of the Ventriculum of Honey Bee (Apis mellifera ligustica Spinola, 1806) Fed with Veterinary Drugs, Dietary Supplements and Non-Protein Amino Acids. *Vet. Sci.* **2020**, *7*, 76. [CrossRef]

Article

Microbial Profile of the *Ventriculum* of Honey Bee (*Apis mellifera ligustica* Spinola, 1806) Fed with Veterinary Drugs, Dietary Supplements and Non-Protein Amino Acids

Giovanni Cilia [1,*], Filippo Fratini [1,2], Elena Tafi [1,3], Barbara Turchi [1], Simone Mancini [1], Simona Sagona [1,4], Antonio Nanetti [5], Domenico Cerri [1] and Antonio Felicioli [1,2]

[1] Department of Veterinary Sciences, University of Pisa, Viale delle Piagge 2, 56124 Pisa, Italy; filippo.fratini@unipi.it (F.F.); elenat93@libero.it (E.T.); barbara.turchi@unipi.it (B.T.); simone.mancini@unipi.it (S.M.); simonasagona@tiscali.it (S.S.); domenico.cerri@unipi.it (D.C.); antonio.felicioli@unipi.it (A.F.)

[2] Interdepartmental Research Center "Nutraceuticals and Food for Health", University of Pisa, Via del Borghetto 80, 56124 Pisa, Italy

[3] Department of Science, University of Basilicata, via dell'Ateneo Lucano 10, 85100 Potenza, Italy

[4] Department of Pharmacy, University of Pisa, via Bonanno 6, 56126 Pisa, Italy

[5] CREA Research Centre for Agriculture and Environment, Via di Saliceto 80, 40128 Bologna, Italy; antonio.nanetti@crea.gov.it

* Correspondence: giovanni.cilia@vet.unipi.it; Tel.: +39-050-2216-969

Received: 21 May 2020; Accepted: 3 June 2020; Published: 6 June 2020

Abstract: The effects of veterinary drugs, dietary supplements and non-protein amino acids on the European honey bee (*Apis mellifera ligustica* Spinola, 1806) *ventriculum* microbial profile were investigated. Total viable aerobic bacteria, Enterobacteriaceae, staphylococci, *Escherichia coli*, lactic acid bacteria, *Pseudomonas* spp., aerobic bacterial endospores and *Enterococcus* spp. were determined using a culture-based method. Two veterinary drugs (Varromed® and Api-Bioxal®), two commercial dietary supplements (ApiHerb® and ApiGo®) and two non-protein amino acids (GABA and beta-alanine) were administered for one week to honey bee foragers reared in laboratory cages. After one week, *E. coli* and *Staphylococcus* spp. were significantly affected by the veterinary drugs ($p < 0.001$). Furthermore, dietary supplements and non-protein amino acids induced significant changes in *Staphylococcus* spp., *E. coli* and *Pseudomonas* spp. ($p < 0.001$). In conclusion, the results of this investigation showed that the administration of the veterinary drugs, dietary supplements and non-protein amino acids tested, affected the *ventriculum* microbiological profile of *Apis mellifera ligustica*.

Keywords: GABA; beta-alanine; oxalic acid; diet effect; microbiota

---

1. Introduction

A decline in the honey bee population is threatening both pollination service and the beekeeping industry globally [1,2]. Honey bee colony losses are related to several causes, including habitat modifications, the massive use of agrochemicals, bacterial and parasitic diseases, climate changes, and multifactorial interactions [2–4].

A key role of gut microorganisms in animal health and welfare has been documented not only in mammals but also in insects [5,6]. The literature provides evidence that the midgut microbiota of eusocial bees, namely honey bees and bumblebees, plays an important role in protecting adults against pathogens [7–11].

The honey bee midgut microbial community was investigated by using culture-dependent methods, resulting in the identification of several microorganisms, including Gram-variable pleomorphic bacteria,

Bacillus spp. and Enterobacteriaceae, together with moulds and yeasts [12–14]. Molecular techniques allowed to repeatedly detect a characteristic gut microbiota of honey bee foragers consisting of nine distinct bacterial phylotypes accounting for the 95% of the total bacterial community [10,15].

The widespread mite *Varroa destructor* (Anderson and Trueman, 2000) and its associated viruses are generally considered among the primary causes of colony collapse [16]. Treatments with acaricides must be administered regularly to reduce the infestation impact on the colony survival. Oxalic and formic acid are natural active compounds commonly used against this mite [17–19]. However, when the acids are accidentally ingested, they are likely to alter the intestinal pH and microbial balance [20].

As for bee products, including beeswax [21], the interest in food supplements like vegetable extracts, microorganisms and vitamins is increasing due to their possible role in promoting colony health, vitality and productivity. Anyhow, these types of products may cause modifications of honey bees midgut microbiota [22].

Adult honey bees use nectar as a primary alimentary source, containing sugars and amino acids that could play a major role in the microbiota modifications [23,24]. Gamma-aminobutyric acid (GABA) and beta-alanine are two of the most frequently and abundant non-protein amino acids of nectar [25,26]. These non-protein amino acids received attention for their potential role in mason bee physiology [27,28].

In light of the increasing use of dietary supplements and veterinary drugs in apiculture, the aim of this study was to investigate the modifications in the viable intestinal microbial community of the worker honey bee fed with (*i*) commercial veterinary drugs containing oxalic acid and formic acid, (*ii*) commercial dietary supplements containing yeasts and vitamins or herbal natural extracts and (*iii*) non-protein amino acids, GABA and beta-alanine.

2. Materials and Methods

2.1. Experimental Design

Seven hundred *Apis mellifera ligustica* (Spinola, 1806) foragers were randomly collected in June 2018 from the same colony in the experimental apiary of the Department of Veterinary Sciences of the University of Pisa (San Piero a Grado, Pisa, Italy). Twenty seven poly(methyl methacrylate) sterilized cages (11 × 13 × 6.5 cm), each including a gravity feeder, one small wax comb, and two transparent walls, were used to host a pool of 25 bees each. The cages were kept between 31 °C and 34 °C with a relative humidity ranging from 50% to 80% until the end of the experiment (one week) [29]. Twenty-five specimens were processed immediately after the sampling and were used as Time 0 control (T0). The remaining honey bees, divided in pool of 25 specimens and reared in cages, were fed with eight different experimental diets and one control diet (C, sterile 1:1 sucrose:water, w:w). Each diet was administrated to three replicate cages ad libitum. Two registered antivarroa veterinary drugs were used: Api-Bioxal® (AB, Chemicals Life, Vigonza, Italy) containing 44.03 g/L of oxalic acid and VarroMed® (VM, BeeVital GmbH, Obertrum am See, Austria) containing 31.42 g/L of oxalic acid and 5 g/L of formic acid. Two commercial dietary supplements containing maltodextrin, yeasts and vitamins (AG, ApiGo®, Chemicals Life, Vigonza, Italy) and essential oils, B vitamins as well as the natural extracts of garlic and cinnamon (AH, ApiHerb®, Chemicals Life, Vigonza, Italy) were also employed. Furthermore, two non-protein amino acids were used at two different concentrations (GABA 1×, GABA 0.75 mM; GABA 20×, GABA 15 mM; BALA 1×, beta-alanine 2.3 mM; BALA 20×, beta-alanine 46 mM).

Diets AB, VM, AG, and AH were formulated according to the producers' instructions. Non-protein amino acids (GABA 1x, GABA 20x, BALA 1x and BALA20x) were mixed with sterile 1:1 sucrose:water (w:w) to reach the aforementioned concentrations. At the beginning of the trial (T0) and after one week (Time7, T7), in order to induce hypothermia and avoid the intestinal microbial profile modification, the honey bees were frozen at −20 °C for 5 min and washed into 95% ethanol to remove the external microbial contamination, right after the *ventriculum* (small intestine and midgut) were dissected for the microbiological analyses. At T7, no mortality was observed in the reared honey bees.

2.2. Microbiological Analyses

The *ventriculi* of the 25 bees belonging to a single cage were pooled (three replicate pools for each diet) and suspended into 1 mL of sterile phosphate buffered saline (PBS) solution. The samples were homogenized for 60 s with a Stomacher® 400 Circulator (VWR International Srl, Milan, Italy). Ten-fold serial dilution series were performed, and different media were used in order to quantify the different target microorganisms. The total viable aerobic count was enumerated on Plate Count Agar (PCA) after incubation at 30 °C for 72 h; the Enterobacteriaceae were quantified on violet red bile glucose agar (VRBGA) after incubation at 37 °C for 24 h; the staphylococci were enumerated on mannitol salt agar (MSA) after incubation at 37 °C for 24–48 h; *Escherichia coli* were enumerated on a tryptone bile X-glucuronide medium (TBX) after incubation at 42 °C for 24 h; de Man–Rogosa–Sharpe agar (MRS) was employed to enumerate the lactic acid bacteria after incubation at 37 °C for 48 h in anaerobic conditions; *Pseudomonas* spp. were quantified on a penicillin pimaricin agar (PPA) plate incubated at 30 °C for 24–48 h; *Enterococcus* spp. were evaluated on kanamycin aesculin azide agar (KAAA) after incubation at 42 °C for 24 h; and the aerobic bacterial endospores were enumerated on PCA (incubation at 37 °C for 48), after a heat-shocking at 80 °C for 10 min of the dilutions. All the culture media and supplements were purchased by ThermoFisher Scientific (Milan, Italy). The bacterial counts were expressed as log Colony-Forming Unit (CFU) per samples.

2.3. Statistical Analysis

Variations that occurred in microbial loads between T0 and T7 were tested by Student's t-tests. Statistical differences between the control groups and the treated ones at T7 were tested by one-way ANOVA. If a significant variation was detected a Tukey's test was performed as post hoc multiple comparisons. The effects were considered significant when $p < 0.05$. Statistical analyses were performed with R free software [30].

3. Results

3.1. Effects of VarroMed® and Api-Bioxal® on the Viable Honey Bee Ventriculum Microbial Profile

The evaluation of the *ventriculum* microbial profile of bees fed with C diet and diets with VM and AB is summarized in Table 1. The bacterial endospores were absent in all the samples; thus, they were not reported.

After one week of experimental diets, both the total viable aerobic counts and Enterobacteriaceae counts increased (both at $p < 0.001$ for all the diets), with no statistical differences within the treatments ($p = 0.798$ and $p = 0.455$ for viable aerobic and Enterobacteriaceae counts, respectively). The *Staphylococcus* spp. and *E. coli* amounts were not affected after one week of the C diet ($p = 0.932$ and $p = 0.887$, respectively), on the contrary, both the tested veterinary drugs statistically influenced both parameters ($p = 0.019$ and $p < 0.001$ for *Staphylococcus* spp. and $p = 0.016$ and $p = 0.022$ for *Escherichia coli*, for, respectively, VM and AB). VarroMed® completely inhibited *E. coli* and *Staphylococcus* spp., on the other hand, Api-Bioxal® doubled the *E. coli* and *Staphylococcus* spp. loads. After one week of treatment, for all the samples, modifications of *E. coli* ($p < 0.001$) and *Staphylococcus* spp. ($p < 0.001$) counts were statistically affected with higher counts in bees fed with AB, followed by those fed the control diet and VM, respectively.

Statistical difference was found between T0 and AB for *Pseudomonas* spp. determination ($p = 0.002$). After one week, the bees fed with AB showed a higher load of these bacteria than at T0. No statistical differences were detected for *Pseudomonas* spp. counts after one week of the control diet and AB compared to VM, nevertheless, the AB samples showed higher amounts than those from the C group ($p = 0.013$).

Table 1. Bacterial counts (log Colony-Forming Unit (CFU) per samples ± s.d.) obtained from honey bees' *ventriculum* after one week of veterinary drugs administration.

Bacteria	Time 0	Diet (Veterinary Drugs and Control) One Week			*p*-Value §
		C	VM	AB	
Total viable aerobic counts	5.36 ± 0.51	7.28 ± 0.71 *	7.03 ± 0.88 *	7.40 ± 0.11 *	0.798
Enterobacteriaceae	3.34 ± 0.77	5.88 ± 0.89 *	5.18 ± 0.04 *	5.59 ± 0.75 *	0.455
Escherichia coli	2.43 ± 1.28	2.52 ± 0.95 [b]	0.00 ± 0.00 *,[c]	4.72 ± 0.44 *,[a]	<0.001
Pseudomonas spp.	2.69 ± 0.96	2.93 ± 0.32 [b]	3.36 ± 0.93 [a,b]	4.38 ± 0.37 *,[a]	0.013
Staphylococcus spp.	1.47 ± 0.69	1.50 ± 0.55 [b]	0.00 ± 0.00 *,[c]	3.48 ± 0.13 *,[a]	<0.001
Enterococcus spp.	1.24 ± 0.34	2.45 ± 1.58	3.88 ± 1.04 *	3.81 ± 0.63 *	0.237
Lactic acid bacteria	2.87 ± 0.67	3.91 ± 1.03 [a,b]	5.26 ± 0.97 *,[b]	2.74 ± 0.05 [a]	0.023

* Means within row different from the Time (T0) ($p < 0.05$). [a,b,c] Means within row lacking a common superscript are different ($p < 0.05$). § p-value of ANOVA between the treatments after one week of the experimental diet (Control (C), VarroMed® (VM) and Api-Bioxal® (AB)).

After one week, both the veterinary drugs induced a significant increase in *Enterococcus* spp. ($p < 0.001$). Anyhow, due to the high variations among the data, no statistical differences were observed between treatments at T7 ($p = 0.237$).

Compared to T0, honey bees treated with VM for one week showed a significant increase in lactic acid bacteria counts ($p = 0.003$). At T7, VM showed the highest values, while AB showed the lowest; moreover, the C diet was statistically comparable to the two diets including veterinary drugs ($p = 0.023$).

3.2. Effects of ApiGo® and ApiHerb® on the Viable Honey Bee Ventriculum Microbial Profile

The microbiological effect of AG and AH on honey bees *ventriculum* microbial profile is reported in Table 2, except for the bacterial endospores which were never detected.

After one week, the total viable aerobic counts increased in both C and the two diets including the dietary supplements AG and AH compared to T0 ($p < 0.001$, $p = 0.008$ and $p = 0.012$, respectively, for C, AG and AH) with no statistical differences between the diets ($p = 0.328$).

Table 2. Bacterial counts (log CFU per samples log CFU per samples ± s.d.) obtained from honey bees' *ventriculum* after one week of dietary supplement administration.

Bacteria	Time 0	Diet with Dietary Supplements, One Week			*p*-Value §
		C	AG	AH	
Total viable aerobic counts	5.36 ± 0.51	7.28 ± 0.71 *	6.76 ± 0.59 *	6.62 ± 0.56 *	0.328
Enterobacteriaceae	3.34 ± 0.77	5.88 ± 0.89 *,[a]	5.61 ± 1.16 *,[a]	3.16 ± 0.13 [b]	0.004
Escherichia coli	2.43 ± 1.28	2.52 ± 0.95 [a]	3.20 ± 0.32 [a]	0.00 ± 0.00 *,[b]	0.001
Pseudomonas spp.	2.69 ± 0.96	2.93 ± 0.32 [a]	3.11 ± 0.49 [a]	0.00 ± 0.00 *,[b]	<0.001
Staphylococcus spp.	1.47 ± 0.69	1.50 ± 0.55 [b]	3.29 ± 0.03 *,[a]	0.00 ± 0.00 [c]	<0.001
Enterococcus spp.	1.24 ± 0.34	2.45 ± 1.58 *	3.68 ± 1.03 *	3.86 ± 0.42 *	0.250
Lactic acid bacteria	2.87 ± 0.67	3.91 ± 1.03	4.51 ± 0.70 *	3.05 ± 0.28	0.160

* Means within row different from the Time (T0) ($p < 0.05$). [a,b,c] Means within row lacking a common superscript are different ($p < 0.05$). § p-value of ANOVA between the treatments after one week of the experimental diet (Control, ApiGo® (AG) and ApiHerb® (AH)).

After one week, the Enterobacteriaceae significantly increased in both C and AG compared to T0 ($p < 0.001$ and $p = 0.009$, respectively), while AH did not differ from T0 ($p = 0.703$). Consequently, Enterobacteriaceae were significantly higher in both C and AG than in AH ($p = 0.004$). Statistical differences were highlighted for *E. coli* ($p = 0.016$) and *Pseudomonas* spp. ($p = 0.002$) only between the T0 and AH treatments. Bees fed with AH showed a significantly lower amount of both *E. coli* ($p = 0.001$) and *Pseudomonas* spp. ($p < 0.001$) than those fed with both C and AG. After AH administration, *Escherichia coli* and *Pseudomonas* spp. were not detected.

Staphylococcus spp. was not detected at T0, and was found only in honey bees fed with AG for one week ($p = 0.012$). Statistical differences were found for lactic acid bacteria, only between the T0 and AG treatment ($p = 0.012$). Indeed, the AG induced a significant increase in lactic acid bacteria. Among the dietary supplement, only the AG differed significantly compared to T0.

After one week, both dietary supplements, as well as the C diet, induced an increase in *Enterococcus* spp. compared to T0 ($p = 0.009$, $p = 0.001$ and $p < 0.001$ for C, AG and AH, respectively). No statistical differences were found among the diets ($p = 0.250$).

3.3. Effects of GABA and Beta-Alanine on the Viable Honey Bee Ventriculum Microbial Profile

The results of the non-protein amino acids on the honey bee *ventriculum* microbial profile are reported in Table 3.

Table 3. Bacterial counts (log CFU per samples log CFU per samples ± s.d.) obtained from the honey bees' *ventriculum* after one week of non-protein amino acid (GABA and β-alanine)-enriched diet administration.

Bacteria	Time 0	Diet with Non-Protein Amino Acids, One Week					*p*-Value [§]
		Control	β-Alanine	β-Alanine 20x	GABA	GABA 20x	
Total viable aerobic counts	5.36 ± 0.51	7.28 ± 0.71 *	7.42 ± 0.8 *	6.97 ± 0.47 *	7.26 ± 0.54 *	6.79 ± 0.90 *	0.793
Enterobacteriaceae	3.34 ± 0.77	5.88 ± 0.89 *	5.48 ± 0.86 *	6.12 ± 0.25 *	5.55 ± 1.08 *	6.12 ± 0.50 *	0.787
Escherichia coli	2.43 ± 1.28	2.52 ± 0.95 [b]	3.43 ± 0.94 [a,b]	5.00 ± 0.55 [*,a]	0.00 ± 0.00 [*,c]	2.73 ± 0.64 [b]	<0.001
Pseudomonas spp.	2.69 ± 0.96	2.93 ± 0.32 [a]	0.00 ± 0.00 [*,b]	0.00 ± 0.00 [*,b]	0.00 ± 0.00 [*,b]	0.00 ± 0.00 [*,b]	<0.001
Staphylococcus spp.	1.47 ± 0.69	1.50 ± 0.55 [a]	0.00 ± 0.00 [*,b]	0.00 ± 0.00 [*,b]	0.00 ± 0.00 [*,b]	0.00 ± 0.00 [*,b]	<0.001
Enterococcus spp.	1.24 ± 0.34	2.45 ± 1.58 [a]	1.39 ± 0.67 [a,b]	1.52 ± 0.23 [a,b]	0.00 ± 0.00 [*,b]	0.73 ± 0.03 [*,a,b]	0.044
Lactic acid bacteria	2.87 ± 0.67	3.91 ± 1.03	3.51 ± 0.59	4.83 ± 1.14 *	4.67 ± 1.36 *	4.39 ± 0.57 *	0.503

* Means within row different from the Time 0 (T0) ($p < 0.05$). [a,b,c] Means within row lacking a common superscript are different ($p < 0.05$). [§] *p*-value of ANOVA between the treatments after one week of the diet (Control, β-alanine, β-alanine 20x, GABA and GABA 20x).

After one week, the total viable aerobic counts increased in C and all the non-protein amino acid-enriched diets, except for GABA 20x ($p < 0.001$, $p = 0.002$, $p = 0.003$, $p = 0.001$ and $p = 0.017$, respectively, for C, β-alanine, β-alanine 20x, GABA and GABA 20x). No statistical differences were found among the diets after one week ($p = 0.793$).

After one week, Enterobacteriaceae significantly increased in both C and all the non-protein amino acid-enriched diets ($p < 0.001$, $p = 0.007$, $p = 0.001$, $p = 0.009$ and $p = 0.001$, respectively, for C, β-alanine, β-alanine 20x, GABA and GABA 20x). No statistical differences were evidenced among the diets ($p = 0.787$).

After one week of all the non-protein amino acid-enriched diets administration, *Staphylococcus* spp. significantly decreased compared to T0 ($p < 0.001$ for all the diets), while, after one week of the C diet administration, *Staphylococcus* spp. was not detected ($p = 0.932$).

Statistical differences were found for *E. coli* between T0 and β-alanine 20x and GABA ($p = 0.014$ and $p = 0.016$, respectively). In both GABA diets, *E. coli* abruptly decreased, while in β-alanine 20x they significantly increased.

Compared to T0, *Pseudomonas* spp. was not detected in the bees fed with non-protein amino acid-enriched diets ($p = 0.002$ for all investigated non-proteins amino acids diets), while it was present after one week of the C diet administration ($p = 0.579$).

No statistical differences were found for lactic acid bacteria among the T0 and all the experimental diets ($p = 0.067$, $p = 0.208$, $p = 0.013$, $p = 0.029$ and $p = 0.028$, respectively, for C, β-alanine, β-alanine 20x, GABA and GABA 20x). Likewise, no statistical differences were found among C and non-protein amino acids for lactic acid bacteria ($p = 0.503$). No statistical differences were found between T0 and C, β-alanine and β-alanine 20x for *Enterococcus* spp. ($p = 0.098$, $p = 0.666$ and $p = 0.244$, respectively). Among the non-protein amino acid-enriched diets, GABA and GABA 20x differed significantly from the T0 ($p = 0.001$ and $p = 0.044$, respectively). Bacterial endospores were never detected in the honey bees treated with non-protein amino acids.

4. Discussion

Concerning the veterinary drugs, Api-Bioxal is composed of oxalic acid, while VarroMed of both oxalic and formic acid, in concentrations of 44 mg/mL and 5 mg/mL, respectively. Oxalic acid usually shows an antibacterial activity [31], as well as formic acid which was reported as a powerful antimicrobial agent even at low concentrations [32–34]. Notably, it seems that formic acid or its synergistic activity with oxalic acid could play a role in the antimicrobial activity of VM. As reported by Raftari et al. (2009), formic acid showed antibacterial activity in vitro against *E. coli* and *S. aureus* [35]. Few research studies reported the *Pseudomonas* spp. tolerance, production and metabolic activity in relation to oxalic acid. Nagarajkumar et al. (2005) reported the ability of several strains of *P. fluorescens* to detoxify soil from oxalic acid produced by fungi, associated to the presence of plasmid genes [36]. Hamel et al. (1999) highlighted the production of oxalic acid as a response to aluminum stress in *P. fluorescens* [37]. As reported in this investigation, the production of oxalic acid in the response to stress suggests that *Pseudomonas* spp. could use it as a substrate. As reported before for *Pseudomonas* spp., *Enterococcus* species could also degrade oxalic acid through metabolic pathways [38]. Even if an antimicrobial activity against *Enterococcus* spp. is reported for formic acid [39,40], probably the combination with oxalic acid in VM may modify or reduce the effect of formic acid. The growth of some bacteria is probably due to their ability to use organic acids as a nitrogen and energy source [41,42]. Oxalic acid, the main component of Api-Bioxal®, showed and antibacterial effect against different strains of *Lactobacillus* spp. [43]. Thus, normal miticide treatments may negatively affect honey bees, since many *Lactobacillus* species are linked to honey bee health, inhibiting some potential bacterial pathogens [44,45]. VM is composed of both oxalic and formic acids, and it seemed that *E. coli* was not inhibited by oxalic acid used alone, while it was strongly inhibited by VM treatment.

Concerning the dietary supplement, AH is composed of garlic, cinnamon, mint and thyme which exert an antimicrobial activity against several bacterial strains. The antibacterial activity of garlic against *E.coli*, Pseudomonadaceae and Enterobacteriaceae was widely discussed in literature [46–48] and attributed to the presence of mono, di, tri and tetra diallyl sulphides which increase with the number of sulphur atoms in the compound [49,50]. Peppermint showed antibacterial activity against several pathogens, including *E. coli* and *Pseudomonas* spp. [51,52]. Its antibacterial activity is mainly due to terpens, namely α-pinene, limonene and α-terpineol [53]. α-pinene, limonene and α-terpineol, together with thymol and linalool [54] which contributed to the antibacterial activity of *Thymus* against *E. coli* and *Pseudomonas* spp. as well [54,55], were reported in this investigation.

Moreover, the antibacterial activity of AH observed in this study could be due to the antibacterial compounds present in cinnamon, as well as cinnamaldehyde, eugenol and cinnamyl acetate, followed by other terpens (including α-pinene, limonene, α-terpineol and linalool) [56]. *Cinnamomun zeylanicum* showed antibacterial effects against many bacteria, such as *E. coli* and *Pseudomonas* spp. [57–59], as highlighted in this investigation with the AH diet.

The garlic and cinnamon present in AH showed an antimicrobial activity against *Staphylococcus* spp. The antibacterial activity of garlic extract has been demonstrated against *S. aureus*, as well as methicillin-resistant and streptomycin-resistant strains, and *S. epidermidis* [60,61]. The antimicrobial activity of garlic is probably due to the action of allicinlin, thiosulfinates, flavonoids and other phenols [62].

It is noteworthy that maltodextrins, present in AG, were metabolized by lactic acid bacteria and used as an energy source, as highlighted in this investigation [63,64].

Although the anti-*Enterococcus* activity of garlic and cinnamon extracts have been widely discussed [65,66], in this investigation the AH diet did not allow the inhibition of *Enterococcus* spp., but caused their increase. On the other hand, the increase in *Enterococcus* spp. observed after the administration of the AG diet was probably due to the capability of this genus to use maltodextrins as a carbon source, as described for lactic acid bacteria [63].

Finally, the action of non-protein amino acids against microorganisms was partially investigated. In this investigation, the effects of GABA on *E. coli* and *Enterococcus* spp. suggested a possible dose-dependent action. A dose-dependent effect of some nectar secondary compounds, including non-protein amino acids, was recognized in some species of *Bombus* [67].

In plants, GABA accumulates in response to various abiotic and biotic stresses, including fungal and bacterial infections [68]. GABA is suggested to mediate the interactions between plants and microorganisms, including bacteria [68,69]. For instance, the *Brassica rapa* plant extracts showed an increase in its in vitro antibacterial activity against *E. coli*, *Pseudomonas aeruginosa* and *Staphylococcus aureus*, after supplementation with GABA [70]. To the best of authors' knowledge, previous studies regarding the in vitro antibacterial effect of GABA are not available. The growth of *E. coli* was not negatively affected by GABA at the concentration present in nectar. Although, some *E. coli* strains could be able to use GABA as a source of nitrogen [71,72].

In beta-alanine-enriched diets, as in GABA, *Staphylococcus* spp. and *Pseudomonas* spp. were not detected. Moreover, a diet enriched with beta-alanine at both concentrations increased the number of *E. coli* colonies. In the biosynthesis of pantothenic acid in *E. coli*, beta-alanine served as a direct precursor [73,74]. Probably, the presence of ready-to-use beta-alanine could promote the growth of *E. coli*. This growth is significantly higher when beta-alanine is administered 20× compared to lower concentrations, suggesting a dose-dependent effect. At the best of authors' knowledge, studies on the in vitro antibacterial effect of beta-alanine are not present in the literature, while there are some on the antibacterial activity of beta-alanine-synthetized derivates [75,76].

5. Conclusions

In conclusion, the reported effects on the microbiological profile of honey bee *ventriculum* may be direct (killing bacteria and inhibiting its growth, etc.) or indirect (improving the bee immunity system, favoring the development of microbiota able to prevent pathogen strain colonization, etc.). All these factors must be considered in the framework of an increasing interest towards the formulation of dietary supplements for honey bee nutrition. In light of the obtained results, non-protein amino acid-enriched nutrition could be used in order to mitigate the possible beneficial microflora imbalance after anti-Varroa and anti-nosemosis treatments.

Author Contributions: Conceptualization, G.C., F.F. and A.F.; methodology, G.C., F.F., A.F. and A.N.; investigation, G.C., E.T. and S.S.; data curation, G.C., F.F., A.F., A.N., E.T., S.M. and B.T.; writing—original draft preparation, G.C., E.T., A.N., B.T.; writing—review and editing, G.C., F.F., A.F., A.N., E.T., S.S., S.M., B.T. and D.C.; supervision, A.F. and D.C.; funding acquisition, F.F., A.F. and D.C. All authors have read and agreed to the published version of the manuscript.

Funding: This research has been supported by Fondi di Ateneo of University of Pisa.

References

1. Gallai, N.; Salles, J.M.; Settele, J.; Vaissière, B.E. Economic valuation of the vulnerability of world agriculture confronted with pollinator decline. *Ecol. Econ.* **2009**, *68*, 810–821. [CrossRef]
2. Potts, S.G.; Roberts, S.P.M.; Dean, R.; Marris, G.; Brown, M.A.; Jones, R.; Neumann, P.; Settele, J. Declines of managed honey bees and beekeepers in Europe. *J. Apic. Res.* **2010**, *49*, 15–22. [CrossRef]
3. Watson, K.; Stallins, J.A. Honey Bees and Colony Collapse Disorder: A Pluralistic Reframing. *Geogr. Compass* **2016**, *10*, 222–236. [CrossRef]
4. McMenamin, A.J.; Genersch, E. Honey bee colony losses and associated viruses. *Curr. Opin. Insect Sci.* **2015**, *8*, 121–129. [CrossRef]
5. Crotti, E.; Balloi, A.; Hamdi, C.; Sansonno, L.; Marzorati, M.; Gonella, E.; Favia, G.; Cherif, A.; Bandi, C.; Alma, A.; et al. Microbial symbionts: A resource for the management of insect-related problems. *Microb. Biotechnol.* **2012**, *5*, 307–317. [CrossRef]
6. Zhao, Y.; Chen, Y.; Li, Z.; Peng, W. Environmental factors have a strong impact on the composition and diversity of the gut bacterial community of Chinese black honeybees. *J. Asia. Pac. Entomol.* **2018**, *21*, 261–267. [CrossRef]

7.   Koch, H.; Schmid-Hempel, P. Socially transmitted gut microbiota protect bumble bees against an intestinal parasite. *Proc. Natl. Acad. Sci. USA* **2011**, *108*, 19288–19292. [CrossRef] [PubMed]

8.   Maes, P.W.; Rodrigues, P.A.P.; Oliver, R.; Mott, B.M.; Anderson, K.E. Diet-related gut bacterial dysbiosis correlates with impaired development, increased mortality and Nosema disease in the honeybee (Apis mellifera). *Mol. Ecol.* **2016**, *25*, 5439–5450. [CrossRef] [PubMed]

9.   Erban, T.; Ledvinka, O.; Kamler, M.; Nesvorna, M.; Hortova, B.; Tyl, J.; Titera, D.; Markovic, M.; Hubert, J. Honeybee (Apis mellifera)-associated bacterial community affected by American foulbrood: Detection of Paenibacillus larvae via microbiome analysis. *Sci. Rep.* **2017**, *7*, 1–10. [CrossRef]

10. Kwong, W.K.; Mancenido, A.L.; Moran, N.A. Immune system stimulation by the native gut microbiota of honey bees. *R. Soc. Open Sci.* **2017**, *4*. [CrossRef]

11. Raymann, K.; Shaffer, Z.; Moran, N.A. Antibiotic exposure perturbs the gut microbiota and elevates mortality in honeybees. *PLoS Biol.* **2017**, *15*, 1–22. [CrossRef] [PubMed]

12. Gilliam, M.; Valentine, D.K. Enterobacteriaceae Isolated from Foraging worker honey bee, Apis mellifera. *J. Invertebr. Pathol.* **1974**, *23*, 38–41. [CrossRef]

13. Gilliam, M.; Morton, H.L. Bacteria belonging to the genus Bacillus isolated from honey bee, Apis mellifera, fed 2,4-d and antibiotics. *Apidologie* **1978**, *9*, 213–222. [CrossRef]

14. Hayden, C.; Road, E.A.; Gilliam, M. Identification and roles of non-pathogenic microflora associated with honey bees. *FEMS Microbiol. Lett.* **2006**, *155*, 1–10.

15. Martinson, V.G.; Moy, J.; Moran, N.A. Establishment of characteristic gut bacteria during development of the honeybee worker. *Appl. Environ. Microbiol.* **2012**, *78*, 2830–2840. [CrossRef]

16. Genersch, E.; Aubert, M. Emerging and re-emerging viruses of the honey bee (Apis mellifera L.). *Vet. Res.* **2010**, *41*. [CrossRef]

17. Underwood, R.M.; Currie, R.W. The effects of temperature and dose of formic acid on treatment efficacy against Varroa destructor (Acari: Varroidae), a parasite of Apis mellifera (Hymenoptera: Apidae). *Exp. Appl. Acarol.* **2003**, *29*, 303–313. [CrossRef]

18. Charriére, J.-D.; Imdorf, A. Oxalic acid treatment by trickling against *Varroa destructor*: Recommendations for use in central Europe and under temperate climate conditions. *Bee World* **2002**, *83*, 51–60. [CrossRef]

19. Giusti, M.; Sabelli, C.; Di Donato, A.; Lamberti, D.; Paturzo, C.E.; Polignano, V.; Lazzari, R.; Felicioli, A. Efficacy and safety of Varterminator, a new formic acid medicine against the varroa mite. *J. Apic. Res.* **2017**, *56*, 162–167. [CrossRef]

20. Anderson, K.; Sheehan, T.H.; Eckholm, B.J.; Mott, B.M.; DeGrandi-Hoffman, G. An emerging paradigm of colony health: Microbial balance of the honey bee and hive (Apis mellifera). *Insectes Soc.* **2011**, *58*, 431–444. [CrossRef]

21. Felicioli, A.; Cilia, G.; Mancini, S.; Turchi, B.; Galaverna, G.; Cirlini, M.; Cerri, D.; Fratini, F. In vitro antibacterial activity and volatile characterisation of organic Apis mellifera ligustica (Spinola, 1906) beeswax ethanol extracts. *Food Biosci.* **2019**, *29*, 102–109. [CrossRef]

22. Czech, A.; Smolczyk, A.; Ognik, K.; Wlazło, Ł.; Nowakowicz-Dębek, B.; Kiesz, M. Effect of dietary supplementation with Yarrowia lipolytica or Saccharomyces cerevisiae yeast and probiotic additives on haematological parameters and the gut microbiota in piglets. *Res. Vet. Sci.* **2018**, *119*, 221–227. [CrossRef] [PubMed]

23. Nicolson, S.W.; Nepi, M.; Pacini, E. *Nectaries and Nectar*; Springer: Berlin, Germany, 2007; ISBN 9781402059377.

24. Heil, M. Nectar: Generation, regulation and ecological functions. *Trends Plant Sci.* **2011**, *16*, 191–200. [CrossRef] [PubMed]

25. Petanidou, T.; Van Laere, A.; Ellis, W.N.; Smets, E. What shapes amino acid and sugar composition in Mediterranean floral nectars? *Oikos* **2006**, *115*, 155–169. [CrossRef]

26. Nepi, M.; Soligo, C.; Nocentini, D.; Abate, M.; Guarnieri, M.; Cai, G.; Bini, L.; Puglia, M.; Bianchi, L.; Pacini, E. Amino acids and protein profile in floral nectar: Much more than a simple reward. *Flora Morphol. Distrib. Funct. Ecol. Plants* **2012**, *207*, 475–481. [CrossRef]

27. Felicioli, A.; Sagona, S.; Galloni, M.; Bortolotti, L.; Bogo, G.; Guarnieri, M.; Nepi, M. Effects of nonprotein amino acids on survival and locomotion of *Osmia bicornis*. *Insect Mol. Biol.* **2018**, *27*, 556–563. [CrossRef]

28. Bogo, G.; Bortolotti, L.; Sagona, S.; Felicioli, A.; Galloni, M.; Barberis, M.; Nepi, M. Effects of Non-Protein Amino Acids in Nectar on Bee Survival and Behavior. *J. Chem. Ecol.* **2019**, 1–8. [CrossRef]

29. Williams, G.R.; Alaux, C.; Costa, C.; Csáki, T.; Doublet, V.; Eisenhardt, D.; Fries, I.; Kuhn, R.; Mcmahon, D.P.; Medrzycki, P.; et al. Standard methods for maintaining adult Apis mellifera in cages under in vitro laboratory conditions. *J. Apic. Res.* **2013**, *52*, 1–36. [CrossRef]

30. R Core Team R: A Language and Environment for Statistical Computing. Available online: https://www.R-project.org (accessed on 5 June 2020).

31. Kwak, A.-M.; Lee, I.-K.; Lee, S.-Y.; Yun, B.-S.; Kang, H.-W. Oxalic Acid from *Lentinula* edodes Culture Filtrate: Antimicrobial Activity on Phytopathogenic Bacteria and Qualitative and Quantitative Analyses. *Mycobiology* **2016**, *44*, 338–342. [CrossRef]

32. Fratini, F.; Cilia, G.; Turchi, B.; Felicioli, A. Insects, arachnids and centipedes venom: A powerful weapon against bacteria. A literature review. *Toxicon* **2017**, *130*, 91–103. [CrossRef]

33. Matsuda, T.; Yano, T.; Maruyama, A.; Kumagai, H. Antimicrobial Activities of Organic Acids Determined by Minimum Inhibitory Concentrations at Different pH Ranged from 4.0 to 7.0. *Nippon Shokuhin Kogyo Gakkaishi* **1994**, *41*, 687–701. [CrossRef]

34. Thompson, J.L.; Hinton, M. Antibacterial activity of formic and propionic acids in the diet of hens on salmonellas in the crop. *Br. Poult. Sci.* **1997**, *38*, 59–65. [CrossRef] [PubMed]

35. Raftari, M.; Jalilian, F.A.; Abdulamir, A.S.; Son, R.; Sekawi, Z.; Fatimah, A.B. Effect of organic acids on Escherichia coli O157:H7 and Staphylococcus aureus contaminated meat. *Open Microbiol. J.* **2009**, *3*, 121–127. [CrossRef] [PubMed]

36. Nagarajkumar, M.; Jayaraj, J.; Muthukrishnan, S.; Bhaskaran, R.; Velazhahan, R. Detoxification of oxalic acid by Pseudomonas fluorescens strain PfMDU2: Implications for the biological control of rice sheath blight caused by Rhizoctonia solani. *Microbiol. Res.* **2005**, *160*, 291–298. [CrossRef]

37. Hamel, R.; Levasseur, R.; Appanna, V.D. Oxalic acid production and aluminum tolerance in Pseudomonas fluorescens. *J. Inorg. Biochem.* **1999**, *76*, 99–104. [CrossRef]

38. Hokama, S.; Honma, Y.; Toma, C.; Ogawa, Y. Oxalate-degrading Enterococcus faecalis. *Microbiol. Immunol.* **2000**, *44*, 235–240. [CrossRef]

39. de Castilho, A.L.; da Silva, J.P.C.; Saraceni, C.H.C.; Díaz, I.E.C.; Paciencia, M.L.B.; Varella, A.D.; Suffredini, I.B. In vitro activity of Amazon plant extracts against Enterococcus faecalis. *Braz. J. Microbiol.* **2014**, *45*, 769–779. [CrossRef]

40. Huang, C.B.; Alimova, Y.; Myers, T.M.; Ebersole, J.L. Short- and medium-chain fatty acids exhibit antimicrobial activity for oral microorganisms. *Arch. Oral Biol.* **2011**, *56*, 650–654. [CrossRef]

41. Dirusso, C.C.; Black, P.N. Bacterial long chain fatty acid transport: Gateway to a fatty acid-responsive signaling system. *J. Biol. Chem.* **2004**, *279*, 49563–49566. [CrossRef]

42. Van Immerseel, F.; Russell, J.B.; Flythe, M.D.; Gantois, I.; Timbermont, L.; Pasmans, F.; Haesebrouck, F.; Ducatelle, R. The use of organic acids to combat *Salmonella* in poultry: A mechanistic explanation of the efficacy. *Avian Pathol.* **2006**, *35*, 182–188. [CrossRef]

43. Diaz, T.; del-Val, E.; Ayala, R.; Larsen, J. Alterations in honey bee gut microorganisms caused by *Nosema* spp. and pest control methods. *Pest Manag. Sci.* **2018**, *75*, 835–843. [CrossRef]

44. Sakamoto, I.; Igarashi, M.; Kimura, K.; Takagi, A.; Miwa, T.; Koga, Y. Suppressive effect of Lactobacillus gasseri OLL 2716 (LG21) on Helicobacter pylori infection in humans. *J. Antimicrob. Chemother.* **2001**, *47*, 709–710. [CrossRef] [PubMed]

45. Halami, P.M.; Chandrashekar, A.; Nand, K. Lactobacillus farciminis MD, a newer strain with potential for bacteriocin and antibiotic assay. *Lett. Appl. Microbiol.* **2000**, *30*, 197–202. [CrossRef] [PubMed]

46. Burt, S.A.; Reinders, R.D. Antibacterial activity of selected plant essential oils against Escherichia coli O157:H7. *Lett. Appl. Microbiol.* **2003**, *36*, 162–167. [CrossRef] [PubMed]

47. Casella, S.; Leonardi, M.; Melai, B.; Fratini, F.; Pistelli, L. The Role of Diallyl Sulfides and Dipropyl Sulfides in the *In Vitro* Antimicrobial Activity of the Essential Oil of Garlic, *Allium sativum* L., and Leek, *Allium porrum* L. *Phyther. Res.* **2013**, *27*, 380–383. [CrossRef] [PubMed]

48. Bjarnsholt, T.; Jensen, P.Ø.; Rasmussen, T.B.; Christophersen, L.; Calum, H.; Hentzer, M.; Hougen, H.-P.; Rygaard, J.; Moser, C.; Eberl, L.; et al. Garlic blocks quorum sensing and promotes rapid clearing of pulmonary Pseudomonas aeruginosa infections. *Microbiology* **2005**, *151*, 3873–3880. [CrossRef]

49. Tsao, S.-M.; Yin, M.-C. In-vitro antimicrobial activity of four diallyl sulphides occurring naturally in garlic and Chinese leek oils. *J. Med. Microbiol.* **2001**, *50*, 646–649. [CrossRef]

50. González-Fandos, E.; García-López, M.L.; Sierra, M.L.; Otero, A. Staphylococcal growth and enterotoxins (A-D) and thermonuclease synthesis in the presence of dehydrated garlic. *J. Appl. Bacteriol.* **1994**, *77*, 549–552. [CrossRef]

51. Moreira, M.R.; Ponce, A.; del Valle, C.E.; Roura, S.I. Inhibitory parameters of essential oils to reduce a foodborne pathogen. *LWT-Food Sci. Technol.* **2005**, *38*, 565–570. [CrossRef]

52. Marino, M.; Bersani, C.; Comi, G. Impedance measurements to study the antimicrobial activity of essential oils from Lamiaceae and Compositae. *Int. J. Food Microbiol.* **2001**, *67*, 187–195. [CrossRef]

53. Hajlaoui, H.; Trabelsi, N.; Noumi, E.; Snoussi, M.; Fallah, H.; Ksouri, R.; Bakhrouf, A. Biological activities of the essential oils and methanol extract of tow cultivated mint species (Mentha longifolia and Mentha pulegium) used in the Tunisian folkloric medicine. *World J. Microbiol. Biotechnol.* **2009**, *25*, 2227–2238. [CrossRef]

54. Rota, M.C.; Herrera, A.; Martínez, R.M.; Sotomayor, J.A.; Jordán, M.J. Antimicrobial activity and chemical composition of Thymus vulgaris, Thymus zygis and Thymus hyemalis essential oils. *Food Control* **2008**, *19*, 681–687. [CrossRef]

55. Turchi, B.; Mancini, S.; Pistelli, L.; Najar, B.; Fratini, F. Sub-inhibitory concentration of essential oils induces antibiotic resistance in *Staphylococcus aureus*. *Nat. Prod. Res.* **2017**, 1–5. [CrossRef]

56. Firmino, D.F.; Cavalcante, T.T.A.; Gomes, G.A.; Firmino, N.C.S.; Rosa, L.D.; de Carvalho, M.G.; Catunda, F.E.A., Jr. Antibacterial and antibiofilm activities of *Cinnamomum* Sp. essential oil and cinnamaldehyde: Antimicrobial activities. *Sci. World J.* **2018**, *2018*, 1–9. [CrossRef] [PubMed]

57. Choi, O.; Cho, S.K.; Kim, J.; Park, C.G.; Kim, J. In vitro antibacterial activity and major bioactive components of Cinnamomum verum essential oils against cariogenic bacteria, Streptococcus mutans and Streptococcus sobrinus. *Asian Pac. J. Trop. Biomed.* **2016**, *6*, 308–314. [CrossRef]

58. Kalia, M.; Yadav, V.K.; Singh, P.K.; Sharma, D.; Pandey, H.; Narvi, S.S.; Agarwal, V. Effect of Cinnamon Oil on Quorum Sensing-Controlled Virulence Factors and Biofilm Formation in Pseudomonas aeruginosa. *PLoS ONE* **2015**, *10*, e0135495. [CrossRef]

59. Utchariyakiat, I.; Surassmo, S.; Jaturanpinyo, M.; Khuntayaporn, P.; Chomnawang, M.T. Efficacy of cinnamon bark oil and cinnamaldehyde on anti-multidrug resistant Pseudomonas aeruginosa and the synergistic effects in combination with other antimicrobial agents. *BMC Complement. Altern. Med.* **2016**, *16*, 158. [CrossRef]

60. Abubakar, E.-M.M. Efficacy of crude extracts of garlic (*Allium sativum Linn.*) against nosocomial *Escherichia coli*, *Staphylococcus aureus*, *Streptococcus pneumoniea* and *Pseudomonas aeruginosa*. *J. Med. Plant Res.* **2009**, *3*, 179–185.

61. Palaksha, M.N.; Ahmed, M.; Das, S. Antibacterial activity of garlic extract on streptomycin-resistant Staphylococcus aureus and Escherichia coli solely and in synergism with streptomycin. *J. Nat. Sci. Biol. Med.* **2010**, *1*, 12–15. [CrossRef]

62. Zhang, Y.; Liu, X.; Wang, Y.; Jiang, P.; Quek, S. Antibacterial activity and mechanism of cinnamon essential oil against Escherichia coli and Staphylococcus aureus. *Food Control* **2016**, *59*, 282–289. [CrossRef]

63. Burns, P.; Borgo, M.F.; Binetti, A.; Puntillo, M.; Bergamini, C.; Páez, R.; Mazzoni, R.; Reinheimer, J.; Vinderola, G. Isolation, Characterization and Performance of Autochthonous Spray Dried Lactic Acid Bacteria in Maize Micro and Bucket-Silos. *Front. Microbiol.* **2018**, *9*, 2861. [CrossRef] [PubMed]

64. Kavitake, D.; Kandasamy, S.; Devi, P.B.; Shetty, P.H. Recent developments on encapsulation of lactic acid bacteria as potential starter culture in fermented foods—A review. *Food Biosci.* **2018**, *21*, 34–44. [CrossRef]

65. Unlu, M.; Ergene, E.; Unlu, G.V.; Zeytinoglu, H.S.; Vural, N. Composition, antimicrobial activity and in vitro cytotoxicity of essential oil from Cinnamomum zeylanicum Blume (Lauraceae). *Food Chem. Toxicol.* **2010**, *48*, 3274–3280. [CrossRef] [PubMed]

66. Buru, A.S.; Pichika, M.R.; Neela, V.; Mohandas, K. In vitro antibacterial effects of Cinnamomum extracts on common bacteria found in wound infections with emphasis on methicillin-resistant Staphylococcus aureus. *J. Ethnopharmacol.* **2014**, *153*, 587–595. [CrossRef] [PubMed]

67. Manson, J.S.; Cook, D.; Gardner, D.R.; Irwin, R.E. Dose-dependent effects of nectar alkaloids in a montane plant-pollinator community. *J. Ecol.* **2013**, *101*, 1604–1612. [CrossRef]

68. Shelp, B.J.; Bown, A.W.; Zarei, A. 4-Aminobutyrate (GABA): A metabolite and signal with practical significance. *Botany* **2017**, *95*, 1015–1032. [CrossRef]

69. Bouché, N.; Fromm, H. GABA in plants: Just a metabolite? *Trends Plant Sci.* **2004**, *9*, 110–115. [CrossRef]

70. Thiruvengadam, M.; Kim, S.-H.; Chung, I.-M. Influence of amphetamine, γ-aminobutyric acid, and fosmidomycin on metabolic, transcriptional variations and determination of their biological activities in turnip (Brassica rapa ssp. rapa). *South Afr. J. Bot.* **2016**, *103*, 181–192. [CrossRef]
71. Soma, Y.; Fujiwara, Y.; Nakagawa, T.; Tsuruno, K.; Hanai, T. Reconstruction of a metabolic regulatory network in Escherichia coli for purposeful switching from cell growth mode to production mode in direct GABA fermentation from glucose. *Metab. Eng.* **2017**, *43*, 54–63. [CrossRef]
72. Dover, S.; Halpern, Y.S. Utilization of -aminobutyric acid as the sole carbon and nitrogen source by Escherichia coli K-12 mutants. *J. Bacteriol.* **1972**, *109*, 835–843. [CrossRef]
73. Cronan, J.E.J. Beta-alanine synthesis in Escherichia coli. *J. Bacteriol.* **1980**, *141*, 1291–1297. [CrossRef] [PubMed]
74. Schneider, F.; Kramer, R.; Burkovski, A. Identification and characterization of the main β-alanine uptake system in Escherichia coli. *Appl. Microbiol. Biotechnol.* **2004**, *65*, 576–582. [CrossRef] [PubMed]
75. Kapoor, A.; Malik, A. Beta-alanine: Design, synthesis and antimicrobial evaluation of synthesized derivatives. *Der Pharm. Lett.* **2016**, *8*, 135–142.
76. Kadhum Shneine, J.; Ziad Qassem, D.; Saad Mahmoud, S. Synthesis, Characterization and Antibacterial Activity of Some Amino Acid Derivatives. *Der Pharma Chem.* **2017**, *9*, 100–104.

Article

On the Importance of the Sound Emitted by Honey Bee Hives

Alessandro Terenzi [†], **Stefania Cecchi** [*,†] **and Susanna Spinsante** [†]

Dipartimento di Ingegneria dell'Informazione, Universitá Politecnica Delle Marche, 60131 Ancona, Italy; a.terenzi@pm.univpm.it (A.T.); s.spinsante@staff.univpm.it (S.S.)
* Correspondence: s.cecchi@staff.univpm.it
† These authors contributed equally to this work.

Received: 22 September 2020; Accepted: 28 October 2020; Published: 31 October 2020

Abstract: Recent years have seen a worsening in the decline of honey bees (*Apis mellifera* L.) colonies. This phenomenon has sparked a great amount of attention regarding the need for intense bee hive monitoring, in order to identify possible causes, and design corresponding countermeasures. Honey bees have a key role in pollination services of both cultivated and spontaneous flora, and the increase in bee mortality could lead to an ecological and economical damage. Despite many smart monitoring systems for honey bees and bee hives, relying on different sensors and measured quantities, have been proposed over the years, the most promising ones are based on sound analysis. Sounds are used by the bees to communicate within the hive, and their analysis can reveal useful information to understand the colony health status and to detect sudden variations, just by using a simple microphone and an acquisition system. The work here presented aims to provide a review of the most interesting approaches proposed over the years for honey bees sound analysis and the type of knowledge about bees that can be extracted from sounds.

Keywords: bee hive monitoring; real-time monitoring; sound measurement; swarming detection; queen bee detection; sound analysis

---

1. Introduction

Among insects, honey bees are well known for their positive effects on a multitude of different scopes. They do not only produce honey, beeswax, royal jelly, and propolis, but they are at the basis of the plants pollination, playing a key role in the proliferation of both spontaneous and cultivated flora. Recent years have seen an increase in bee mortality, due to several different factors. One of the most recent and dangerous syndromes affecting honey bee colonies is the Colony Collapse Disorder (CCD), which is characterized by a sudden disappearance of honey bees from the hive [1–3]. Many bee scientists agree that the decline of honey bee colonies is the result of multiple negative factors working independently, in combination, or synergistically to impact honey bee health [2,4]. Large bee mortality can result in a loss of pollination services with negative ecological and economic [5] impacts that could significantly affect the maintenance of wild plant diversity, wider ecosystem stability and crop production [6,7]. Solutions such as robotic pollination were proposed in recent years, but they were already shown to be technically and economically unacceptable solutions, currently posing substantial ecological and ethical risks [8]. In this context, the necessity of a continuous and intensive monitoring of bee hives' status emerges clearly, in order to safeguard and protect these important insects. Over the years, several monitoring systems have been proposed in the literature: most of them are based on the measurement of several hive parameters such as temperature, humidity, carbon dioxide, and weight [9,10]. In recent years, a great improvement has come from the integration of web technologies and smart sensors. In [11], a web-based monitoring system built upon sensors and a cloud architecture, to monitor and follow bees' behavior, is described. In [12–15], different approaches for heterogeneous wireless sensor networks technologies to gather data unobtrusively from a bee hive have been described. Approaches based on computer vision have been proposed as well: in [16], a system to track bees in a 50 Hz frame rate 2D video of the hive entrance close view is presented. In [17], a real-time imaging system for multiple honey bees tracking and activity monitoring, by counting the honey bees entering and exiting the bee hive, is proposed. Despite all these approaches having shown good results, they still require remarkable computational power, several different types of sensors, and sometimes they impose some modifications to the hive physical structure. A non-invasive technique to monitor the bees' status relies on sound analysis [18]. Vibration and sound signals are used by bees to communicate within the colony [19,20]. Modalities of vibroacoustic signal production include gross body movements, wing movements, high-frequency muscle contractions without wing movements, and pressing the thorax against the substrates or another bee [21–23]. Vibroacoustic signals modulate behaviors that affect swarming, and the queen's behavior during swarming. In fact, it has been proved that there is a strict correlation between the frequencies of vibroacoustic signals, the amplitudes detected inside the honey bee hives, and the forecasting of events like swarming. The sound can be recorded by means of microphones or accelerometers placed in specific spots inside or outside the hives, and then it can be analyzed to detect the colony health status according to a workflow schematically represented in a general fashion in Figure 1.

This paper aims to provide an up-to-date review on honey bees' sound analysis approaches, discussing the pros and the cons of each technique, following their historical evolution. The paper is organized as follows: Section 2 provides a review of the scientific literature and how the research on sound emitted by bees evolved along time. Patents related to the analysis of sounds from bees are presented in Section 3, while Section 4 focuses on recent results. Finally, Section 5 provides some concluding remarks and insights for future investigations and developments.

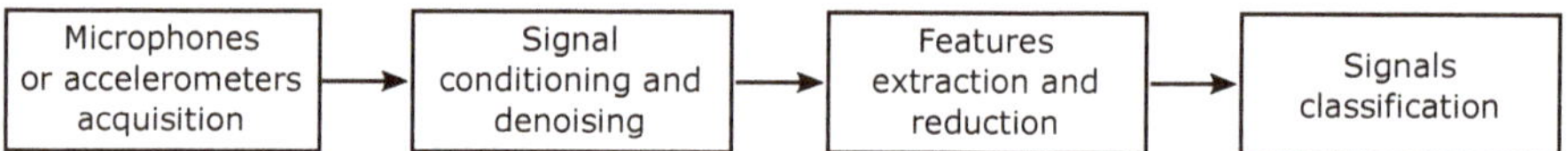

Figure 1. Typical workflow for vibroacoustic signal analysis and classification. First, the sound is acquired inside or outside the hive using microphones or accelerometers. Then, the recorded signal is usually filtered or resampled to remove noise and unwanted frequencies; then, features are extracted from the signal, exploiting different algorithms. If necessary, a features reduction process is applied, based on algorithms such as Principal Component Analysis (PCA). Finally, the data are passed to a classification algorithm to detect the colony health status: typical classifiers are based on Support Vector Machine (SVM), Linear Discriminant Analysis (LDA), and Machine Learning (ML).

2. Literature Review

2.1. The Early Works

The topic of analysing the sound emitted by honey bees has always raised the interest of many scientists from different fields. Among the first ones to make some observations on honey bees' sound was Aristotle who lived in the third century BC. In one of his writings [24], he observed that, when a swarming is imminent, a particular sound is produced continuously by the bees, several days before the swarming event. Later on, one of the early works came from the 17th century AD, in particular one of the first descriptions of queen piping was given by Charles Butler [25], who reported in his book that the bees produce two different sounds. A century later, Francis Huber [26], studying the birth of a new queen, discovered that the first young queens born emit a sound called tooting, and that other young queens, who are still sitting in their cells, respond with another sound called quacking.

2.2. The 20th Century and the First Modern Study

During the second half of the XX century, thanks to modern electronics, deeper and more accurate studies on the sound emitted by honey bees were carried out. In 1957, Frings et al. [19]

published an article showing that bees react to different sounds. In particular, it was demonstrated that signals of certain frequencies and amplitude produce an almost total cessation of movement of workers bee and drones. Some years later, in 1964, Wenner et al. [27] performed one of the first spectral analyses on the honey bee's sound. Using a spectrograph and a microphone placed inside the colony, he discovered information about the sound produced during the waggle dance and after the birth of a new queen. The waggle dance is used by bees to communicate within the colony the information about the distance and position of flowers. The work shows that, during the dance, bees emit a specific pulsed sound, in which the number of pulses is proportional to the distance between the colony and the flowers. Wenner also deepened the study about tooting, showing that this sound is composed of two different pulses, the first is a long pulse of 1 s duration followed by shorter pulses, with a fundamental frequency of 500 Hz and some overtones. The quacking has a lower fundamental frequency and starts with shorter pulses. In 1974, Simpson et al. [28] performed a study to better understand the queen bee piping sound [29]. Using a modified speaker attached to a hive, a sinusoidal 650 Hz vibration signal was generated, in order to obtain a signal similar to the one produced by queen bees. When the signal was applied to four hives containing very small colonies with unmated queens, all the four colonies swarmed. The second experiment was carried out with sixteen bigger colonies, each with a young queen bee. This time the sound was applied only to eight colonies which swarmed, leaving a portion of their bees in the hive. The other eight colonies which did not receive the sound didn't swarm. Despite the simplicity of the generated sound, the experiments show that bees react to sound, even if the limited technology available at the time didn't allow the authors to generate more complex signals that could also work for colonies with an older queen bee.

In 1985, Dietlein et al. [30] performed a longer study acquiring and analyzing the sound from three different colonies for over a year. The setup used involved two different microphones: one placed inside the hive (for the recording of bees' sound) and the other one placed between the inner cover and the telescoping outer cover of the hive (to capture the background noise). The microphones were both connected to an analog circuit, which, thanks to a differential amplifier, subtracts the background noise, acquired from the outer microphone, from the honey bees' sounds. The amplitude of the signal was recorded hourly for a year, showing that there are some significant variations in the amplitude and frequencies of the sound produced in different seasons. Moreover, a frequency analysis has also shown that the spectral content of the signal changes with the seasons and the hour of the day. This study shows an interesting prospective on long-term sound analysis inside the colony, but one of the main drawbacks is probably the analog circuitry used to process the signal, i.e., the analog subtraction between the signals generated by the microphones and the amplification could introduce an additive noise on the signal. One year later, in 1986, Michelsen et al. [20] performed a study more focused on the sound emitted during the waggle dance, analyzing both the vibration and the airborne sound produced. The idea was to better understand which mechanisms are used to communicate during the waggle dance: since it is performed inside the hive in the dark, bees cannot rely on their sight to see the dance, and they must use other communication means, such as vibration, airborne sound, and chemical transmission. Using both a laser vibrometer and microphones placed inside the colony, Michelsen discovered that no vibration is produced during the dance, and that all of the information is passed through the airborne sounds. In particular, as already discovered by Wenner [27], the waggle sound is a pulsed sound, with 20 ms—long pulses at the frequency of 250–300 Hz. It was also discovered that the waggle sound at 2 cm from the bee who is dancing produces a sound wave with a Sound Pressure Level (SPL) of 73 dB. The second part of Michelsen's work is more focused on the mechanical aspects of the propagation of sound and vibration inside the colony. In this case, the experiment involves the generation of synthetic comb vibration to better understand the freezing phenomena already described in [19]. Again, as already discovered, bees seem to freeze at specific frequencies, and this led Michelsen to speculate that these particular signals are used by bees to quiet the colony, in order to allow a better transmission of important messages. About ten years later, Eren et al. [31] carried out a study in order to better understand and try to emulate the communication between worker bees and the queen bee. From this study, it results that worker bees are not able to produce sound with frequencies higher than 500 Hz, while queen bees are able to produce almost all workers frequencies and many higher ones. Researchers tried to acquire the sound from over 150 queen bees confined in standard queen bee shipment cases, arranged in a matrix form. The acquired sound has been analyzed and then synthesized thanks to a computer. However, the results seem quite inconsistent, exhibiting a weak response from both queen bee and workers to the synthesized sounds that are more complex than the one proposed in [20,27]. This research topic seems quite promising and interesting: the inconsistent results could be mainly due to the experimental setup used for the queen bee sound acquisition. In fact, putting together more than 150 queen bees is not a normal situation, and the sound produced could be very different from the real one generated inside the colony. To achieve better results, it is desirable to acquire the signals in a more natural situation, trying to avoid stressors acting on the bees.

2.3. The 21st Century and the Technological Advancement

Within the last twenty years, technological progress has allowed the application of new solutions such as digital signal processing, ML, and low cost smart sensors, enabling new discoveries. In 2005, Ferrari et al. [32] focused their studies on the swarming phenomena. In particular, researchers acquired signals from three hives for over 270 h, observing nine swarming events. Each hive was equipped with an omnidirectional microphone, placed on top of a hive's looms under the cover, and one humidity and temperature sensor placed inside the hive in between the looms. In the paper, authors do not provide specific motivation for the positioning of the sensors, but we can speculate about the need for keeping the number of microphones used as low as possible-thus adopting a single omnidirectional one-and for monitoring in-the-hive ambient conditions which could as well be correlated to a swarming event. Acquired data were synchronized and then manually labeled in order to identify the swarming events, but authors do not specify which of the three hives was involved in the process. Each event has been analyzed both in frequency and time domain, in order to find any correlation between sound, temperature, and humidity during swarming. Sound spectrograms have shown that there is a quick change in the frequency content during swarming. Before the swarming, most of the energy content of the signal is around 150 Hz, while, during swarming, it jumps to 500 Hz. Moreover, the joint analysis of sound, temperature, and humidity has shown that, during swarming, there is an increase in the sound amplitude, and, meanwhile, temperature and humidity decrease. According to the authors, this behavior could be due to the ventilation produced by bees ready for swarming.

In 2008, Papachristoforou et al. [33] made an interesting discovery, showing that the sound emitted by honey bees could reach high frequencies, even up to 15–16 kHz. The experiment carried out was focused on sound emitted by the bees during a hornet attack. A hornet was artificially introduced inside the hive and the bees' reaction was recorded. A frequency analysis based on spectrograms revealed that, during a hornet attack, guard bees produce hissing sounds with a fundamental frequency at around 5 kHz, with several high order harmonics which can reach 15 and 16 kHz. The generated sound seems to have a precise structure and this led to thinking that this is a true signal used for communication and not a simple noise produced under stress. Some studies seem to show that honey bees are capable of picking up sound at over 10 kHz with the sub-genual organ, a chordotonal sensor localized in the proximal part of the tibia of each leg [34]. Such a high-frequency structure of the signal could be motivated by the fact that it makes the sound very distinctive from the background noise of the colony, allowing a clear transmission of the alarm message to the whole colony. This study has an important consequence, since it seems to show that the sound produced by the bees reaches frequencies that many other studies seem to ignore completely. In 2011, Eskov et al. [35] proposed a totally different approach, based on statistical sound analysis. The authors considered the recorded signal as a noise with a specific spectral content. As other types of colored noise, the one produced from the bees also has a precise statistical behavior which could be measured. Data from twelve hives were acquired every night from midnight to 3:00 a.m. within the spring and summer periods.

Each recording was divided into one second long fragments which could be analyzed by estimating the relative fluctuations of the signal. Each fragment was first normalized and then smoothed using the Procedure of Optimal Linear Smoothing (POLS) [36] with Gaussian kernel, applied to the integrated sequence of original data. Finally, the sequences of ranked amplitudes of relative fluctuations have been estimated giving a statistical indicator of sound behavior. The paper shows that, when bees are going to swarm, the fluctuation changes and this could be used to predict the swarming event several days before it could happen. This work is quite interesting as it proposes an original point of view and could be used for swarming detection. The main drawback is related to the fact that the authors assume that most of the recorded signal is only noise, and this assumption could be a limit in some cases. In 2013, Howard et al. [37] proposed a classification algorithm for queenless colonies, based on the Stockwell Transform (S-Transform) [38] as a features extraction tool. Four colonies of two different species have been monitored, with two normal colonies and two orphaned ones. Data were first analyzed using the S-Transform, which is a time-frequency representation of the signal derived from the wavelet transform with a modification in the phase of the mother wavelet. A qualitative analysis was carried out, comparing the S-Transform, with spectrograms and Fourier transform of the signal. For the classification and representation of the signals, a Self-Organising Maps (SOM) [39] was used. The SOM is trained with two neural networks and allows a clear representation and classification of datasets featuring high dimensionality. Results show that the SOM approach is able to classify the two states with some issues, probably related to the use of two different bee species, or to the classification algorithm. In 2014, Quandour et al. [40] went beyond the simple sound analysis, using the recorded sound to automatically detect if the colony was healthy or infected by the varroa mite. The varroa destructor is one of the most dangerous honey bees parasites and is considered as one of the factors which is contributing to the higher levels of bee losses around the world [41]. The proposed approach was based on the extraction of four statistical indicators from the sound: peak frequency, spectral centroid, bandwidth, and root variance frequency. These four features were then passed to a Principal Component Analysis (PCA) algorithm to reduce the data dimensionality, and finally the data were input to a classifier to detect if the colony was healthy or infected. Two types of classifiers were used, the former based on SVM and the latter on LDA. Both show good performances allowing the discrimination of the two colony statuses. A notable aspect is the reduced computational costs of the proposed algorithm which was implemented on a low-cost single board computer, but, on the other side, the dataset used in the experiments was too limited to give a reliable benchmark on algorithm performances. One year later, Murphy et al. [42] proposed a complete platform for honey bees' monitoring. In this work, the microphone is only one of the sensors installed within the hive, and the other parameters acquired include CO_2, temperature, humidity, acceleration data. Additionally, an infrared camera inside the hive and a thermal camera outside the hive are installed. Sound is used to detect possible swarming events in a very simple way: the system filters and then estimates the envelope of the recorded sound, and, if the amplitude of the signal rises quickly, the system sends a message to the beekeeper. This simple monitoring system could be used for swarming detection on a very simple hardware with also a reduced power consumption, but, due to its simplicity, it could easily generate false positive warnings. In 2017, Ramsey et al. [43] focused their work on the whooping signal, trying to detect this particular signal in different conditions. Vibracoustic signals of three colonies were acquired in different periods of the year and in different geographical locations, with two colonies placed in the UK and one in France. Each colony was equipped with a high precision accelerometer placed in the center of a brood frame. The whooping signal was detected with a two-step process. First, the spectrogram of the recorded signal is matched to the spectrogram of a template pulse, then the ratio of the cross-correlation product and the Euclidean distance among the two is used to find the pulsed signal. Then, in order to discriminate between a whooping signal and non-whooping signal, PCA and LDA are used. The number of whooping signals recorded shows variations which seem related to weather conditions, geographical positions, and time of the day. The whooping sound monitoring could be used as an indicator for hive monitoring;

however, the type of sensor used and its precise positioning inside the colony limit this approach, avoiding a possible large scale implementation of the system. In 2018, Kulyukin et al. [44] published an article where they exploit ML techniques to analyze the sound of the hives. In particular, the main goal was to distinguish the honey bee sound, from the background noise and the cricket chirping noise. Four microphones were placed outside the entrance of six Langstroth bee hives. A sound frame of 30 s was recorded every hour, from May to July. Bee hives were placed in different locations, with many different background noises. Data have been manually labeled into three categories: honey bee, cricket, and background noise. With the obtained dataset, several approaches were tested to classify the data; in particular, an ML approach based on a Convolutional Neural Network (CNN) has been compared to traditional classifiers such as Logistic Regression, k-Nearest Neighbors (k-NN), SVM with a linear kernel, one vs. rest classification, and (M4) Random Forests. The results show that, for these types of problems, an ML approach could be used with success, but, since it wasn't directly applied to the honey bees' sound, a similar algorithm could be used only as a preprocessing technique to remove unwanted sounds. In the same year, Cejrowski et al. [45] proposed an algorithm to detect the presence of the queen bee based on sound analysis. The system involves a custom brood frame placed inside the colony, with a microphone, and a temperature and humidity sensor. Data have been acquired from a single hive from February 2017 to August 2017, forcing a critical situation for the bees, by removing the queen bee from the hive. Data acquired with the normal colony and the orphaned colony have been analyzed using a Linear Predictive Coding (LPC) for features extraction. This algorithm is based on the source-filter model of a speech signal, i.e., first, the input signal is used to produce a linear model with a number of poles; then, the resulting set of coefficients is used to model the unknown system. Following the LPC coefficients estimation, the t-distributed stochastic neighbor embedding (t-SNE) [46] algorithm has been used to reduce the data dimensionality and, finally, an SVM algorithm was used to classify the results. The algorithm seems to work well with a good ability of detecting changes in the colony sound; however, the data used are limited, with a single colony and a single orphaned event analyzed. Moreover, with the new queen bee, the algorithm seems to need a new training process while a similar approach should be able to detect orphaned and normal colony situations with no need for such a step.

3. Patents

Several patents have also been issued over the years. One of the first was presented in 1957 proposing a particular instrument called the Apidictor [47]. The system presented is composed of a microphone, a vu-meter and a series of tunable analog filters. The idea is to adjust the filters to the specific frequency bands at which bees emit specific sounds (i.e., the piping sound) and then to monitor the activity. In 2007, Bromenshenk et al. [48] submitted a patent to exploit the sound emitted by honey bees to monitor air pollution. The main idea is that, when bees are exposed to sub-lethal concentrations of various airborne toxicants, the generated sounds are different and specific for each pollutant. These sounds can be acquired and stored into a database. Then, when a new sound is acquired inside a colony, it is possible to make a comparison with the database, in order to detect pollutants near the colony. For the comparison, first, the spectrograms of various sounds are extracted and then a classification by means of linear discriminant functions is carried on. In 2012, Brundage et al. [49] proposed a system to detect bees productivity based on sound analysis. The idea was to analyze the fundamental harmonic produced by the bees' flight, which is in the range between 180 Hz to 260 Hz. The presence of a downward frequency shift in the fundamental frequency corresponds to a flying bee launching from locations around the bee hive entrance. By counting the number of frequency shifts, it is possible to estimate the number of bees which have left the hive during a day, and then it is possible to use these data to monitor the hive productivity. One of the most recent patents is from Bencsik et al. [50], who proposed, in 2015, a solution based on previous research on sound emitted by honey bees [43]. The system involves the use of one or more accelerometers, placed at the center of the brood frame. The acquired signal is then transformed into the frequency domain producing a spectrogram. The spectrogram is then processed with PCA to reduce the data dimension. Finally, a linear discriminant function is used to analyze the signal at a reduced computational cost in order to detect events such as swarming or a Varroa mite infection.

4. Recent Results

In 2019–2020, several works have been published on bee sound analysis. Nolasco et al. [51] applied ML techniques to analyze honey bees' sound and detect the queen bee presence. Data from the Nu-Hive project [52] were used, analyzing sound from two different colonies in normal and orphaned situations. Two features' extraction techniques were used: Mel Frequency Cepstral Coefficients (MFCC) and the Hilbert–Huang Transform (HHT) [53]. MFCC is a widely known technique for signal representation, in which the coefficients are generated starting from the square of signal spectrogram; then, a triangular filterbank is applied to the signal; finally, the Discrete Cosine Transform (DCT) is applied to the logarithmic output of the filterbank. Regarding HHT, the features extraction technique is based on a combination of two algorithms, i.e., the Empirical Mode Decomposition (EMD), which decomposes the signal into a set of basis functions, and the Hilbert Transform (HT) which transforms each basis function into a time-frequency representation of the signal. Using the SVM classifier, a comparison of the performance of both approaches was presented, showing that the best results were achieved with a combination of MFCCs and HHT coefficients. Some experiments with CNNs were carried out as well, using MFCCs as input data and providing good performances. In the mentioned work, authors exploit ML for orphaned colony analysis; however, the most innovative part i.e., the HHT, has not been used in combination with ML, leaving this aspect to a future development. Within the same year, another work was presented involving the queen bee presence detection: Robles-Guerrero et al. [54] acquired sensor data from five hives, and each hive was equipped with a single microphone placed a the center of a brood frame. Data were acquired from March to April, recording 30-s long frames every ten minutes with a Raspberry Pi 2. MFFCs were first computed and then statistical moments for each mel coefficient were calculated involving: mean, standard deviation, variance, skewness, median, and kurtosis. Lasso regularization was then used to reduce again the dimensionality and finally a Logistic Regression algorithm was used for classification. Some experiments were also proposed using Singular Value Decomposition (SVD) and scatter plots to analyze the behavior and separability of the datasets. Among the five colonies, one lost its queen naturally, while, in other two families, the queen bee was removed intentionally. Two experiments were carried out, the former comparing one orphaned colony with the four healthy ones, and the latter removing two queen bees and comparing the data with two normal colonies, and the queenless one. The classifier was able to separate an orphaned colony from the healthy one with a good accuracy. In the same year, another study on sound based swarming detection was published [55]. Data from the open source bee hive project [56] were taken, using MFCCs and LPC as features. As classifiers, two different algorithms were used, the former based on Hidden Markov Model (HMM) and the latter on a Gaussian mixture model (GMM). Several experiments with different combinations of classifiers and features have been carried on, showing that the best solution is MFCCs in combination with an HMM classifier. The work also shows some experiments focusing on the performance degradation due to a high level of background noise in the recorded signals. Again, as in other works, the main drawback is the limited dataset size, since only three swarming events were available and were used for the experiments with only 90 min of recordings for the training phase. Two other studies were published in the same year, which more deeply analyze another point, i.e., how to extract the information from the recorded sound inside the colony. In [57,58], the authors proposed some innovative approaches for the features extraction problem. Many previous research works focused their work on the classifier, using only spectrograms or at least MFCCs to extract important information from the sound. In these two works, some innovative approaches for features extraction are discussed, focusing on orphaned colony situations [57] and swarming [58]. Well known approaches such as spectrograms and MFCCs are compared with innovative algorithms, based on Hilbert–Huang transform, and wavelet transform [59].

Wavelet transform (WT) allows the generation of a time-frequency representation of the signal, and it is based on the decomposition of the signal with specific basis functions called *mother wavelet*. Similar to the Fourier transform, the mother wavelet is translated in time and compressed in amplitude to represent the original signal. Two types of wavelet transforms are available, i.e., the discrete wavelet transform and the continuous one. The results presented seem to show that these approaches are able to better distinguish different signals, and this will be more deeply investigated in a future work where these approaches will be combined with classification algorithms. Another study proposed in 2019 is from [60]. Here, the authors proposed a Raspberry Pi-based acquisition system, able to collect the following parameters: images from the hive entrance, sound, weight, external temperature and humidity, internal and external light level, internal temperature, and air quality. Data are analyzed to detect changes in different periods; weight is also used for swarming detection. A simple analysis of recorded sound is performed by means of a Fourier transform. One of the most recent contributions in this field is from Ramsey et al. [61]. In this work, the authors improve the results of their previous work [43]. In particular, using the same dataset, they propose two different methodologies to predict swarming. One is based on the extraction of one hour logarithmic averaged spectra; then, the cross-correlation product between the spectra and three discriminant functions is calculated. Finally, a linear discriminant function is used to detect swarming. The discriminant functions are estimated by a specific algorithm which analyzes spectrograms of swarming events to estimate the discriminant curves. The second approach is based on a long-term estimation: the algorithm analyzes ten days of vibrational data, acquired from midnight to 5:00 a.m. Ten spectrograms are firstly estimated and then the Fast Fourier Transform (FFT) of each spectrogram is performed on the time course of the magnitude of all the uploaded spectral frequencies for each day, yielding two-dimensional Fourier transform images (2DFT). The FFT is then further calculated over each pixel of the series of 2DFTs found in the preceding days, obtaining a three-dimensional Fourier transform (3DFT). The 3DFT spectrograms are cross correlated with specific discriminant functions and eventually the swarming is predicted using a linear discriminant function. The results seem to show that the first approach is more reliable, providing less false positive swarming events. The second part of the article is focused on the structure of tooting and quacking signals. The results are quite interesting, opening to the possibility of having a reliable swarming prediction. The only problem in the proposed approach is the type of sensor used, which has a high cost and is not suitable for a large scale application. Finally, another recent work is presented in [62], where a combined study is considered by crossing the results of the analysis of the sound generated inside the hive by the honey bees, with other common parameters such as temperature, humidity, CO_2 and hive weight, and weather conditions, acquired with several sensors. In this work, the changes in the weight produced during the swarming phenomena are related to the changes in the sound amplitude. Moreover, the other sensors' variations are analyzed in a normal day and during longer periods, showing trends related to the honey production and the colony activity.

5. Future Works and Conclusions

In this paper, following the historical path, a complete overview of the state-of-the-art of approaches and techniques for honey bees sound analysis has been presented: Table 1 shows a summary of all the proposed approaches analyzed in this work, while, in Figure 2, the microphones and accelerometers placement used in several works are visible. Signal analysis techniques, such as HHT, wavelet transform, and multidimensional FFT and LPC have been applied for the first time to the analysis of honey bees' sound and have been compared to classical approaches based on Fourier transform and MFCCs. Classification algorithms have been discussed as well: linear discriminant approaches have been improved using more sophisticated discriminant functions; other classifiers such as CNNs, SVM, GMM, and HMM have been applied and compared with more traditional approaches. Future works will more deeply analyze these recently proposed approaches, combining new features extraction techniques with innovative classification algorithms.

Table 1. Summary of the state-of-the-art of approaches discussed in this work.

Approach	Description	Applications	References
Spectrograms	The sound is recorded and then analyzed using spectrograms, searching for changes in the harmonic content of the signal.	Analysis of waggle dance, analysis of piping and tooting. Swarming detection. Measuring bees reaction to hornet attack.	[20,27,30,32,33]
Tone based sound synthesis	A loudspeaker or a shaker is placed inside the hive, different tones at different amplitudes and frequencies are generated and bees reaction is monitored.	Find the frequency at which bees react with movement cessation. Reproduce the harmonic generated by the queen bee, to stimulate a swarming.	[19,20,28]
Amplitude monitoring	Amplitude and envelope of the recorded signal is used to detect different behaviors.	Changes of the amplitude in different seasons and conditions. Measuring of SPL during waggle dance. Swarming detection.	[20,30,42]
Bees sound synthesis	The bees sound is firstly recorded and then analyzed and synthesized by means of a computer. The synthesized sound is then reproduced inside the colony and bees reaction is monitored.	Measure the response of worker bees to the synthesized queen bee sound.	[31]
Noise analysis	The recorded sound inside the colony is considered as a noise with a specific statistical behavior. Some statistical indicators are extracted from the sound, changes in the statistical indicators are related to specific colony behaviors.	Swarming detection and prediction.	[35]
Statistical indicator analysis	From the recorded sound, peak frequency, spectral centroid, bandwidth and root variance frequency are extracted. PCA is used to reduce the dimensionality of the indicators and finally SVM or LDA is used to classify the signals.	Detect the presence of Varroa destructor inside the colony.	[40]
Whooping detection	Precision accelerometer inside the colony are used to record the bees vibrations. Spectrograms of vibrations are cross correlated with a pulse signal to detect pulsed signals, LDA and PCA are then used to isolate whooping signals.	Measuring the variation of the whooping signal during different seasons and geographical locations.	[43]
Bees sound detection	The sound is acquired at the hive entrance. Spectrograms of the recorded sound are classified using different algorithms such as, CNNs, logistic regression, SVM, k-NN, one vs. rest and random forest.	Distinguishing the honey bee sound, from the background noise and the cricket chirping noise	[44]
LPC sound analysis	Sound acquired inside the hive is analyzed using LPC as features extraction algorithm. T-SNE algorithm is then used to reduce dimensionality, and finally SVM is used to classify the signals.	Queen bee presence detection.	[45]
HHT and MFCC analysis	Recorded sound inside the colony is analyzed using MFCCs and HHT as features. CNNs and SVM are then applied to classify the signals.	Queen bee presence detection, swarming detection.	[51,57,58]
MFCC analysis	MFCCs are estimated from the recorded signal. Lasso regularization is then used for dimensionality reduction and finally logistic regression algorithm is used for classification.	Queen bee presence detection.	[54]
Wavelet analisys	Wavelet transform is applied to the recorded signal to analyze the sound and detect different behavior.	Queen bee presence detection, swarming detection	[57,58]
MFCC and LPC analysis	MFCC and LPC are used as features, HMM and GMM are used as classifier.	Swarming detection.	[55]
Multimensional FFT	Two and three-dimensional spectrograms are generated starting from the sound recorded using accelerometers placed inside the colony. A discriminant function is then used to classify the signals and detect specific events using two different algorithms.	Swarming detection and swarming prediction.	[61]

Figure 2. Different microphones and accelerometers placement inside the colonies, based on different approaches. In particular: in (**a**) from [27], B is the microphone used and it is placed above the queen cage. (**b**) shows the microphone placement of [63]: the sensors are placed upon the brood frames inside a cage to protect them from propolization. (**c**) shows the solution adopted in [45], where a custom frame with the sensors inside has been chosen. (**d**) refers to the approach proposed in [44], with microphones placed outside the hive, so that a cage to protect against propolization is not necessary. In (**e**), accelerometers placement proposed in [43] is presented: the sensors exploit vibrations and do not suffer from propolization problems. (**f**) belongs to approaches presented in [51,52,57,58]: the microphones are hidden inside the hive walls, and a grid is used to protect them. (**g**) from [54] shows a custom brood frame. Finally, (**h**) from [61] shows accelerometers positioning similar to the one proposed in [43].

The issue of collecting datasets of significant dimension and having them available should also be more deeply analyzed: indeed, many proposed works lack dataset dimensionality. Future works should take into consideration this aspect since bigger and labeled datasets could improve significantly the quality, reliability, and significance of the obtained results. Finally, once the best combination between features extraction and classifier has been found, the development of a low-cost system able to monitor the beehive health is a crucial implementation aspect.

Author Contributions: All the authors contributed equally to the idea of this review. A.T. prepared the original draft with considerable contributions from S.C. and S.S. for the analysis of the literature, and the editing of the manuscript. All authors have read and agreed to the published version of the manuscript.

Funding: This research was internally funded by Università Politecnica delle Marche with the project "NU-Hive: New technology for bee hive monitoring".

Conflicts of Interest: The authors declare no conflict of interest.

References

1. Faucon, J.; Mathieu, L.; Ribière, M.; Martel, A.; Drajnudel, P.; Zeggane, S.; Aurieres, C; Aubert, M.F.A. Honey bee winter mortality in France in 1999 and 2000. *Bee World* **2002**, *83*, 14–23. [CrossRef]
2. Oldroyd, B. What's killing American honey bees? *PLoS Biol.* **2007**, *5*, e168. [CrossRef]
3. Van der Zee, R.; Pisa, L.; Andonov, S.; Brodschneider, R.; Chlebo, R.; Coffey, M.; Crailsheim, K.; Dahle, B.; Gajda, A.; Gray, A. Managed honey bee colony losses in Canada, China, Europe, Israel and Turkey, for the winters of 2008–9 and 2009–10. *J. Apic. Res.* **2012**, *51*, 100–114. [CrossRef]
4. Van Engelsdorp, D.; Hayes, J., Jr.; Underwood, R.M.; Pettis, J.S. A survey of honey bee colony losses in the United States, fall 2008 to spring 2009. *J. Apic. Res.* **2010**, *49*, 7–14. [CrossRef]
5. Gallai, N.; Salles, J.M.; Settele, J.; Vaissière, B.E. Economic valuation of the vulnerability of world agriculture confronted with pollinator decline. *Ecol. Econ.* **2009**, *68*, 810–821. [CrossRef]
6. Biesmeijer, J.C.; Roberts, S.P.; Reemer, M.; Ohlemüller, R.; Edwards, M.; Peeters, T.; Schaffers, A.; Potts, S.G.; Kleukers, R.; Thomas, C.; et al. Parallel declines in pollinators and insect-pollinated plants in Britain and the Netherlands. *Science* **2006**, *313*, 351–354. [CrossRef] [PubMed]
7. Potts, S.G.; Biesmeijer, J.C.; Kremen, C.; Neumann, P.; Schweiger, O.; Kunin, W.E. Global pollinator declines: Trends, impacts and drivers. *Trends Ecol. Evol.* **2010**, *25*, 345–353. [CrossRef]
8. Potts, S.G.; Neumann, P.; Vaissière, B.; Vereecken, N.J. Robotic bees for crop pollination: Why drones cannot replace biodiversity. *Sci. Total Environ.* **2018**, *642*, 665–667. [CrossRef]
9. Voskarides, S.; Josserand, L.; Martin, J.P.; Novales, C.; Micheletto, D. Electronic Bee–Hive (E–Ruche) Project. In Proceedings of the Conference VIVUS-Environmentalism, Agriculture, Horticulture, Food Production and Processing Knowledge and Experience for New Entrepreneurial Opportunities, Naklo, Slovenia, 24–25 April 2013; pp. 24–25.
10. Braga, A.R.; Gomes, D.G.; Rogers, R.; Hassler, E.E.; Freitas, B.M.; Cazier, J.A. A method for mining combined data from in-hive sensors, weather and apiary inspections to forecast the health status of honey bee colonies. *Comput. Electron. Agric.* **2020**, *169*, 105161. [CrossRef]
11. Debauche, O.; El Moulat, M.; Mahmoudi, S.; Boukraa, S.; Manneback, P.; Lebeau, F. Web monitoring of bee health for researchers and beekeepers based on the internet of things. *Procedia Comput. Sci.* **2018**, *130*, 991–998. [CrossRef]
12. Murphy, F.E.; Magno, M.; O'Leary, L.; Troy, K.; Whelan, P.; Popovici, E.M. Big brother for bees (3B)—Energy neutral platform for remote monitoring of beehive imagery and sound. In Proceedings of the 2015 6th International Workshop on Advances in Sensors and Interfaces (IWASI), Gallipoli, Italy, 18–19 June 2015; pp. 106–111.
13. Chazette, L.; Becker, M.; Szczerbicka, H. Basic algorithms for bee hive monitoring and laser-based mite control. In Proceedings of the 2016 IEEE Symposium Series on Computational Intelligence (SSCI), Athens, Greece, 6–9 December 2016; pp. 1–8.
14. Gil-Lebrero, S.; Quiles-Latorre, F.J.; Ortiz-López, M.; Sánchez-Ruiz, V.; Gámiz-López, V.; Luna-Rodríguez, J.J. Honey bee colonies remote monitoring system. *Sensors* **2017**, *17*, 55. [CrossRef]
15. Kviesis, A.; Zacepins, A.; Durgun, M.; Tekin, S. Application of wireless sensor networks in precision apiculture. *Eng. Rural. Dev.* **2015**, *20*, 440–445.

16. Yang, C.; Collins, J. A model for honey bee tracking on 2D video. In Proceedings of the 2015 International Conference on Image and Vision Computing New Zealand (IVCNZ), Auckland, New Zealand, 23–24 November 2015; pp. 1–6.

17. Ngo, T.N.; Wu, K.C.; Yang, E.-C.; Lin, T.T. A real-time imaging system for multiple honey bee tracking and activity monitoring. *Comput. Electron. Agric.* **2019**, *163*, 104841. [CrossRef]

18. Ntalampiras, S.; Potamitis, I.; Fakotakis, N. Acoustic Detection of Human Activities in Natural Environments. *J. Audio Eng. Soc.* **2012**, *60*, 686–695.

19. Frings, H.; Little, F. Reactions of honey bees in the hive to simple sounds. *Science* **1957**, *125*, 122. [CrossRef]

20. Michelsen, A.; Kirchner, W.H.; Lindauer, M. Sound and vibrational signals in the dance language of the honeybee, Apis mellifera. *Behav. Ecol. Sociobiol.* **1986**, *18*, 207–212. [CrossRef]

21. Kirchner, W.H. Acoustical communication in honeybees. *Apidologie* **1993**, *24*, 297–307. [CrossRef]

22. Hrncir, M.; Barth, F.G.; Tautz, J. 32 vibratory and airborne-sound signals in bee communication (hymenoptera). In *Insect Sounds and Communication: Physiology, Behaviour, Ecology, and Evolution*; CRC Press: Boca Raton, FL, USA, 2005; p. 421.

23. Hunt, J.H.; Richard, F.J. Intracolony vibroacoustic communication in social insects. *Insectes Sociaux* **2013**, *60*, 403–417. [CrossRef]

24. Aristoteles. *The Works of Aristotle the Famous Philosopher: Containing His Complete Masterpiece and Family Physician; His Experienced Midwife, His Book of Problems and His Remarks on Physiognomy*; Library of Alexandria: Ashland, OH, USA, 1870.

25. Sarton, G. The Feminine Monarchie of Charles Butler, 1609. *Isis* **1943**, *34*, 469–472. [CrossRef]

26. Huber, F. New observations on bees, I and II. *Transl. CP Dadant Dadant Sons Hamilt.* **1792**, *111*, 230.

27. Wenner, A.M. Sound communication in honeybees. *Sci. Am.* **1964**, *210*, 116–125. [CrossRef]

28. Simpson, J.; Greenwood, S.P. Influence of artificial queen-piping sound on the tendency of honeybee, Apis mellifera, colonies to swarm. *Insectes Sociaux* **1974**, *21*, 283–287. [CrossRef]

29. Simpson, J. The mechanism of honey-bee queen piping. *Z. Vgl. Physiol.* **1964**, *48*, 277–282.

30. Dietlein, D. A method for remote monitoring of activity of honeybee colonies by sound analysis. *J. Apic. Res.* **1985**, *24*, 176–183. [CrossRef]

31. Eren, H.; Whiffler, L.; Manning, R. Electronic sensing and identification of queen bees in honeybee colonies. In Proceedings of the IEEE Instrumentation and Measurement Technology Conference Sensing, Processing, Networking, IMTC Proceedings, Ottawa, ON, Canada, 19–21 May 1997.

32. Ferrari, S.; Silva, M.; Guarino, M.; Berckmans, D. Monitoring of swarming sounds in bee hives for prevention of honey loss. In Proceedings of the International Workshop on Smart Sensors in Livestock Monitoring, Gargnano, Italy, 22–23 September 2006.

33. Papachristoforou, A.; Sueur, J.; Rortais, A.; Angelopoulos, S.; Thrasyvoulou, A.; Arnold, G. High frequency sounds produced by Cyprian honeybees Apis mellifera cypria when confronting their predator, the Oriental hornet Vespa orientalis. *Apidologie* **2008**, *39*, 468–474. [CrossRef]

34. Kilpinen, O.; Storm, J. Biophysics of the subgenual organ of the honeybee, Apis mellifera. *J. Comp. Physiol. A* **1997**, *181*, 309–318. [CrossRef]

35. Eskov, E.; Toboev, V. Changes in the structure of sounds generated by bee colonies during sociotomy. *Entomol. Rev.* **2011**, *91*, 347. [CrossRef]

36. Maybeck, P.S. Chapter 8 Optimal smoothing. In *Stochastic Models, Estimation, and Control: Volume 2*; Mathematics in Science and Engineering; Elsevier: Amsterdam, The Netherlands, 1982; Volume 141, pp. 1–22. [CrossRef]

37. Howard, D.; Duran, O.; Hunter, G.; Stebel, K. Signal Processing the acoustics of honeybees (APIS MELLIFERA) to identify the "queenless" state in Hives. *Proc. Inst. Acoust.* **2013**, *35*, 290–297.

38. Stockwell, R. A basis for efficient representation of the S-transform. *Digit. Signal Process.* **2007**, *17*, 371–393. [CrossRef]

39. Kohonen, T. The Self-Organizing Map. *Proc. IEEE* **1990**, *78*, 1464–1480. [CrossRef]

40. Qandour, A.; Ahmad, I.; Habibi, D.; Leppard, M. Remote Beehive Monitoring using Acoustic Signals. *Acoust. Aust. Aust. Acoust. Soc.* **2014**, *42*, 204–209.

41. Goulson, D.; Nicholls, E.; Botías, C.; Rotheray, E.L. Bee declines driven by combined stress from parasites, pesticides, and lack of flowers. *Science* **2015**, *347*, 1255957. [CrossRef] [PubMed]

42. Murphy, F.E.; Popovici, E.; Whelan, P.; Magno, M. Development of an heterogeneous wireless sensor network for instrumentation and analysis of beehives. In Proceedings of the 2015 IEEE International Instrumentation and Measurement Technology Conference (I2MTC) Proceedings, Pisa, Italy, 11–14 May 2015; pp. 346–351. [CrossRef]

43. Ramsey, M.; Bencsik, M.; Newton, M.I. Long-term trends in the honeybee 'whooping signal' revealed by automated detection. *PLoS ONE* **2017**, *12*, e0171162.

44. Kulyukin, V.; Mukherjee, S.; Amlathe, P. Toward Audio Beehive Monitoring: Deep Learning vs. Standard Machine Learning in Classifying Beehive Audio Samples. *Appl. Sci.* **2018**, *8*, 1573. [CrossRef]

45. Cejrowski, T.; Szymański, J.; Mora, H.; Gil, D. Detection of the bee queen presence using sound analysis. In Proceedings of the Asian Conference on Intelligent Information and Database Systems, Dong Hoi City, Vietnam, 19–21 March 2018; pp. 297–306.

46. Maaten, L.v.d.; Hinton, G. Visualizing data using t-SNE. *J. Mach. Learn. Res.* **2008**, *9*, 2579–2605.

47. Woods, E.F. Means for Detecting and Indicating the Activities of Bees and Conditions in Beehives. U.S. Patent 2806082A, 10 September 1957.

48. Bromenshenk, J.J. Honey Bee Acoustic Recording and Analysis System for Monitoring Hive Health. U.S. Patent 7,549,907, 8 January 2007.

49. Brundage, T.J. Acoustic Sensor for Beehive Monitoring. U.S. Patent 8,152,590, 10 April 2012.

50. Bencsik, M. Beehive Monitoring. GB2015/051580, 10 December 2015.

51. Nolasco, I.; Terenzi, A.; Cecchi, S.; Orcioni, S.; Bear, H.L.; Benetos, E. Audio-based identification of beehive states. In Proceedings of the ICASSP 2019-2019 IEEE International Conference on Acoustics, Speech and Signal Processing (ICASSP), Brighton, UK, 12–17 May 2019; pp. 8256–8260.

52. Cecchi, S.; Terenzi, A.; Orcioni, S.; Riolo, P.; Ruschioni, S.; Isidoro, N. A Preliminary Study of Sounds Emitted by Honey Bees in a Beehive. In Proceedings of the 144th Audio Engineering Society Convention, Milan, Italy, 23–26 May 2018.

53. Huang, N. Introduction to the Hilbert–Huang transform and its related mathematical problems. In *Hilbert–Huang Transform and Its Applications*; World Scientific: Singapore, 2005; pp. 1–26.

54. Robles-Guerrero, A.; Saucedo-Anaya, T.; González-Ramírez, E.; De la Rosa-Vargas, J.I. Analysis of a multiclass classification problem by lasso logistic regression and singular value decomposition to identify sound patterns in queenless bee colonies. *Comput. Electron. Agric.* **2019**, *159*, 69–74. [CrossRef]

55. Zgank, A. Bee Swarm Activity Acoustic Classification for an IoT-Based Farm Service. *Sensors* **2020**, *20*, 21. [CrossRef]

56. Open Source Beehives Project. 2019. Available online: https://www.osbeehives.com/pages/about-us (accessed on 30 October 2020).

57. Terenzi, A.; Cecchi, S.; Orcioni, S.; Piazza, F. Features Extraction Applied to the Analysis of the Sounds Emitted by Honey Bees in a Beehive. In Proceedings of the 2019 11th International Symposium on Image and Signal Processing and Analysis (ISPA), Dubrovnik, Croatia, 23–25 September 2019; pp. 03–08.

58. Cecchi, S.; Terenzi, A.; Orcioni, S.; Piazza, F. Analysis of the Sound Emitted by Honey Bees in a Beehive. In Proceedings of the 147th Audio Engineering Society Convention, New York, NY, USA, 16–19 October 2019.

59. Daubechies, I. *Ten Lectures on Wavelets*; Society for Industrial and Applied Mathematics: Philadelphia, PA, USA, 1992.

60. Hunter, G.; Howard, D.; Gauvreau, S.; Duran, O.; Busquets, R. Processing of multi-modal environmental signals recorded from a "smart" beehive. *Proc. Inst. Acoust.* **2019**, *41*, 339–348.

61. Ramsey, M.T.; Bencsik, M.; Newton, M.I.; Reyes, M.; Pioz, M.; Crauser, D.; Delso, N.S.; Le Conte, Y. The prediction of swarming in honeybee colonies using vibrational spectra. *Sci. Rep.* **2020**, *10*, 9798. [CrossRef]

62. Cecchi, S.; Spinsante, S.; Terenzi, A.; Orcioni, S. A Smart Sensor-Based Measurement System for Advanced Bee Hive Monitoring. *Sensors* **2020**, *20*, 2726. [CrossRef]

63. Ferrari, S.; Silva, M.; Guarino, M.; Berckmans, D. Monitoring of swarming sounds in beehives for early detection of the swarming period. *Comput. Electron. Agric.* **2008**, *65*, 72–77. [CrossRef]

Publisher's Note: MDPI stays neutral with regard to jurisdictional claims in published maps and institutional affiliations.

Communication

Histopathological Findings in Testes from Apparently Healthy Drones of *Apis mellifera ligustica*

Karen Power *, Manuela Martano, Gennaro Altamura and Paola Maiolino

Department of Veterinary Medicine and Animal Productions, University of Naples "Federico II", Via Delpino, 1, 80137 Naples, Italy; manuela.martano@unina.it (M.M.); gennaro.altamura@unina.it (G.A.); maiolino@unina.it (P.M.)

* Correspondence: karen.power@unina.it

Received: 23 July 2020; Accepted: 31 August 2020; Published: 2 September 2020

Abstract: It is well known that factors acting on the decrease of population of honeybees, can act on the male and female reproductive system, compromising the fertility of queens and drones. While there are many studies on female fertility, only a few studies have focused on male fertility and the possible alterations of the reproductive system. The testes of 25 samples of adult drones of *Apis mellifera ligustica* were analyzed by histopathology using an innovative histological processing technique and the alterations that were found are here described. Most of the samples showed unaltered testes but, in some cases, samples showed degenerated seminiferous tubules, while others appeared immature. Although a limited number of samples were analyzed, the results obtained displayed that histopathological alterations of the testes exist also in honeybees and that more interest should be put to the matter, as honeybees could be considered as bioindicators for endocrine disruptors. Future studies on a larger number of samples are necessary to analyze how different environmental factors can act and induce alterations in the honeybee reproductive system.

Keywords: histopathology; honeybee; testes

1. Introduction

Honeybees are of proven importance for the agricultural economy and the conservation of biodiversity [1], as given by their global distribution and generalist foraging behavior. Honeybees can be considered the most important single pollinator species of a wide variety of wild flora, livestock pastures, and private gardens [2,3]. Pathogens, agrochemicals, and climate alterations are only a few of the environmental biotic and abiotic elements that, acting singly or synergistically, are able to threaten the fitness of honeybees and lead to colony losses [4,5]. When a drastic reduction in the number of individuals of a species occurs, effective reproduction is potentially the key to perpetuation and conservation of the species. Ineffective reproduction, caused by numerous factors such as heavy metals, chemicals, and diseases, could lead to a reduction in the offspring, increasing the risk of a further decrease in the population [6]. The stressors mentioned above are able to directly or indirectly have pathological effects on the reproductive system of honeybees, impairing their fertility and often diminishing the number of future offspring that should perpetuate the species [7–9]. To date, many studies have focused on alterations of queen fertility, as outputs, such as low egg deposition or a high prevalence of drone brood, are easily recognizable in hives [10]. On the contrary, hypofertility of drones is often subclinical and therefore less studied, but, as the reproductive activity is strictly connected to the success of mating, all the elements that could invalidate drones' fertility, could consequently threaten colony fitness [11].

In colonies of *Apis mellifera*, drones typically represent 5 to 10% of the adult population, however the production and maintenance are regulated by the colony in accordance with several environmental factors, namely food availability, size of colony, number of drones already present in the colony, queen presence/absence, and season [12]. The main role of drones in the colony organization is to reach the Drone Congregation Areas (DCAs) and mate with queens from other colonies, spreading the genetic material of the colony from which they belong to new colonies [12]. Therefore, ability to fly, copulate, and a great amount of high-quality sperm, are required to fulfill the precious role of drones [13].

The reproductive process in honeybees shows unique features as queens only mate in the early stages of their life with multiple drones and acquire on that occasion the whole amount of spermatozoa that will be stored in the spermatheca and used during their whole life [14]. Therefore, drone fertility is strictly connected to the queen's reproductive capacity as mating with unfit drones, which are not able to produce high-quality fertile semen, could lead to queen failure, which occurs when queens stop efficient egg laying or start laying haploid male eggs [10]. Queen failure causes queen replacement by colony workers or by beekeepers, with a consequent increase in production time and costs [15].

The male reproductive system of *A. mellifera* consists of a pair of bean-shaped testes, composed of 150 or more seminiferous tubules per testis, from which originate two vasa deferentia that enlarge distally forming the seminal vesicles, which open in the mucus glands [16]. Mucus glands are elongated accessory glands that produce white mucus, a proteic substance used to produce a mating sign in the queen after successful copulation [17]. The mucus glands open into paired, lateral ejaculatory ducts, which in turn convey into a long, slender, common ejaculatory duct. The common ejaculatory duct opens in the bulb of the endophallus, the copulatory organ. The honeybee copulatory organ is located internally, in the ventral region of the abdomen and it is composed of three main elements: the bulb, provided with the chitinous plates, the cervix, and the vestibulum, presenting two yellow cornua.

Two more accessory sex glands, found near the endophallus bulb, are recognized: the bulbous gland [16], and the cornual glands. The cornual glands secrete an orange-colored secretion that reinforces the attachment of the mating sign in the queen's reproductive tract [18].

In *A. mellifera* drones, the formation of the male reproductive system starts during the first stages of embryonic development. Testes are formed soon before the larva hatches from the egg while spermatogenesis starts on the third day of the larval stage. Spermatogonia, the undifferentiated germ cells, undergo multiple mitoses, developing in primary spermatocytes [16]. The primary spermatocytes are subjected to a reductional meiosis where a secondary haploid spermatocyte and one cell containing only cytoplasm, are formed. The secondary spermatocytes then undergo a non-reductional meiosis, giving origin to two spherical spermatids. Spermatid multiplication stops prior to pupation [16] and honeybee drones seem to be the only insects in which spermatogenesis occurs only during their developmental stages, therefore, drones have a predetermined quantity of sperm in adult life [19,20]. Spermiogenesis, which is the morphological differentiation resulting in spermatozoa maturation, starts two to three days after pupal molting, and sperm migrate from the testes to the seminal vesicles, where they absorb nutrients to become fully functional [21]. Drones are sexually mature twelve days after emergence when they are able to evert their endophallus and creamy colored semen containing spermatozoa, which can be located at the posterior extremity of the ejaculate, on top of the white mucus, which is void of spermatozoa [17]. Drones produce an average of 1.5 to 1.7 µL semen with approximately 7.5 million spermatozoa/µL [22].

Considering the peculiarity of the male reproductive system development, factors affecting the larval and pupal stages are potentially able to impair the whole reproductive life of a drone.

Colonies in which a poor protein diet was fed, raised drones with lower body and thorax mass and lower ejaculate volumes, compared to colonies in which pollen administration was guaranteed [23]. On the other hand, by feeding colonies with sucrose syrup and protein supplements during the early spring, the semen quality improved [24].

It has been proven that low levels of *Varroa destructor* during drone pupal development can directly cause flight reduction and reduce sperm production down to 45%, making them unlikely to reach the DCAs, chase the queen, copulate with her, and transfer a sufficient amount of spermatozoa [25].

However, it has also been shown that drones infected by *Nosema apis* during adult life, particularly shortly after hatching, face substantial fitness costs that impair the production and maintenance of high-quality sperm, reducing fertility [26].

An active debate is still open on the influence of age on sperm quality, quantity, and viability: some have shown that sperm viability decreases with age [27], while others observed an increased viability with age [28]. According to Rousseau et al. [22] age has no effect on spermatozoa viability and motility, while for Stürup et al. [11] senescence negatively influenced sperm viability only from 20 to 25 days after emergence, but the length of viability decrease is influenced by colony factors, especially genetics.

Previous studies have shown that various pathogens can be found in the reproductive organs of drones [26,29] and that stressors like pesticides [7], miticides [30], and high temperatures [11] can directly impair drones' fertility by reducing spermatozoa concentration, viability as well as ATP concentration, necessary for spermatozoa motility [8].

To date, many studies have focused on the viability and motility of spermatozoa, given the increasing interest of beekeeping in instrumental insemination, and little importance has been given to the study of possible alterations of spermatozoa [31] and not much is known about pathological changes in the microscopic structure of reproductive organs which could lead to the formation of altered spermatozoa.

The aim of this preliminary study was to analyze, through the use of an innovative histologic processing technique, the testes of apparently healthy *Apis mellifera ligustica* drones collected in different apiaries across the Campania region (Italy) and to describe the presence of the alterations unexpectedly found.

2. Materials and Methods

Twenty-five adult drones of *A. m. ligustica* were collected in 5 different apiaries located in Campania (Italy) (5 drones/apiary) from March to June 2019, a time-span that includes the natural reproductive season of local honeybees. Apiaries were located in small beekeeping farms (less than 30 hives) surrounded by orchards and tomato crops. Drones were individually collected from apparently healthy hives, with low levels of *V. destructor* infestation (<2%) and absence of clinical signs of viral and *N. apis* infection. Insects were selected according to size, vitality, and absence of visible clinical signs of pathologies such as trembling, deformed wings, and swollen abdomens. They were manually caught from the comb and subsequently transported in sterile tubes to the laboratory of Veterinary General Pathology and Anatomical Pathology of the Department of Veterinary Medicine and Animal Productions, University of Naples "Federico II". All drones were anesthetized with chilling for 3 min at −20 °C [32], observed at the stereomicroscope (Zeiss Stemi 305 trino, New York, NY, USA) to detect the presence of macroscopic alterations, and whole-body mass was weighted to assess age according to Metz and Tarpy [13]. Samples were then placed in tubes containing 10% buffered formalin for 1 h for histopathological examination. Subsequently, they were individually injected with 10 µL of 10% buffered formalin using a micropipette (Accumax Smart, Dantali, India) and a 10 µL tip. While holding the thorax of the drone between the thumb and the index finger of one hand, the injection was performed laterally between the 3rd and 4th tergite, holding the tip parallel to the tergite in order to avoid puncturing of the gut and contamination of specimens with pollen and feces [32]. After injection, drones were moved to a tube containing 10% formalin. After 24 h, each sample was cut lengthwise, placed in an embedding cassette, and then in an automatic embedding processor (VTP300, Bio-Optica, Milan, Italy). For each half, paraffin blocks were manually created using an embedding console system, and 3 µm thick sections were obtained with a microtome. In order to facilitate sectioning, a disposable stainless-steel blade (N35, Feather, Osaka, Japan) for fine cuts of hard tissues was used. Single sections were placed on the surface of hot water and then collected on a slide and dried at room temperature for 12 h. Slides were mechanically stained with hematoxylin and eosin (H-E) using an automatic tissue slide stainer (ST5010 Autostainer XL, Leica, Germany) and finally mounted. Tissue preparations were observed by light microscopy (Nikon Eclipse E-600, Nikon, Tokyo, Japan). Although observation focused on the testes, all tissues were analyzed for the presence of visible pathogens, i.e., *Nosema* spp.

3. Results

Macroscopically, samples showed no visible alterations and were aged 10 to 27 days old. Samples processed with the protocol described above appeared well preserved since tissues were well stained, and no artifacts were seen. Microscopically, 17/25 samples showed healthy testes that appeared as elongated, bean-shaped structures with numerous seminiferous tubules, surrounded by an external epithelial layer (seminiferous epithelium). The seminiferous tubules contained follicular and germ cells that encapsulate to form a thin wall cyst; the tubular lumen was filled with many coiled spermatozoa (Figure 1).

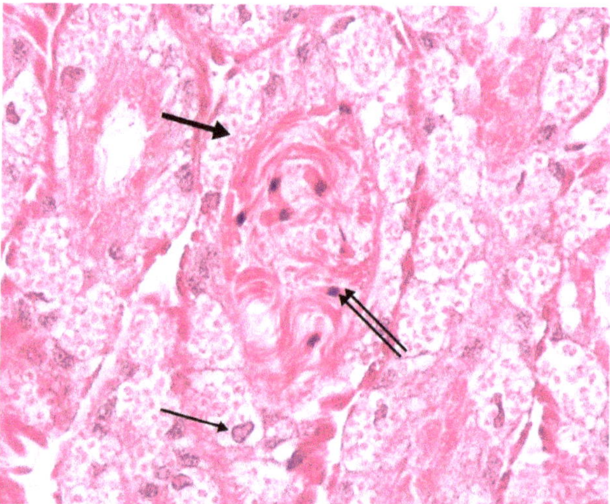

Figure 1. Testes. Normal seminiferous tubules. Follicular (thin arrow) and germ cells (thick arrow); coiled spermatozoa in the lumen of the tubules (double arrow). H-E 40×.

In 5/25 samples, the seminiferous tubules presented severe and widespread degenerative phenomena characterized by the appearance of necrosis of follicular and germ cells, disappearance of the external epithelial layer and of the tubular lumen, until the complete disruption of the normal tissue structure. It was possible to distinguish numerous spermatozoa and the complete absence of spermatogonia and spermatocytes (Figure 2).

In 2/25 samples, the seminiferous tubules were characterized by a small number of tubules of reduced dimensions, absence of tubular lumen and numerous spermatogonia but no spermatozoa, suggesting they had not reached complete maturation. Between the tubules it was possible to observe the presence of trophocytes and eosinophilic granules, attributable to residues of trophocytes (Figure 3).

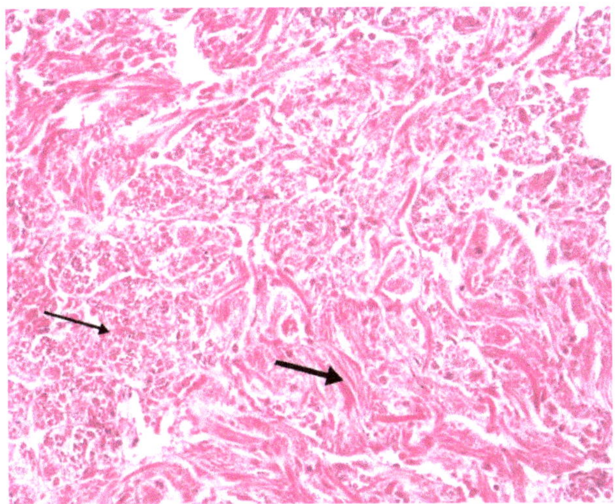

Figure 2. Testes. Altered seminiferous tubules. Disappearance of the seminiferous epithelium and the lumen; necrosis of follicular and germ cells (thin arrow); numerous spermatozoa (thick arrow); absence of spermatocytes and spermatogonia. H-E 20×.

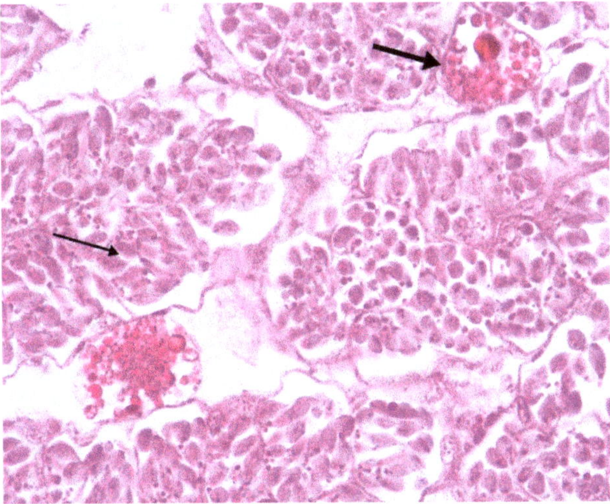

Figure 3. Testes. Altered seminiferous tubules. Absence of tubular lumen; presence of numerous spermatogonia (thin arrow); absence of spermatozoa; trophocytes between the tubules (thick arrow). H-E 40×.

In 1/25 samples it was possible to observe the detachment of the germ cells from the basal lamina of the tubules and the rupture of the membranes of the spermatogonia, probably a consequence of severe and diffuse intratubular edema (Figure 4).

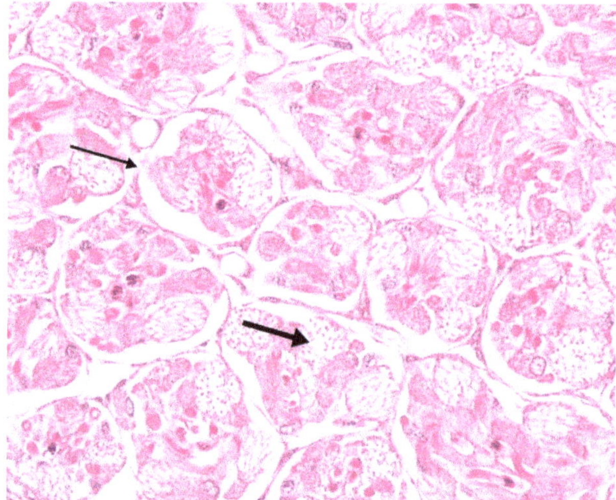

Figure 4. Testes. Altered seminiferous tubules. Detachment of germ cells from the basal lamina (thin arrow); rupture of the membranes of spermatogonia (thick arrow). H—E 20×.

4. Discussion

Compared to the past, in recent years, male hypofertility/infertility has been arousing greater interest, since malformations, genetic alterations, infectious diseases, food shortages, and managerial errors have been identified as responsible for a decreased reproductive capacity of many zootechnical species [33,34]. In a previous study [35], we had highlighted the existence of spermatozoa showing visible defects such as broken, split, and double tails. In this study, we describe the presence of alterations of the reproductive system in *A. m. ligustica* drones, regardless of the absence of macroscopic alterations, and we suggest that altered testes could probably be the cause of altered spermatozoa, that are unfit to swim up the oviducts, reach the spermatheca, and subsequently fertilize the egg. Therefore, a queen who would have mated with these drones would have very likely preserved in her spermatheca fewer or abnormal spermatozoa, reducing the potential to lay diploid eggs.

The presence of alterations in apparently healthy drones becomes particularly relevant for instrumental insemination. Considering the peculiar reproductive behavior of honeybees, instrumental insemination is a valuable tool to control the source of males and avoid undesired mating with drones that could negatively influence the genetics of the colony. Donor drones are chosen mainly on genetic characteristics and the absence of clinically visible signs of disease, while actual health status and semen analysis is not always performed prior insemination. Unhealthy and unfertile semen can, therefore, erroneously be used causing a reduction in queens' reproductive performance, as well as disease spreading.

Most of the samples did not show any pathological alterations, however, conversely to previous descriptions, the lumen of the seminal tubules appeared filled with coiled spermatozoa. Spermatozoa maturation and migration to the seminal vesicles have been often described as completed during the first week of adult life [16,19], however, empirical data supporting this theory appears old and limited, and no histological study has ever been performed, therefore comparable results are currently unavailable. On the contrary, a study by Metz and Terpy [13] found that the transfer of spermatozoa from the testes to the seminal vesicles can actually begin in the first week, but no data is reported for the end time of the migration.

Five samples showed clear degenerative phenomena affecting the seminiferous tubules. In other species, degenerative alterations have been associated with high levels of heavy metals and pesticides, such as organophosphates or neonicotinoids [36,37], in the environment which could induce oxidative stress in tissues. Oxidative stress occurs following the accumulation in organisms of reactive oxygen species (ROS) which can determine high molecular damage, degeneration of tissues, and premature aging [38]. It has been shown that drones are able to survive acute oxidative stress due to individual tolerance and resistance, and not to repair of oxidative damage of lipids and of cells [39], thus leading to a subclinical disease.

For this study, only adult drones ready for mating were collected. Nonetheless, the microscopic examination highlighted samples with testes that showed reduced maturation, as can be inferred from the presence of many spermatogonia and degenerated trophocytes in the intertubular space.

Testis development and spermatogenesis of drones of *A. mellifera* have been precisely described by Lago et al. [40] based on histological sections. Changes in whole testicular architecture, as well as of the seminiferous tubules, are described from the first-instar larvae to the pharate-adult stage. According to the histological descriptions, our findings correspond with changes occurring in testes of a fifth-instar stage larvae, however, the presence of trophocytes is not described in the cited research.

Neonicotinoids and other pesticides are considered as endocrine disruptors able to induce both hormonal and morphological alterations of the reproductive system, by mimicking the effects of estrogens and inducing signs of feminization and demasculinization [41,42]. Although endocrine disruptors are found in minimal quantities, in the long run, their chronic accumulation can interfere with honeybee health and the correct function and development of the reproductive system of drones [43,44]. Furthermore, endocrine disruptors seem to induce the production of vitellogenin (Vg) in male specimens of many animal species, and probably also in honeybees [45]. Vg is a protein present in the fat body and in the hypopharyngeal glands of workers, queens, and drones which plays a key role in phenomena related to egg laying, immunity, and longevity [46,47]. It cannot be excluded that in the drones analyzed in this study there may be an up-regulation of Vg during the developmental stages which could have influenced the correct maturation of the male reproductive system or induce alterations.

Degeneration or delayed or incomplete maturation could also have been related to the presence of subclinical viral and parasitic diseases.

Deformed Wing Virus (DWV) as well as *N. apis* have been localized in the testes of mature drones [26,29] suggesting a possible action of these pathogens in drones' fertility impairment although no histopathological findings in the reproductive tissues have been described in previous studies.

In the present study, drones did not show any clinical signs of either disease, but, while we can exclude the presence of *N. apis* and *Nosema ceranae*, as no spores were identified in the gastrointestinal tissue neither in the reproductive tissue, the presence of low levels of DWV cannot be excluded.

Samples were collected from colonies infested with low infestation levels of *V. destructor*, which, as previously stated is correlated with a low number of spermatozoa [25] and oxidative stress [48] but also to the spreading of DWV [49].

5. Conclusions

In conclusion, the results obtained, although carried out on a limited number of samples, have allowed us to display that the morphological alterations of the testes also exist in honeybees and that these could cause the formation of altered spermatozoa. Moreover, we have shown that the macroscopic appearance does not always reflect the actual health status of drones and this information appears particularly important when selecting donors for instrumental insemination, therefore, semen analysis should always be performed. We can also hypothesize that the alterations found here can be ascribed to the same causes as those in humans and other animals. Unfortunately, due to the techniques used to process the samples (formalin fixation and paraffin embedding) and to the unexpectedness of the results, it was not possible to use the same samples for further studies and correlate a specific cause for the histological alterations found. Therefore, more studies are needed to identify the etiology of testicular lesions. If the hypothesis of a major role of endocrine disruptors should be confirmed, honeybees could have the potential of becoming bioindicators of the presence of endocrine disruptors in the environment that could also affect fertility in male humans. The alterations found in honeybee testes and spermatozoa could be a red flag for similar issues affecting humans.

Author Contributions: Conceptualization, K.P.; methodology, K.P. and G.M.; validation, M.M. and P.M.; writing—original draft preparation, K.P.; writing—review and editing, M.M., G.A., and P.M. All authors have read and agreed to the published version of the manuscript.

Funding: This research received no external funding.

Conflicts of Interest: The authors declare no conflict of interest.

References

1. Potts, S.G.; Biesmeijer, J.C.; Kremen, C.; Neumann, P.; Schweiger, O.; Kunin, W.E. Global pollinator declines: Trends, impacts and drivers. *Trends Ecol. Evol.* **2010**, *25*, 345–353. [CrossRef] [PubMed]
2. Hung, K.J.; Kingston, J.M.; Albrecht, M.; Holway, D.A.; Kohn, J.R. The worldwide importance of honey bees as pollinators in natural habitats. *Proc. Biol. Sci. USA* **2018**, *285*, 20172140. [CrossRef]
3. Garibaldi, L.A.; Steffan-Dewenter, I.; Winfree, R.; Aizen, M.A.; Bommarco, R.; Cunningham, S.A.; Kremen, C.; Carvalheiro, L.G.; Harder, L.D.; Afik, O.; et al. Wild pollinators enhance fruit set of crops regardless of honey bee abundance. *Science* **2013**, *339*, 1608–1611. [CrossRef] [PubMed]
4. Le Conte, Y.; Navajas, M. Climate change: Impact on honey bee populations and diseases. *Rev. Sci. Tech.* **2008**, *27*, 485–510. [CrossRef]
5. Goulson, D.; Nicholls, E.; Botías, C.; Rotheray, E.L. Bee declines driven by combined stress from parasites, pesticides, and lack of flowers. *Science* **2015**, *347*, 1255957. [CrossRef] [PubMed]
6. Pettis, J.S.; Rice, N.; Joselow, K.; van Engelsdorp, D.; Chaimanee, V. Colony failure linked to low sperm viability in honey bee (*Apis mellifera*) queens and an exploration of potential causative factors. *PLoS ONE* **2016**, *11*, e0147220. [CrossRef]
7. Straub, L.; Villamar-Bouza, L.; Bruckner, S.; Chantawannakul, P.; Gauthier, L.; Khongphinitbunjong, K.; Retsching, G.; Troxler, A.; Vindondo, B.; Neuman, P.; et al. Neonicotinoid insecticides can serve as inadvertent insect contraceptives. *Proc. Biol. Sci. USA* **2016**, *283*, 20160506. [CrossRef]
8. Kairo, G.; Provost, B.; Tchamitchian, S.; Abdelkader, F.B.; Bonnet, M.; Cousin, M.; Sénéchal, J.; Benet, P.; Kretzschmar, A.; Belzunces, L.P.; et al. Drone exposure to the systemic insecticide Fipronil indirectly impairs queen reproductive potential. *Sci. Rep.* **2016**, *6*, 31904. [CrossRef]
9. McAfee, A.; Pettis, J.S.; Tarpy, D.R.; Foster, L.J. Feminizer and doublesex knock-outs cause honey bees to switch sexes. *PLoS Biol.* **2019**, *17*, e3000256. [CrossRef]
10. Brutscher, L.M.; Baer, B.; Niño, E.L. Putative drone copulation factors regulating honey bee (*Apis mellifera*) queen reproduction and health: A review. *Insects* **2019**, *10*, 8. [CrossRef]
11. Stürup, M.; Baer-Imhoof, B.; Nash, D.R.; Boomsma, J.J.; Baer, B. When every sperm counts: Factors affecting male fertility in the honeybee Apis mellifera. *Behav. Ecol.* **2013**, *24*, 1192–1198. [CrossRef]
12. Boes, K.E. Honeybee colony drone production and maintenance in accordance with environmental factors: An interplay of queen and worker decisions. *Insectes Soc.* **2010**, *57*, 1–9. [CrossRef]
13. Metz, B.N.; Tarpy, D.R. Reproductive Senescence in Drones of the Honey Bee (*Apis mellifera*). *Insects* **2019**, *10*, 11. [CrossRef] [PubMed]
14. Boomsma, J.J.; Baer, B.; Heinze, J. The evolution of male traits in social insects. *Annu. Rev. Entomol.* **2005**, *50*, 395–420. [CrossRef] [PubMed]
15. Amiri, E.; Strand, M.; Rueppell, O.; Tarpy, D. Queen Quality and the impact of honey bee diseases on queen health: Potential for interactions between two major threats to colony health. *Insects* **2017**, *8*, 48. [CrossRef] [PubMed]
16. Snodgrass, R.E. *The Anatomy of the Honey Bee*; USDA Bureau of Entomology: Washington, DC, USA, 1910; Volume 18.
17. Koeniger, G.; Koeniger, N.; Jamie, E.; Lawrence, C. *Mating Biology of Honey Bees (Apis Mellifera)*; Wicwas Press: Kalamazoo, MI, USA, 2014.
18. Colonello, N.; Hartfelder, K. Protein content and pattern during mucus gland maturation and its ecdysteroid controlin honey bee drones. *Apidologie* **2003**, *34*, 257–267. [CrossRef]

19. Bishop, G.H. Fertilization in the honey-bee I. The male sexual organs their histological structure and physiological functioning. *J. Exp. Zool.* **1920**, *31*, 225–265. [CrossRef]

20. Page, R.E., Jr.; Peng, C.Y. Aging and development in social insects with emphasis on the honey bee, *Apis mellifera* L. *Exp. Gerontol.* **2001**, *36*, 695–711. [CrossRef]

21. Ruttner, F. *The Instrumental Insemination of the Queen Bee*; Apimondia International Beekeeping Technology and Economy Institute: Bucharest, Romania, 1976.

22. Rousseau, A.; Fournier, V.; Giovenazzo, P. *Apis mellifera* (Hymenoptera: Apidae) drone sperm quality in relation to age, genetic line, and time of breeding. *Can. Entomol.* **2015**, *147*, 702–711. [CrossRef]

23. Czekońska, K.; Chuda-Mickiewicz, B.; Samborski, J. Quality of honeybee drones reared in colonies with limited and unlimited access to pollen. *Apidologie* **2015**, *46*, 1–9. [CrossRef]

24. Rousseau, A.; Giovenazzo, P. Optimizing Drone Fertility With Spring Nutritional Supplements to Honey Bee (Hymenoptera: Apidae) Colonies. *J. Econ. Entomol.* **2016**, *109*, 1009–1014. [CrossRef] [PubMed]

25. Duay, P.; De Jong, D.; Engels, W. Decreased flight performance and sperm production in drones of the honey bee (*Apis mellifera*) slightly infested by Varroa destructor mites during pupal development. *Genet. Mol. Res.* **2002**, *1*, 227–232. [PubMed]

26. Peng, Y.; Baer-Imhoof, B.; Millar, A.H.; Baer, B. Consequences of Nosema apis infection for male honey bees and their fertility. *Sci. Rep.* **2015**, *5*, 10565. [CrossRef] [PubMed]

27. Locke, S.J.; Peng, Y.S. The effects of drone age, semen storage and contamination on semen quality in the honey bee (*Apis mellifera*). *Physiol. Entomol.* **1993**, *18*, 144–148. [CrossRef]

28. Czekonska, K.; Chorbinski, P.; Czekońska, K. The Influence of Honey Bee (*Apis Mellifera*) Drone Age on Volume of Semen and Viability of Spermatozoa. *J. Apic. Sci.* **2013**, *57*, 61–66. [CrossRef]

29. Fievet, J.; Tentcheva, D.; Gauthier, L.; De Miranda, J.; Cousserans, F.; Colin, M.E.; Bergoin, M. Localization of deformed wing virus infection in queen and drone *Apis mellifera* L. *Virol. J.* **2006**, *3*, 16. [CrossRef] [PubMed]

30. Johnson, R.M.; Dahlgren, L.; Siegfried, B.E.; Ellis, M.D. Effect of in-hive miticides on drone honey bee survival and sperm viability. *J. Apic. Res.* **2013**, *52*, 88–95. [CrossRef]

31. Gatimel, N.; Moreau, J.; Parinaud, J.; Léandri, R.D. Sperm morphology: Assessment, pathophysiology, clinical relevance, and state of the art in 2017. *Andrology* **2017**, *5*, 845–862. [CrossRef]

32. Human, H.; Brodschneider, R.; Dietemann, V. Miscellaneous Standard Methods for *Apis mellifera* Research. *J. Apic. Res.* **2013**, *52*. [CrossRef]

33. Kastelic, J.P. Male involvement in fertility and factors affecting semen quality in bulls. *Anim. Front.* **2013**, *3*, 20–25. [CrossRef]

34. Kunavongkrit, A.; Suriyasomboon, A.; Lundeheim, N.; Heard, T.W.; Einarsson, S. Management and sperm production of boars under differing environmental conditions. *Theriogenology* **2005**, *63*, 657–667. [CrossRef]

35. Power, K.; D'Anza, E.; Martano, M.; Albarella, S.; Ciotola, F.; Peretti, V.; Maiolino, P. Morphological and morphometric analysis of the Italian honeybee (*Apis mellifera ligustica*) spermatozoa: A preliminary study in Campania region. *Vet. Med. Anim. Sci.* **2018**, *6*, 2. [CrossRef]

36. Babazadeh, M.; Najafi, G. Effect of chlorpyrifos on sperm characteristics and testicular tissue changes in adult male rats. *Vet. Res. Forum.* **2017**, *8*, 319–326. [PubMed]

37. Ebrahimi, M.; Taherianfard, M. The effects of heavy metals exposure on reproductive system of cyprinid fish from Kor River. *Iran. J. Fish. Sci.* **2011**, *10*, 13–26.

38. Finkel, T.; Holbrook, N.J. Oxidants, oxidative stress and the biology of ageing. *Nature* **2000**, *408*, 239–247. [CrossRef] [PubMed]

39. Li-Byarlay, H.; Huang, M.H.; Simone-Finstrom, M.; Strand, M.K.; Tarpy, D.R.; Rueppell, O. Honey bee (*Apis mellifera*) drones survive oxidative stress due to increased tolerance instead of avoidance or repair of oxidative damage. *Exp. Gerontol.* **2016**, *83*, 15–21. [CrossRef]

40. Lago, D.C.; Martins, J.R.; Rodrigo, P.D.; Santos, D.E.; Bitondi, M.M.; Hartfelder, K. Testis development and spermatogenesis in drones of the honey bee *Apis mellifera* L. *Apidologie* **2020**. [CrossRef]

41. Hayes, T.B.; Anderson, L.L.; Beasley, V.R.; De Solla, S.R.; Iguchi, T.; Ingraham, H.; Kestemon, P.; Kniewald, J.; Kniewald, Z.; Longlois, V.S.; et al. Demasculinization and feminization of male gonads by atrazine: Consistent effects across vertebrate classes. *J. Steroid Biochem. Mol. Biol.* **2011**, *127*, 64–73. [CrossRef]

42. Huang, G.Y.; Liu, Y.S.; Chen, X.W.; Liang, Y.Q.; Liu, S.S.; Yang, Y.Y.; Hu, L.X.; Shi, W.-J.; Tian, F.; Zhao, J.-L.; et al. Feminization and masculinization of western mosquitofish (*Gambusia affinis*) observed in rivers impacted by municipal wastewaters. *Sci. Rep.* **2016**, *6*, 20884. [CrossRef]

43. Sandroc, C.; Tanadini, L.G.; Pettis, J.S.; Biesmeijer, J.C.; Potts, S.G.; Neumann, P. Sublethal neonicotinoid insecticide exposure reduces solitary bee reproductive success. *Agric. For. Entomol.* **2014**, *16*, 119–128. [CrossRef]

44. Baines, D.; Wilton, E.; Pawluk, A.; de Gorter, M.; Chomistek, N. Neonicotinoids act like endocrine disrupting chemicals in newly-emerged bees and winter bees. *Sci. Rep.* **2017**, *7*, 10979. [CrossRef]

45. Tufail, M.; Nagaba, Y.; Elgendy, A.M.; Takeda, M. Regulation of vitellogenin genes in insects. *Entomol. Sci.* **2014**, *17*, 269–282. [CrossRef]

46. Amdam, G.V.; Norberg, K.; Hagen, A.; Omholt, S.W. Social exploitation of vitellogenin. *Proc. Natl. Acad. Sci. USA* **2003**, *100*, 1799–1802. [CrossRef] [PubMed]

47. Colonello-Frattini, N.A.; Guidugli-Lazzarini, K.R.; Simões, Z.L.; Hartfelder, K. Mars is close to venus–female reproductive proteins are expressed in the fat body and reproductive tract of honey bee (*Apis mellifera* L.) drones. *J. Insect Physiol.* **2010**, *56*, 1638–1644. [CrossRef] [PubMed]

48. Lipiński, Z.; Żółtowska, K. Preliminary evidence associating oxidative stress in honey bee drone brood with Varroa destructor. *J. Apic. Res.* **2005**, *44*, 126–128. [CrossRef]

49. Di Prisco, G.; Annoscia, D.; Margiotta, M.; Ferrara, R.; Varricchio, P.; Zanni, V.; Caprio, E.; Nazzi, F.; Pennacchio, F. A mutualistic symbiosis between a parasitic mite and a pathogenic virus undermines honey bee immunity and health. *Proc. Natl. Acad. Sci. USA* **2016**, *113*, 3203–3208. [CrossRef]

MDPI

St. Alban-Anlage 66

4052 Basel

Switzerland

Tel. +41 61 683 77 34

Fax +41 61 302 89 18

www.mdpi.com

Veterinary Sciences Editorial Office

E-mail: vetsci@mdpi.com

www.mdpi.com/journal/vetsci